Radio Photonics

Techniques and enabling technology

SERIES EDITOR

Professor Rajpal S Sirohi Consultant Scientist

About the Editor

Rajpal S Sirohi is currently working as a faculty member in the Department of Physics, Alabama A&M University, Huntsville, Alabama (USA). Prior to this, he was a consultant scientist at the Indian Institute of Science Bangalore, and before that he was chair professor in the Department of Physics, Tezpur University, Assam. During 2000–11, he was academic administrator, being vice chancellor to a couple of universities and the director of the Indian Institute of Technology Delhi. He is the recipient of many international and national awards and the author of more than 400 papers. Dr Sirohi is involved with research concerning optical metrology, optical instrumentation, holography, and speckle phenomenon.

About the series

Optics, photonics and optoelectronics are enabling technologies in many branches of science, engineering, medicine and agriculture. These technologies have reshaped our outlook, our way of interaction with each other and brought people closer. They help us to understand many phenomena better and provide a deeper insight in the functioning of nature. Further, these technologies themselves are evolving at a rapid rate. Their applications encompass very large spatial scales from nanometers to astronomical and a very large temporal range from picoseconds to billions of years. The series on the advances on optics, photonics and optoelectronics aims at covering topics that are of interest to both academia and industry. Some of the topics that the books in the series will cover include bio-photonics and medical imaging, devices, electromagnetics, fiber optics, information storage, instrumentation, light sources, CCD and CMOS imagers, metamaterials, optical metrology, optical networks, photovoltaics, freeform optics and its evaluation, singular optics, cryptography and sensors.

About IOP ebooks

The authors are encouraged to take advantage of the features made possible by electronic publication to enhance the reader experience through the use of colour, animation and video, and incorporating supplementary files in their work.

Do you have an idea of a book you'd like to explore?

For further information and details of submitting book proposals see iopscience.org/books or contact Ashley Gasque on Ashley.gasque@iop.org.

Radio Photonics

Techniques and enabling technology

Le Nguyen Binh
Huawei Research Institute, Munich, Germany
and
Exatronix, Schwabing West, Muenchen, Germany

IOP Publishing, Bristol, UK

ISBN 978-0-7503-3511-9 (ebook)
ISBN 978-0-7503-3509-6 (print)
ISBN 978-0-7503-3512-6 (myPrint)
ISBN 978-0-7503-3510-2 (mobi)

DOI 10.1088/978-0-7503-3511-9

Version: 20211201

IOP ebooks

British Library Cataloguing-in-Publication Data: A catalogue record for this book is available from the British Library.

Published by IOP Publishing, wholly owned by The Institute of Physics, London

IOP Publishing, Temple Circus, Temple Way, Bristol, BS1 6HG, UK

US Office: IOP Publishing, Inc., 190 North Independence Mall West, Suite 601, Philadelphia, PA 19106, USA

To my mother T H Nguyen and the memory of my father Mr Le Nguyen.

Contents

Preface

Optics has been an important ingredient in silicon integrated circuit technologies over the last 70 years, allowing the technological evolution of software and reliable automated systems for realizing mass-scale and 5G infrastructure deployments. That is why several industrial sectors invest so heavily in the optical communications sector. For example, CISCO Inc acquired Acacia Communications Inc for coherent optical communication products for extensive deployment in access, long-haul, and metro networking, and the distribution of ultra-high speed data channels. When we state that photonics and optical fibers will be at the heart of the next generation of mobile networks, we may sound biased towards optical technologies because of our background in the community and our research focus on optical transmission and networking. However, when one of the largest networking software companies makes the same claim we think that optical networking has reached the point at which global communities will benefit from unlimited bandwidth, and hence an information capacity that humans have never experienced thus far.

The capacity demands for modern wireless 5G technologies and beyond have increased progressively. Figure 0.1 depicts the evolution towards data-center-based networking. This leads to significant expansion of the delivery of data to access networking, in both the optical and wireless domain, including those within the data centers and their networking. Furthermore, distributed dc networking is essential so as to minimize the latency time, as required by 5G channel processing.

The frequency regions (FR1–FR*n*) for 5G wireless networks are specified in many radio frequency (RF) spectral windows, including those in the ultra-high frequency

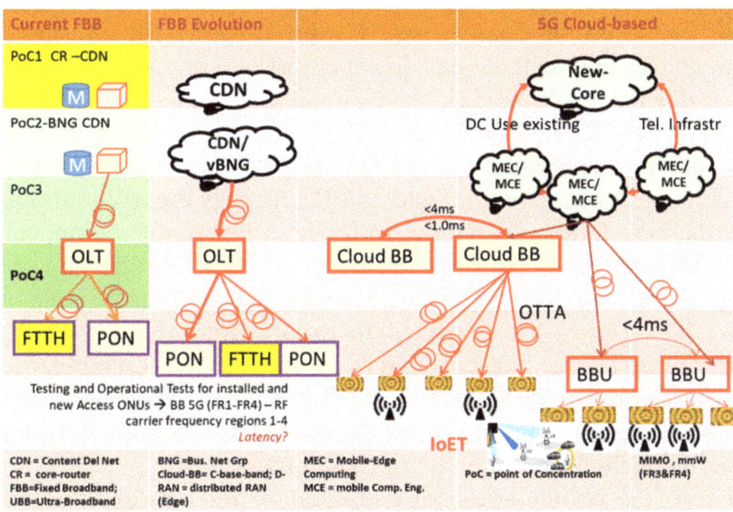

Figure 0.1. Evolution of current transport to 5G cloudified networks. MIMO = multiple input multiple output; PON = passive optical network; FFTx = fiber to the x (x = house, building, premises...); OTTA = optical to the antenna; RAN = radio access network; BBU = business broadband unit; OLT = optical lightwave terminal; PoC = proof of concept (levels 1, 2, 3).

(UHF) low–high microwave (μW), and thence to the millimeter wave (mmW), so that wide wireless RF bands can be employed in order to increase the information capacity. The upper most frequency can reach 90 GHz (FR2/3), and even in the terahertz region (x-rays). Under ultra-high frequency the attenuation becomes appreciable and thus the transmitted power, and hence the power amplifiers and their weights, add additional loads on the antenna tower structures which house the antennae and their amplifiers, plus the coaxial cables interconnecting them to the information boxes connected not far from the antenna tower.

It should be noted that the μW and mmW frequencies are much smaller than those of lightwaves of about 193.1 THz for low-loss silica fibers through which the information and the RF carriers are transported. Therefore, we can consider that the wireless waves and channels are up-converted to the optical spectral regions, then transported through the optical guided media, commonly single mode optical fibers, then down-converted back to the wireless domain frequency via the photodetection process, and then recovered in the electrical domain. Therefore in this book we address the technologies for optical–wireless and wireless–optical conversion technologies and the transmission and transport technologies using both analog and digital techniques.

Figure 0.1 shows the evolution of the current fixed broadband networks to 5G cloudified networks, in which the delivery of broadband data to wireless access antennae is critical, as is the extension of data centers into distributed configurations for high speed deployment. This also includes the evolution of the traditional carrier networks, commonly consisting of a seven-layer network which cannot respond to the high speed operation of massive data delivery due to the significant delay times through the whole network. Thus telecoms carrier networks must evolve into data-center cloudified networking.

Under the evolution of networking, 5G broadband wireless access must be established to deliver 5G technologies, in which the two first frequency regions, FR1 (2–6 GHz) and FR2 (6–40 GHz), then 40–60 GHz, then 90 GHz, need to be designed to carry and distribute broadband data channels to an antenna and vice versa. It is noted that the RF carriers in FR1 are still low enough that we can classify this FR1 as a low frequency region, and as FR2 reaches the μW and mmW regions we term this the high frequency region. In the high frequency region, carrying such RF carriers is impossible via coaxial cables, but can be achieved using optical fibers, namely single mode optical fibers, commonly the standard single mode fiber (SSMF). The RF regions are expected to expand into the 100 GHz region, in particular in the free-license band of 7 GHz centered at 56.8 GHz. Hence the use of the terms radio frequency over fiber (RFoF) and radio photonics to indicate the convergence of RF radiation into guided optical transmission lines and the employment of photonic technology to assist in these processes.

The chapters of the book address the transmission and networking techniques and the enabling technologies for delivering broadband data channels for 5G, 6G, and 7G transport and networking, including the following subjects:

- A basic introduction to radio photonics over fiber (RPoF).
- RPoF enabling techniques and technologies.

- Long-haul RPoF transmission techniques.
- Photonics-assisted front end reception for RPoF systems.
- Ultra-high speed RPoF transmission techniques, including both direct modulation and external modulation, employing double sidebands, single sidebands, and digital signal processing.
- Techniques for RPoF with intermediate frequency (IF) processing and multiple band antenna sensing techniques for RPoF via plasmonic electro-optic effects, and passive and active modulation optical transmission for remote RPoF. Furthermore, a future vision for RPoF in 5G, 6G, and 7G is provided.
- The generation of RF oscillators using electro-photonic techniques in a closed loop using electro-photonic modulation and integrated in a harmonic mode-locked lasing loop.
- 5G, 6G, and 7G optical to wireless and vice versa techniques and technologies.
- The main features of single mode and few-mode fibers and the corresponding transmittance transfer function for fibers operating in the linear dispersive and nonlinear regions. This chapter is considered to be an appendix. Some of the fundamental diagrams are familiar.

Dr Dr Eng Le Nguyen Binh
Huawei München Forschungzentrum, 80992 and
Exatronix, 80796, Munich, Germany
Summer 2021

Acknowledgements

I wish to thank my colleagues at the Huawei European and Munich Research Centers, Ye Zhengqing, Long Xingyun, Zhang Zheng, Haroon Khan, Hong Fei, Eswara Rao, Maxim Kushernov, Christian Blum, Ivan Cano, Jinlong Wei, Hans Bunner, and Stephano Bettell for many fruitful exchanges and interactions in the laboratories and workshops. Without their help it would have been impossible for me to complete a number of the experimental platforms and results presented herein.

I would also like to thank Ashley Gasque and Robert Trevelyan of IOP Publishing, UK, for their encouragement and help.

Finally, I thank my wife Phuong and my son Lam for their understanding of the number of hours spent writing instead of family time.

Author biography

Le Nguyen Binh

Le Nguyen Binh received the degrees of B.E. (Electronic Eng.-Summa cum Laude) and PhD and Dr. Eng. in Electronics Engineering and Integrated Photonics, respectively, all from the University of Western University of Western Australia, Chartered Engineer.

Prof.Dr.Dr.Eng. Le Binh has authored and co-authored of more than 300 papers in leading journals and refereed conferences and fourteen books in the field of photonics and digital optical communications, all published by CRC Press, Taylors and Francis Group LLC. of Boca Raton, Florida (USA) and IOP publishing House, UK.

While in academic institutions, he has developed and read several courses in engineering such as Fundamentals of Electrical Engineering, Physical Electronics, Small Signal Electronics, Large Signal Electronics, Signals and Systems, Signal Processing, Digital Systems, Micro-Computer Systems, Electromagnetism, Wireless and Guided Wave Electromagnetics, Communications Theory, Coding and Communications, Optical Communications, Advanced Optical Fiber Communications, Advanced Photonics, Integrated Photonics, Fiber Optics. He also led several courses and curricula developments in electrical and electronic engineering (EEE) and joint courses in physics and EEE.

Le Binh worked for Siemens AG Central Research Centre in Neu Perlach, Munich Germany, Nortel Networks Advance Technology Center, UK and Huawei EU R&D Center, Munich. As a technical director of Huawei Technologies, he has directed and collaborated on several engineering aspects of programmable 100G DWDM Tbps long haul and metro coherent optical transmission networks, 1,6Tbps per lambda of 75 GHz bandwidth, ultra-fast 110 GHz and 256 GSa/s digital signal analyzer, integrated photonics and electronics technology for effective packaging of sub-systems, and currently over last few years, on radio photonics over fiber for 6G/7G networks including transmission and automatic testing engineering, especially the generation of ultra-high frequency oscillation via radio-photonic resonance.

He was acting as a Series Editor *Optics and Photonics* for CRC Press and chaired the Commission D (Electronics and Photonics) of the National Committee of Radio Sciences of the Australian Academy of Sciences.

Chapter 1

Introduction to RPoF techniques

1.1 Introductory remarks

5G systems and networking demonstrate the breadth and depth of innovative and comprehensive mobile wireless communications technology. The potential of 5G is extremely large. Potential applications include interactive virtual reality over the entire globe, autonomous auto-driving in crowded environments, interactive communications of vehicle-to-vehicle and everything (V2V&E), the Internet of Things (IoT), the Internet of Everything (IoET), multi-Gb wireless mobile broadband communications, and fixed ultra-broadband wireless access. Figure 1.1 shows typical home uses of 5G wideband ultra-high-speed Internet and associates IoET/IoT, etc.

The RF spectral regions of 5G wireless technology have evolved to offer faster data rates and wider bandwidth allocations using an extended spectral window to 2–6 GHz and then to the microwave (μW) and millimeter wave (mmW) spectral regions, to 90 GHz and even beyond, pushing into the terahertz region. In these extended spectral regions the carriers and associated baseband data channels need to be carried or transmitted and distributed over long distances of more than a few hundreds of meters to several hundreds of kilometers. It is thus impossible for metallic coaxial cables to transport these RF carrier modulated data channels. Therefore, optical guided media, namely single mode optical fibers, are the choice for transporting RF carrier data channels. The most commonly used fiber is the standard single mode fiber (SSMF) ITU-G652 or coded as Corning SMF-28/SMF-28e.

This chapter introduces the fundamental aspects of radio-photonics over fiber (RPoF), that is the conversion of the RF carrier and data channels to the optical domain, transmitting them over the optical transmission line (the optical fiber), and the reception to convert them back to the RF domain and radiation to free-space via broadband antenna sites, and hence the recovery of the data channels. In the opposite direction, radiated electromagnetic (EM) waves can also be induced via a reception antenna or arrays of antennae and then converted to the optical domain

doi:10.1088/978-0-7503-3511-9ch1

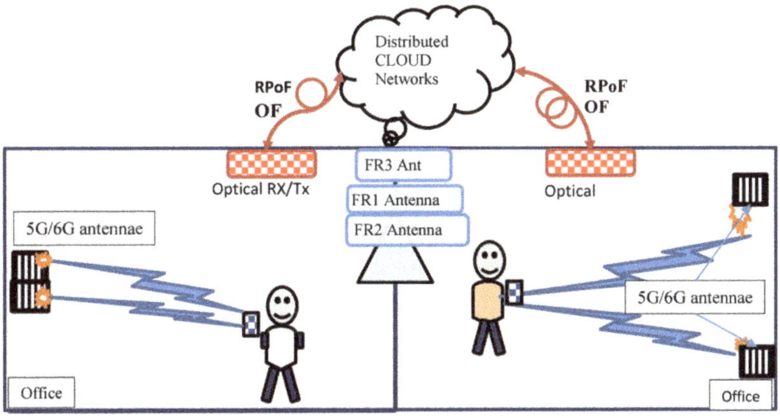

Figure 1.1. Typical user access to broadband wireless channels in a close environment. FR = frequency region in 5G networking; OF = optical fiber (single mode/multi-mode); RPoF = radio-photonics over fiber.

and transported over optical networks to the receiving sites for further distribution to specific ends of 5G and 6G application systems. The basic principles representing the physical processes of the RPoF using a 2 × 2 transfer matrix are illustrated in figure 1.2(a), in which two-port electrical networks can be used from the generation to reception. The intermediate stages of optical conversion, transmission, and reception can be estimated using the normal optical transmission and or networking techniques. Therefore, a scattering matrix can be used, as shown in figure 1.2(b).

1.2 5G and 6G frequency bands and spectrum allocations

This section gives a brief outline of the frequency bands in 5G networks with the essentials of using optical guided waves for such transmission, distribution, and networking for optical-to-wireless and vice versa [1].

1.2.1 The 5G, 6G and 7G RF spectral windows

The fifth-, sixth-, and seventh-generation (5G, 6G, and 7G) wireless systems and networks began to be deployed in 2018, and continuing into the decade 2020–2030. Initially, 5G networking may use the existing 4G and then the newly specified 5G radio frequency (RF) carrier frequency bands (2 GHz to 6 GHz) to operate, then 6G and 7G in the higher frequency windows in the μW and mmW frequency bands. The primary technologies, including the μW and mmW bands (10–20 GHz and 26, 28, 38, and 60–90 GHz, respectively) offer performance as high as 20 gigabits per second (Gbps). Massive multiple input multiple output (MIMO, such as 64–256 matrix element antennas) with a 2 × 2 matrix array offers performance up to ten times the current 4G/5G networks due to the significant antenna gain of the MIMO and processing capabilities of digital signal processing (DSP) advanced technology. Some prefer the more rigorous International Telecoms Union (ITU) spectral region towards the frequency regions of the International Mobile Telecommunication (IMT) definition, that only includes the μW and mmW bands for much higher

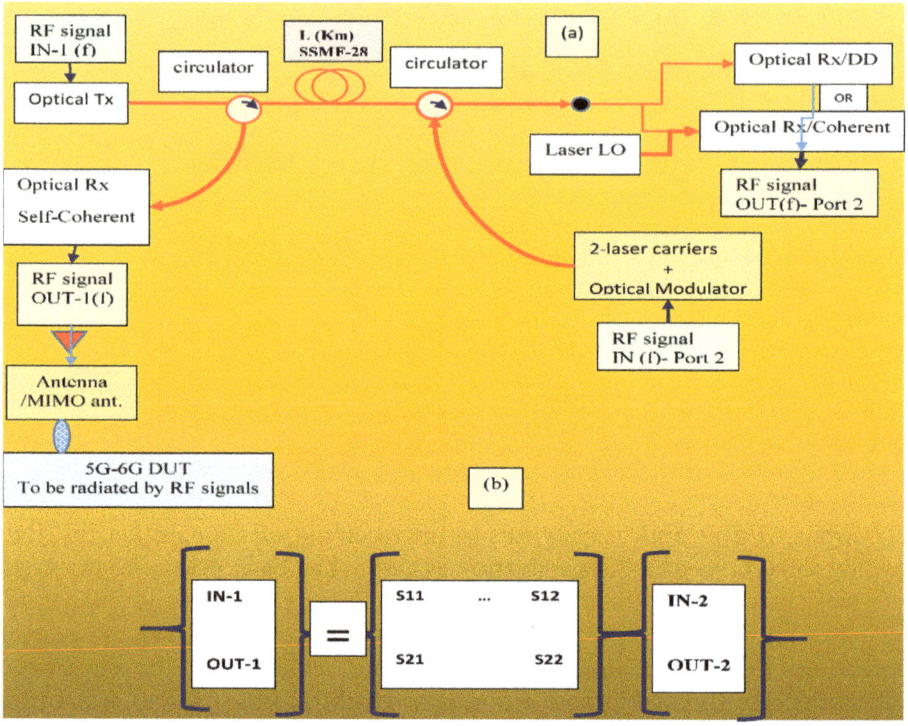

Figure 1.2. (a) Typical bi-directional uplink and downlink RPoF for optical-to-antenna and antenna-to-optical networking. (b) Two-network representation of the transfer function, either from wireless-to-optical or optical-to-wireless, with multiplication of the linear function matrices of different transfer sub-systems of the RPoF link. MIMO = multiple input multiple output; LO = local oscillator; DUT = device under test; SSMF = standard single mode fiber; RF = radio frequency; DD = direct detection; Rx = receiver; Tx = transmitter.

speeds, in which the allowable bandwidth would reach 7–10 GHz for data channels. Under these RF carriers and broadband data channels, it is essential that the optical fibers are the transmission media and that the distribution systems are close to the wireless antennae. This chapter and subsequent chapters give an introduction to and details of RPoF technological development. Figures 1.3 and 1.4 illustrate, respectively, such transmission of RPoF in a local area and the wider long-haul RPoF networking with ultra-broadband fiber networks.

Table 1.1 lists in brief the frequency bands of 5G and the channel bandwidths which the band can support (see [2] for more details of the frequency band allocations). The frequency and bandwidth, including guard-bands, are illustrated in figure 1.4. The frequency region in this new radio (NR) falls into the sub-6 GHz region and is jammed with several NR frequency bands. Multiple channels in the 5G NR can be transported over the wider network via digital RPoF technologies, in particular using the Cloud radio access network (CRAN) in a cloudified Internet, as shown in figure 0.1 of the preface. A list of the frequencies and their bands for data channels are listed in table 1.1. Table 1.2 shows the ultra-high frequency bands,

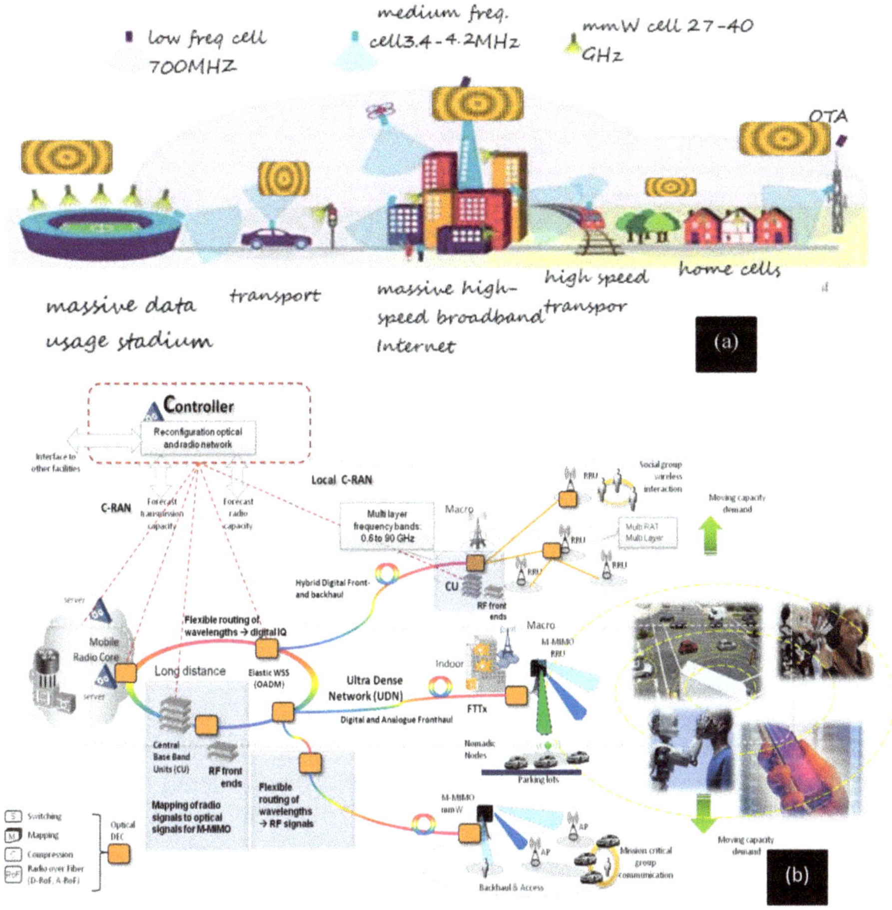

Figure 1.3. Expected 5G frequency band usage from optical-to-antenna to massive data channels in stadium, transport, and business/domestic environments. (a) Planar serial modes of wireless. (b) Wireless and optical distribution and networking.

the μW and mmW regions or FR2 and FR3, for the 28–40 GHz spectral window. Tables 1.3 and 1.4 outline the allocations of the frequency regions for different areas of the continents.

1.2.1.1 RF spectral range 1 (FR1)

The frequency regions of FR1 are tabulated in table 1.1. The RF frequencies can be scaled accordingly for RPoF networking. In addition, the region 2, FR2 (detailed in the next section) and higher regions can rely on hardware technologies, in particular optical systems, thus the convergence between RF and optical technologies.

1.2.1.2 RF spectral range 2 (FR2)

Table 1.2 tabulates the RF spectral ranges in the mmW and μW ranges, which offer a wider baud-rate with effective bandwidth, and thus a higher total bit rate per

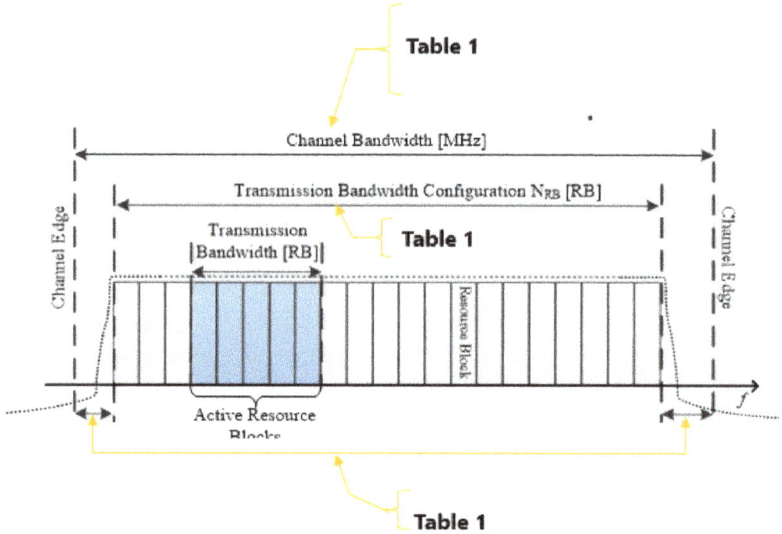

Figure 1.4. The wireless channel and bandwidth in the modulated rectangular form through pulse shaping modulation. Data channel BW = maximum transmission bandwidth plus the guard-bands on both sides of the channel spectral regions.

carrier to transport and distribution over wireless and optical media through optical–wireless networks, in particular optical access networking. Even higher RF frequency ranges, reaching 58.6 and 90 GHz are considered, in particular the 56.8 GHz free-license band with a 7 GHz band will allow 500 MBd 1.0 Gbps per RF carrier with 500 GHz spacing to transport massive wideband high bit rate radio channels.

1.2.1.3 Continental RF spectral bands

1.2.1.3.1 Continental 5G RF spectral bands

Table 1.3 lists the bands which have been or are available and fast cleared for 5G transmission and receiving over different geographical/continental regions so that equipment manufacturers can conform to requirements.

1.2.1.3.2 Continental 6G and 7G RF spectral bands (mmW)

Table 1.4 lists the μW and mmW spectral bands for the deployment of hotspots or micro-cells providing a very high throughput data rate. The most common baud-rate is now 500–1000 MBd. Thus, a high-order modulation format of RF carriers, for example 128–256QAM, leads to multi-gigabits per second (Gbps) per data channel.

Even higher frequency windows in the 58.6 GHz band with the 7 GHz free-license band and the 90 GHz RF band are considered and proven to be highly useful in the 5G or 7G network. These RF spectral bands are also being considered to be increased to the terahertz region, in particular for bio-imaging in the field of security.

Table 1.1. Frequency spectral region 1 (FR1) using RF carrier frequencies and the bandwidth of the baseband data channel for up/downlink or single uplink only.

Band No.	Up/downlink (full) or uplink only	Central RF frequency region (MHz)	Uplink bad(MHz)	Downlink band (MHz)	Up–down frequency spacing (MHz)	Data channel BW (MHz)
N1–N5	Full	850–2100	824–1980	2110–2170	45–190	5, 10, 15, 20
N6–N20	Full	800–2600	815–2570	758–2690	40–20	5, 10, 15, 20, 25, 30, 40, 50
N25–N59	Full	1900–2300	1850–3700	1930–2360	45–80	5, 10, 15, 20
N65–N80	Full	2100–4700	1700–5000	2110–2200	190	5, 10, 15, 20
N81–N90	Uplink only	900–2500	700–2700	N/A	N/A	5, 10, 15, 20, 40

Table 1.2. RF spectral region 2 (FR2) using carriers in the mmW band for RF carriers and the data channel bandwidth for up/downlink or single uplink over an optical guided medium—optical fibers.

Band No.	f (GHz)	Up/downlink (GHz)	Data channel bandwidths (MHz)
	Ku-band 12–18	12–18	50, 100, 200, 400
N257–N261	Ka-band 26.5–40	26.50–40	50, 100, 200, 400
M(NATO)–EHF (ITU)	W-band 75–110	75–110 GHz	400–1.0 GHz

Table 1.3. Frequency region 1 (FR1) for RF carriers allocated/used in global systems.

Continental regions	5G RF spectral regions
Europe (NATO)	3400–3800 MHz (awarding trial licenses)
China	3300–4990 MHz (ongoing trial)
Japan	3600–4900 MHz
Korea	3100–3700 MHz
USA (ITU)	3100–4200 MHz

Table 1.4. RF spectral regions 2 and 3 (FR2 and FR3) in the mmW region for different geographical areas.

Continental regions	6G/7G frequency band
Europe (M-band NATO)	24.25–27.5 GHz, 56.8 GHz + 7 GHz free-license band
China	24.25–27.5 GHz and 37–43.5 GHz
Japan	27.5–28.28 GHz and 90 GHz band
Korea	26.5–29.5 GHz
USA (ITU)	27.5–28.35 GHz and 37–40 GHz

1.3 RPoF and transfer functions

When considering the RPoF link or any other electrical or optical network, one can normally express the optical link by an equivalent two-port or N-port network in which there are a number of input ports and output ports. In the case of RPoF we can represent it by a two-port network or four-port network if we also consider the reverse direction. However, via the electrical-to-optical conversion and then the optical-to-electrical conversion the signal flow cannot be in such direction, thus a two-port network is sufficient for such link. We can assume that all the devices of the link operate in the small signal linear range such that the total transfer function can be equivalent to a cascade of all the sub-system transfer functions. In the frequency domain this overall transfer function is a multiplication of the frequency transfer

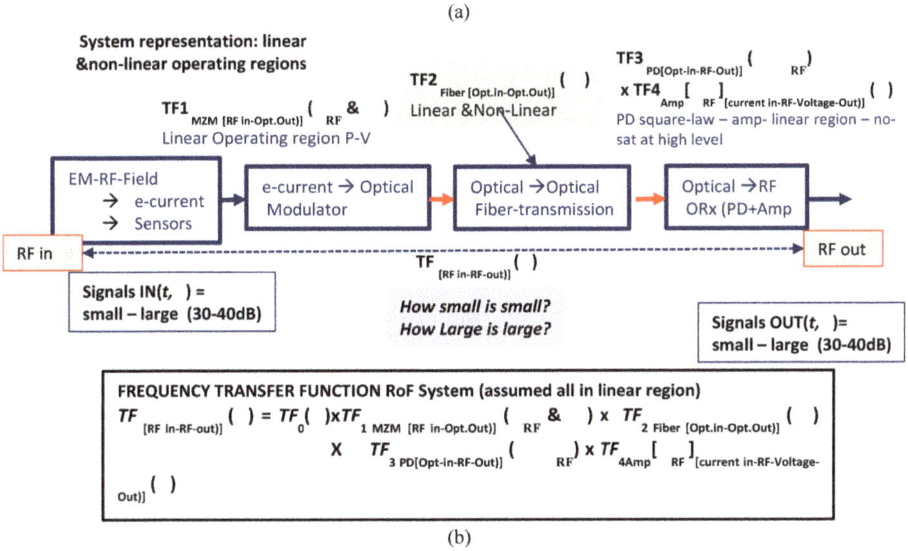

Figure 1.5. Representing an RPoF by a transfer function: (a) overall transfer function RF in → RF out and (b) detailed representation of the transfer function of each stage of the RPoF link.

functions of each sub-system, that is, the transfer of the EO sub-system or RF signals modulating the optical waves to convert them to the RF to optical transfer function, the optical-to-optical transmittance transfer via an optical fiber, and then the optical-to-electrical (OE) transfer via a photodetection (PD) or optical receiving (ORx) device. These representations for the overall transfer function and its cascade of different transfer functions are given in figure 1.5.

In figure 1.5, the transfer function 1 (on the left) represents an electromagnetic (EM) field, a wireless signal field sensor, a transducer from the wireless to current of its output and 50 Ω line, or the input impedance of the optical modulator (OM). The OM converts, by modulation of the lightwaves, the RF waves into the optical domain or the sidebands of the optical waves, which are then propagated through an optical guided medium whose loss/attenuation is low. The modulated lightwaves also suffer typical multi-path fading due to the interference of the modes in a multi-mode fiber, or the phase delay and group delay in single mode fiber, as well as nonlinear dispersion effects when its power is sufficiently high.

In order to satisfy the cascade representation of the overall transfer function of an RPoF, the operating regions of each sub-system must satisfy the small signal operating conditions, that is, the signals need to be in the linear region of the sub-system transfer characteristics. Section 4.1 provides outlines of all the sub-system operating principles and their transfer characteristics for readers to appreciate the linear and nonlinear operating regions in such cascaded sub-systems, such that the theoretical representation of the overall transfer function of RPoF can be satisfied.

Section 1.5 provides the physical design and techniques for sensing or inducing wireless high frequency signals into electrical current to modulate the optical modulator without any amplifier such that they can be completely passive but provide active modulation on optical guided lightwaves.

1.4 RPoF transmission techniques

1.4.1 Analog RPoFs

Analog RPoF modules convert RF signals to optical signals via the modulation of an optical modulator with a data channel modulated RF carrier signal, which are then received by an optical receiver that converts them back to the electrical RF domain with demodulation of the data channel whose frequency is much lower than that of the RF carriers.

The transmitter sub-system consists of an optical modulator which converts the signal from RF to optical signals, and then a second sub-system, the opto-electronic (OE) reception unit, converts the optical signal back to the RF electrical region. The data channels of the RF envelope can then be demodulated back to the original shape. Both sub-systems are linked via an optical fiber transmission line, the standard single mode fiber (SSMF). The RPoF link can operate over three RF spectral windows of (i) a low frequency of 0.8 MHz–6 GHz, (ii) the μW region of 8 GHz –18 GHz, and (iii) the mmW RF spectral region, 18–60 GHz and 90 GHz with the 7–10 GHz band for data channels.

The low frequency programmable RPoF modules are commonly based on direct modulation and include 2.5, 3, 4, and 6 GHz, the FR1 of 5G networking. This is due to the certainty of direct modulation of the laser diode (LD) current and the linearity of modern LDs without much distortion. This offers a low cost solution, unlike the use of external modulation of the LD via the employment of an optical modulator which is placed externally to the LD. In this case the LD is operating in the continuous wave (CW) mode, thus the laser phase noise is minimum due to the non-disturbance of the LD lasing dynamics.

The optical operating wavelength can either be in the coarse wavelength division multiplexing (CWDM) or dense wavelength division multiplexing (DWDM) assignment, that is the wavelength spacing between the RPoF wavelength can be of the order of less than 1.0 nm to a few nanometers, in particular in the O-band of the 1295 to 1320 nm optical spectral window of the SSMF. The RPoF technology can offer RPoF links in an operating frequency of range of 0.5 MHz to 6 GHz with a passband flatness within ±1 dB, and greater than 40 dB gain links, with a noise figure (NF) and 5 dB with a linear RF amplifier integrated opto-electronic (OE) receiver. Applications for such RPoF links include the up/downlink of remote antenna systems, the Global System for Mobile Communications (GSM) and cellular communications, ultra-broadband broadcast, ultra-broadband satellite, and 5G, 6G, and 7G network access. The packing of several RF modulated optical carriers is commonly known as sub-carrier multiplexing (SCM) on the optical carrier.

The RPoF transport of μW and mmW of the RF spectral regions is essentially based on external modulation, as the optical modulators are the most efficient in the RF spectral windows of 8, 15, 18, 20, and 40 GHz, in particular the lithium niobite based modulators and Si-photonic integrated modulators, as well as the InP-based modulators. In this 6G and 7G FR2 transport networking the external modulation technique offers the best performance. However, due to the use of the data channel band, narrow resonant type modulators are preferred. These optical modulators are not yet commercially available, but are found in research laboratories [3, 4]. The provide a frequency range of 10 GHz up to and including 40 GHz with good flatness and good phase noise. The noise figure (NF) gain can be improved by cascading an electrical RF amplifier with the optical receiver output. Parameters such as gain, P1dB, and NF can be designed. The main feature and figure of merit of such an RPoF is the transfer ratio (in dB) of the RF power in and RF power out and the ability of the optical and opto-electronic sub-systems to handle the maximum power of the RF carriers such that when the antenna RF signals operate at a high power level the signal can be transported over the fiber without distortion.

Applications for such high frequency FR2 RPoFs include distributed and long-distance delivery of data channels to antennae from a central control center, remote antenna networking, such as the Square Kilometre Array for radio physics, radio radar astronomy, and military communications.

The overall transfer function or the ratio of RF output over RF input is critical. In this section, we consider an RPoF consisting of the following sub-systems:

- *Optical modulators*: conversion of the RF signal in into an optical carrier modulated by an RF signal with optical-out–RF-in electro-optical transfer characteristics.
- *Optical transport or propagation*: optical-to-optical transfer, fiber propagation, or transmittance characteristics.
- *Photodetection (PD)*: the conversion of an optical field into electrical current via a square law conversion of optical power to electric current.
- *Electronic amplification*: electrical-to-electrical transfer with a gain, that is current received from a photocurrent to electrical current or the voltage output of the amplifier. It is possible to use an automatic gain control (AGC) state to increase the dynamic range of the RPoF link.

1.4.2 Optical modulation techniques

Modulation of lightwaves by RF signals can be implemented by:

- Direct modulation by superimposing the RF signals on the biasing voltage or current biased for the operating point at which the RF signals fall into the linear operating region of the device. Lasers such as distributed feedback grating (DFB) lasers or vertical cavity surface emitting lasers (VCSELs) are used.

- External modulation by keeping the laser lasing in the continuous wave (CW) mode or unmodulated, which involves coupling to an optical modulator through which the RF signal is input into its RF port. The optical waves are modulated via electro-optic effects.

1.4.2.1 Direct modulation

Direct modulation means manipulating the laser current using the RF channel signals when the LD is biased at an operating point thus allowing continuous modulation of the lasing lightwaves. The direct modulation of a laser is illustrated in figures 1.6(a) and (b) for, respectively, a DFB laser and a VCSEL. The DFB laser consists of an active layer sandwiched between two cladding layers. InGaAsP is a typical compound semiconductor (SC) material for the 1550 nm C-band and L-band of a silica fiber. The electrodes for the injection of current into the DFB laser and VCSEL are placed parallel to the optical waveguide cavity and both sides of the vertical cavity of the VCSEL, respectively. The small signal capacitances of these electrodes determine the frequency responses of these direct modulation lasers. The typical operating exponential characteristics for such lasers are show in figures 1.6(b) and 1.7. The detailed transfer characteristics with the *I–V* and *P–I* characteristics for a VCSEL are shown in figure 1.7. The biasing point must be located at the most symmetrical point of the linear line of the transfer curve. Due to the exponential transfer property of the laser, the linearity normally suffers some nonlinearity. Figure 1.8 shows a typical frequency response of the transmittance or optical power

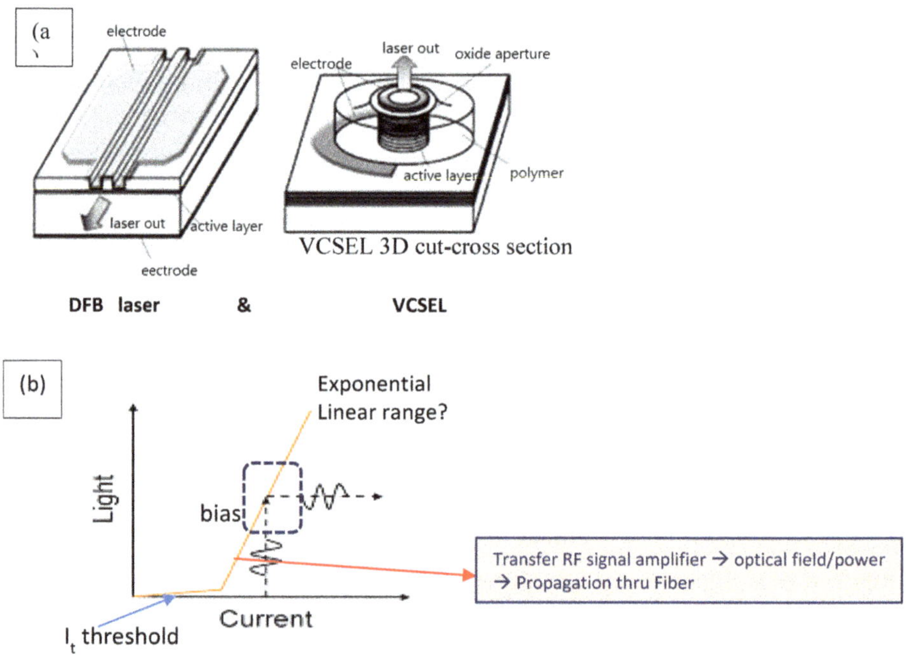

Figure 1.6. Direct modulation of an RPoF link using a DFB laser or VCSEL. (a) Schematic structure of a DFB laser and VCSEL. (b) Lightwave generated power versus driving current for a VCSEL.

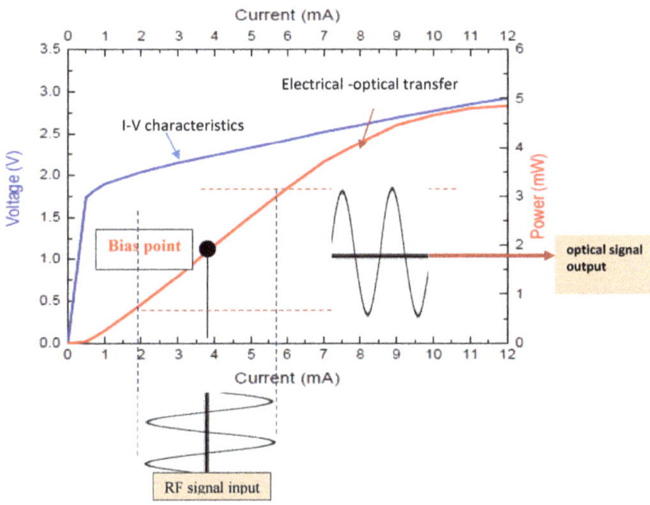

Figure 1.7. Direct/external modulation of an RPoF using a VCSEL, showing the modulation and transfer characteristics (optical power versus applied current) and *I–V* (junction voltage versus driving current) driving characteristics.

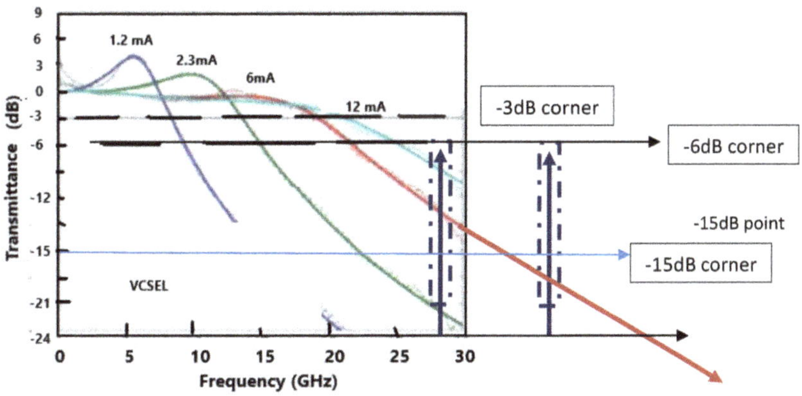

Figure 1.8. VCSEL small signal modulation characteristics: optical relative power output in transmittance (dB) versus RF carrier frequency input excitation with a driving RF current as a parameter. A VCSEL under different biasing current with a small signal. Some specific RPoF signals can be in the range of –20 dBm to +10 dB, thus it can fall into the small signal to large signal operating range. It is noted that the projected passbands at 3 dB and 6 dB, thus the operating bandwidths, can be extended from 22 GHz (3 dB) to 43 GHz (6 dB).

of a high frequency VCSEL. We can use the 3 dB bandwidth or 6 dB bandwidth to reach close to 40 GHz with a loss of 3 dB optical power (figure 1.9).

1.4.2.2 External modulation

In an RPoF external modulation the laser is kept lasing in CW mode so that the linewidth of the laser is kept constant. A temperature thermal electric cooler (TEC)

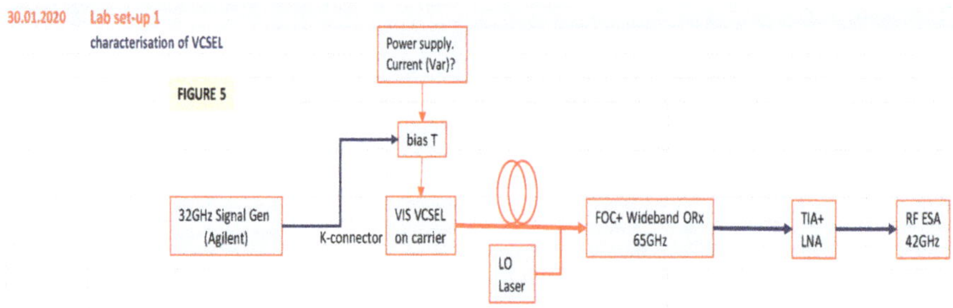

Figure 1.9. A VCSEL to carry RPoF signals: small (−20 dBm −70 mV p-p) to very large (+10 dBm −2000 mV p-p) and coherent reception with mixing of a local oscillator (LO) laser.

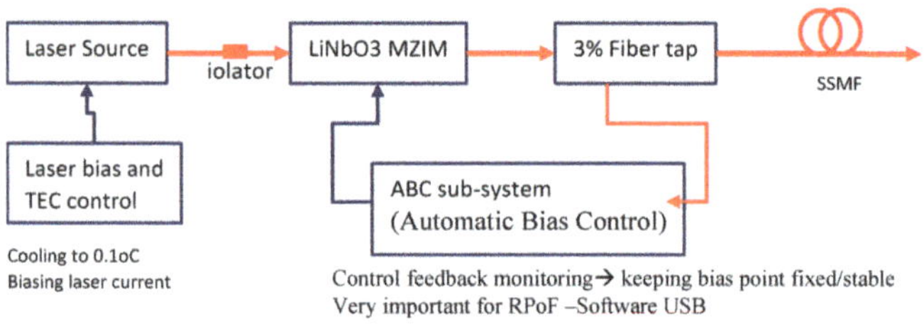

Figure 1.10. Schematic of an external modulation sub-system for an RPoF analog link using an MZIM.

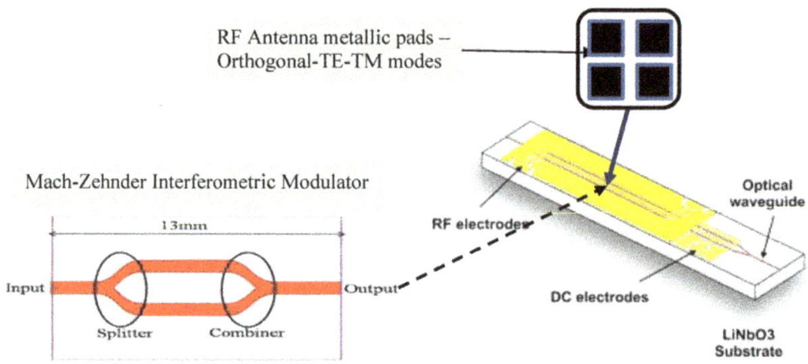

Figure 1.11. External modulation via an RF field induced from antenna pads to electrodes and the subsequent electro-optic effects in the guided optical waveguides of the interferometric optical modulator, shown with the typical dimensions of electrodes and overall length indicated.

is used to keep the temperature of the laser constant within ±0.1 °C. The output of the CW laser is normally coupled to an isolator so that any reflection feedback from other connected optical devices does not create unwanted oscillation of the laser due to this positive optical feedback to the lasing cavity, as shown in figure 1.10. The modulator is normally a Mach–Zehnder interferometric modulator (MZIM) whose structure is shown in figure 1.11. The input waveguide is coupled with the lightwaves generated from an LD operating in the CW mode. This waveguide is split into two branches, one of which is placed between an electrode and its ground plane over which the lightwaves are modulated by RF signals, and the other branch can be under no effects from any electric field. These two branches are then combined and the lightwaves are interfered to produce a constructive or destructive lightwave, thus the term interferometric modulator. For high frequencies the electrode would follow a co-planar structure (CPS) traveling type or asymmetric co-planar structure (ACPS) type. The main difficulty of this type of modulator for high frequency operation is the matching of the velocities of the optical wave and the μW or mmW wave along the optical guiding and that of the electrode. Figure 1.11 gives more details of the induction of the RF radiation induced by the antenna pads into the electro-optic modulation via the electrodes of an MZIM. Figure 1.12 gives the transfer characteristics of the MZIM and the operating points on which one can apply the biasing and then apply or superimpose the RF waves to generate RF modulated optical waves so as to propagate through the optical guided medium, the SMF or most commonly the standard SSMF.

An advanced lightwave laser (source) integrated on the same substrate with an external modulator, an electro-absorption type. The structure and its cross section of such an integrated external modulation laser is shown in figure 1.13.

1.4.3 Optical guiding media: optical fiber

Figure 1.14 shows the propagation of the optical rays through optical fibers of types (a1) step-index multi-mode fibers (MMF); (a2) graded-index MMF and (a3) single mode graded- or step-index fiber (SMF). We can observe that for step-index MMF the ray path is straight as there is no change of the refractive index of the core region.

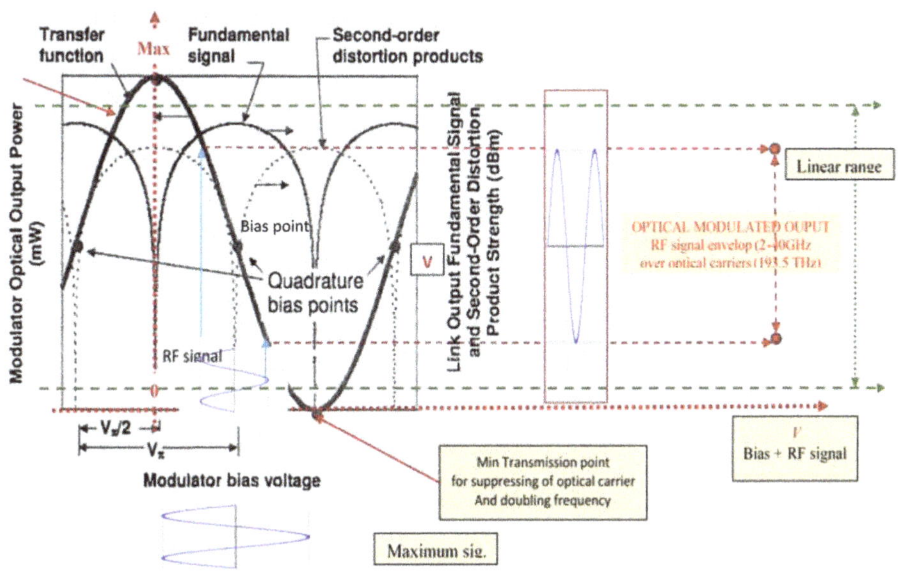

Figure 1.12. Modulation principles of an external modulator using an MZIM: Transfer RF signal amplifier → optical field/power → propagation through the fiber. Also shown is the second harmonic generation curve of the optical power of the second harmonic as a function of applied voltage.

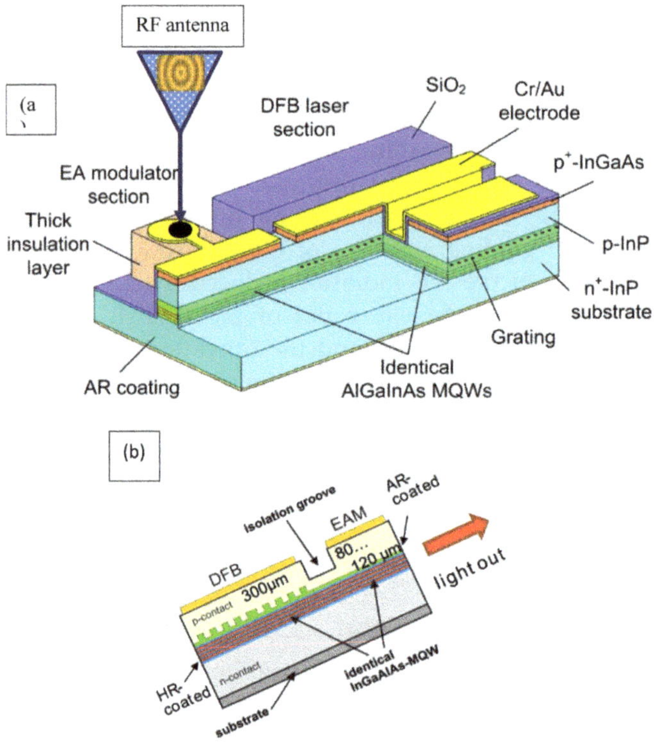

Figure 1.13. Direct/external modulation of the RPoF using an external modulation laser. (a) Image of an integrated external modulation laser (IEML) with the antenna as an induced current driving the electro-absorption (EA) modulator. (b) Schematic (in reverse direction) structure of a DFB laser integrated with an EA modulator [5]. EAM = electro-absorption modulator; DFB = distributed feedback grating; AR = anti-reflection (~30% reflection coefficient); HR = high reflection (~100% reflection); MQW = multi-quantum-well.

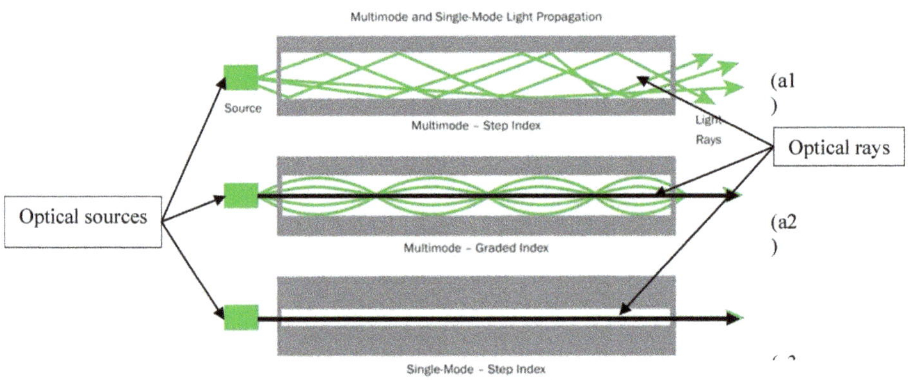

Figure 1.14. Optical ray transmission over optical fiber types. Note the multipaths for multi-mode step-index and graded-index fibers and straight (or slightly bent angled) rays for a single mode fiber: (a1) multi-mode step-index fiber; (a2) multi-mode graded-index fiber; (a3) and single mode step-index or graded-index core.

There is some penetration of the ray into the cladding region, the Goos–Hanchen shift phenomena). On the other hand the rays propagating in graded-index MMF follow a curved path and focus to a common point. This behaviour can thus reduce the time difference of the rays which are generated from the same source point. Thus this reduces the spreading of optical pulse signals. Finally for the SMF shown in figure 1.14(a3). There is only 'one' ray going straight through the fiber. Not quite only one ray, but it should be represented this way, but the optical field of either horizontal or vertical polarizations. For SMF the propagation and evolution of the optical fields are represented by the Schrödinger equation, in which the linear and nonlinear effects on the optical phases of the lightwaves are under the envelope of optical pulses. The transmittance of such propagation over the SMF, the optical transfer function characteristics, under small signal and nonlinear large signal operation is shown in figure 1.15. For RFoF using SMF, we should select the frequency band over which the transfer function is flat so that minimum distortion can be achieved, that is, avoiding the regions of the dips of the transfer function in figure 1.15. In chapter 8 we give a number of techniques to minimize the 'dips' so that RFooF can be highly effective.

1.4.4 Optical reception in the analog section of an RPoF

Figure 1.16 shows the equivalent circuit of an opto-electronic receiving sub-system which consists of a photodetector (PD) followed by an electronic pre-amplifier and then a main amplifier. This structure can be represented by a small model including noise sources. The generated photo-current is presented by a current source, the noises are presented by a double circle. Noises include quantum shot noises due to the residual laser energy (in dc or direct current level) and pulse small signal source. In addition there is a total equivalent noise current referred to the input port which is the total of all electronic noise current of the pre-amplifier and the main amplifier as they are referred to the input port of the equivalent model. Figure 1.17 thus shows

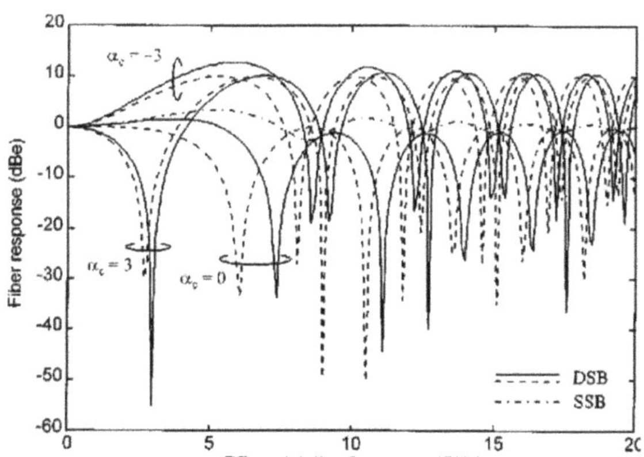

Figure 1.15. Transfer function characteristics of a single mode fiber under small signal and nonlinear large signal operation: transmittance (dB) versus input RF signal frequency with α as a parameter. dBe = the logarithmic scale in electrical quantity or detected electrical RF signal. Note the band-rejection points.

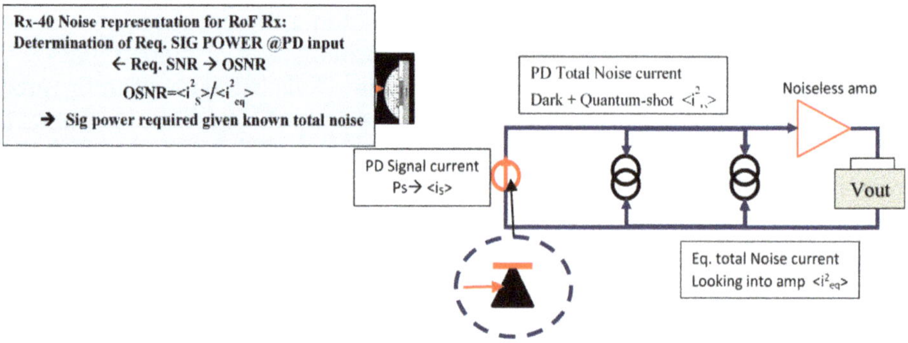

Figure 1.16. Schematic and small signal model of photodetection.

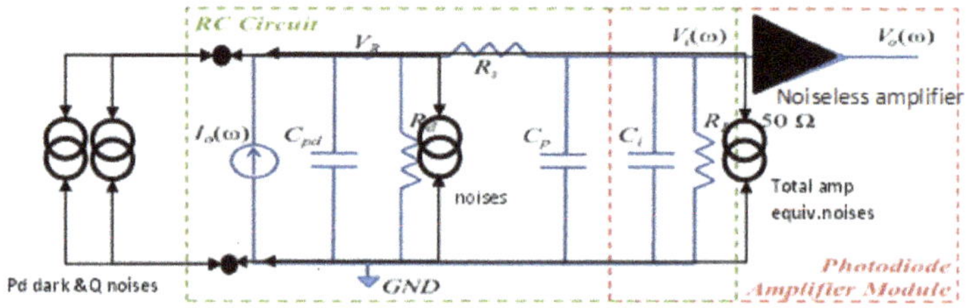

Figure 1.17. Photodetection and amplifier noise model of an optical reception sub-system for an RPoF. Q-30noise is the quantum shot noise generated by the photodetector (PD) of optical power. PD dark current noise is generated by the random thermal noise of the PD p–n junction of either p–i–n (PIN-PD) or avalanche (APD) type.

the details of the small signal model in which all noise sources are given referred to the input port. Thus, the amplifier is a noiseless device.

1.5 EM field sensing

Figure 1.18 shows the sensing devices which are used to monitor the RF radiation or receiving RF wave energy of an antenna. The RF energy absorbed by the sensor is used to modulate the optical waves which are guided on a substrate and then transmitted to a control centre. Optical resonators can be one possibility so that selected frequency bands can be selected to obtain specific RF band embedded in the optical carrier window. Alternatively, the middle section of figure 1.18 shows a plasmonic modulator in which the lightwaves are concentrated and guided through a narrow section over which the plasmonic surface waves derived from the RF waves are launched . The RF and optical waves are interacted and the lightwaves are modulated, thence transmitted over the fiber to a central centre. The optical fields over different types of optical wave guides and the plasmonic RF fields are shown, respectively, in figure 1.20(a), (b), and (c). Figure 1.19 shows an optical sensor for sensing RF radiation and modulating the optical waves which are transmitted from

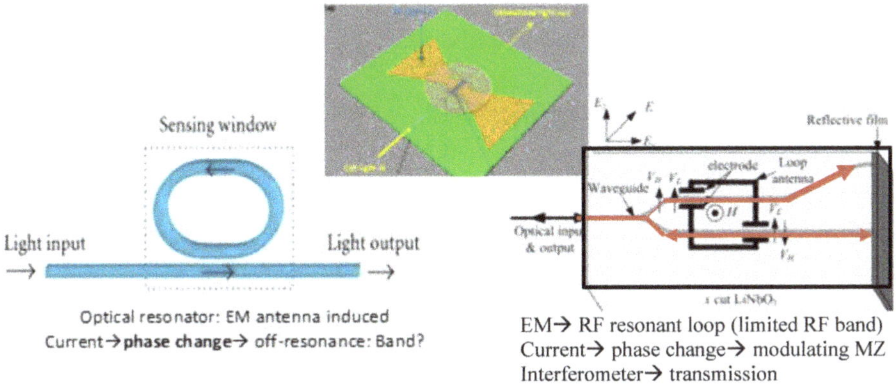

Figure 1.18. EM sensing I: electro-optic effects and resonance structure.

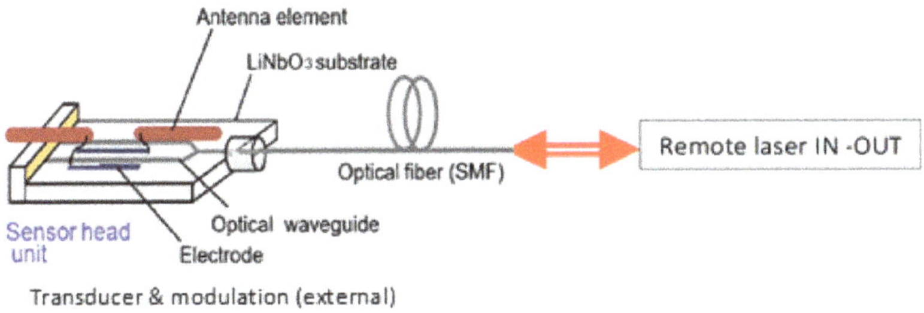

Figure 1.19. EM sensing 2: electro-optic effects and remote structure.

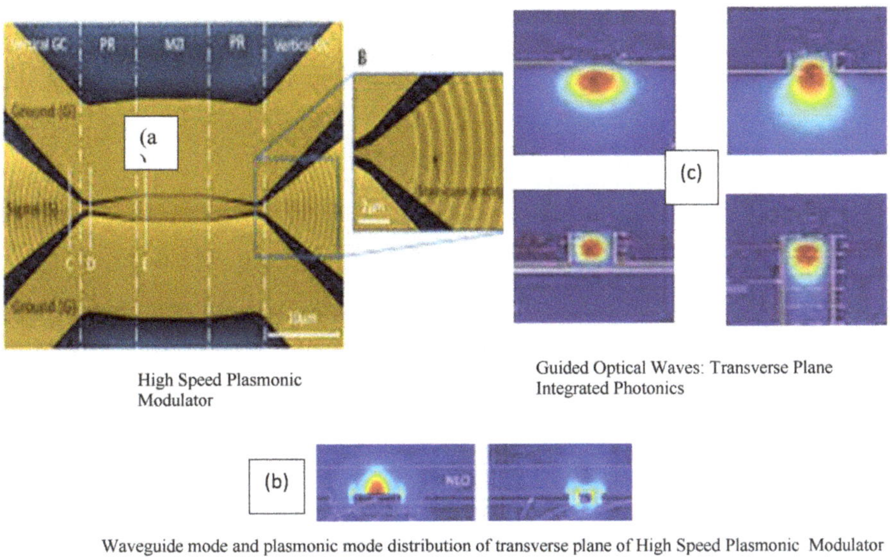

Figure 1.20. EM sensing using plasmo-antenna and electro-optic modulation. (a) Planar view of the plasmo-antenna-modulator. (b) Optical waveguide mode and plasmonic mode. (c) Optical waveguide modes in the transverse plane of different optical waveguide structures.

a remote lightwave source, passing through the optical waveguide and reflected from a reflective film, returning to the same fiber and traveling back to the control centre.

1.6 Distortion effects in RPoF

See figures 1.21 and 1.22 for details.

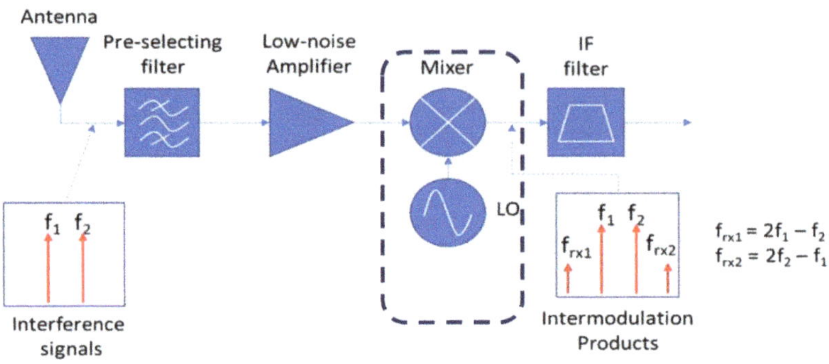

Figure 1.21. Nonlinearity and intermodulation (IMD) cross-talks in RF and RPoF systems. IMD is the interaction between closely spaced RF channels packed in the frequency region band via the nonlinear effects of the transmission medium or the nonlinear power amplifier for radiation of a wireless field. Third- and higher-order mixing of the two interfering RF signals produces IMD signals embedded in the desired band in an optical receiver, due to nonlinearity of the transmitter, the nonlinear effects of the fiber (see chapter 8), and the nonlinearity of the PD and the signal electrical RF amplifier. The IMD signals may degrade the receiver's performance in terms of sensitivity. The IMD frequencies can be set such that one of the third-order IMD products ($f_{rx1} = 2f_1$–f_2 and $f_{rx2} = 2f_2$–f_1) falls within the passband of the receiver.

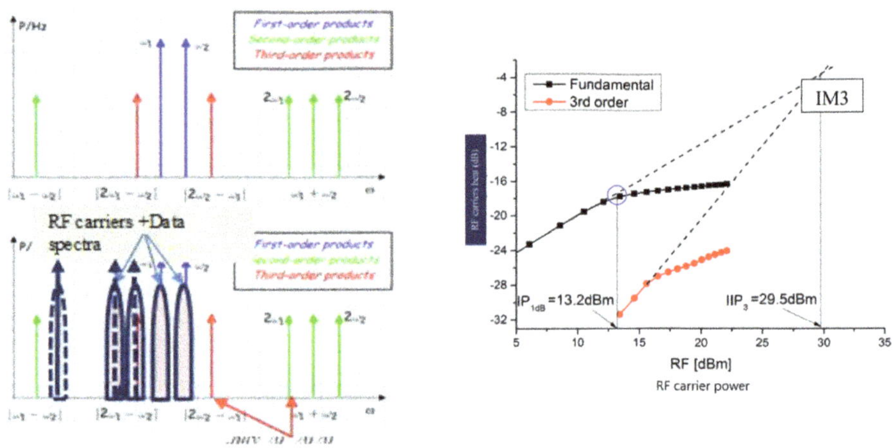

Figure 1.22. Intermodulation (IMD) effects in an RPoF due to fiber nonlinear Kerr effects; Tx and PD + Elect. Amp. → IMD noise. IMD3 = the third-order nonlinear RF component due to IMD of the RF sub-carriers which carry information channels.

1.7 Optical transport guided media

The dielectric waveguides proposed as optical guided media by Kao and Hockham [6] in 1966 led to the first research phase that attracted significant interest in the 1970s. Then, through the demonstration of fiber optics, optical communications have progressed tremendously over the last five decades.

The first-generation lightwave systems were deployed commercially in 1983 and operated in the first optical spectral window of 800 nm over multi-mode optical fibers (MMF) due to the availability of LD operation in this region, at transmission bit rates of up to 45 Mb s^{-1} over only a few kilometers. The introduction of ITU-G652 SSMFs in the late 1970s was due to the significant development and invention of both theoretical and practical demonstrations of single mode guiding phenomena, in particular the weakly guiding phenomena.

The second generation of lightwave transmission systems, transporting modulated signals over a single mode fiber, was then possible in the 1980s. The operating wavelengths were shifted to the O-band window of 1300 nm in which the loss is about 0.5 dB K^{-1} m^{-1} with almost no material dispersion compared to that in the 800 nm spectral region. This is because the fiber chromatic dispersion (CD) factor is almost zero in this region. The second-generation systems could operate at bit rates of up to 1.7 Gb s^{-1} and had a repeaterless fiber span of about 40 km. Thus there were several infrastructural buildings to house the repeaters at this distance along the transmission line. Later, when fibers could be designed and fabricated to operate using the lowest loss silica fiber, in the C, C+, and L bands, the repeaterless distance was longer, up to 80 or 100 km. These housings could also be reused for economic reasons. It is also noted that inline optical amplifiers are used in these spans to compensate for the fiber propagation loss. Furthermore, dispersion compensating fibers are used to compensate for SSMF dispersion. They are commonly housed at the same site as the inline optical amplifiers.

Once single mode optical fibers are fabricated and employed they preserve the polarization directions of the linearly polarized guided mode. Further efforts were also made for the improvement of the receiver sensitivity using coherent detection techniques. The repeaterless distance could reach 60 km in installed systems with a bit rate of 2.5 Gbps, in particular the electronic transimpedance amplifiers, to offer wideband characteristics for an operation bandwidth of 50–60 GHz, then later to 110 GHz using distributed amplification design techniques in SiGE or InP integrated microelectronics platforms. However, due to the laser linewidth and its stability, these coherent optical transmission systems were limited in the determination of the frequency difference between the optical carriers and that of the local oscillator laser. Thus there was instability of the recovery of the data channels until advanced digital processing techniques were employed recently [7].

Optical fiber communications systems then evolved to third-generation transmission systems which utilized the lowest attenuation 1550 nm wavelength window and operated up to a 2.5 Gbps bit^{-1} rate in single mode optical fibers. These systems were deployed commercially in 1990 with a repeaterless span of around 60 km. Coherent transmission optical systems started to attract attention in long-haul systems, but were limited by the difficulty of frequency matching in homodyne and heterodyne reception due to the analog features of the coherence, thus producing a significant error rate.

At this stage, the generation of optical signals was based mainly on either direct modulation of the semiconductor laser source or direct detection. External modulators are used with a bandwidth limited to about 8 GHz. Direct modulation of the DFB laser can also reach the 8 GHz maximum bandwidth level. The bandwidth of the $LiNbO_3$ integrated optic modulator gradually reaches 35 GHz with diffused waveguides and traveling wave electrodes. The insertion loss of these external modulators of about 5 dB limited the extension of the span length of long-haul optical transmission system until the invention of the Er-doped fiber amplifier (EDFA).

Since the invention of optical amplifier devices, the EDFAs, in the early 1990s, lightwave systems evolved rapidly to multiple carriers (wavelength division multiplexing (WDM)) and shortly after that to higher density packed optical channels or dense WDM, optically amplified transmission systems which can increase the transmission capacity per wavelength carrier to 10 Gbps, thus multiple $N \times 10$ Gbps channels. N can reach 110 with a spectral spacing of 50 GHz in the fiber C-band. This is contributed mainly by the overcoming of fiber transmission loss by optical amplification. Further, the loss due to the insertion loss of external optical modulators, ~4 dB in total, can be compensated by an optical amplification booster. The external modulation permits the continuous operation of the laser, hence no disturbance of the narrow linewidth of external cavity lasers. Thus laser coherence can be used. Furthermore, the advancement of digital signal processing (DSP) systems has allowed tremendous progress with sampling rates reaching 200 GSa s^{-1} with long and fast storage sequences. These DSP systems have contributed processing algorithms to deal with the shortfalls of analog coherent systems in homodyne reception, compensating for fiber nonlinear and linear distortion effects. Thus these DSP-based coherent and non-coherent reception optical transmission systems can lead to high speed and ultra-high capacity optical systems which exploit external modulation extensively for coherent transmission and reception optical fiber communication systems. The present optical transmission systems are considered as the fifth generation, having a transmission capacity of a few terabits per second (Tbps) per wavelength carrier. Thus over the entire C-band the total transmission capacity can reach several tens of Tbps.

Due to the preservation of the polarization of the optical guided modes in a single mode fiber, coherent detection, homodyne or heterodyne, became possible during the 1980s and early 1990s, which offered more sensitive optical reception due to the mixing of the signal and a high power local optical oscillator. However, these coherent systems operated in the analog mode and the matching of the optical frequencies of the RF shifted optical carrier could not be matched to those of the optical local oscillator, so coherent transmission systems were not exploited. At that time the main driving force of the development of coherent optical systems was to improve the receiver sensitivity, commonly by 3 to 6 dB or 15 to 30 km of fiber loss. The repeaterless transmission distance could thus be extended to more than 80–100 km of SSMF (with a 0.15 dB km^{-1} fiber attenuation factor). However, coherent optical systems suffer severe performance degradation due to fiber dispersion impairments. In addition, the phase coherence for lightwave carriers of the laser source and the local laser oscillator was very difficult to maintain.

In contrast, the incoherent detection technique minimizes the linewidth broadening of the laser source as well as that of the local laser oscillator and thus minimizes the stringent requirements on their phase coherence. In addition, incoherent detection mitigates the problem of polarization control in the mixing of transmitted lightwaves and the local laser oscillator in the multi-THz optical frequency range.

The invention of EDFAs in the late 1980s offered optical gains of the order of 20 dB and above, revolutionizing both the incoherent and coherent digital optical transmission systems, pushing the total capacity into the multi-Tbps range and the transmission distance could reach several thousand of kilometers.

In recent years, the huge increase in the demand for information via wireless networks and smart access mobile devices has led to the significant development of broadband communications. The rapid growth of multi-media services, machine-to-machines networks, the IoET, and IP streaming services has placed great demands on both significant transport techniques for global optical networks and the transport of RF modulated optical signals over widespread and dense media. It is most likely that such tremendous growth will continue in the coming 5G, 6G, and 7G generations of networking. These 5G–7G network demands are driving local and global telecommunications service carriers to develop high-performance and high capacity next-generation optical transmission systems and data center networking. The overall capacity of WDM or DWDM optical systems can be boosted by increasing the base transmission bit rate of each optical channel and then combining them with the multiplexing techniques of several packed channels in a DWDM system. However, these radio-optical transmission systems encounter a number of challenging issues which can be outlined as described in figure 1.23.

Currently, 100 Gbps and higher, up to 800 GBps, transmission systems employ intensity modulation (IM), and complex signals are possible. The complex signals are made of two single amplitude shift keyings (ASKs) so that the phase of the signals can be a dimension of the modulation scheme. These systems are now

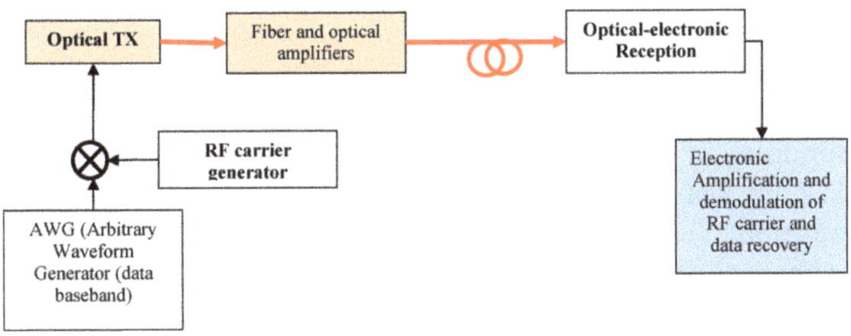

Figure 1.23. An optical transmitter using external modulation, plus opto-electronic direct reception and demodulation of an RPoF using a digital/analog modulation format for the data channel and analog RF signals or oscillation signals. Commonly M-QAM for RF carrier (in the micro- or millimeter wave region) optical modulation and the narrowband data channel (bandwidth < 1 GHz).

employed commonly in short distance access networks. The term OOK can also be used interchangeably with ASK, but mainly for the case of direct detection of the optical communication system. The OOK format simply implies the on-off states of the lightwaves where only the optical intensity is considered. On the other hand, the ASK format is a digital modulation technique representing the signals in the constellation diagram by both the amplitude and phase components. The ASK is normally used for coherent techniques.

Moving towards high bit rate transmission such as 40 Gbps, the performance of OOK photonic transmission systems is severely degraded due to fiber impairments, including fiber dispersion and fiber nonlinearities. The fiber dispersion is classified into chromatic dispersion (CD) and polarization mode dispersion (PMD), causing the rotation of the optical signals which are shifted by the RF carriers and the IMD effects. On the other hand, severe deterioration of the system performance due to fiber nonlinearities leads to interference into other optical channels and the IMD as well. These distortions come from the high power spectral components at the carrier and signal frequencies of the ASK-modulated RF-optical signals.

The total transmission capacity can be enhanced by increasing the number of multiplexed DWDM optical channels. The dense packing of channels can be implemented by reducing the spectral spacing between these optical channels, e.g. from 100 GHz down to 50 GHz, or even 25 GHz and 12.5 GHz. The packing of the channels with narrower spacing also leads to narrower data channel bandwidths, hence the requirement for narrow passband optical multiplexers (mux) and demultiplexers (demux) which is much more difficult for optical devices. Passing through these narrowband optical filters, the signal waveforms are distorted and the optical channels suffer the problem of inter-channel cross-talk. The narrowband filtering problems are becoming more severe at a high data bit rate, e.g. 40 Gbps, thus degrading the system performance significantly.

Together with the demand for boosting the total system capacity, another challenge for service carriers is to find cost-effective solutions for the upgrading process. These cost-effective solutions should require minimum renovation to the existing photonic and electronic sub-systems, i.e. the upgrading should only take place at the transmitter and receiver ends of an optical transmission link. Another possible cost-effective solution is to extend the uncompensated reach of optical transmission links significantly, i.e. without using dispersion compensation fibers (DCFs), thus reducing by a considerable number the required inline EDFAs. This network configuration has received much interest recently from both the photonic research community as well as service carriers.

Over the last few years, extensive development and research work has proven that DSP-based coherent reception techniques can enable the bit rates per wavelength channel to increase to higher than 800 Gbps, employing 200 GBd and polarization multiplexing with advanced multi-level modulation formats such as QPSK (2 bits/symbol) and/or Mary-QAM (*M* bits per symbol) to aggregate to 400 or 800 Gbps. Further, the pulse sequences can be shaped and distributed statistically using advanced digital-to-analog converters (DAC) to pack the channels into super-channels to generate Tbps per channel with a number of optical sub-carriers. The advanced ultra-high sampling rate analog-to-digital converters (ADCs) and DACs at 200 GSa s^{-1} (giga samples per second) allow the DSP to recover the clock, hence the sampling rate and time, combating the linear and nonlinear impairments due to various linear and nonlinear fiber impairments such as the CD, PMD, self-phase modulation (SPM), and cross-phase modulation (XPM) effects. The transmission for 112 Gbps per optical carrier could reach 3500 km in a field trial and 1750 km for 200 Gbps over an optically amplified and non-DCF fiber span transmission distance.

Thus the main motivations of this section are to describe the employment of the RF modulated digital optical transmission technique. The fundamental principles of digital communications, for RPoF links employing both coherent and incoherent reception techniques, are given.

1.8 Modulation formats and optical signal generation

1.8.1 Basic modulation

Radio-photonic modulation is a transfer process of information from the radio domain to an optical medium with an appropriate format that would allow transmitting and reception with minimum distortion. In radio-photonic communications, the radio-modulated electrical data channel can be converted to the optical domain via an optical modulation process, the optical modulator/transmitter, and can then be transmitted through the guided medium, the optical fiber. This is optical modulation.

There are four basic digital modulation types, (i) amplitude shift keying (ASK), (ii) frequency shift keying (FSK), (iii) phase shift keying (PSK), and (iv) orthogonal frequency division multiplexing (OFDM), in which the amplitude, frequency, or phase of the parameter or carrier can be varied to represent the information. Further, the data channels are also modulated with RF sub-carriers. Digital modulation maps the digital information symbols of '1' and '0' into carrier specific properties such as amplitude, phase, both amplitude and phase, or frequency in the orthogonality of the optical carriers which are then transmitted over the guided optical media. The modulated and transmitted lightwave carrier is then re-mapped at the reception systems back to a nearest recovered copy of the information.

Modulation is a process that facilitates the transport of information over a medium. In this book our medium is the optical guided fiber and associate photonic components. In digital communications there are three basic types of digital modulation techniques: ASK, PSK, and FSK. All these techniques vary a parameter of a sinusoidal carrier to represent the information '1' and '0'.

In ASK the amplitude of the lightwave RF carrier, normally generated by a narrow linewidth laser source and modulated by an RF oscillation signal, is changed in response to the digital data and everything else is kept fixed. That is bit '1' is transmitted by a bright lightwave amplitude of the RF modulated carrier. The state '0' can be represented be the lightwave dim amplitude. The optical frequency is kept unchanged, as shown in figures 1.24(a) and (b).

Both the non-return to zero (NRZ) and RZ formats can be generated with or without the suppression of the optical RF carrier (RZ-CS or NRZ-CS) by a dc bias with a π phase shift difference of the two optical guided paths in an MZIM.

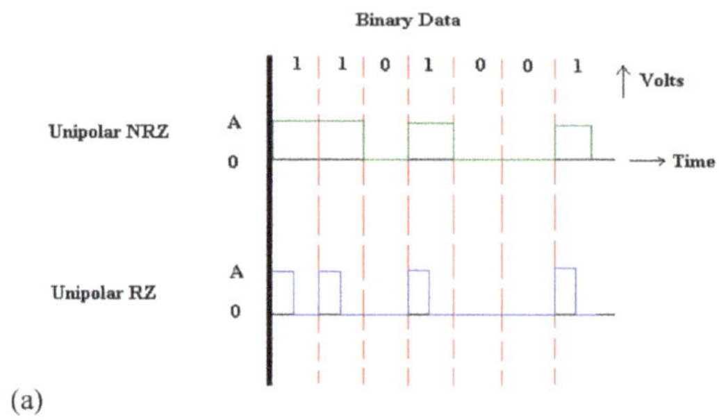

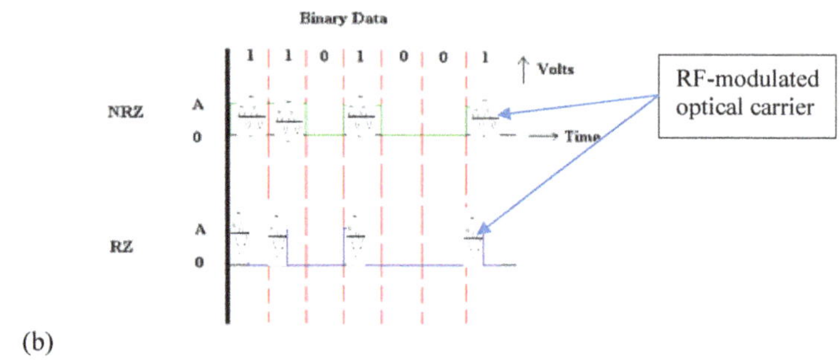

Figure 1.24. NRZ and RZ pulse amplitude modulated formats for a sequence of {1 0 1 0 1 0 1 0 1 0 1 0}: (a) non-RF carrier modulated OOK signals and (b) RF modulated optical carrier NRZ and NZ formats.

In PSK RPoF the phase of the lightwave RF modulated carrier is changed to represent the information, thus the amplitude of the signals remains constant. Thus the RF modulated signal can preserve significant energy, offering high sensitivity at the receiver. The phase in this context is the shift of angle at the point at which the sinusoidal carrier starts. To transmit a '0' the phase would be shifted by π and a '1' with no change of phase. The phase angle can be changed and take a value of a set of phases corresponding to the mapping of the symbols, as shown in figure 1.25.

In FSK the frequency of the RF shifted optical carrier represents the digital information. One particular frequency is assigned to the '1' state and another frequency is assigned as the '0', state as shown in figure 1.26. FSK can be considered as a continuous phase modulation (CPM) process. An example of FSK is the minimum phase shift keying (MSK) whose frequency separation f_d *is* selected such that the signals carried by these frequencies are orthogonal and continuous at the boundary of the two states.

ASK can be combined with PSK to create a hybrid modulation scheme such as quadrature amplitude modulation (QAM), where both the phase and amplitude of the carrier are changed at the same time. The carriers are expected to follow a similar pattern as that of the DPSK in figure 1.25(d) but with different frequencies of the carrier under the envelop of the bit '0' and '1'. For MSK signals the carrier frequency is chirped up or down depending on the '0' or '1'. That is, the phase of the carrier is continuously varied during the bit period, and the carrier frequencies of the bits are such that there is an orthogonality of the carriers and the signal envelope.

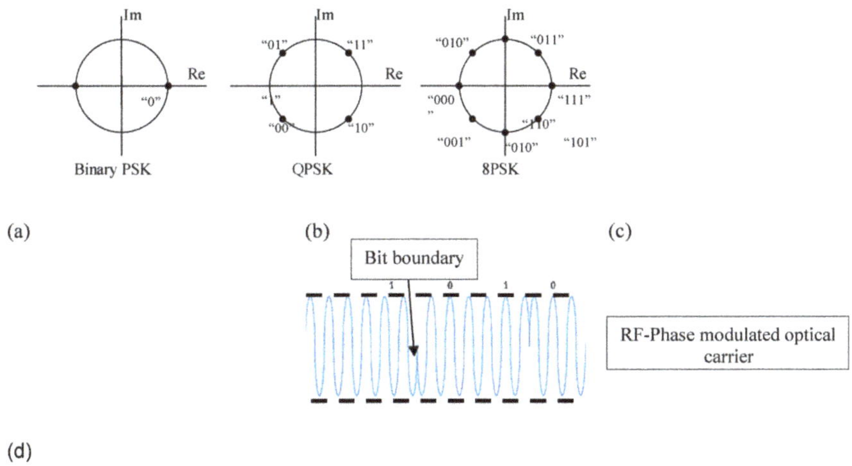

Figure 1.25. Signal-phase space constellation of discrete phase modulation: (a) binary PSK (BPSK); (b) quadrature binary PSK (QPSK); and (c) 8-PSK. (d) Carrier-phase under modulation with π phase shift of the BPSK at the edge of the pulse periods.

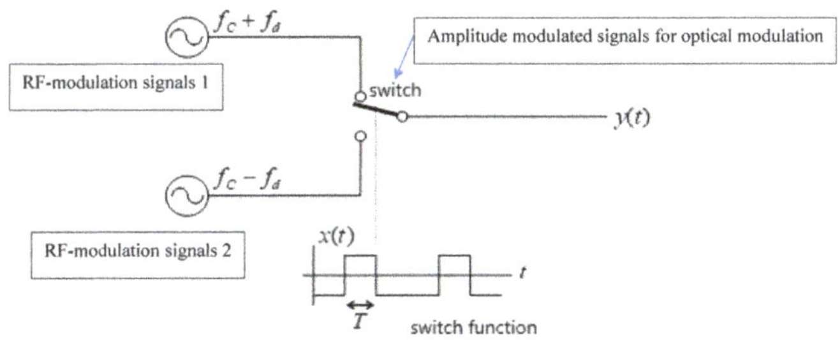

Figure 1.26. Structure of an FSK transmitter. f_d is the deviation frequency from the RF modulated optical carrier frequency of f_c.

1.8.2 Tri- and multi-level modulation formats

Additional degrees of modulation freedom and demodulation can be used to increase the information capacity by multi-level signaling with the same symbol/baud-rate. Under these modulation schemes the receiving sub-systems commonly incorporate DSP. The correction of the phase or frequency difference of the RF modulated optical carrier and that of the local oscillation can be resolved and can overcome many of the difficulties faced by the analog coherent RPoFs in the 1980s.

Tri-level modulation formats and pseudo-multi-level, tri-level, poly-binary signals are given in figure 1.27. These tri-level signals can be generated in terms of the phase or frequency of the RF modulated lightwave carrier. The modulator with RF input signals can offer such modulation of the optical carriers. This can normally be implemented by using a separate optical modulator for the RF signals and another separate modulator in cascade optical modulators for the data channel. This optical modulator can be narrowband because the data channel bandwidth is normally less than 1.0 GHz.

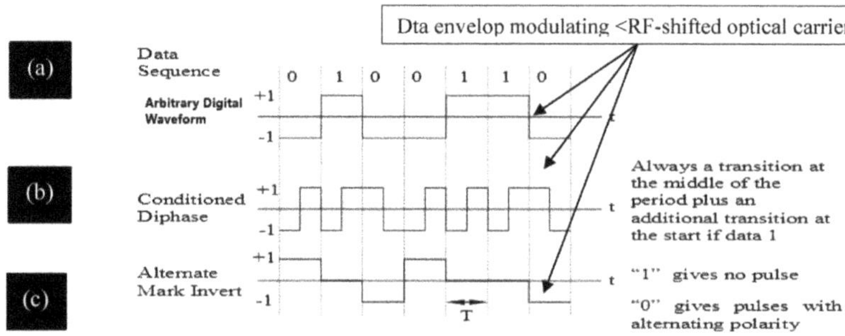

Figure 1.27. Baseband pseudo-multi-level or poly-binary baseband signals for modulating RF shifted optical carriers. (a) Binary data sequence digital signals, and (b) diphase RZ and (c) alternate mark inversion (AMI) formats. The RF carriers are not shown.

More than two symbols can be used to encode a single bit of information, enhancing the information bits per symbol, B_S, and hence the capacity. Under optical transmission -1 and $+1$ can be coded in terms of the variation of the phase of the RF modulated optical carrier as there is no negative intensity representation unless the field of the lightwaves is used. The tri-level uses $\{+|E|, -|E|, \text{ and } 0\}$ which with its equivalent phase representation can be mapped to $\{0, |E|^2\}$ at the optical coherent receiver, for example the duo-binary format. A phase difference of π and 0 between the three levels would minimize the pulse dispersion as they are propagating along the fiber due to the relative phase difference of π, hence destructive interference of any pulse spreading due to dispersion. These tri-levels can be implemented in the optical domain or in the electrical domain by the half bandwidth filtering and DSP-processing of a seven-level eye diagram.

This tri-level should not be confused with multi-level signaling, in which $\log_2 M$ bits are encoded on N symbols, and then transmitted at a reduced rate $B_s/\log_2 N$. Both multi-level amplitude or amplitude-phase shift keying and differential quadrature phase shift keying (DQPSK) can be considered as multi-level RF-optical modulation techniques.

1.8.3 Complex in-phase and quadrature phase (IQ or QPSK) modulation

Another form of modulation that would enhance the capacity of the transmission is the use of orthogonality or complex signaling. The information can be coded in both the in-phase and quadrature components, as shown in figure 1.28.

QPSK is the most commonly used approach in the differential and non-differential phase modulation in which the I and Q components are used extensively due to its efficiently high energy per bit, very similar to that of PSK, and doubling the capacity of PSK. QPSK is a summation of two orthogonal binary ASK signals. Thus two optical modulators are necessary to create such QPSK or IQ signals.

1.8.4 RPoF optical signal generation

External modulation is an essential technique for modulating lightwaves such that their linewidth preserves its narrowness and only the sidebands of the modulation

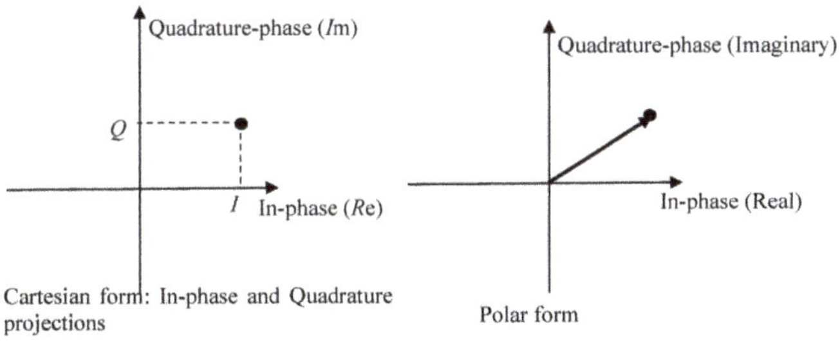

Figure 1.28. QPSK signal vectors represented in (a) Cartesian coordinates and (b) polar form; I = in-phase (real) component and Q = quadrature phase component.

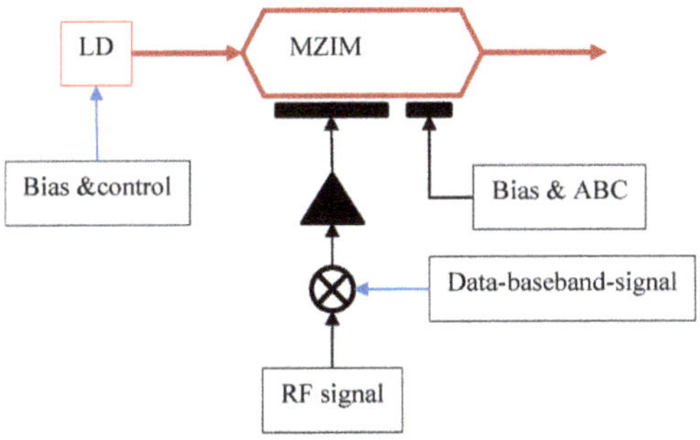

Figure 1.29. Generation of RF modulated optical NRZ using an external MZIM. ABC = automatic bias control.

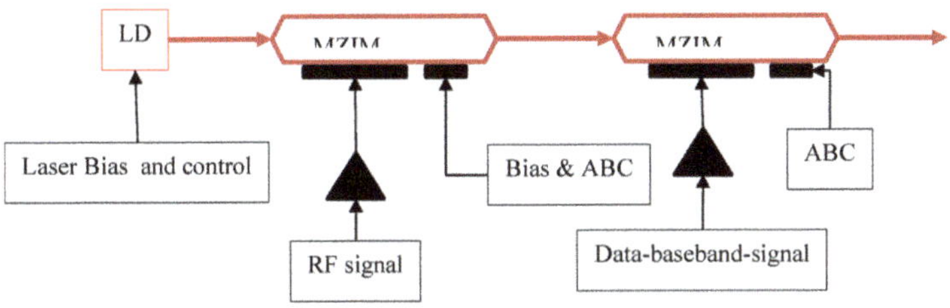

Figure 1.30. Generation of RF modulated optical RZ using two MZIMs, one to generate the suppression of the RF-optical carrier and one for NRZ data formats. ABC = automatic bias control.

scheme dominate the spectral properties of the generated passband characteristics. Figures 1.29 and 1.30 show the typical structure of optical transmitters for the generation of NRZ and RZ optical signals. The laser is always switched on and its lightwaves are modulated via the electro-optic modulator using the principles of interferometric constructive and destructive interference to represent the ON and OFF states of the lightwaves. The RZ can then similarly be generated but with an additional optical modulator that would generate periodic optical pulses whose width is half that of the bit period. The phase and frequency modulation can also be generated using these electro-optic modulators by biasing conditions and control of the amplitudes of the electrical digital waveforms. We note here that the fiber that connects the two modulators of figure 1.30 must be of polarization maintaining (PM) type. Otherwise there would be polarization fluctuation and hence reduction of the coupling of one lightwave power to the other.

The laser source would normally be a narrow linewidth laser that is turned on at all times to preserve its narrowness characteristics. The lightwaves are generated and coupled to the optical modulator via the pigtails of both devices. The modulator would be driven by the data pulse sequence output of a bit pattern generator condition to appropriate the driving level required by the V_π and the phase variation of the carrier if phase modulation is needed. When a modulation format is necessary then an electronic pre-coder is required to code the serial sequence to an appropriate coding. The pre-coder can be a differential coding, multi-level coding or inverse fast Fourier transform to generate orthogonal data sub-channels in the case of OFDM.

The pulses of all modulation formats can take the form of NRZ or RZ. For the RZ format an additional optical modulator is required to generate or condition the '1' NRZ to RZ, as shown in figure 1.30. The second modulator can exchange its position with that of the other modulator without affecting the generation of the modulation formats. This second modulator is usually called the pulse carver.

It is noted that if the RZ modulator is biased such that the phase difference is the π phase difference then a carrier suppression of the RF modulated optical carriers can be generated under the envelop of the ASK data channel. The bandwidth of the modulator determines the rise time and fall time of the edges of the pulse sequence in NRZ or RZ with suppression of the RF modulated optical carrier modulation formats, shown in figures 1.29 and 1.30, respectively.

1.9 Opto-electronic reception: incoherent and self-coherent

1.9.1 The reception analog front end

RF modulated optical differential PSK (DPSK) and MSK based formats can be demodulated self-coherently by using an optical balanced receiver which employs a Mach–Zehnder delay interferometer (MZDI). In the case of the optical DPSK format, MZDI is used to detect differentially coded phase information between every two consecutive symbols [8, 9]. This reception is carried out in the photonic domain as it is beyond the speed of the electrical domain, in particular at very high bit rates of 40 Gbps or above. Similarly, this technique can be used to generate an RF oscillator or reception to power intermediate frequency, as described in chapter 8. An MZDI assisted self-homodyne balanced receiver is also used for optical MSK signals by detecting the differential phase of MSK pulses.

As a subset of the continuous phase shift keying (CPFSK) family, MSK-RF modulated lightwaves can also be incoherently detected based on frequency discrimination. Thus, an optical frequency discrimination receiver (OFDR) using dual optical filters cooperating an optical delay line (ODL) is possible.

1.9.2 DSP-coherent optical reception of RPoF

Coherent and self-coherent detection and transmission techniques were exploited extensively over the last decade of the twentieth century so as to extend the repeaterless distance a further 20–40 km for SSMF of non-optically amplified optical transmission systems due to an improvement of the receiver sensitivity in the order of 10–20 dB, depending on the signal modulated format and coherent receiver structure.

In general, a coherent receiver can operate based on the beating of the received optical signals in the PD, with the optical field of a local laser oscillator (LO). This results from the square law detection of the PD, the total field of the LO, and the RF modulated optical signals. The beating optical signals give the phase of the carrier which is preserved and thus the detection of the phase of the carriers and its envelop can be used to recover the data information of a RPoF. Hence the phase modulation and continuous phase or frequency modulation signals can be processed in the electronic DSP domain. With the advancement of DSP, the processing of the received signals can be either in the intermediate frequency (IF) or baseband of the heterodyne and/or homodyne detection, respectively, and can be processed to determine the phase of the modulated and transmitted signals, in particular those under the QPSK complex-signal formats.

There are three type of coherent receivers, the homodyne, heterodyne, and intradyne types, and these are dependent on a frequency difference of zero, an IF greater or smaller than the passband of the signals between the LO and that of the RF modulated optical signal carrier.

Under advanced optically amplified fiber communications for RPoF, broadband amplified stimulation emission (ASE) noise always exists and under coherent detection the beating between the local laser source and ASE dominates the electronic noise of the receiver at the front end. The advanced aspects of coherent reception systems are currently pushed to a new generation by incorporating DSP sub-systems and naturally the ultra-high sampling rate ADC and DAC, permitting the bit rate per channel to reach 100 Gbps and higher to 800 Gbps using polarization multiplexing (PM) PM-QPSK or PM-M-QAM modulation schemes with M of 16 or 32 or higher for the RF modulated optical carriers.

Figure 1.31 illustrates an example of the baseband data channels in orthogonal frequency division multiplexing (OFDM) formats carried by an RF carrier on the optical domain, then transmitted, detected, and recovered by a DSP-based receiver with a DAC generating data channels via the use of an arbitrary waveform generator (AWG) and an optical IQ optical modulator (IQ = in-phase and quadrature phase). The OFDM format gives a pulse spectral sharp spectrum, hence the high spectral efficiency.

The schematic of an RF-256-QAM OFDM RPoF is given in figure 1.32. Figures 1.33(a)–(d) show the simulated RPoF using OFDM and a 256-QAM (quadrature amplitude modulation or M-IQ signaling) under coherent transmission and reception using a balanced receiver and polarization multiplexing. The spectra of the optical and electrical signals indicate block-wall type OFDM channels and the

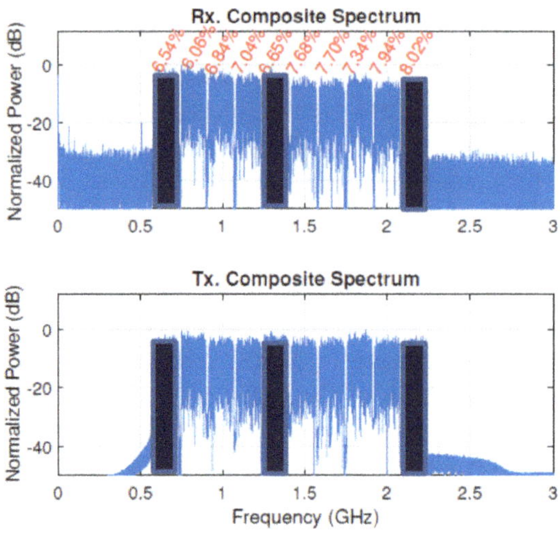

Figure 1.31. Typical DSP generated waveforms of channels and RF carriers modulating optical lightwaves and DSP-based reception output as processed by a DSP-based optical receiver.

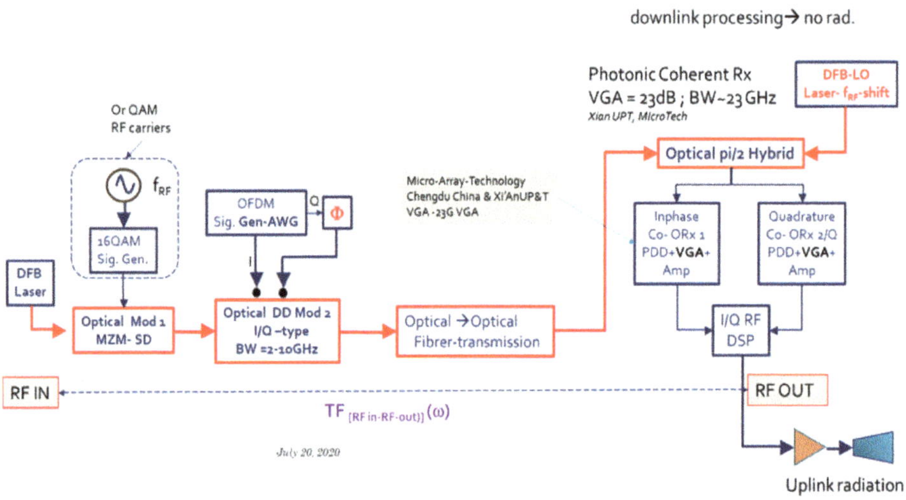

Figure 1.32. 16QAM-OFDM RPoF test-system QAM (RF)–OFDM (baseband): digital RPoF. DFB = distributed feedback; I/Q = IQ modulation; QAM = quadrature amplitude modulation; AWG = arbitrary waveform generator; VGA = variable gain amplifier; LO = local oscillator; f_{RF} = RF frequency; DD = dual drive.

constellation shows low distortion of the RF carriers after the reception. The VPI TransmissionMaker model is used as the simulation platform.

1.10 Organization of the chapters

The rest of the chapters of this book are organized as follows.

Chapter 2 outlines the essential enabling technologies for RPoF, in particular the optical modulation techniques, the demands on the phase noise and stability of the

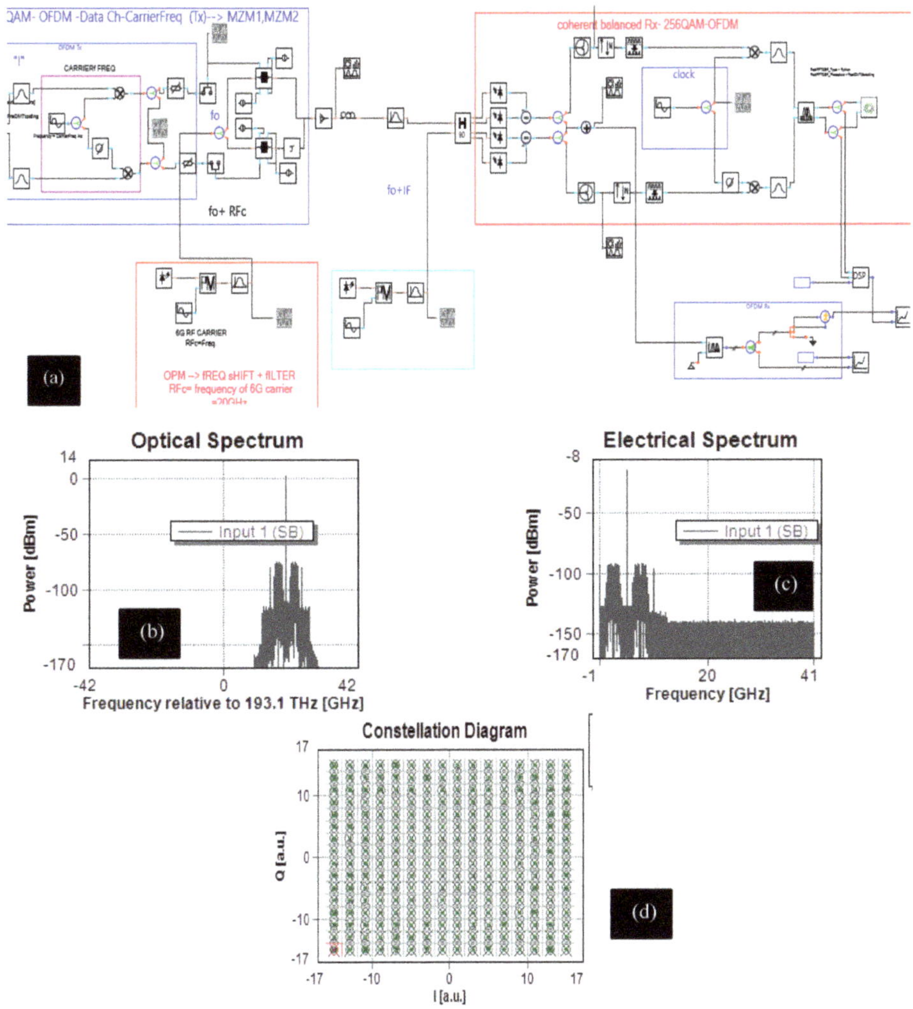

Figure 1.33. (a) VPI TransmissionMaker [10] simulated model for OFDM RF-optical carrier channels under coherent balance reception. (b) Optical spectra of two OFDM channels. (c) Electrical spectrum at the receiver. (d) Constellation of an RF modulated carrier by m-QAM.

narrow linewidth lasers, the propagation of radio-photonic signals over a fiber, and its nonlinearity effects on the distortion and intermodulation on information channels.

Chapter 3 presents techniques for the generation and transport of multi-channel RPoF of RF modulated optical carriers and data channels.

Chapter 4 presents the channels as well as techniques for linearization of RPoF via novel linear modulation devices and by suppression of intermodulation (IMD) via biasing methods and by parallel optical modulators. Furthermore, the processing of the RF modulated optical channels is demodulated and processing at lower sampling rates via self-homodyne and coherent reception techniques is given in order to minimize the requirement for a high sampling rate and thus high power consumption.

Chapter 5 gives an account of the access transmission system in both digital broadband and multi-carrier RF shifted radio-photonics over a fiber in the optical domain. Networking in 5G, 6G, and 7G is important and critical for high speed user access to technology. These are the bandwidth hungry areas and they drive the innovations in global information communication technology (ICT). Thus this chapter only outlines the essential features of the broadband access techniques.

Chapter 6 gives an account of the radio-photonic networking for 5G, 6G, and 7G networks, in particular for testing under a production environment or remote testing of antennae which operate under different frequency regions.

Chapter 7 gives an account of the RF generation using techniques of electro-photonic oscillation, in particular the parallel and differential resonance structures. This allows the possibility of the design and fabrication of a universal RF oscillator for the production of some structures which is tunable to any operating region of 6G and 7G RPoF systems. Further integration of an electro-photonic oscillator in a harmonic mode locked laser can generate both optical comb lines and RF oscillation.

Chapter 8 can be considered as an appendix in which a detailed account of the guiding properties of radio-photonic signals over fiber under linear and nonlinear dispersion and intermodulation effects. The transmittance transfer function is given so as to observe the effects of fiber linear dispersion and nonlinear distortion on single sideband or double sideband RF modulated signals.

References

[1] Cable Free 10+ Wireless Networks https://cablefree.net/wirelesstechnology/4glte/5g-frequency-bands-lte/

[2] Share Tech Note https://sharetechnote.com/html/5G/5G_FR_Bandwidth.html

[3] Popović M A 2010 Resonant optical modulators beyond conventional energy efficiency and modulation frequency limitations *Integrated Photonics Research, Silicon and Nanophotonics and Photonics in Switching OSA Technical Digest* (Washington, DC: Optical Society of America) IMC2, p 364

[4] Roger H L and Howerton M M 2001 Investigations on short-path-length high-speed optical modulators in $LiNbO_3$ with resonant-type electrodes *IEEE J. Lightw. Technol.* **19** 1287

[5] Binh L N 2017 *Optical ModulationOptics and Photonics* (Boca Raton, FL: CRC Press/Taylor and Francis)

[6] Kao K C and Hockham G A 1986 Dielectric-fibre surface waveguides for optical frequencies *Proc. IEE* **133** 191

[7] Binh L N 2013 *Digital Processing: Optical Transmission and Coherent Receiving Techniques* (Boca Raton, FL: CRC Press/Taylor and Francis)

[8] Binh L N 2010 *Digital Optical Communications* (Boca Raton, FL: CRC Press)

[9] Browning C, Gazman A, Abrams N, Bergman K and Barry L P 2018 256/64-QAM multicarrier analog radio-over-fiber modulation using a linear differential drive silicon Mach–Zehnder modulator *Proc. Int. Topical Meeting on Microwave Photonics (Toulouse, France, 22–25 Oct.)*

[10] VPI Photonics https://vpiphotonics.com/index.php

Chapter 2

Multi-carrier radio photonics

2.1 Historical remarks

In the years since the introduction of the concept and phenomena of using lightwaves and their guiding in a dielectric waveguide [1], there has been a tremendous increase in the demonstration and deployment of optical transmission systems and networks throughout the world. The bit rate distance and bandwidth distance products have increased over the last five decades.

In the early 1970s Corning demonstrated the guiding of lightwaves through a very short length of glass fiber. These guided optical waves were guided at 633 nm through a core–cladding silica fiber [2, 3]. Since then there has been an explosion in guiding phenomena using highly pure optical glass as the core and cladding using chemical vapor deposition and appropriate core–cladding radii such as to support either multi-mode or single mode use in the transverse plane. Thus the effective transmission of lightwave signals over long distances has increased from a few meters in 1970 to few million meters in 2020 with subsystems such as the transmitter and receiver in low loss spectral regions. This astounding medium, with a core diameter of about 8 μm and a cladding outside diameter of 125 μm, thinner than a human hair, has revolutionized global communications technology and enabled the Internet, with extensive interconnections bringing global human communities closer and closer together. Thus we have a much smaller world thank to information technology and information usage.

Transmission systems and optical networks began attracting the interest of several academic institutions, corporations, and research laboratories for the deployment of this technology and connecting human communities via terrestrial or underseas links. The developments of laser sources, low loss optical fibers of single mode type, optical and receiver amplifiers, optical modulators, and optical reception techniques and components accelerated the development of such optical

doi:10.1088/978-0-7503-3511-9ch2

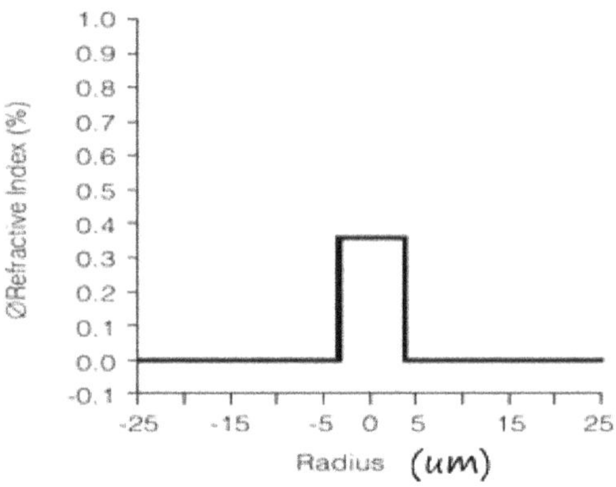

Figure 2.1. Refractive index profile of the SSMF or G.652 type by Corning SMF-28.

systems and networks that we have witnessed to-date. Integrated optical components have also been developed and now the integrated optical component size can be much smaller and more compact. They can now have a form factor of about 10 mm × 20 mm with a capacity a few million times higher than the 1970 version, both in terms of distance and bit rates.

The single mode fiber (SMF) was invented in association with the theoretical concept of weakly guiding phenomena, that is the lightwaves are guided through the fiber core whose refractive index is minutely different (higher) with respect to that of the cladding layer, about 0.3%, as shown in figure 2.1. Hence this allows much deeper penetration of the guided optical mode field into the cladding region, the evanescent field of the guided mode. This avoids higher loss due to core–cladding interface reflection. Furthermore, the group delay and the dispersion factors are also smaller for only one guided mode. Thus longer distance transmission is possible with a minimum dispersion factor, hence low distortion.

The guided mode distribution profile in the transverse plane of the SMF is circular. However, the polarizations of the mode can be randomly rotated as there is no deterministic direction of the guide cross section in either the geometrical or refractive index profile. This leads to the possibility of coherent transmission, that is, using the coherence properties of the guided mode and mixing with a local oscillator. Complex modulation and coherent reception can be employed so that the total capacity can be increased. More details on optical fibers and guided mode properties are provided in chapter 8.

Optical modulation techniques and modulators employing guided wave structures have been researched and developed extensively throughout the last few decades of the twentieth century [4, 5]. At the initial stage various material systems such as thin film deposition technology and diffusion or proton exchange technology were used to introduce titanium into $LiNbO_3$ or out-diffusion in $LiNbO_3$ substrates to create regions of higher refractive index due to the stresses and strain effects in these crystal structures, thus forming waveguiding regions or planar or channel waveguides. Electrodes were then deposited such that electric fields can be established across the guiding region and electro-optic (EO) effects can be created to manipulate the change of the phases of the optical waves. These phase changes can then be used to generate optical modulation using interferometric phenomena. In these decades the coupling losses in guided $LiNbO_3$ were high (about 4–5 dB total insertion loss or equivalent to 16–20 km of fiber losses). Thus prior to the invention of optical amplifiers these losses were too high and significant improvements had to be made. However, such coupling losses of integrated optic modulators cannot be reduced much further due to the nature of the coupling of the two different mode sizes of the circular fiber and planar or channel waveguide structures. Since then, the optical fiber amplifiers using erbium doped fibers were developed in the early 1990s, and revived the interest in and use of guided wave modulators in current advanced optical transmission systems. These integrated modulators are indispensable in today's photonic technology.

Parallel with the invention and development of optical fibers and guided photonic devices, wireless [6] and computing technologies have also been explored for higher and higher digital communication speeds. The high speed computing processors based on advanced micro-electronic technology have a linewidth resolution that can now go down to 7 nm, or about 1/10 the diameter of an average human hair. Thus the sampling rate can now reach 200 giga-samples per second (GSa s^{-1}), allowing significant improvement in the processing power of ultra-wideband signals. The photodetector bandwidth can reach 110 GHz with distributed amplification to offer a bandwidth of electronic amplification near the transition frequency of the transistors (~470 GHz for SiGe technology and ~270 GHz for InP technology) [7, 8].

Digital signal processing (DSP) systems with high sampling rates and fast memory allow the processing of samples of the received signals as well as the processing and formation of higher order complex modulation formats [9]. Thus we have now witnessed that the transmission rate can reach 1.6 Tbps, with 128 GBd 64QAM coherent reception and DSP generating format per wavelength. The transmission reach distance can vary according to the transmission link with or without optical amplification and dispersion compensation.

Furthermore, optical modulation by external optical modulators has been developed using $LiNbO_3$ [10–12] as the substrate and integrated Si-photonics with a doped p–n junction as the guided region under reverse bias. The 3 dB bandwidth can reach 65 GH. In addition, the electronics can also offer an amplifier reaching 110 GHz using distributed amplification and SiGe technology.

The fifth to seventh generation wireless networks allow several type of technological exploitation of information for communications, not only human-to-human but also machine-to-machine, vehicle-to-vehicle, and to the Internet of Everything (IoET) etc. The bandwidth becomes wider with the increase in RF, reaching the microwave (μW) and millimeter wave (mmW) regions.

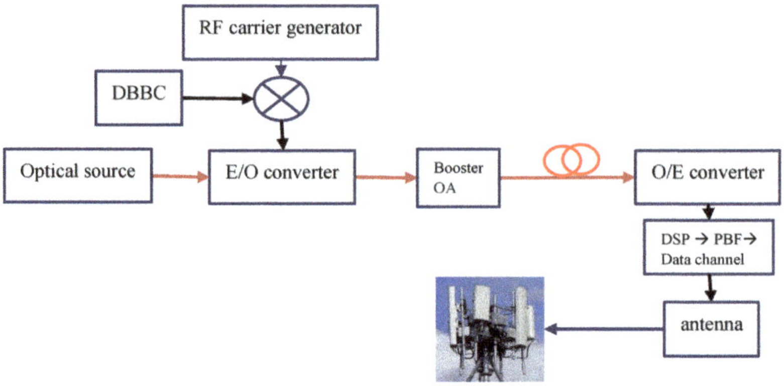

Figure 2.2. Schematics of an RF optical transmission system. OA = optical amplifier; DBBC = data baseband channel; DSP = digital signal processor; PBF= digital band pass filtering and data baseband algorithm; O/E = optical-to-electrical converter; E/O = electrical-to-optical converter.

Optical communications technologies have also enabled the transport of wideband high speed data channels and thus offer access media for the transport of wireless channels over global networks and ease of access by users whose personal device bandwidth can reach the wideband information. In order to make the transport of multi-channels of wideband information channels tenable, ultra-wideband optical modulation must be available to transfer the wireless channels into the optical domain.

These advanced development in the fields of optical modulators, ultra-broadband electronics in small signal amplification and large signal drivers, and ultra-high sampling rates in DSP have allowed advances in ultra-wideband transmission and radio photonic over fiber (RPoF) systems. That means that the RF carriers in the μW or mmW regions superimposed by data channels can modulate the lightwave carrier and then transport signals over the fiber. We thus term this physical mechanism as RPoF rather than the term used commonly in previous years, namely radio over fiber (RoF) or radio frequency over fiber (RFoF).

A basic schematic of the transmission of RF signals, including the RF carrier and the data baseband channels (DBBC), is depicted in figure 2.2. An optical source is the most essential lightwave generator for the generation of optical waves so that the optical modulation can be implemented to impede the RF carriers and the data baseband channels. Hence they can be transported over the optical fiber. The optical fiber suffers attenuation effects or propagation spectral loss so the spectral windows are normally selected so that minimum losses can be obtained. For long-reach transport the C-band (1520 to 1565 nm) and L-band (1565 to 1620 nm) are commonly used. The optical power at the output of the electrical-to-optical (E/O) converter can be optically amplified by a booster and then received at the other end of the fiber by an optical-to-electrical (O/E) converter and then digitally processed via the DSP, in which an analog-to-digital converter (ADC) is incorporated with a high speed sampler. Processing algorithms can be used to recover the transmission rate of the DBBC, the optical carrier, and the low frequency offset (LOFO) in the case of coherent reception.

Depending on the receiver sensitivity or the level of signal required at the output of the O/E converter the reception can be direct detection or coherent reception, which can be the homodyne, heterodyne, or intra-dyne method depending on the frequency offset (FO) between the local oscillator and the carrier of the RF in the optical domain.

This chapter provides details of some of basic advances of optical modulation technologies to carry wireless channels in the μW and mmW regions.

2.2 Optical and RF spectra

2.2.1 Optical fiber spectrum and low loss spectral windows

In order to appreciate the signal propagation characteristics of optical waves and RF waves in optical fibers and free-space media, this section gives a brief outline of the propagation loss and dispersion characteristics for signals across the whole spectrum, which is the most feasible and extensively exploited option for optical networks. We also cover the attenuation and propagation characteristics of RF carriers and their embedded data baseband channels so that we can refer to and appreciate the motivations for propagation in either optical or RF media.

2.2.2 Optical fibers

Since the introduction and demonstration of silica glass based circular optical waveguides [1, 2], extensive work has been carried out over the last 50 years on several components and transmission media, the single mode optical fibers (SMFs) and then the most common standard SMFs (SSMFs). The C-band spectral window (1520 to 1565 nm) and wavelength with 100 GHz channel spacing can be seen in table 2.2 and [65]. For 50 GHz spacing the frequency of the optical channels can be found by finding the middle frequency of the two adjacent channels. The optical frequency is to be selected and then the corresponding wavelength. This C-band was extended recently to the C+- and then the L-band. Five spectral windows, called the O-, E-, S-, C-, and L-bands, indicate the operating wavelength regions. The C+-band can now be extended so that the total number of channels of 50 GHz spacing can reach 110. Over these spectral grids the RF sub-carrier multiplexed channels are used when integrating such RPoF channels and the broadband optical channels.

Furthermore, the dispersion characteristics of SMFs can be given by equation (2.1) in which we can see that the dispersion can be zero at a certain wavelength. For the Corning SSMF type or SMF-28 this zero-dispersion wavelength is in the region of 1290–1310 nm. The variation is due to the variation of the fabricated uniformity or variation of the geometrical dimension of the core of the fiber during the fiber drawing process, and the minute fluctuation of the doping concentration of impurities in SiO_2, hence the core refractive index. Figure 2.2 shows the dispersion factor as a function of wavelength for silica doped GeO_2. The zero-dispersion wavelength is within 4 nm of 1300 nm. In principle the total dispersion factor is composed of the dispersion due to material, the refractive index of the core material varying with respect to wavelength, and due to the geometrical structure and the refractive index profile of the core and the cladding. Thus one can tailor the core refractive index, the geometrical structure, and the doping material to obtain a specific dispersion, such as zero-dispersion shifted fibers (DSFs) or dispersion flattened fibers (DFFs). The equation is

$$D(\lambda) = \frac{S_0}{4}\left\{\lambda - \frac{\lambda_0^4}{\lambda^3}\right\}^2, \tag{2.1}$$

where $D(\lambda)$ is the dispersion factor as a function of wavelength λ, in units of ps nm^{-1} km^{-1}, S_0 is the zero-dispersion slope (<0.092 ps nm^{-2} km^{-1} for SMF-28), that is, the slope of the dispersion curve versus λ_0 at which the dispersion factor is zero, and λ is the operating wavelength. Figure 2.3(a) shows the variation of the the attenuation factor with respect to the wavelength across the spectral windows in which optical channels are transmitted 1200 nm to 1625 nm of optical fibers of Corning type G652A and G.652D of the family of SMF-28. Note that G652D is in blue and the resonant loss around 1370 nm is eliminated as compared to G.652A, that is, the 'water OH' absorption is eliminated during the fabrication process. Figure 2.3(b) depicts the suppression of the water OH absorption by special fiber fabrication process during the fiber drawing, this is the G652D fiber type. Figure 2.4 depicts the dispersion factor versus wavelength for a silica core doped with GeO_2. The dispersion zero wavelength at 1312 nm. Different slopes show different impurity doped materials in the fiber core. So we can note that for silica fiber, no dispersion at wavelength around 1300 nm depending on the doped impurity in the fiber core. So if the absorption can be suppressed then the spectral window 1280 nm to 1320 nm can be highly attractive for transmission of optical channels. This is the O-band. However, the spacing between the central wavelength must be not uniform, otherwise four wave mixing would occur to crate cross talk.

Figure 2.5 depicts the transmittance versus RF frequency, that is the frequency response of the SSMF under the propagation of RF modulated lightwaves whose narrow linewidth is very narrow. It is noted that there are a number of notched frequencies of the transfer transmittance function. This is due to the destructive interference of the two sidebands of the modulated optical waves propagating along the fiber. Within the 20 GHz band there are three spectral dips so if the 5G or 6G RF carriers are to be carried along the SSMF by a double sideband (DSB) signal, there would be non-flat frequency responses and hence severe distortion [13]. This is even more severe if the transmission distance is much longer than 1.0 km. Thus, single sideband (SSB) signals can be applied by modulating the optical carrier using proper biasing and levels of applied RF signals [14, 15]. Chapter 3 will address this technique for high carrier frequency RF carriers in RPoF systems.

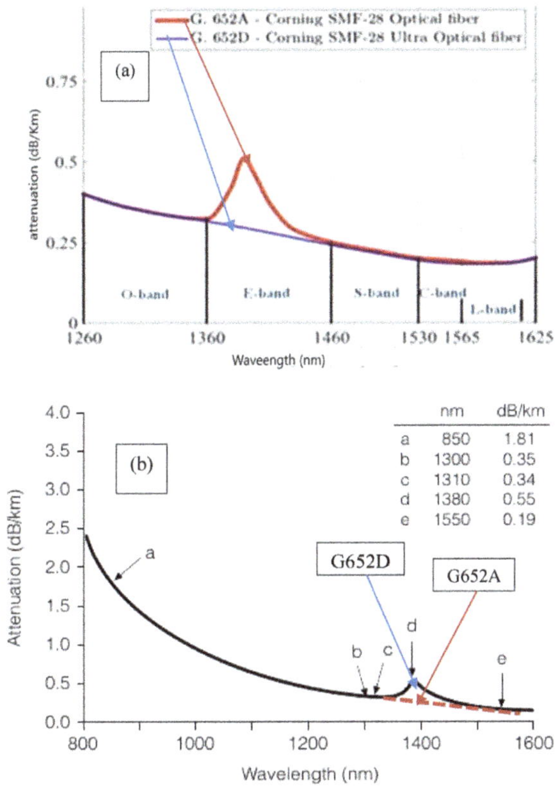

Figure 2.3. The attenuation factor (dB K^{-1}) of an optical fiber of Corning type G652A and G.652D of the family of SMF-28. Note that G652D is in blue and the resonant loss around 1370 nm is eliminated as compared to G.652A, that is the 'water OH' absorption is eliminated during the fabrication process. (a) The general and magnified spectral regions around the 1550 nm region. (b) The Corning fiber. In the G652D the water OH absorption is removed via a special process during fiber drawing.

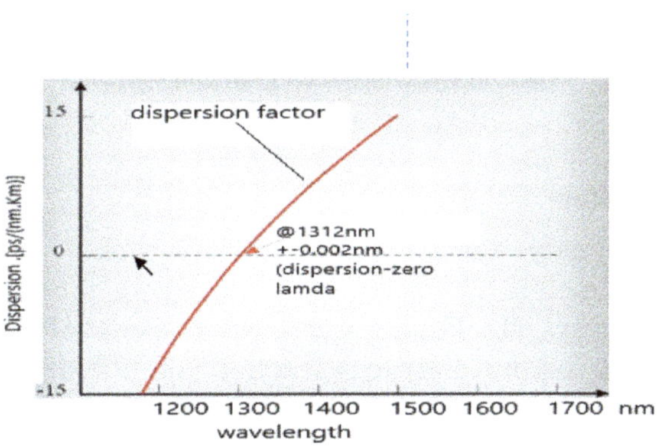

Figure 2.4. Total dispersion factor versus wavelength for a silica core doped with GeO_2. The dispersion zero wavelength at 1312 nm. Different slopes show different impurity doped materials in the fiber core.

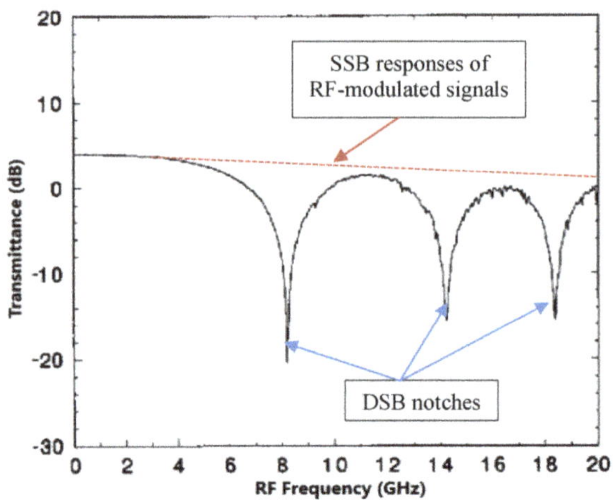

Figure 2.5. Linear fiber transfer transmittance (in dB) as a function of RF frequency whose signal modulates the lightwave and transmission over a 100 km long SSMF. Plots show the transmittance (dB) or optical power out/optical power in as a function of the RF frequency oscillator. Lightwave source = single mode DFB (distributed feedback) type. SSB = single sideband; DSB = double sideband. The laser is a narrow linewidth type.

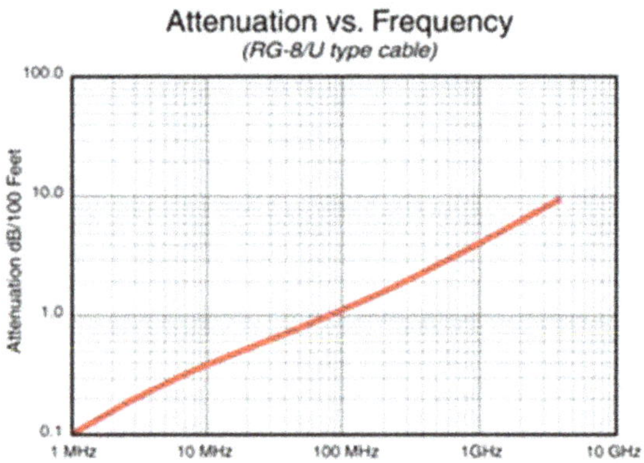

Figure 2.6. Attenuation factor as a function of RF frequency over a length of 30 meters. For up to 20 GHz, a linear project can be made for higher frequency. Cable type: RG-8/U coaxial.

2.2.3 Coaxial RF medium

Regarding the coaxial cable the attenuation factors cannot provide sufficient low attenuation over 30 meters characteristics (see figure 2.6) for interconnection between 5G or 6G microcells. Higher frequency coaxial cables such as V- or K-type can also lower the loss but only by a fraction, about 5 dB lower, but the phase fluctuation can also severely distort the signals. Thus the SSMF attenuation curve (figure 2.3) is much preferred. Hence the RPoF technology is needed urgently for 5G and 6G networking.

2.3 Optical transmitter

2.3.1 Narrow linewidth lasers for RPoF

2.3.1.1 Definition and uses

A narrow linewidth laser can be defined as a single frequency laser whose linewidth is very narrow, in the order of 10 to 100 kHz. Thus the coherence length of the laser is very long. These lasers are essential for modern coherent long-haul communications transmission, in particular RPoF technology.

A number of laser applications, such as coherent transmission, low dispersion transmission under direct detection and external modulation, and RPoF links, require lasers with a very small optical linewidth, that is with a narrow linewidth optical spectrum. The term 'narrow linewidth lasers' refers to single frequency lasers, that is the laser cavity resonates only at a particular desired wavelength with low phase noise, and hence high spectral purity. Typically, such lasers also exhibit low intensity noise with respect to the carrier, the carrier-to-noise ratio (CNR) being less than -130 dB Hz^{-1}.

2.3.1.2 Narrow linewidth lasers

The most important types of narrow linewidth lasers (NLWL) are the following:

- Distributed laser diodes (DFB lasers) and distributed Bragg reflector lasers (DBR lasers) operating e.g. in the 1.5 or 1.0 μm wavelength region, are the most common. Typical operation characteristics are an output power of some tens of milliwatts and a linewidth of a few megahertz.
- Significantly smaller linewidths (3 dB bandwidth in the spectral domain of the frequency scale) from semiconductor lasers are implemented by using an external cavity or resonator with a single mode fiber with an inline grating, a narrow bandwidth fiber Bragg grating, or with a free-space grating with a high reflection coefficient and narrow passband. Linewidths of a few kilohertz or even below 1.0 kHz can be achieved.
- Small, compact erbium doped fiber lasers of the resonator form a pair of fiber Bragg gratings as a cavity. Pump energy from a semiconductor laser (SCL) can be injected to create lasing action. The fiber laser can generate tens of milliwatts of output power with a linewidth in the region of a few kilohertz, e.g. NKT Photonics fiber lasers [16].

2.3.1.3 Conditions for narrow linewidth lasing

To achieve narrow bandwidth (linewidth) emission, it is required that:

1. The laser cavity has a high gain medium with a resonant cavity of narrow gain bandwidth. The laser resonator must has a short length such that a large free spectral range (FSR) can be obtained. Longitudinal side mode suppression is high.
2. The external noise is minimized. This requires a highly stable resonator cavity, normally a single polarized cavity so that there is no polarization fluctuation of the optical mode. If using an electrically pumped laser then a

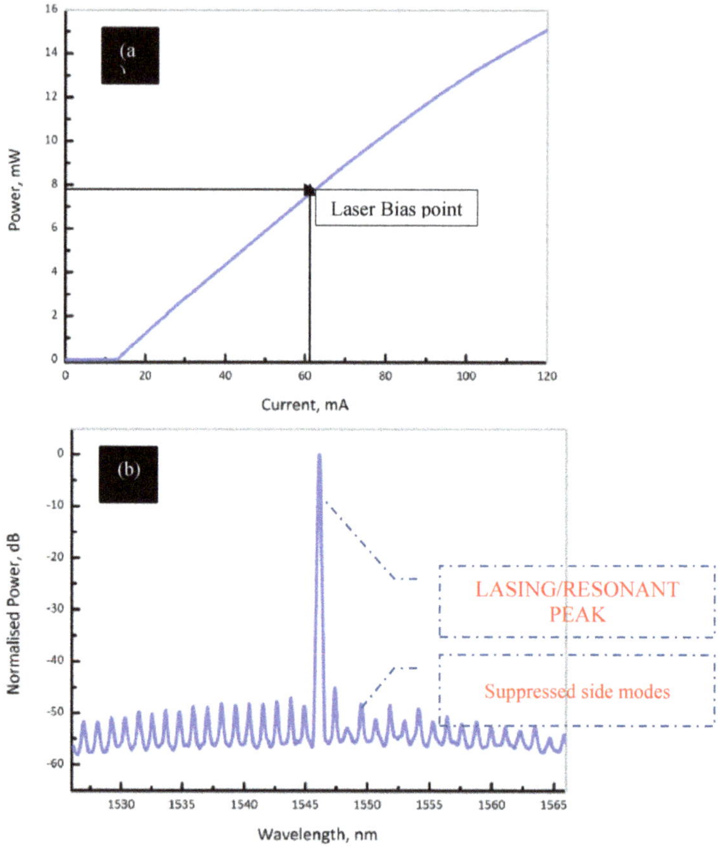

Figure 2.7. Narrow linewidth laser bias and emission characteristics of an ECL laser. (a) Lasing power versus cavity driving current. (b) Laser spectral emission versus generated wavelength.

low noise voltage or current source should be used. Any optical positive feedback is to be avoided, using a Faraday isolator in the resonant cavity.

3. The laser design is optimized so that the laser phase noise is minimized.

A typical NLWL (Eblana Co Ltd) can have the following properties: (i) a linewidth as narrow as 100 kHz; (ii) a stable single cavity without an external cavity or feedback; (iii) tunable by either temperature or current; (iv) very low mode partition noise, low relative intensity noise (RIN) = 160 dB Hz^{-1}; and (v) low sensitivity to environmental and mechanical vibration. The external cavity can be etched so that the cavity can contain an optical filter, a Fabry–Pérot filter, so that only discrete modes can be amplified [17] (see figure 2.7).

2.3.2 Comb lasers: generation of multi-sub-carriers by recirculating frequency shifting

Comb lasers could play an important role in RPoF due to their property that several equally spaced laser lines exist which can be demultiplexed into individual lines and

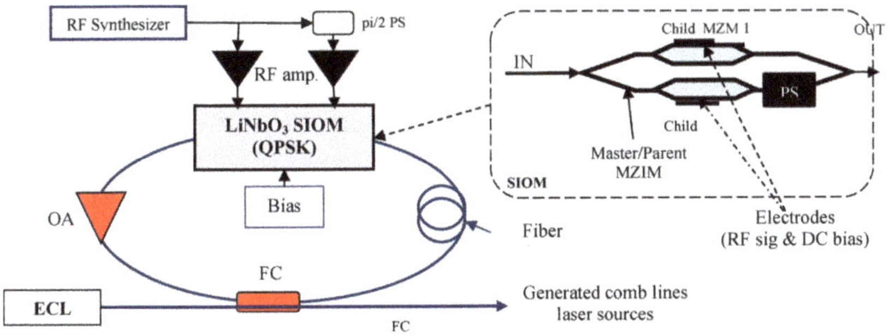

Figure 2.8. Schematic structure of the RCFS comb generator. ECL = external cavity laser; OA = optical amplifier; EDFA = Er-doped fiber amplifier; FC = fiber coupler; PS = phase shifter; SIOM = interferometric synchronous optical modulator or QPSK modulator.

modulated with RF waves of different frequencies of the 5G and 6G bands. This sub-section provides details of a fiber ring comb laser for comb line generation for RPoF system applications.

2.3.2.1 Generic structure

The structure of a comb generator employing the recirculating frequency shifting (RCFS) technique is depicted in figure 2.8. An optical recirculating loop consists of an inline optical modulator which transfers the RF into an optical RF modulated carrier, an optical amplifier to compensate for the loop optical insertion losses. Furthermore, a loop inline optical adjustable attenuator is incorporated to adjust the loop's overall gain so that a slightly less than unity gain can be achieved so as to avoid self-oscillation or lasing. Recently, such a recirculating comb source has been demonstrated in [18, 19]. A 3 dB optical coupler couples into and outputs continuous lightwaves and the generated RF shifted optical sub-carrier lines, respectively. This fiber coupler can be replaced by an optical switch and associate couplers for generation of shifted carriers to both sides of the main lightwave carrier—the dual shift recirculating comb generator which is described in this section and illustrated in figure 2.8. An electrical synthesizer generating sinusoidal waves whose frequency determines the frequency shifting of the optical lines of the generated comb is used to apply into the electrodes of the modulator, hence via the electro-optic effect to modulate the phase of the guided lightwaves and thus optical modulation.

2.3.2.2 Complex optical modulator for multi-lasing line generation

The synchronous interferometric optical modulator (SIOM) or complex optical modulator is the main component in the loop that can be termed as synchronous type, as depicted in figure 2.8. Indeed, the SIOM is the same as the QPSK modulator or complex modulator, in which real and imaginary modulated lightwave components are combined at the optical output. These synchronize the propagation of the modulated lightwaves and the loop delay so that no collusion of many optical comb laser lines occurs. The complex optical modulator is composed of a master/parent Mach–Zehnder interferometric modulator (MZIM) whose two arms are embedded with two slave/child Mach–Zehnder modulator (MZMs) electro-optic interferometric modulators. The lightwaves guided through the two child MZMs can be biased and activated such that a constructive interference mode can be generated at the combined output of the parent MZM. Optical waveguides are formed by diffusion of thin Ti-film deposited on top of the Z-cut or X-cut $LiNbO_3$ substrate. Optical waveguides can support only a single mode of both TE and TM polarizations at the input and output waveguide branches of the parent MZM.

Only one polarized mode is excited in the circulating ring. In an MZM, thin SiO_2 film is deposited on the electrode area prior to the deposition of the electrode film to obtain a matching of the velocity of the RF traveling wave and that of the guided optical waves. This allows wideband transfer $S_{21}(\omega)$ characteristics of the MZM. Traveling wave electrodes (TWEs) are independent with respect to the dc electrodes (i.e. unconnected) so that independent biasing to create the initial phase states of the guided optical waves can be performed.

In addition to the two dc electrodes deposited in one light path of the child MZIMs, another dc electrode is on the phase modulator of one child MZM (the Q-branch) so that a $\pi/2$ delay can be generated to create complex optical signals at the parent MZM output. Synchronization of the two RF waves is to be achieved. Thus this modulator can be called a synchronous interferometric optical modulator (SIOM) or synchronous QPSK modulator.

As depicted in the inset of figure 2.8, the modulators in the two arms of the SIOM are known as child MZMs are excited with $\pi/2$ phase shift RF oscillation signals with respect to each other so that single-sideband (SSB) lightwaves can be generated at the output of the parent MZM. Such implementation by a $\pi/2$ phase shift is indeed performing a Hilbert transformation so that the imaginary components are canceling each other out, hence the SSB carrier.

The principles of the shifting of the optical carrier frequency to the right or to the left of the original lightwave carrier are shown in figure 2.9. Applying sinusoidal or co-sinusoidal RF signals to the electrodes of the child MZMs will shift the lightwave

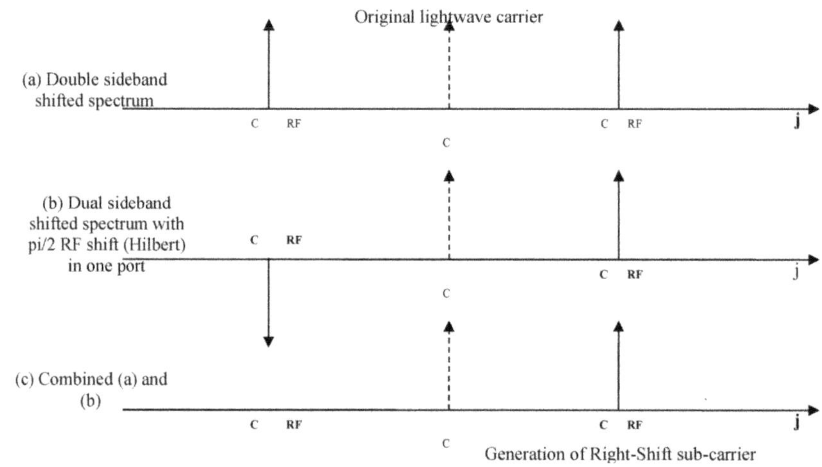

Figure 2.9. The spectra of (a) an RF modulated carrier and (b) a $\pi/2$ phase shift or Hilbert transformed RF wave driven signal modulated lightwave. (c) The constructive and synchronous spectrum of the generated right shift sub-carrier. Reproduced with permission from [32].

carrier to the right or left, respectively, due to the summation of the two modulated lightwaves emerging at the output of the child MZM being constructively or destructively combined in either band of the spectrum, thus resulting in SSB frequency shifting. The SSB shift is critical so that multiple lines or interference of the other sideband sub-carriers with the shifted frequency sub-carriers generated in the previous modulation can be avoided. It is noted here that the SSB principle was first demonstrated by Smith *et al* [20] to transport RF signals over long length fibers. In our recirculating comb generator, the optical wave is shifted and then circulates through the loop and is then shifted again and again to generate other multiple shifted sub-carriers.

Thus the SSB modulation serves two purposes, first to shift the original carrier only to either the right or the left side of the original carrier, and second to minimize the interference of the other sideband generated carriers if double sideband (DSB) modulation is employed. This SSB modulation technique was also employed by several research groups for transmission over a long-haul fiber to carry radio channels [21–23] or digital channels using orthogonal frequency division multiplexing (OFDM) [24–26] and in the comb generation for composite terabits per second (Tbps) transmission but not in synchronization [27–30].

Therefore, for the case of dual band shifting of the lightwave main carrier, described later in section 2.3, two comb generators would be required combined with an optical switch and optical couplers to generate the shift left and shift right sub-carriers with respect to the central original lightwave following the principles illustrated in figures 2.9(b) and (c) with both the normal and the reverse $\pi/2$ phase shift.

2.3.2.3 Implementation of the comb generator

The comb generator is implemented using fiber-based devices. A ring of fiber of length of about 25 m is formed with one input and one output port of a 3 dB optical fiber coupler (FC). This means that there is always a 50% intensity of an RF shifted lightwave coupled back into the ring and 50% coupled to the output port. A $LiNbO_3$ SIOM is incorporated in the fiber ring. The ring length is accounted for by all the pig-tailing of optical components associated within the ring. The overall gain of the ring must be less than unity so as to avoid any lasing. Er-doped fiber amplifiers (EDFA) are included so as to compensate for the optical loss of the fiber, the modulator, and the coupler FC. The open loop is constructed first and the modulation of the lightwave carrier by a purely sinusoidal wave from a reference synthesizer is applied. The shifting of the lightwave carrier to the left or to the right can be observed, as shown in figures 2.10(a) and (b), respectively. This frequency shifting can be shifted to the right or to the left side of the optical central carrier depending on the relative phase shift applied to the two slave electro-optic intensity modulators (EOIMs) embedded in the two arms of the master MZIM. As observed, the suppression of the residual lightwaves can reach higher than 25 dB through the tuning of the biasing voltage applied to the arms of the EOIM, in particular the dc bias of the master MZIM. This is why synchronization of the optical modulator is essential for efficiency and a high carrier-to-noise ratio (CNR) to be achieved.

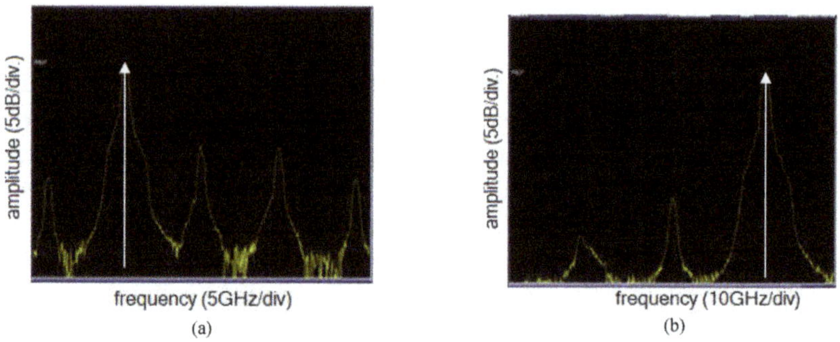

Figure 2.10. The spectrum of the SSB lightwave at the output of the synchronous optical modulator: (a) shifting left and (b) shifting right in the frequency scale. Resolution of the horizontal scale for (a) is 40 pm div^{-1} and for (b) is 80 pm div^{-1}, and is 5 dB div^{-1} in the vertical scale for power. Reproduced from [33, 34] by permission of Taylor & Francis Ltd.

Under the synchronization scenario a minimum phase noise level can be obtained. This is very important for these comb lines to be employed as low noise sources or the CNR for long-haul and coherent reception of the modulated sub-carriers over many thousands of kilometers of SSMF. These residual sub-carriers must be at this level so that noise due to interference with other generated sub-carriers can be considered negligible.

The comb stability can be controlled by a control feedback circuit. The feedback signals come from the synchronization of the jittering of the referenced RF synthesizer, the injection optical power level into the recirculating ring, and the pumping level of the pump laser into the loop inline optical amplifying devices. The RF frequency determines the spectral spacing between the comb lines. The RF signals can be generated from a micro- or millimeter wave synthesizer, and applied to the two arms of the EOIMs. RF amplifiers can be used to obtain appropriate RF signals so that a maximum modulation depth can be achieved with minimum nonlinear distortion effects. It is noted that although the modulator 3 dB bandwidth can be at 23 GHz, it is still possible to operate the modulator at 33 GHz of the 6 dB level or even to 40 GHz at 10 dB with some RF signal loss, hence requiring higher RF signal power. This is because the driving signal is purely sinusoidal and exhibits a very narrow band such that the nonuniformity can be affected so much. We have generated comb lines from 28 GHz to 33 GHz successfully for channel spacing between sub-carriers, as shown in figure 2.11. Note also in figure 2.11(b) the residual original carrier in the far right of the spectrum. This appears because of the use of a moderate gain optical amplifiers (OA of EDFA type) and it takes some time for the optical signals to reach a saturation state in the amplifier. Depending on the loop and the gain of the OA with respect to those components embedded in the recirculating ring, there may be one or two residual frequency shifted lines in the spectrum, as is observable in figure 2.10.

The spectrum is flat over some spectral range, about 5.5 nm or the 687 GHz band which is sufficient to support 13 sub-carriers of 50 GHz frequency spacing in the 1550 nm spectral region. With these 13 sub-carriers and the symbol rate of 28 GBd s^{-1} per sub-

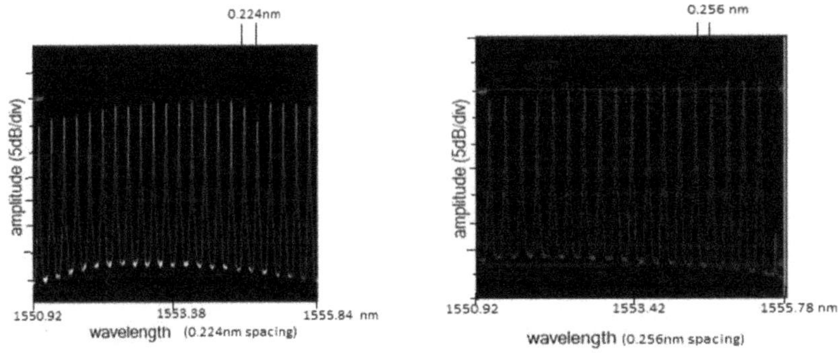

(a) (b)

Figure 2.11. The generated multi-sub-carriers and comb lines of an RCFS loop. Vertical scale: arbitrary power amplitude. Horizontal scale: 50 GHz div^{-1} frequency.

carrier channel, the aggregate bit rate is 100 Gbps per sub-carrier if using two polarized channels and QPSK or 2 bits per symbol. Similarly, this can support an RPoF system with 40 sub-RF carrier multiplexing of the 1 GHz data channel band and guard band per comb line. Altogether over 13 comb lines a total capacity for an RPoF system of 13 × 40 GHz or 520 GHz and 4.16 Tbps with 8 bits Hz^{-1} is achieved if the 256 QAM format is used for RF carriers and OFDM data modulation format is utilized.

The comb lines can also be equalized by using a wavelength selective switch (WSS) to obtain amplitude equalized sub-carriers. The ripple level of the comb lines depends on the flatness of the optical amplifier incorporated in the ring, thus a flat spectral OA would normally be selected for this type of recirculating RF shifted comb generator. On both the left and right shift spectra a CNR is achieved in the order of 27 dB. This is quite remarkable compared to the other comb generated lines reported so far [31, 32]. We believe that this is contributed by the synchronization of the modulated lightwaves passing through the master MZIM and slave EOIs.

A narrow linewidth external cavity laser (ECL) used as the primary lightwave is launched into the SSB frequency shifted fiber ring to generate comb sub-channels as described. Non-resonance indicates that the resonance condition for lasing is not satisfied by ensuring that the total optical loop gain is less than unity. The ECL output power is set at its highest level of 13 dBm. A higher power level can be used if an EDFA booster amplifier is employed.

The comb line spectra are shown in figures 2.11(a) and (b). It is noted that the comb line power levels at −10 dBm and −4 dBm are due to the lightwaves passing through an EDFA operating under the saturation condition of an output power of 16 dBm. This total power is distributed to all sub-carrier lines.

Both sides of the primary laser lines can also be generated by employing a dual ring structure with the SSB shifted to the right or the left of the primary line, as described in the next section. This ultra-wide spectral comb line source has also been generated in our laboratory.

The comb generator described in this section is employed in a multi-Tbps transmission system [34], but not yet in RPoF.

2.3.2.4 A full band frequency shifting recirculating comb generator

Further to the possibility of generating comb sub-carriers over the whole C-band in the region of 1520 to 1565 nm for optical communication systems with the referenced central lightwave carrier located at the center of the band, the frequency shifting of the original carrier must be into both the left and right sides of the carrier. We thus propose a dual band comb generator structure in which two recirculating frequency shifting ring comb generators share 'a common optical switching structure', as shown in figure 2.12.

Further detailed principles of the experimental generation of the dual band comb lines can be described as follows. First, the continuous laser line is fed into a 2 × 2 fiber coupler (FC) (one port left unused) whose outputs can then be fed to 2 × 2 optical switches, as shown in figure 2.12. The integrated optical guided wave switch is shown as an inset in this figure. The optical switch is of $LiNbO_3$ type and fabricated using a Ti in-diffusion technique [35, 36]. The electrical switching signals are fed into the electrical port of the OSWs as shown by the dotted line in the figure. All optical paths are shown as continuous lines. The bandwidth of the OSW is 18 GHz which allows the switching of 100 ns to be achieved without any difficulty. The switching electrical signals are complementary to each other, indicating that one switch is on and the other is off. The switching period of the electrical switching signal is just less than that of the circulating delay time which is estimated to be 100 ns for the 20 m long fiber loop of each comb generator. Each output of the 1 × 2 OSW is fed into a comb generator whose RF signals are fed into an SIOM with the $\pi/2$ shifts in reverse to each other in the RF signals applied to the electrical ports so that the shift left and shift right of the sidebands can be obtained.

The lightwaves are injected at a power level such that the optical amplifiers incorporated in the ring, similarly to that given in figure 2.8, can operate in the

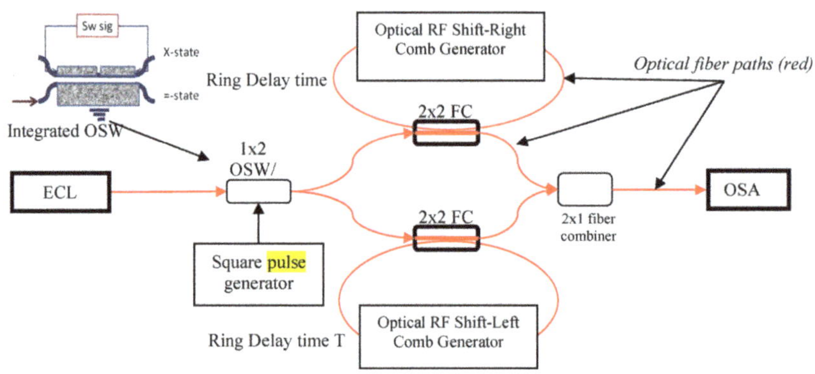

Figure 2.12. Schematic of the experimental dual band comb generation. OSW = optical switching device with a switch time much shorter than that of the ring propagation delay time *T*; FC = fiber coupler; OSW = optical switch; OSA = optical spectrum analyzer. All continuous lines are optical paths. Dotted lines are electrical lines. The inset is the structure of an integrated OSW. Reproduced with permission from [32].

saturation mode. The outputs of the two ring comb generators are then combined via another passive FC. The output of the dual band comb generator is monitored by an optical spectrum analyzer (OSA). The switching time of the optical switch must be much less than that of the ring delay time so that no interference will occur. It is obvious that the delay time of the lightwaves circulating through a loop is sufficiently long, about 75 ns for a 25 m long ring. Thus a moderate bandwidth optical switch can be employed. Such electro-optic switches with a 3 dB bandwidth of 20 to 30 GHz in $LiNbO_3$ are commercially available. These switches can thus provide the switching time between shifting left and right within 50 ps to 33 ps [37–39].

The observed dual band spectral comb lines are shown in figure 2.13(c). We note here that due to the switching efficiency of the OSW being not 100% but about 95%, there are some residual optical carriers which are embedded in the shifted optical sub-carriers, which could cause some interference noise in subsequent comb lines. Thus we expect some degradation in the performance bit error rate (BER) versus optical signal-to-noise ratio (OSNR) as compared to those depicted in [32]. We also observed some undulation of the amplitudes of the comb lines of the shift to the left side due to unequal transmission loss in the recirculating loop. Indeed, we have also set up the dual ring comb generator using a passive FC instead of the OSW shown figure 2.12. The optical spectrum of the comb sub-carriers can also be observed in figure 2.13. We note here that the optical amplifiers embedded in both comb rings of the dual ring comb generator are implemented with a high output power of about

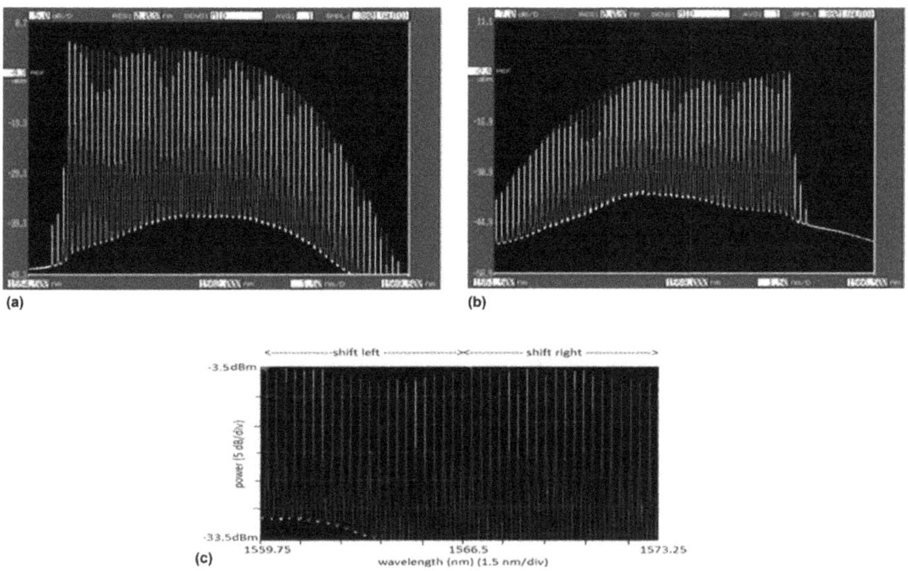

Figure 2.13. Observed optical spectra of a dual ring comb generator with double sideband shifting. The sub-carriers are shown as combined RF frequency left shifted and right shifted sub-carriers. (a) Shift right. (b) Shift left. (c) Combined right and left comb generation, a section of the spectrum only. Reproduced with permission from [33] by permission of Taylor & Francis Ltd.

20 dBm and they are set such that they are always operating in the saturated region under the optical launched power. Hence the comb lines reach the saturated power level in the first circulation. The difference between the switched configuration using OSW and the passive FC is that some residual lines at switching speed would appear in the spectrum when it is monitored by a high-resolution spectrum analyzer. However, under the spectrum monitored by the OSA of minimum resolution of 0.1 nm these residual lines cannot be seen. The continuous original laser lightwaves are fed into the two rings alternately so that the first order shifting of the optical sub-carrier can be generated at all times and hence higher order shifted lines are generated due to the recirculating and frequency shifting via the SIOM. The two SIOMs are modulated with the $\pi/2$ phase shift in opposite ways, as described in section 3.2. In figure 2.13 the shifting left and shifting right are referred to in frequency terms, so if specified in terms of wavelength then the remarks must be in reverse.

2.3.3 Integrated quantum dot comb laser

A compact comb source with reasonable power per comb line is critical for transmission system line cards and applications. We have demonstrated a comb laser in an integrated structure in which a quantum dot multi-layer is formed as the gain medium, and the cavity length and its reflector are fabricated so that the cavity length for a Fabry–Pérot mode spacing is equal to the frequency spacing of the International Telecoms Union (ITU) grid. Indeed the comb laser is an ultra-short pulse sequence whose spectrum is equivalent to a comb spectrum under Fourier transform. Figure 2.14 shows the structure of the Fabry–Pérot mode locked laser in

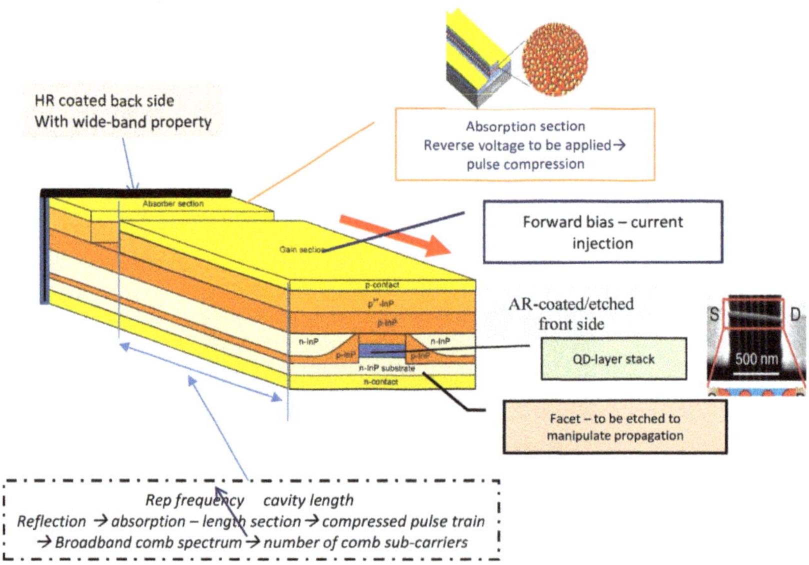

Figure 2.14. Quantum dot (QD) mode locked cavity laser (MLCL) structures and principles without a distributed Bragg grating (DBG).

which an etched slot is created so that the exact length of the cavity can be accurately determined, and hence the mode spacing.

The condition is proportional to the resonance proportional to the effective refractive index of the guided wave in the buried waveguide structure in which the lasing occurs. The waveguide is a single mode type. The field can be represented by a complex field with total internal reflection and complex phase (see [40, 41]). Thus, by etching the facet of the laser with a slanted angle the resonance condition will change or shift the initial wavelength mode to a desired spectral window. The frequency spacing remains almost the same. See the curve depicted in figure 2.15. The measurement of the mode spectra and the linewidth of the mode are depicted in figures 2.16 and 2.17.

The quantum dot gain region is significantly higher due to the randomness of the dots and greater thermal stability. The cavity can be determined exactly to the required mode gain and mode spacing by an etched slot as indicated in figures 2.15 and 2.14, respectively.

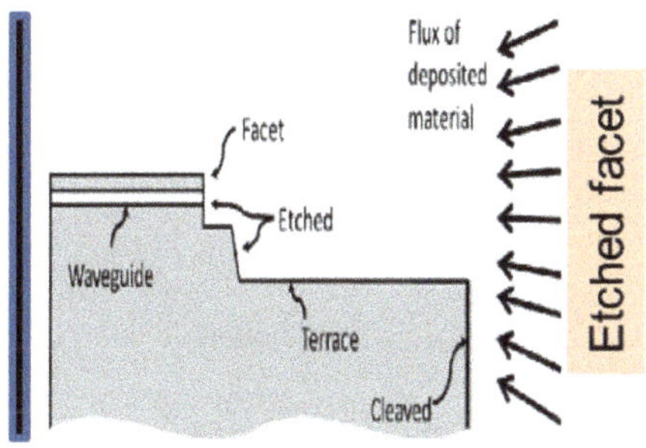

Figure 2.15. Mode spacing of a mode locked laser cavity as a function of the length of the cavity and the refractive index of the guided region.

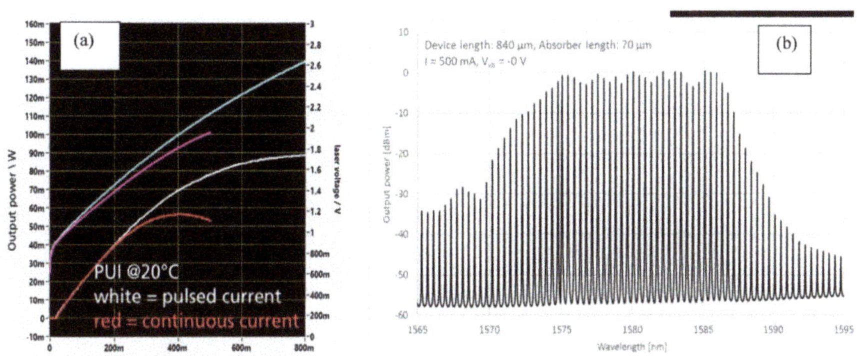

- 90 mW output power @ 800 mA → Lower as compared to 2%/90% but still much larger than targeted
- Spectral comb measurement only possible under continuous current → Results thermally limited

Figure 2.16. Comb laser sub-carrier spectrum without using the etching of the facet. (a) Output power as a function of driving current. (b) Comb spectrum with 50 GHz mode spacing.

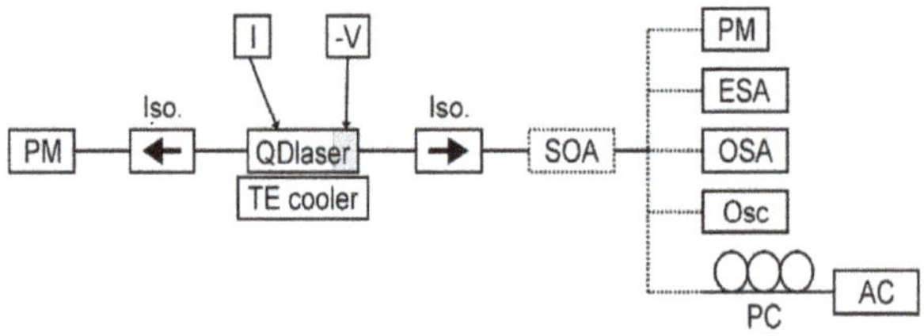

(a) schematic

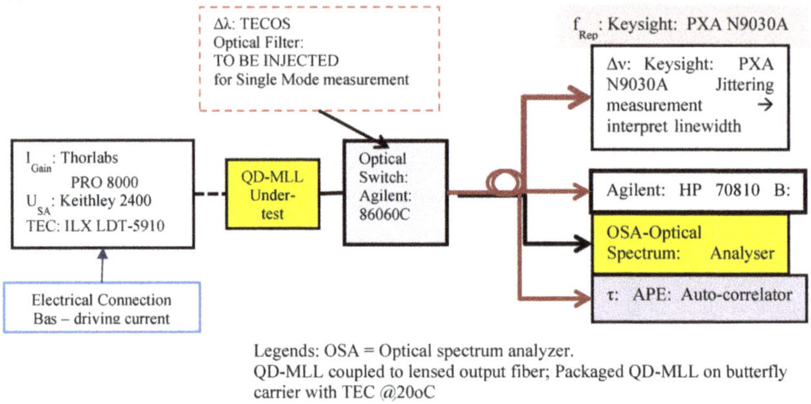

(b) detailed schematic with measurement equipment.

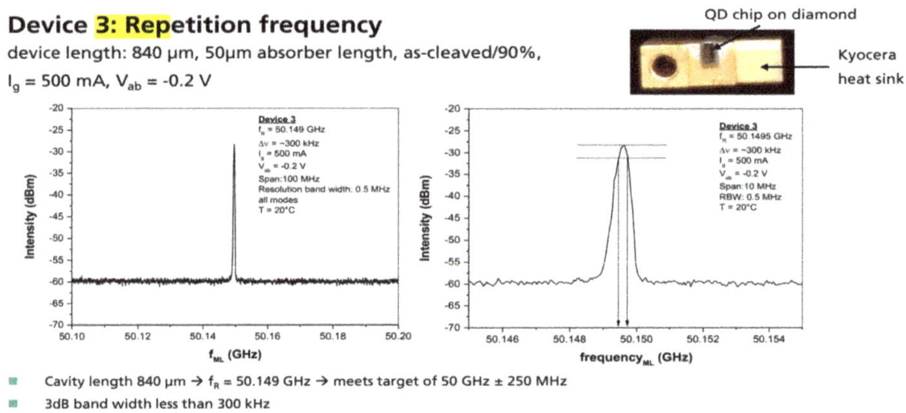

(c) measured linewidth for two QD-MLL lasers, insert is the mounting of laser on cooler.

Figure 2.17. Linewidth measured by self-homodyne modulation over a 1 km long SSMF and direct detection by an optical receiver. (a) Schematic structure. (b) Detailed set-up for linewidth measurement. (c) Linewidth and noise level. The inset shows a device under mounting and cooling. TEC = thermal electric cooler; I = current; U = voltage supply unit; QD = quantum dot; MLL= mode locked laser.

Figure 2.16 shows the comb spectrum whose total number is about 4 over the whole telecoms C-band with the mode spacing equal to 50 GHz or the ITU grid (see table 2.2) and the power is about −2 dBm with 1 dB amplitude variation. The mode line RIN and linewidth of each mode are within 100 KHz as measured under the set-up depicted in figure 2.17.

2.3.4 Optical frequency combs in RPoF networks

In later chapters of this book we will address the details of RPoF networking and transmission systems. This section gives a brief outline of the networking technique for transport RF channels and RF carriers over 5G and 6G networks with multi-level RF carrier modulation and carrying data baseband channels. These techniques are suitable for either distribution or networking of data channels to multiple frequency antenna sites or antenna on-site testing for minimizing maintenance costs or testing in production lines of 5G handheld devices or antennae in multiple structures.

RPoF link technology is expected to support multiple RF carriers and channel signals over SSMFs for both mobile back- and front-haul 5G, 6G, and 7G networks. RPoF link systems utilizing lightwaves to carry multiple analog RF signals through optical networks can extend the cellular perimeter while remaining transparent to the bit rates, modulation formats, and protocols.

As a result the complexity of a wireless system to deliver multi-band, multi-gigabit wireless services can be simplified by taking advantage of microwave photonics in optical wireless network design and system integration. Comb lasers are the multi-carrier light sources which can be employed in an integrated system in which RF carriers of different frequencies are distributed over the whole microwave to millimeter wave range to be carried by one lightwave comb line.

The end-to-end RPoF systems, from the generation schemes of 1–100 GHz optical wireless signals in the central office to the design of transceivers for base stations and radio access units, can be employed and illustrated as an exemplar of the ultra-high capacity optical wireless transmission and distribution investigated in this chapter. Two typical RPoF networking technologies and the optical wireless interface budget specifications employing comb lasers are required to build a converged multi-service, gigabit wireless access network.

Figure 2.18(a) depicts the RPoF networking employing mode-locked laser (MLL) multi-carrier optical sources and RF optical modulators for access optical networks or metropolitan multi-antennae site access and distribution. The integrated lasers of MLL shown in the inset provide compact network cards or antenna on-site system cards. Both polarizations of the optical modes can be employed and thus polarization maintaining (PM) fiber couplers are used throughout. Figure 2.18(b) lists the three tables for properties of MLL and the minimum properties required for meeting the power budget for error-free transmission in an RPoF environment.

Figure 2.19 illustrates a transmission technique with time division multiplexing of the polarized mode for increasing the total capacity of RPoF channels. This is highly feasible because the channel rate of the RPoF is normally less than 500 MHz and the time delay is long enough for an optical delay line for time multiplexing. A comb laser can be employed here to demonstrate the applications of a multi-carrier comb laser. Indeed, an optical demultiplexer and modulation with RF carriers and data baseband and multiplexer can be used to combine all channels of different polarizations and transmission over the optical networks as shown in figure 2.20.

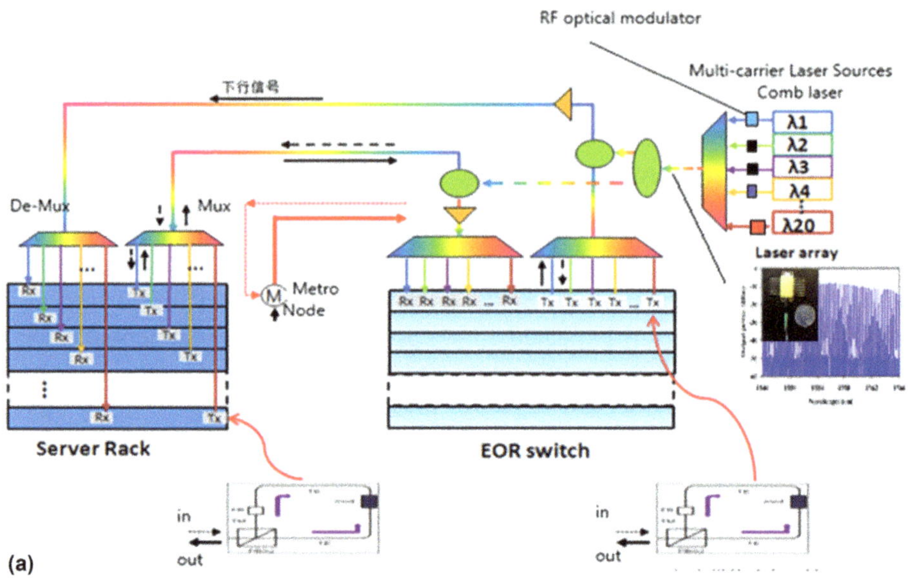

Components	Optical power/loss
QD comb laser	-2dBm/ch
Optical link budget	-26dB
EDFA in optical link	20dB
Sensitivity of PD	-11dBm (w/ FEC 25GBaud PAM4)
Power margin	2dB

Components	Optical power/loss
QD laser	-15dBm/ch
EDFA after laser	15dB
Optical link budget (incl. mod and AWG etc.)	-26dB
EDFA in optical link	20dB
Sensitivity of PD	-10dBm (w/o FEC, 25G PAM4@BER= 10^{-12})
Power margin	4dB

Components	Optical power/loss
QD comb laser	-15dBm/ch
EDFA after Laser	15dB
Ring modulator	-5dB
Chip-to-fiber Coupling	-2dB
Optical link budget	-1dB
Fiber-to-chip coupling	-2dB (ORx)
Ring filter	-2dB
Required power to GeSi APD	-18dBm @56Gbps w/ FEC
Power margin	6dB

(b) Typical laser property for RPoF

Low-cost PID for short-reach RPoF access

Figure 2.18. Schematic of optical networking for RPoF to multiple point antenna sites for 5G and 6G networks, showing the comb lasers, demultiplexer, and modulator for multi-RPoF transmission and networking. The inset shows the spectrum of an integrated MLL. (a) Schematic. (b) Property of the MLL comb laser and the minimum for access RPoF transmission. ch = channel; EDFA/OA = erbium doped fiber optical amplifier; FEC = forward error correction; PD = photodetector; QD = quantum dot; PAM = pulse amplitude modulation.

2.3.5 Optical modulators

2.3.5.1 Electro-absorption modulators

The modulation of lightwaves via an external optical modulator can be of three types, depending on the special EO effects that alter the lightwave properties, in particular the intensity or the phase of the lightwave carrier. In an external modulator the intensity is normally manipulated by manipulating the phase of the carrier lightwaves guided in one path of an interferometer. The Mach–Zehnder interferometric structure using an integrated optical waveguide is the most common type.

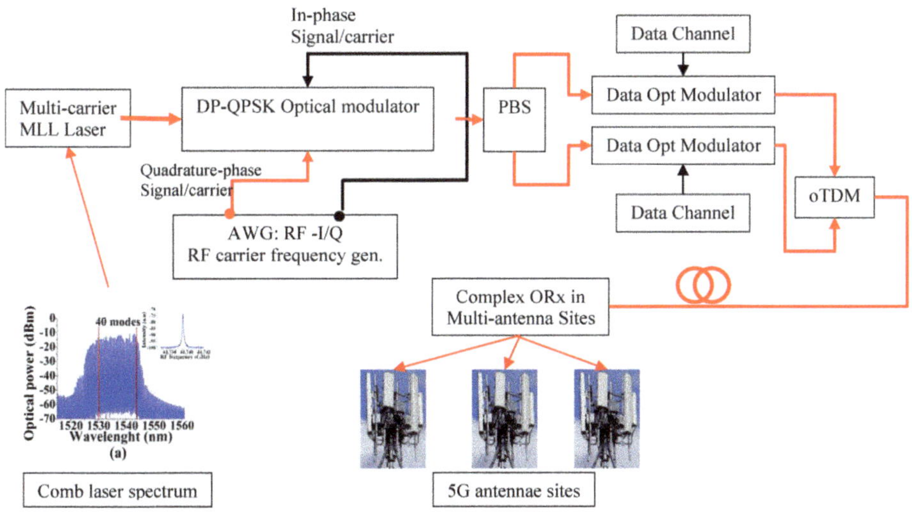

Figure 2.19. Multi-frequency RF antenna networking using an MLL and optical IQ modulator with or without polarization multiplexing for doubling capacity. AWG = arbitrary waveguide grating.

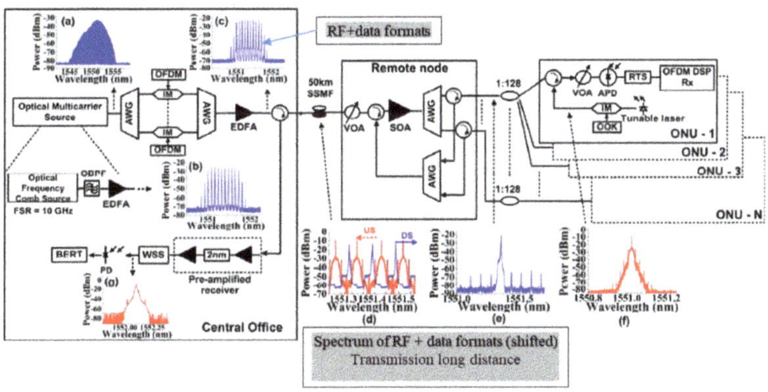

Figure 2.20. Multi-carrier modulation for RPoF transmission and networking for massive frequency antenna delivery networking. See also figure 2.18. (Adapted with permission from [42] for RPoF). OBPF = optical band pass filter—optical array waveguide gratings (AWG) for multi-carrier modulated demultiplexing; ONU = optical network unit, located at the antenna site; EDFA = optical amplifier–erbium doped fiber amplifier; OFDM = orthogonal frequency division multiplexing; SOA = semiconductor optical amplifier; SSMF = standard SMF; OOK = on–off keying modulation; IM = intensity modulator; VOA = variable optical attenuator; WSS = wavelength selective switch.

An electro-absorption (EA) modulator employs the Franz and Keldysh effect which is observed as lengthening of the wavelength of the absorption edge of a semiconductor medium under the influence of an electric field [27–29]. In a quantum structure such as the multi-quantum well structure this effect is called the Stark effect, or the EA effect. The EA modulator can be integrated with a laser structure on the same integrated circuit chip as described in sections 3.3 and 3.8. For a

$LiNbO_3$ modulator, the device is externally connected to a laser source via an optical fiber.

The total insertion loss of a semiconductor intensity modulator is about 8–10 dB, including fiber-waveguide coupling loss which is quite high. However, this loss can be compensated by a semiconductor optical amplifier (SOA) that can be integrated on the same circuit. Compared to $LiNbO_3$ its total insertion loss is about 3–4 dB, which can be allowable as Er-doped fiber amplifiers (EDFAs) is now readily available.

The driving voltage for an EA modulator is usually lower than that required for $LiNbO_3$. However, the extension ratio is not as high as that of the $LiNbO_3$ type, which is about 25 dB compared to the 10 dB for an EA modulator. This feature contrasts the operating characteristics of the $LiNbO_3$ and EA modulators. Although the driving voltage for an EA modulator is about 2–3 V while it is 5–7 V for $LiNbO_3$, the former type would be preferred for intensity or phase modulation formats due to the extinction ratio that offers a much lower 'zero' noise level and hence a high quality factor.

2.3.5.2 Phase modulators

The phase modulator is a device which manipulates the 'phase' of optical carrier signals under the influence of an electric field created by an applied voltage. When a voltage is not applied to the RF electrode, the number of periods of the lightwaves, n, exists in a certain path length. When a voltage is applied to the RF electrode, one or a fraction of one period of the wave is added, which now means $(n + 1)$ waves exist in the same length. In this case, the phase has been changed by 2π and the half voltage of this is called the driving voltage. In the case of long-distance optical transmission, the waveform is susceptible to degradation due to nonlinear effects such as self-phase modulation, etc. A phase modulator can be used to alter the phase of the carrier to compensate for this degradation. The magnitude of the change of the phase depends on the change of the refractive index created via the electro-optic effect that in turn depends on the orientation of the crystal axis with respect to the direction of the established electric field by the applied signal voltage.

An integrated optical phase modulator operates in a similar manner, except that the lightwave carrier is guided via an optical waveguide with a titanium diffused or ion-exchanged confined region for $LiNbO_3$, and rib waveguide structures for the semiconductor type. Two electrodes are deposited such that an electric field can be established across the waveguiding cross section so that a change of the refractive index via the electro-optic or EA effect is achieved, as shown in figure 2.21. For ultra-fast operation one of the electrodes is a traveling wave type or hot electrode and the other is a ground electrode. The traveling wave electrode must be terminated with matching impedance at the end to avoid wave reflection. Usually a quarter wavelength impedance is used to match the impedance of the traveling wave electrode to that of the 50 Ω transmission line.

A phasor representation of a phase modulated lightwave can be achieved by circular rotation at a radial speed of ω_c. Thus the vector with an angle ϕ represents the magnitude and phase of the lightwave.

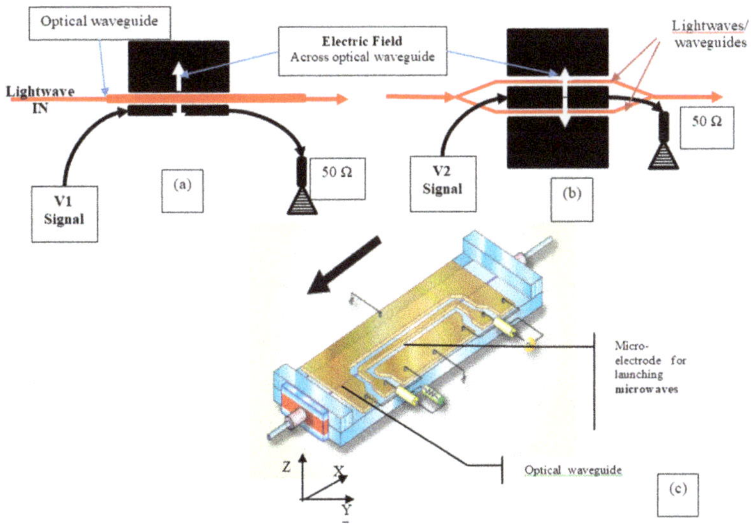

Figure 2.21. EO phase modulation in a planar integrated modulator using Ti-diffused $LiNbO_3$. Electrode impedance matching is not shown. (a) Schematic diagram. (b) Integrated optic Mach–Zehnder modulator structure (split and combining two phase modulation sections). (c) Photograph of a packaged modulator.

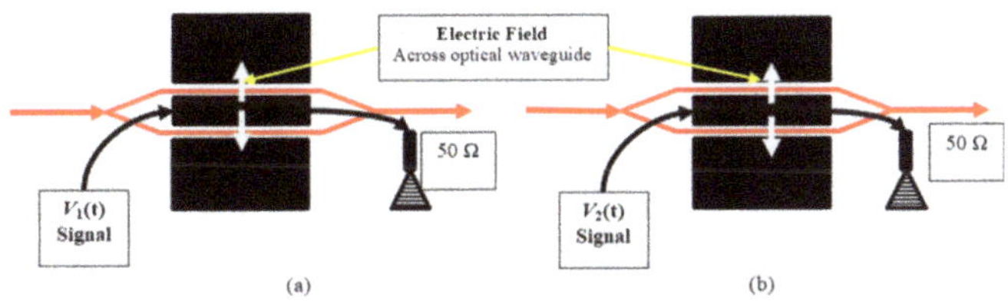

Figure 2.22. Intensity modulation using interferometric principles in guidewave structures in $LiNbO_3$ (a) ON—constructive interference mode and (b) OFF—destructive interference mode. Optical guided wave paths 1 and 2. An electric field is established across the optical waveguide section both on the surface and in the crystal crossing the guided region of the optical waveguide. The electrode is not to scale. The typical width of a high speed optical modulator is only about 5 μm.

2.3.5.3 Intensity/interferometric modulators

The basic structure of a $LiNbO_3$ modulator comprises (i) two waveguides, (ii) two Y-junctions, and (iii) RF/dc traveling wave electrodes (figure 2.22). The optical signals coming from the lightwave source are launched into the $LiNbO_3$ modulator through the polarization maintaining fiber, and are then split equally into two branches at the first Y-junction on the substrate. When no voltage is applied to the RF electrodes, the two signals are re-combined constructively at the second Y-junction and coupled into a single output. In this case the output signals from the $LiNbO_3$ modulator are recognized as 'ONE'. When a voltage is applied to the RF electrode, due to the electro-optic effects of the $LiNbO_3$ crystal substrate, the waveguide refractive index is changed and hence the carrier phase in one arm is advanced although retarded in the other arm. Then the two signals are re-combined destructively at the second Y-junction, transformed into a higher order mode, and radiated at the junction. If the phase retarding is in a multiple odd factor of π, the two signals are completely out-of-phase, the combined signals are radiated into the substrate and the output signal from the $LiNbO_3$ modulator is recognized as a 'ZERO'. The voltage difference which induces this 'ZERO' and 'ONE' is called the driving voltage of the modulator, and is one of the important parameters in deciding the modulator's performance. The design of traveling wave electrodes for broadband operation is given in section 2.6.

2.3.6 Modulator details

2.3.6.1 Phasor representation and transfer characteristics

Consider an interferometric intensity modulator that consists of an input waveguide that is split into two branches and then recombines into a single output waveguide. If the two electrodes are initially biased with voltages V_{b1} and V_{b2} then the initial phases exerted on the lightwaves would be $\phi_1 = \pi V_{b1}/V_\pi = -\phi_2$, which are indicated by the bias vectors shown in figure 2.23(b). From these positions the phasors are swinging according to the magnitude and sign of the pulse voltages applied to the electrodes. They can be switched to the two positions that can be constructive or destructive. The output field of the lightwave carrier can be represented by

$$E_0 = \frac{1}{2} E_{i\text{RMS}} e^{j\omega_c t}(e^{j\phi_1(t)} + e^{j\phi_2(t)}), \tag{2.2}$$

where ω_c is the carrier radial frequency, $E_{i\text{RMS}}$ is the root mean square value of the magnitude of the carrier, and $\phi_1(t)$ and $\phi_2(t)$ are the temporal phases generated by the two time-dependent pulse sequences applied to the two electrodes. With the voltage levels varying according to the magnitude of the pulse sequence one can obtain the transfer curve, as shown in figure 2.23(a). This phasor representation can be used to determine exactly the biasing conditions and magnitude of the RF or digital signals required for driving the optical modulators to achieve 50%, 33%, or 67% bit period pulse shapes.

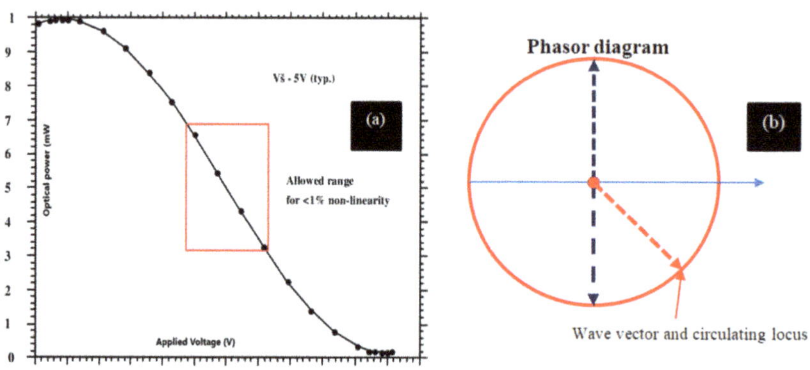

Figure 2.23. (a) E/O transfer curve of a Mach–Zehnder interferometric intensity modulator. (b) Phasor diagram.

The power transfer function of a Mach–Zehnder modulator is expressed as[1]

$$P_0(t) = \alpha P_i \cos^2 \frac{\pi V(t)}{V_\pi}, \tag{2.3}$$

where $P_0(t)$ is the output transmitted power, α is the modulator total insertion loss, P_i is the input power (usually from the laser diode), $V(t)$ is the time-dependent signal applied voltage, V_π is the driving voltage so that a π phase shift is exerted on the lightwave carrier. It is necessary to set the static bias on the transmission curve through a bias electrode. It is common practice to set a bias point at a 50% transmission point or a $\pi/2$ phase difference between the two optical waveguide branches, the quadrature bias point, or the Q-point. There are two Q-points, one with the rising slope and one with the negative descending slope. Figure 2.23(b) shows this destructive interference at the dc bias point. Then if a sinusoidal wave voltage is assumed on this point then the modulated optical RF frequency is doubling with suppression of the optical carrier.

2.3.6.2 Automatic bias control

One factor that affects the modulator performance is the drift of the bias voltage which happens after a few hours of operation due to the accumulation of charges on the surface of the crystal, i.e. the electro-effects of a high field. For the Mach–Zehnder interferometric modulator (MZIM or MZM) type this is critical when it is required to bias at the quadrature point or at the minimum or maximum locations on the transfer curve. dc drift is the phenomenon occurring in $LiNbO_3$ due to the build up of charges on the surface of the crystal substrate. Under this drift the transmission curve gradually shifts in the long term [30, 31]. In the case of a $LiNbO_3$ modulator, the bias point control is vital as the bias point will shift long term. To compensate for the drift it is necessary to monitor the output signals and feed them back into the bias control circuits to adjust the dc voltage so that the operating points stay at the same point, as shown in figure 2.24, for example the quadrature point. It is the manufacturer's responsibility to reduce the dc drift so that the dc voltage is not beyond the limit throughout the lifetime of a device.

2.3.6.3 Detailed physical structures

Figures 2.25(a) and (b) show the structure of a Mach–Zehnder intensity modulator using a single and dual electrode configuration, respectively. The thin line electrode is called the 'hot' electrode or traveling wave electrode. RF connectors are required for launching the RF data signals to establish the electric field required for electro-optic effects. Impedance termination is also required. Optical fiber pig-tails are also attached to the end faces of the diffused waveguide. The mode spot size of the diffused waveguide is not symmetric and hence some diffusion parameters are

[1] Note that this equation represents a single drive MZIM—it is the same for a dual drive MZIM provided that the bias voltages applied to the two electrodes are equal and opposite in sign. The transfer curve of the field representation would have half the periodic frequency of the transmission curve shown in figure 2.23.

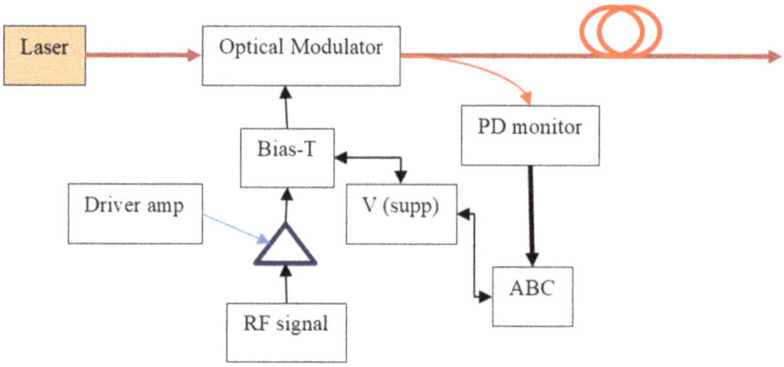

Figure 2.24. Arrangement of the bias control of integrated optical modulators to achieve a stable operating point. ABC = automatic bias control; PD = photodetector.

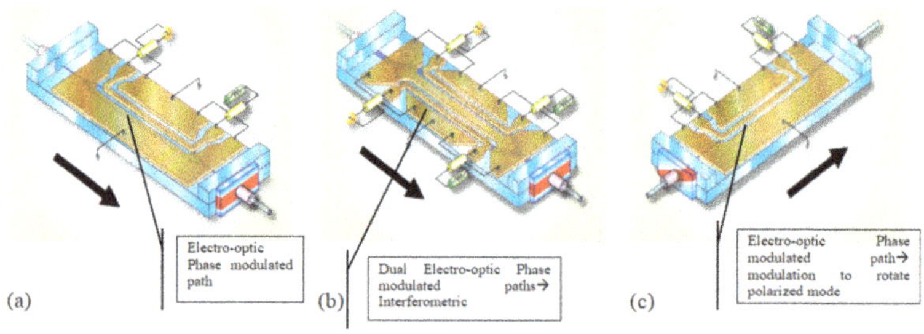

Figure 2.25. MZM using $LiNbO_3$: (a) single drive type, (b) dual drive structure, and (c) EO polarization scrambler using a Ti-diffused $LiNbO_3$ substrate.

controlled so that maximizing the coupling between the fiber and the diffused or rib waveguide can be achieved. Due to this mismatch between the mode spot sizes of the circular and diffused optical waveguides coupling loss occurs. Furthermore, the difference between the refractive indices of the fiber and $LiNbO_3$ is quite substantial and thus Fresnel reflection loss would also be incurred.

Table 2.1 shows the typical operational characteristics of a MZM in lithium niobate substrate by titanium diffusion. It is noted here that the operation bandwidth is in the small linear region. The bandwidth at 6 dB can be extended to about 40 GHz, thus the RF carrier and its superimposed data channel can be used with an extra 3 dB loss.

Figure 2.25(c) shows the structure of a polarization modulator which can be essential for polarization multiplexing of two polarized data sequences so as to double the transmission capacity, e.g. 40 Gbps to 80 Gbps. Furthermore, this type of polarization modulator can be used as a polarization rotator in a polarization mode dispersion (PMD) compensating RPoF sub-system [32].

2.3.6.4 Typical operational parameters

See table 2.1.

Table 2.1. Typical operational parameters for a $LiNBO_3$ optical MZM.

Parameters	Typical values	Definition/comments
Modulation speed and bandwidth	40 Gbps and 23 GHz	Digital and RF sub-carrier multiplexed signals
Insertion loss (lithium niobate)	<5 dB	Coupling loss and propagation losses in a diffused waveguide
Driving voltage	<4 V	The RF oscillation voltage to achieve 100% modulation depth
Optical bandwidth	Min. 23 GHz	3 dB roll-off of the $S_{21}(f)$ parameter
Extinction ratio (ER)	Min. 20 dB, typical 30 dB	The ratio of max. (ON) and min. (OFF) optical power
Polarization extinction ratio (PER)	Min. 20 dB	The extinction ratio between two polarized guided modes (the TM and TE guided modes)

2.3.6.5 Traveling wave electrodes (TWEs) for integrated modulators

An integrated EO modulator comprises optical waveguides and electrodes. The integrated optical waveguides will follow the structure of 3D waveguides, as described in chapter 4. The electrodes must be in either a coplanar structure (CPS) or asymmetric coplanar structure (ACPS) so that ultra-broadband operation can be achieved. This section provides the detailed modeling and design of such electrodes as the modulating electrodes for ultra-high speed broadband optical modulators. The modulation efficiency is also examined with consideration of the overlap-integral between the fields of the optical guided mode and that guided by RF waves.

Optical transmission at an ultra-high bit rate up to 112 and 212 Gbps and multiplexing of several optical channels are reaching a high standard deployment of dense division wavelength multiplexing (DWDM) access optical networks. Various ASK and PSK formats, in particular the quadrature phase shift keying (QPSK) and M-QAM formats, and RPoF high RF frequency data channel multiplexed channels are demonstrated to increase the transmission capacity with DSP based mitigation of linear dispersion and nonlinear dispersion effects. External modulators of the Si photonic type offer such capability of modulation. The phase modulation can be achieved with single, asymmetric, or dual (balanced or symmetric) electrode structures that require precision in the design and fabrication.

2.3.7 Complex IQ/QPSK modulators

The in-phase and quadrature phase (IQ) optical modulators are modulation devices which have attracted significant interest for their potential for doubling the transmission capacity in optical systems by coherent reception. This is possible and has been demonstrated over the first two decades of the twenty-first century, as it allows complex modulation under coherent reception and thus digital signal processing under ultra-high sampling rate DSP subsystems. Furthermore, polarization multiplexing can also be implemented easily in integrated photonic structures, and will double the capacity with two polarized channels. This section gives a brief description of such IQ modulators or QPSK optical modulators.

2.3.7.1 Schematics of a QPSK modulator

Figure 2.26(a) shows the schematic of a QPSK (or IQ) optical modulator and the operating transfer curve for the conversion of electrical signals into optical modulated signals. The single polarization IQ modulator consists of two 'child' (inner) MZMs which are placed in the optical paths of the 'parent' (outer) MZM. An optical phase shifter via EO effects also integrated on one path of the parent MZM is used to shift the $\pi/2$ phase on one path, thus generating the quadrature phase component. The data channels in the baseband are inserted into the electrodes of the child MZMs to create the I- and Q-channels in the optical domain for transmitting along the optical fiber transmission lines. The biasing of the child MZMs and parent MZM is a bit more complicated and is achieved commonly via the use of an automatic biasing control (ABC) unit. This should follow a sequence of feedback control via the monitoring of photodetectors integrated in the same optical device. A typical commercial IQ modulator is shown in figures 2.26(b) and (c) where the arrangements of the biasing and RF electrodes are indicated.

Some typical parameters of QPSK modulator parameters for high speed optical modulation and transport over fiber transmission lines can be stated as:

- V_π (I-branch MZM) dc < 10 V; V_π (Q-branch MZM) dc < 10 V; V_π (phase) < 15 V.
- V_π (1 GHz) < 3 V (3.3 V max.).
- X-cut chirpless 3 dB at 25 GHz 6 dB projected to 45 GHz for each MZM (I) and MZM (Q).

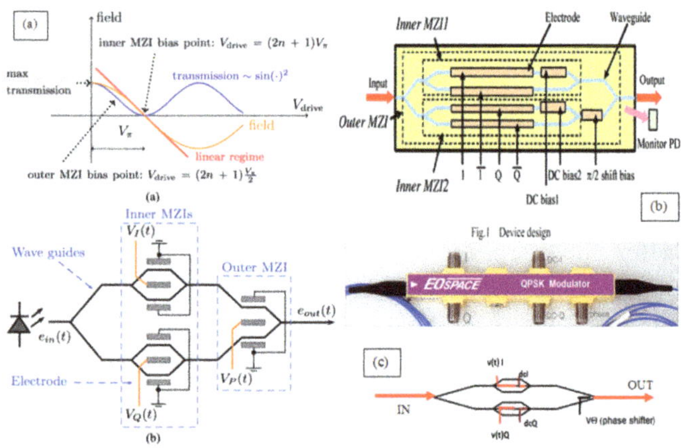

Figure 2.26. (a) Schematics of the QPSK modulator for EO space. Biasing voltages V_b(I), V_b(Q), and V_{phase} (PS) are the dc applied voltages. This structure needs bias *T*. Types of general IQ modulator (single polarization): (b) integrated photonic structure of IQ modulator (single polarization) with phase shifter on one branch of the parent MZM and (c) IQ modulator with a phase shifter at the parent MZM (outer) of the EOSpace company. The schematic (below the image) shows the typical waveguide structure of the modulator.

- Insertion loss (IL) < 5 dB; IQ modulator QPSK.
- Reflection S_{11} < 10 dB (dc 20 GHz); extinction ratio (ER) > 20 dB.
- Bias: recommended automatic biasing control (ABC).

2.3.7.2 PD-QPSK modulators

Figure 2.27 shows the schematic of a dual polarization IQ modulator which consists of two IQ modulators and polarization beam splitters and combiners at the optical input and output ports. The optical waveguides of the integrated optic modulators must be designed and fabricated such that they support two polarized optical modes, the transverse electric (TE) and transverse magnetic (TM) modes so that each polarized mode is split and entering into each IQ modulator section. The RF signals are inserted via the traveling wave electrodes deposited on top of the optical waveguide paths of the PDM-IQ modulator, as indicated in figure 2.27. Thus in the RPoF the use of a PDM-IQ modulator can double the transport capacity of the RF channels in each polarization through the fiber polarized modes. The 90o polarization rotator rotates

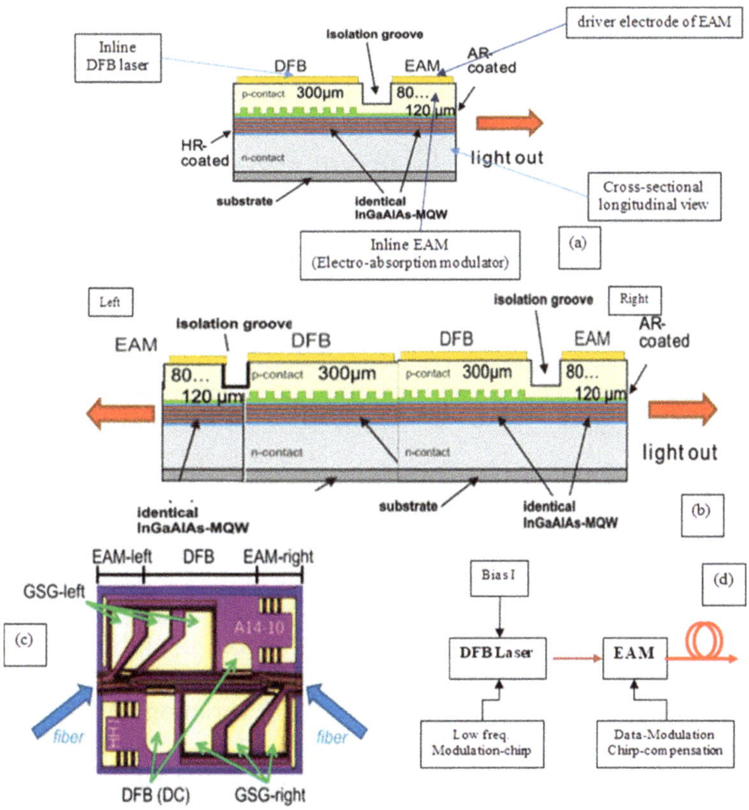

Figure 2.27. (a) Schematic of the cross-sectional view of an integrated EML, with an integrated field effect microwave transistor FET for driving the EAM. (b) Dual back-to-back EML structure. (c) Integrated circuit plane view of a dual EML with electrode deposited on top of the EA modulator. (d) Modulation of the laser and EAM. Al = aluminum; GaAs = gallium arsenide; MQW = multi-quantum well.

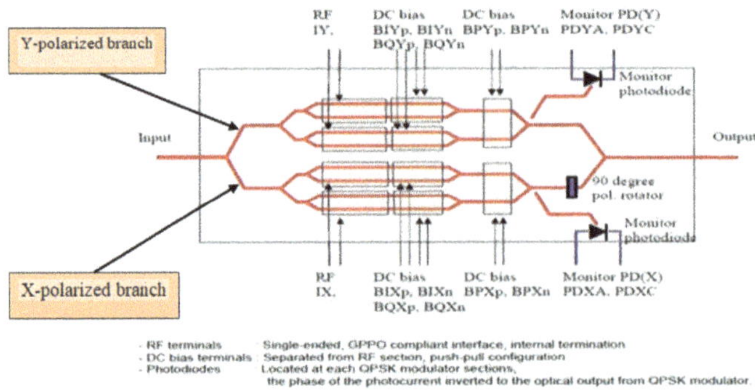

Figure 2.28. Schematic of a PDM-QPSK integrated optic modulator with H-polarized and V-polarized branches and including biasing electrodes and ports. The H-polarized QPSK modulator is in the upper branch and the V-polarized QPSK modulator is in the lower branch. PDM = polarization division multiplexing; BIYp, BIYn = bias Y-polarized 'child' modulators; p and n = positive and negative electrodes, respectively; BPY p, n = bias phase Y-branch with positive and negative electrodes. Similar notations for the X-branch.

the X-polarized mode to a horizontal TE-polarized mode so that both modes of the two branches can be aligned with the polarization of the output waveguide and coupled to the fiber for transmission to the optical fiber transmission line.

We note that there are eight parallel optical waveguides in the central section of the PDM-IQ modulator, so when biasing it is necessary to implement a sequence of biasing the electrodes one by one such that we can set the conditions and states of the biasing of the branch which we would like to operate.

The uses of the PDM-IQ modulator are given in chapter 3, in particular for the polarization multiplexing of data channels or RF channels of different RF modulation frequencies in complex planes, thus M-QAM modulation of the RF carriers to transport wideband data channels to an effective bit rate reaching tens of gigabits per second (figure 2.28).

2.3.8 EA modulator and integrated structures

2.3.8.1 Electro-absorption modulator

An electro-absorption modulator (EAM) is a semiconductor based device. An EAM can be used for modulating the intensity based on the variation of the absorption coefficients at different electrical driving levels. EAM operation is based on the Franz–Keldysh effect [43], that is, that the medium, the guided optical mode field, absorption coefficient is varied with respect to the variation of an applied electric signal.

EAMs are commonly fabricated using primary materials of III–V compound semiconductors or recently using group IV materials such as Si and Ge. Ge exhibits a good absorption coefficient in the telecoms spectral window of 1550 nm and is integrated inline Si photonic devices. The structures of two types of EAMs are shown in figure 2.29. The EAM can be in the p–i–n (p-doped/intrinsic/p-doped regions) photodetection structure. The transmission type does not lead to a high

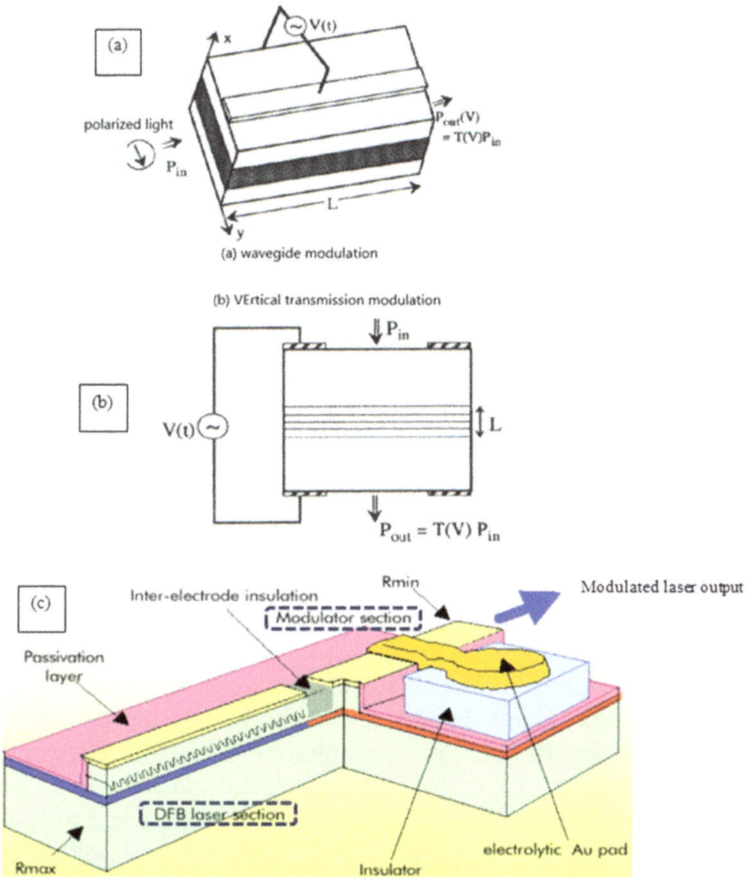

Figure 2.29. The structure of EAMs: (a) waveguide transmission modulation; (b) vertical transmission modulation; and (c) cross-sectional view of a waveguide EAM laser (inline EAM laser structure) type. R_{min} = minimum resistance; DFB = distributed feedback. Copyright (2013), reprinted by permission of Taylor & Francis Ltd.

enough extinction ratio (ER) and thus is not much used in compact transmitters. The waveguide type is more commonly used and has higher optical confinement, and hence highly efficient modulation efficiency.

In contrast to the electro-optic (EO) MZM, EAMs can operate under much lower voltages with good linearity and offer bandwidths of several gigahertz, thus making these devices useful for RPoF systems. An EAM can be integrated with a DFB laser on a single integrated circuit to form an RPoF transmitter, as shown in figure 2.29(c). Compared to direct modulation of the laser diode, a higher bandwidth and reduced chirping effects can be obtained. An EAM can be integrated on branches of an MZM so that two EAM modulated branches can be formed and combined in a time division multiplexed modulation device to increase the transmission capacity.

2.3.8.2 Si photonic EAM modulator

A SiGe EAM [44] can be fabricated with a Ge optical waveguide that is 600 nm wide and 80 μm long. Lightwaves are coupled via an optical taper to the Ge waveguides or using photonic grating couplers on a Ge substrate. The grating coupler loss is about ~6 dB/coupler. An electrical RF probe can be used to apply the bias voltage to the electrodes deposited across the Ge waveguide and high speed or RF carrier modulated baseband data (RPoF) signal. A cross section of the modulator is shown in figure 2.30. The 3 dB bandwidth of this EAM can reach 70 GHz and thus several sub-carrier multiplexed radio photonic channels can be generated and transmitted. Ge is preferred as its absorption coefficient is high in the C- and L-bands of the silica spectral low loss regions. Furthermore, it can be integrated with electronic circuits fabricated in such a SiGe substrate as well as other silicon-on-insulator (SOI)-based MZM modulators.

For RPoF systems SiGe EAM linearity can be achieved when biased at −1.8 V, which is lower that for $LiNbO_3$ based MZM. However, the ER of the EAM is not as high as we can achieve in $LiNBO_3$ MZMs, which is about 30 dB. Figures 2.32(a) and (b) depict the measurement set-up and the input electrical voltage–optical power transfer characteristics of a semiconductor EAM in which a linear range region is indicated for uses in RPoF. In the next chapter further details of the operation characteristics of the EAM will be described, in particular the nonlinear distortion and generation of harmonic components which create cross-talk or intermodulation effects in adjacent radio channels carried by the RF carrier under an EAM or external modulated laser (EML). The packaging of an EAM for testing is depicted in figure 2.31, in which the thermal electric controlled platform (a 14 pin butterfly mount platform) the EAM chip is mounted and a coplanar waveguide (CPW) electric electrode is shown to couple the RF input signal and the T-bias to

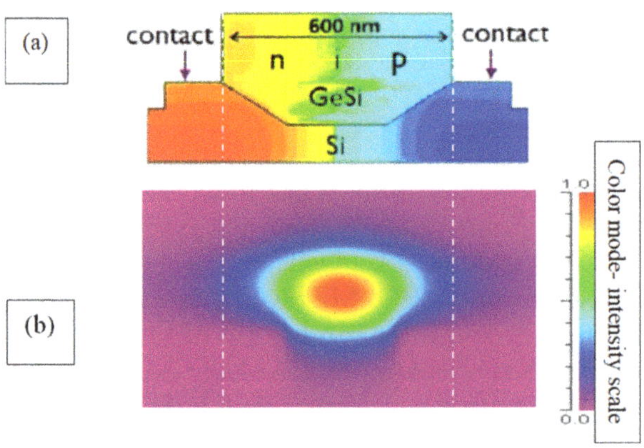

Figure 2.30. (a) Cross-sectional structure of an EAM based on a SiGe material system structure with a p–n junction as the absorption compound group IV material film deposited on a Si waveguide on an insulator substrate. (b) Profile of the transverse guided fundamental mode with an effective refractive index of 3.259O74 estimated by finite element simulation [45]. Electric field of the RF signals developed across the optical guided mode and dropped across the two electrodes deposited on top of the p and n regions is shown.

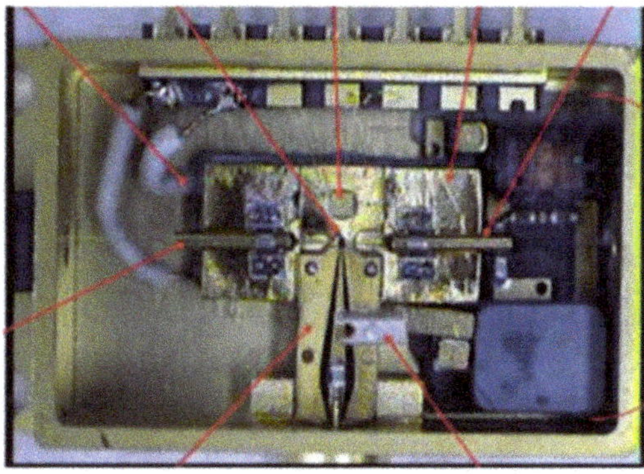

Figure 2.31. EAM integrated die and mounting platform for testing. The EAM devices and thermal electric Peltier controller TEC and associate electronics are housed in a 16 pin butterfly.

superimpose the biasing voltage according to the transfer characteristics given in figure 2.32.

We note that one can bias the EAM so that there would be a chirp effect on the lightwaves and then the laser can also be chirped in the opposite direction so that the EAM chirp can compensate and thus the lightwaves at the EAM output suffer no chirping and thus can be transmitted over longer distance. From the transfer function of the EAM depicted in figure 2.32, the total voltage range is about 3.5 V in which the linear range would be estimated to about 1.5 to 2.5 V. The driving range of the EAM offers lower loss than the coupling MZM and external laser, thus lower costs and the lower loss offers significant advantages in RPoF transmission systems.

2.3.8.3 Integrated external modulation laser structure and direct modulation

The integration of an external modulator such as the EAM can offer the significant advantages of compactness and thus a form which can be installed easily in data centers or intra-DCs (DCIs). Thus the term EML is used if a DFB laser can also be integrated in the compound semiconductor based on InP (the EAM) fabricated on an InP substrate by molecular organic chemical vapor epitaxy (MOCVD) or molecular beam epitaxy (MBE) [46, 47]. Some advantages of using an EAM in RPoF are the large absorption coefficients with reasonable low driving sinusoidal signal and the fact that it is integrable with an inline laser using the flip chip technique, on the same chip. The E/O transmission bandwidth for a typical EAM is limited by the time constant $\tau = C_s/Z_o/2 + R_s$, where Z_o is the output impedance of the driving amplifier, C_s is the EAM junction capacitance, and R_s is the series resistance of the modulator. To reach a S_{21}, the scattering parameter of port 2 to port 1 with a bandwidth of 70 GHz for RPoF, the electronic amplifier's output impedance must be matched to the input impedance of the EAM. The integration of an inline DFB laser and the EAM offer significant advantages when the chirp

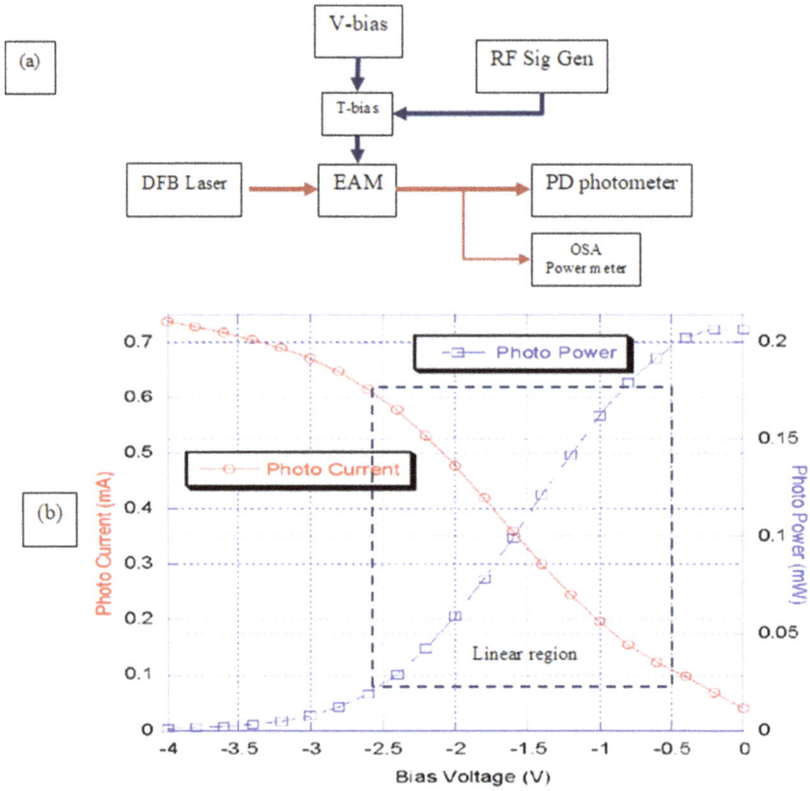

Figure 2.32. A typical electrical to optical power or photo-current of an EAM, with applied bias voltage versus output optical power/current with an input optical power of 0 dBm at a 1550 nm operating wavelength. The dotted rectangle indicates the linear operating region in the power output which is important for RPoF transport. The red circles and blue squares are plots of for measured optical power and detected PD currents plotted, respectively, as functions of the applied biased voltage across the electrodes of the EA modulator. (a) Schematic of the measurement set-up. (b) Measured transfer characteristics at 1.0 mW optical power coupled into an EAM and detected by an OSA or power meter versus applied bias voltage. Sig Gen = signal generator; EAM = electro-absorption modulator.

characteristics of one device can be compensated by the other, such as when the DFB is modulated directly.

The main motivation for an EAM is that the laser can be modulated directly to generate chirp lightwaves. This chirp is then compensated by the counter-chirp effects generated via the modulation of the EAM, thus allowing the generation of chirpless data lightwaves and hence possible longer distance transmission. Monolithic integration of an InGaAsP laser and an EAM in this compound semiconductor can be fabricated to form an EAM laser structure (an EML) which offers potential applications as a 50 GHz RPoF. This section describes an EML and its application in RPoF applications to frequencies up to a 40 GHz RF carrier, as illustrated in figure 2.27.

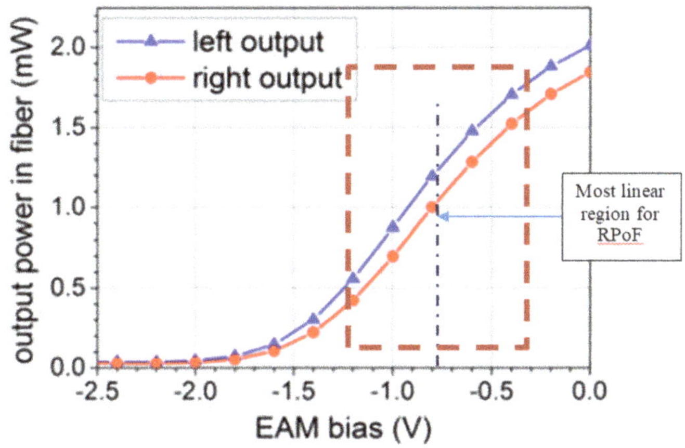

Figure 2.33. Output power of two EM lasers versus biasing voltage applied to the EAM section of a dual EML (DEML) structure and the linear region for biasing the DEML for RPoF.

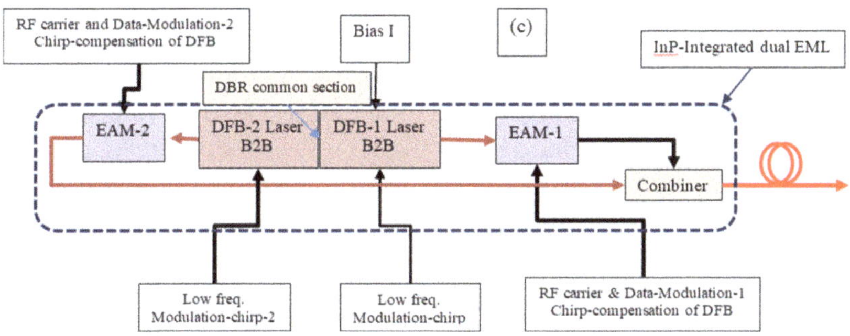

Figure 2.34. Schematic of a dual EML with B2B DFB lasers coupled via an optical coupler (combiner) and thus an integrated doubling capacity transmitter.

Figure 2.27(b) shows the plane view of an EML fabricated by the Heinrich Hertz Institute of Berlin [48, 49], which can offer 28 GBd and 53 GBd. They can be optimized for 100 Gb (4 × 25 GBd) data telecoms and 5G RPoF front-, middle-, and back-haul transmission systems (figure 2.33).

2.3.8.4 Integrated dual EMLs

The EML can be integrated as shown in figure 2.34 into a dual structure with a shared back-to-back (B2B) DBR region so that reflections can be formed two cavity lasers. The outputs of the EAM1 and EAM2 can be coupled via an integrated optical coupler so that two RF carriers and data channels can be transmitted through a common SSMF [50].

2.3.9 Plasmonic modulator to 500 GHz

The frequency response and modulation efficiency can be extended to the several hundreds of gigahertz region by employing plasmonic modulation in which the

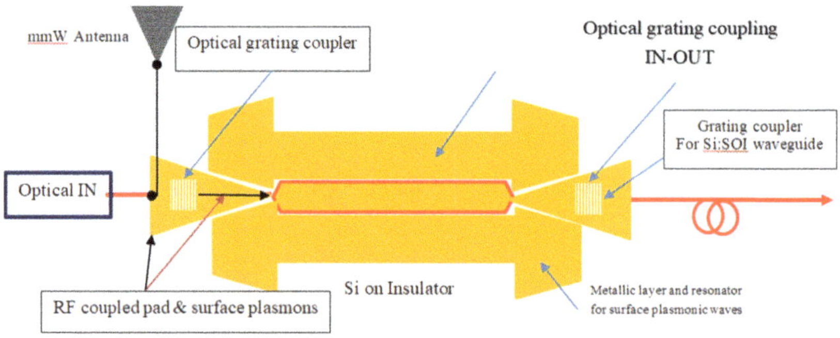

Figure 2.35. Schematic of a plasmonic QPSK modulator to offer 500 GHz 3 dB bandwidth modulation for RPoF.

plasmonic electromagnetic (EM) wave is created by a metallic layer on top of a nonlinear (NL) organic layer. The nonlinear layer is essential so that the electric fields can travel like a soliton wave. An example of such a plasmonic modulator is shown in figure 2.35. The resonance of the plasmonic electrode focuses the electromagnetic waves into the nanowire optical waveguides made by the SOI waveguide based MZM structure, the high field established in the nano-gap between the active electrode and the ground electrode region under biases, leading to high electro-optic interaction, and thus a high modulation index. The operating bandwidth of the plasmonic modulator can reach many hundreds of gigahertz [51]. Furthermore, the required driving electric field is low enough that the antenna signals can be applied directly on the plasmonic electrodes without re-equipment of an electronic amplifier. This allows potential uses in EM antenna field sensing devices. The transmittance characteristics and the effects of nonlinear distortion, an intermodulation product (IMP) of the intersection point of third order and fundamental order or the intersection point of the third order IMD and the first order harmonic, the IIP3, of 19 dBm and a SFDR of 110 dB, can be obtained with a 3 dB bandwidth of 350 GHz [49].

2.4 Linear and nonlinear testing methods for RPoF

For RPoF systems the linearity of the operation region when the RF carrier is swinging over the minimum to maximum range is important. This section gives a brief account of the testing of optical modulators so that the linear and nonlinear distortion can be evaluated for such modulators used for RPoF system. The following parameters can be measured, including the spurious free dynamic range (SFDR).

2.4.1 Scattering parameters S_{21}, S_{11}

The frequency responses of an optical modulator, in particular for RPoF systems, are commonly tested to obtain the transfer function in the frequency domain using the scattering parameter test set. The optical $S_{21}(\omega)$ gain parameter of an RPoF transmission link is normally conducted using the set-up shown in figure 2.32 with the ports 1 and 2 of the network analyser as shown in figures 2.36(a) and (b). Both

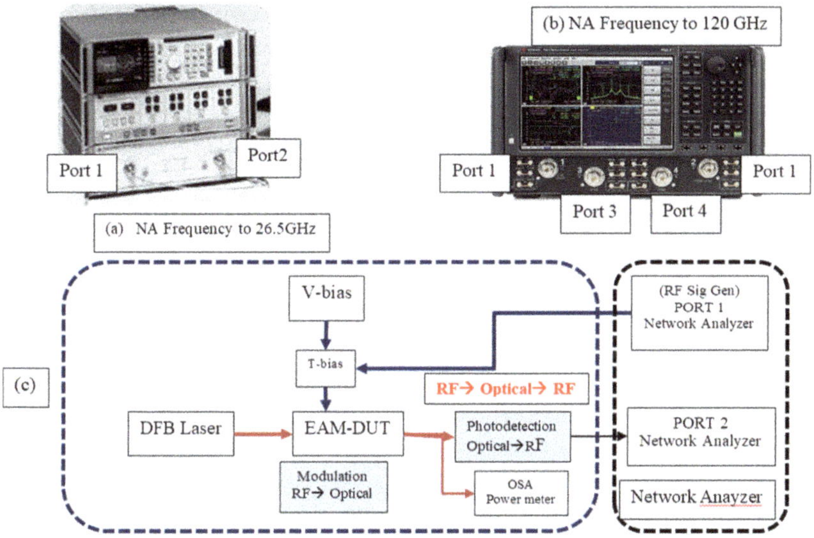

Figure 2.36. Commercial network analyzers (NAs): (a) 45 M to 26.5 GHz; (b) dc to 120 GHz. The scattering parameters $S_{11}(\omega)$, $S_{12}(\omega)$, $S_{21}(\omega)$, and $S_{22}(\omega)$ with subscripts indicate the order or the ports. (b) An RPoF as the DUT with an EAM with a biasing circuit and ports for connecting to the NA, a digital signal analyzer (DSA) of ultra-high sampling rate. (c) The two-step conversion process in the NA connection for the characterization of the RPoF.

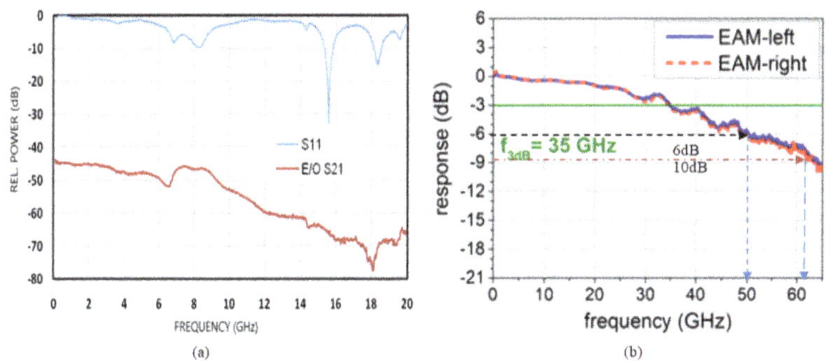

Figure 2.37. Scattering parameters and amplitude as a function of the frequency of an EAM. The blue curve is the reflection scattering parameter S_{11} (input port 1 and measured reflected power from port 1) and the red curve shows the output RF power as measured at the output of a PD or an optical to electrical converter, commonly termed as the optical S_{21} of the DUT. (a) Single EAM and (b) S_{21} of the left and right EML given in figure 2.27.

the amplitude and phase of the device under test (DUT) can be measured. A measured scattering parameter plot of the S_{11} and S_{21} parameters is shown in figure 2.37. $S_{11}(\omega)$, the reflection coefficient, is a bit higher than the normal acceptable level which should be −10 dB or less. $S_{21}(\omega)$ is also measured with the output RF power versus RF frequency of modulation fed into the EAM RF port. The output port is monitored by a photodetector (PD). Obviously, the RF signal applied to the EAM electrical port is small enough so that the electrical network analyser can be calibrated with the scattering matrix components. Note the dip in $S_{11}(\omega)$ at around 8 GHz due to a good match of the RF and the optical traveling velocity. The input optical power into the EAM from the inline laser is estimated at the driving current to be about 3 mW and the total insertion loss of the EAM is about 10 dB or about 0.3 mW power received at the PD. The scattering parameters $S_{11}(\omega)$, $S_{21}(\omega)$, and $S_{22}(\omega)$ are obtained over a wide range of RF spectral regions. The reflected RF power from the EAM electrode is fed into the input port of the RPoF and the scattering parameter S_{11} is obtained.

2.4.2 SFDR

If a number of RF signals/channels are transmitted through a nonlinear (NL) medium there would be intermodulation effects, hence the generation of IMPs between these signals [52]. These are the spurious frequency components and interference with other RF channels located in these bands, hence the intermodulation interference (IMI) noise, which is a function of the signal power at the operating NL region. The generated IMP can be located in the other channel spectral bands, thus generating channel IMP cross-talk. Thus, we can define the SFDR parameter (in dB) as the dynamic range (DR) in which no IMP spurious products are generated for RPoF systems.

The SFDR is estimated as the ratio of the fundamental oscillation component to the strongest spurious oscillation component of the output of an RPoF system. The SFDR was originally defined as a measure to specify the linearity and IMP in analog-to-digital converters (ADCs) and digital-to-analog converters (DACs), an in particular in audio amplifiers to determine the system distortion level.

In the case of an RPoF, the SFDR determines the range of the 'RF tone' so that the linearity of the RF tone can be employed such that there is a minimum intermodulation cross-talk [53] noise interference from other RF channels which can be co-transmitted. The SFDR of an RPoF system is determined by determining the output oscillation component obtained at the output of a PD after transmitting through a length of optical fiber and conversion of the RF signal via the optical modulator and the lightwave carrier of the laser, as shown in figures 2.38 (a) and (b).

For the testing and measurement of the SFDR we require two combined signal generators whose frequencies are different by an amount equal to that of the channel spacing. In 5G and 6G networking this frequency spacing between data channels can be from 250 to 1000 MHz. These RF carriers are coupled via a broadband directional coupler, as shown in figure 2.39. The spacing and power launched of these generators including an internal band pass filter (BPF) are monitored via an electrical spectrum analyzer (ESA1). The coupled RF carriers are fed into the RF port of the DUT and are coupled with a biasing voltage via a T-bias device, as shown in figure 2.31. It is noted that almost all the commercial network analyzers operate in the 'small signal' regions, that is the RF signals are operating in the linear region of the transfer impedance characteristics of the RPoF system. Thus, in order to evaluate the nonlinear distortion and intermodulation cross-talk of an RPoF we must set up a measurement system to determine these nonlinear parameters, first with a B2B link or a short fiber length, commonly a fiber pig-tail length of less than 1.0 m, thus characterizing the transfer transmittance of the optical transmission line.

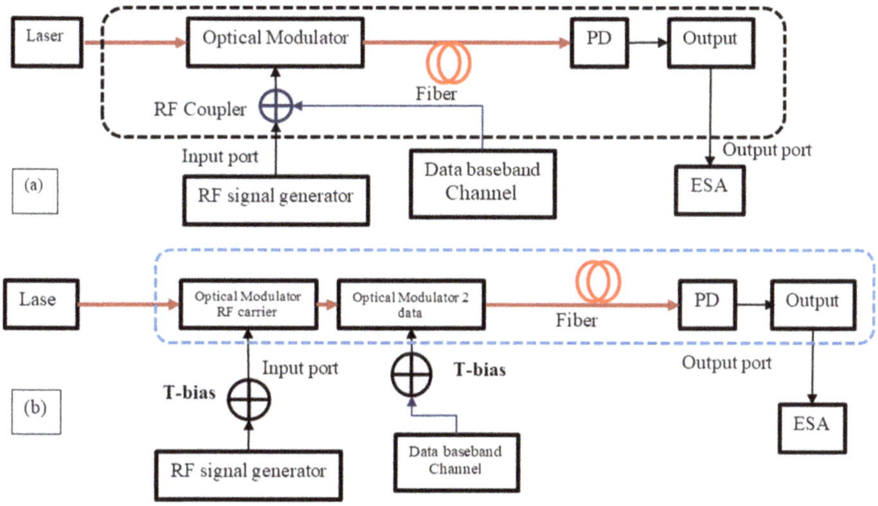

Figure 2.38. Schematic of a typical RPoF. (a) An RPoF with an RF carrier and data baseband channel coupled through an RF coupler. (b) A database channel coupled through a second optical modulator. The optical receivers that can be employed for the RPoF system are described in sections 1.5 and 1.9 of chapter 1. The DUT can be a complete packaged device or a mounted die, such as the EAM die shown in figure 2.31.

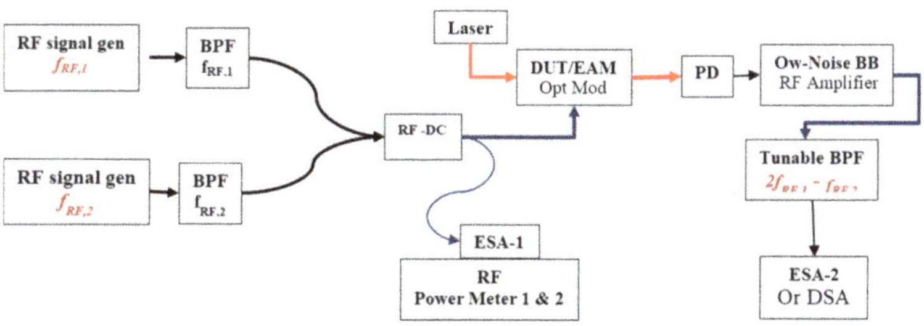

Figure 2.39. Measurement set-up for SFDR. dc = directional coupler; DUT = device under test; BPF = band pass filter; PD = photodetector; ESA = electrical spectrum analyzer; BB = broadband; DSA = digital signal analyzer (see figure 2.36(b)).

2.4.2.1 Intermodulation and IIP3

Figure 2.40 shows the basic concept of the intermodulation effects or the interaction between the frequency components of the modulation and transmission over the RPoF, which can be in a B2B structure and transmitted over an optical transmission line. As introduced in section 1.3 and figure 1.5 of chapter 1, the efficiency of the transfer of the data modulated RF carrier channel at the input to the RF-carrier modulated channel at the output of the fibre channel, is highly critical. The transfer is performed via two-step conversions, a conversion from RF to optical via an optical modulation and then transmission over the optical transmission line, and then the second conversion from optical to RF via a photodetection process of either

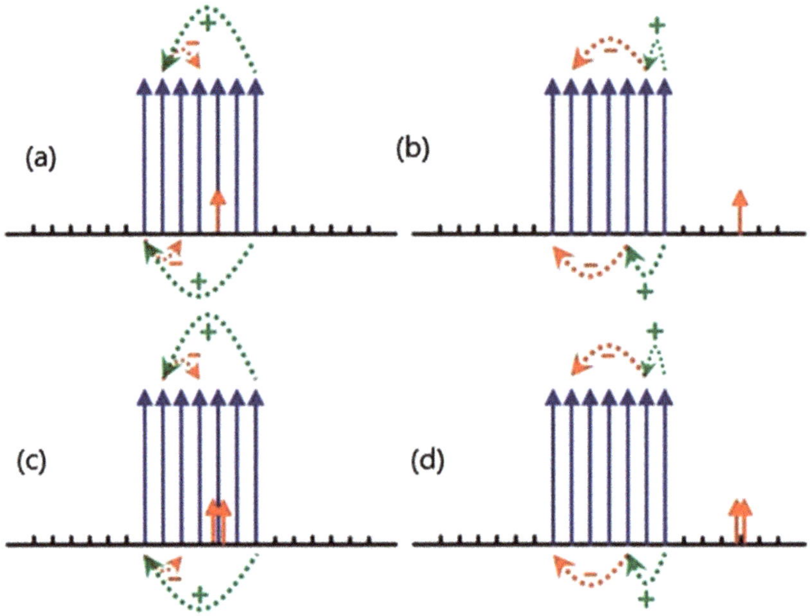

Figure 2.40. The formation of intermodulation products from multi-tone carrying data channels (blue vertical arrows). The frequency spacing between tones ranges from 250 to 1000 MHz in 5G and 6G networks.

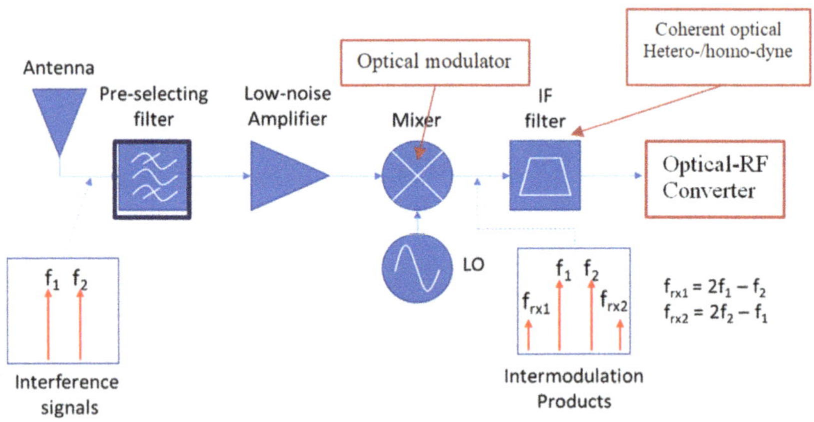

Figure 2.41. Basic schematic for the evaluation of the higher order intermodulation concept in a pure RF domain. The inset shows the equivalent optical–RF and RF–optical conversions. Adjacent RF carriers f_1 and f_2 are launched into the RPoF and then pass through a two-step conversion, and then are detected and converted back to the RF domain with intermodulation products f_{rx1} and f_{rx2}.

a coherent or incoherent receiver. Thus we can represent these two-step processes by a transfer function in the linear domain, as shown in figures 2.41 and 2.39.

Third- and higher order mixing of the two RF signals produces IMP signals in the desired band of the channel at the receiver due to the transmitter and fiber nonlinearity, fiber nonlinear Kerr effect, and the PD and amplifier nonlinearity.

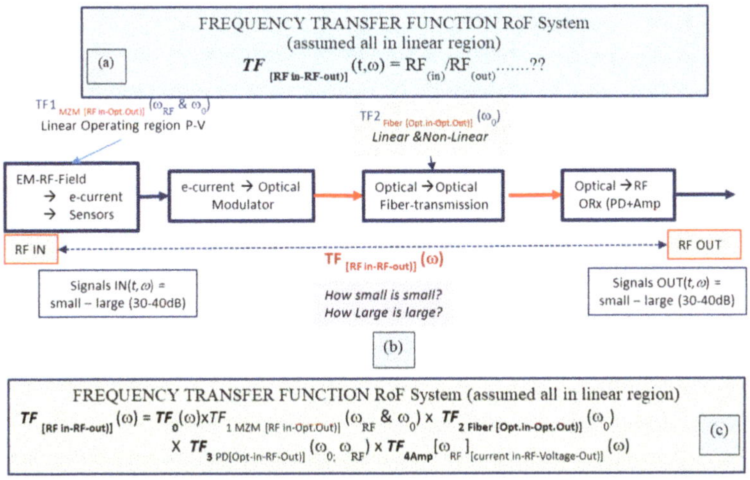

Figure 2.42. Transfer function of an RPoF link as a cascade of transfer functions/matrix of different sub-sections of the system. System representation: linear and nonlinear operating regions. (a) Overall transfer function. (b) The cascade of the sub-system of an RPoF input can be RF signals from a wireless sensor. (c) Transfer functions (TF(ω)) of the subsystems in a cascade. The subscripts indicate the functionalities of the subsystems.

The intermodulation distortion products of the excitation signals degrade the receiver's sensitivity performance. The IMP frequencies can be set such that one of the third-order IMPs, $f_{rx1} = 2f_1 - f_2$ and $f_{rx2} = 2f_2 - f_1$, fall into the receiver passband, as shown in figure 2.41. The generation of the IMP spectral components is shown in figure 2.40. Figure 2.40(a) shows the in-band IMP of multi-RF tones. Figure 2.40(b) shows the out-of-band IMPs from different combinations of the input RF tones. Figure 2.40(c) shows the in-band IMPs generated from the offset multi-RF inputs. Finally, figure 2.40(d) shows the out-of-band IMPs from the offset multi-RF mixing of distinct spectral components. The dotted arrows show the dynamics of the evolution of the frequencies which combine to form third-order IMPs (IMP3 or IIP3), hence the cross-talk in the other active channels. Figure 2.39 shows the summation of two RF outputs via an RF directional coupler (DirC). This is then launched into the DUT RF port. The power of another dc output is monitored by an RF power meter or via an electrical spectrum analyzer (ESA). After the two-step (E/O) RF to optical conversion (via modulation) and then the O/E optical to RF conversion (via an optical reception), the RF frequency components are measured and projected onto the IM3 (or IMP3), the third-order intermodulation product (figure 2.42).

2.4.2.2 *Third-order IMP3 measurement of an EML device*

The fundamental and third-order components of an EML are plotted in figure 2.43. This shows the linearity of such EML structures. It is seen later that two EMLs can be connected B2B to create a powerful lasing and modulation device for RPoF systems.

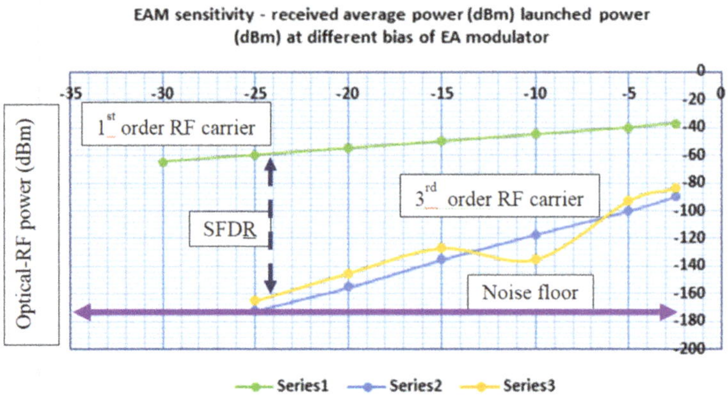

Figure 2.43. Linear and nonlinear behavior of an EAM laser which consists of an inline DFB laser and an inline integrated EAM. IM = intermodulation; IM3 = third-order IMP point; V_b = bias voltage applied to the VCSEL.

2.5 Reception subsystems

2.5.1 High speed photodetector

Over the first two decades of the twenty-first century the employment of digital signal processing (DSP) has attracted significant attention from optical communications technology communities around the world [54]. Both the transmission of wideband data and RPoF have exploited the powerful capability of DSP extensively in mitigating the impairments of transmission over fiber [55]. Multiple input multiple output (MIMO) technology is also integrated with DSP systems to further improve the transmission quality of channel signals [56]. However, the front-end of the optical receiver for RPoF is critical to recover the RF carrier as well as the digital data channels which are transported in the RF modulated optical domain.

The DSP algorithm is a programmable software installed in the random access memory (RAM) to be loaded in the processing unit of the DSP so that specific algorithms and functions can be executed and output signals can be sent to an ADC. Thus several signal mitigation algorithms can be used to improve the signal quality. The sampling rate of the ADC, DAC, and DSP now reach > 200 GSa s^{-1}, hence allowing the recovery of the RF carrier frequency to < 100 GHz or to the upper limit of the mmW (W-band) region [57, 58]. It is noted that the data rates of the channels which are transported by the RF shifted optical carrier for wireless transmissions are lower than 500 MHz. The sampling rate is required in DSP mainly for sampling and recovery of the RF carriers. In the reverse direction or in the generation of the RF carrier carrying the data channels an arbitrary waveform generator (AWG) using DAC and DSP can be formed to generate such RF channels. Figure 2.44 depicts the schematics for these DSP based transmitters and receivers. Furthermore, figure 2.45 shows the schematic for an MIMO RPoF system.

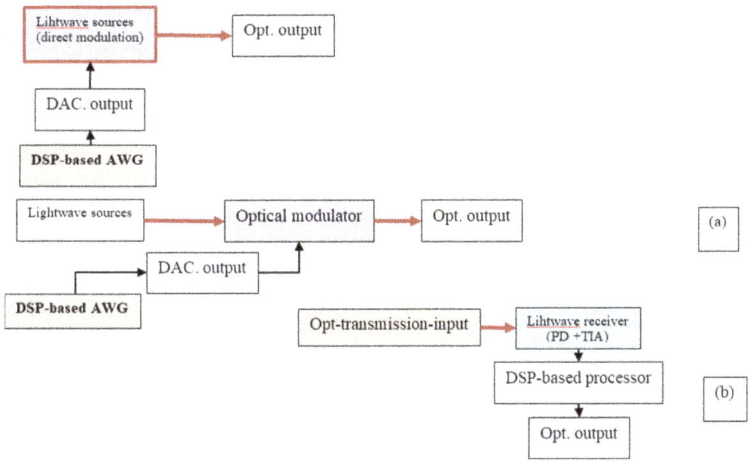

Figure 2.44. Schematics of the generator and receptor using DSP systems. (a) AWG, direct modulation, and external modulation system. (b) DSP RF optical based receiver. AWG = arbitrary wave generator; PD = photodetector; TIA = transimpedance amplifier.

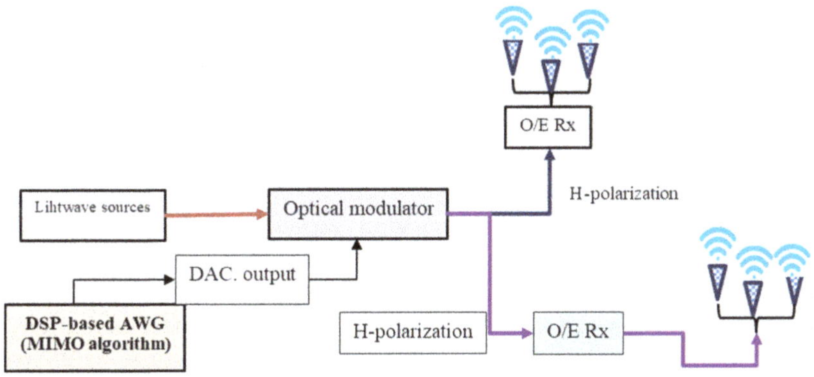

Figure 2.45. MIMO processing for the generation of multiple channels or polarized optical streams in a DSP based optical receiver of RPoF or radio photonic systems.

2.5.2 Advanced DSP at the transmitter and receiver for RPoF

Single-sideband (SSB) and direct detection (DD) offer promising transmission techniques for RPoF links due to their moderate complexity and electronic dispersion compensation. In addition, many effective signal–signal beat interference (SSBI) cancellation schemes for SSB-DD systems, including Kramers–Kronig (KK) and iterative SSBI cancellation, have been reported, pushing the net data rate per channel of SSB-DD transmission systems beyond 400 Gbps and thus are excellent for use in RPoF systems [59, 60]. Figure 2.46 shows the structure of a 4 × ADCS unit for processing of the correlation of the samples so as to increase the bit rate above the capability of a unit ADC. Thus, in this sense with the use of an ultra-wideband front amplifier such as the distributed amplifier of 110 GHz several RPoF channels can be processed.

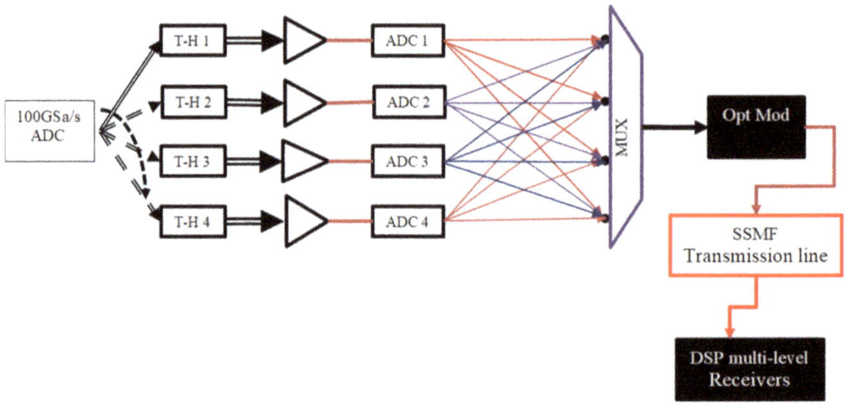

Figure 2.46. (a) ADC front-end test for multi-level reception up to 200 GBd. (b) Photograph of the wire-bonded ADC front-end. (c) Electrical SNR of the complete multi-IC B2B PAM4 test.

2.5.3 Advances in transmission technologies for RPoF access networking

Recent advances in achieving an ultra-high sampling rate and ultra-high speed DSP have allowed the baud rate or symbol rate to be implemented and hence the polarization multiplexing and complex modulation formats M-QAM to result in the aggregate bit rate reaching higher than 1.0 Tbps per wavelength with a bandwidth no wider than 75 GHz. This is critical and can fit into the 50 GHz or 7 GHz spectral grid of the ITU standard.

Such an increase in capacity is an important dimension to exploit to achieve high capacity RPoF systems with DSP based coherent technology. High sampling rate DACs and ADCs are essential to generate signals beyond 200 GBd. The sampling rates of DACs can reach 128 GSa s^{-1} based on SiGe integrated technology [61, 62] and 120 GSa s^{-1} for complementary metal-oxide semiconductor (CMOS) integrated circuits [63] due to the fast transition velocity in SiGe.

Furthermore, the wideband GaAs based optical modulator, in particular the dual polarization inphase quadrature (DP-IQ) phase modulator based on this compound semiconductor, which has been improved recently in the propagation loss and integration technology. Pittala *et al* [64] have demonstrated 220 GBd signal generation enabled by a two-channel 256 GSa s^{-1} arbitrary waveform generator (AWG) and advanced DSP. The output waveforms of the DACs are time multiplexed via a digital multiplexer in the electronic domain to double the baud-rate. These ultra-high capacity transmissions for access and inter-communications between data centers are given in chapter 4.

2.6 Plasmonic ultra-wideband technology for RPoF EM sensing and transmission

Data-hungry global communities have driven the enormous progress in the Si electronics industry. We have witnessed a continuous progression towards smaller,

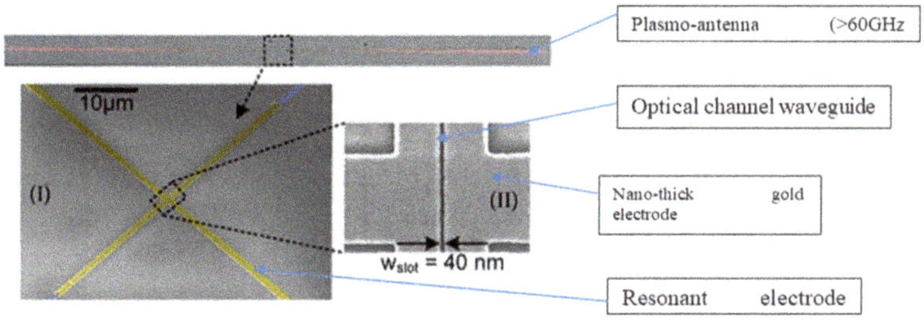

Figure 2.47. A plasmo-antenna incorporated on an MZM optical structure for RPoF. The optical waveguide has Si as the insulator and a Au electrode deposited across the optical waveguide coupled to the plasmo-antenna. The frequency response is shown for a 60 GHz center and 5 GHz bandwidth.

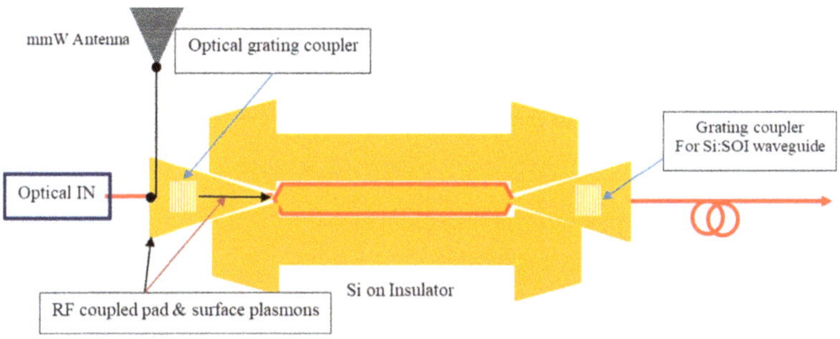

Figure 2.48. Planar view of a plasmonic SOI modulator which consists of a MZM (red) and the plasmonic electrode (gold) whose excitation is the RF waves from a mmW antenna. The plasmonic waves can in in turn modulate the lightwaves in the two branches of the Mach–Zehnder interferometer. The output of the MZM is the transmitted to the remote centre via optical fibre. Note: the width of the optical waveguide is in nanometers, which is narrower that that of the lightwave wavelength in vacuum. The input and output lightwaves can be coupled into and out of the waveguide via a pair of grating couplers.

faster, and more efficient optical transmission system components. The transmission of RPoF over short distances can reach more than 1.0 Tbps. Electromagnetic (EM) radiation sensing is important for monitoring and servicing of RF antenna sites, in particular the wireless micro-cell.

Currently, two of the largest technological obstacles are the DSP speed and front-end ultra-broadband amplifiers, whose thermal dissipation and signal delay times are associated with the electronic interconnection. Optical interconnections of RPoF systems can offer solutions for massive RPoF access networking.

Dielectric photonic devices are limited in size by the fundamental laws of diffraction to about half a wavelength of light. This obvious size mismatch between the two technological solutions, the electronic and optical components, presents challenges for ultra-high speed and ultra-high capacity RPoF. New chip-scale device technology is required that can facilitate information transport between nanoscale devices at optical frequencies and bridge the gap between the world of nanoscale electronics and microscale photonics.

Table 2.2. DWDM channels with numbers for the ITU-T G.694.1 100 GHz spacing C-band frequency grid including wavelength and frequency. Note 100 GHz ~ 0.8 nm in the C-band around 1550 nm [65].

Channel (#)	Frequency (GHz)	Wavelength (nm)	Channel (#)	Frequency (GHz)	Wavelength (nm)
1	190 100	1577.03	37	193 700	1547.72
2	190 200	1576.2	38	193 800	1546.92
3	190 300	1575.37	39	193 900	1546.12
4	190 400	1574.54	40	194 000	1545.32
5	190 500	1573.71	41	194 100	1544.53
6	190 600	1572.89	42	194 200	1543.73
7	190 700	1572.06	43	194 300	1542.94
8	190 800	1571.24	44	194 400	1542.14
9	190 900	1570.42	45	194 500	1541.35
10	191 000	1569.59	46	194 600	1540.56
11	191 100	1568.11	47	194 700	1539.77
12	191 200	1567.95	48	194 800	1538.98
13	191 300	1567.13	49	194 900	1538.19
14	191 400	1566.31	50	195 000	1537.4
15	191 500	1565.5	51	195 100	1536.61
16	191 600	1564.68	52	195 200	1535.82
17	191 700	1563.86	53	195 300	1535.04
18	191 800	1563.05	54	195 400	1534.25
19	191 900	1562.23	55	195 500	1533.47
20	192 000	1561.42	56	195 600	1532.68
21	192 100	1560.61	57	195 700	1531.9
22	192 200	1559.79	58	195 800	1531.12
23	192 300	1558.98	59	195 900	1530.33
24	192 400	1558.17	60	196 000	1529.55
25	192 500	1557.36	61	196 100	1528.77
26	192 600	1556.56	62	196 200	1527.99
27	192 700	1555.75	63	196 300	1527.22
28	192 800	1554.94	64	196 400	1526.44
29	192 900	1554.13	65	196 500	1525.66
30	193 000	1553.33	66	196 600	1524.89
31	193 100	1552.52	67	196 700	1524.11
32	193 200	1551.72	68	196 800	1523.34
33	193 300	1550.92	69	196 900	1522.56
34	193 400	1550.12	70	197 000	1521.79
35	193 500	1549.32	71	197 100	1521.02
36	193 600	1548.52	72	197 200	1520.25

RPoF, in particular, requires both up-link and down-link transport of ultra-high frequency RF carriers over a fiber guided optical transmission line. The back-hauling of RF signals is critical and much more efficient optical modulators which can accept RF radiation signals and convert these lightwaves are essential.

Plasmonic resonant modulators have been reported to offer an amplifier-less conversion of a 60 GHz carrier with a 5.0 GHz bandwidth employing an MZM structure with a plasmonic electrode, as depicted in figure 2.47. These plasmonic modulators would bridge the gap between electronics and photonics enabling the interconnection of data centers for Internet working at terabits per second, thus serving multi-exabyte global networks. RPoF back-haul transport mmW and μW carriers will require plasmonic technology. The 3 dB bandwidth of the plasmonic modulator can reach more than 500 MHz, leading to exciting technologies not only in RPoF but several other potential applications such as x-ray imaging, etc.

The extension of this into broadband modulators can also be employed to fabricate MZM structures, as shown in figure 2.48. The plasmonic MZM consists of (i) two metallic electrodes forming a metal–insulator–metal (MIM) slot waveguide, with nanometer width Si-on-insulator, (ii) a slot filled with an NL metallo-organic material whose refractive index can change linearly with an applied electric field via the EO effect, (iii) a Si-ridge waveguide (450 nm wide, 220 nm thick) to guide lightwaves to the plasmonic slot waveguide, where a linear taper can transform the photonic mode in the silicon nanowire for the surface plasmon interaction along the slot, and (iv) an optical phase modulated by the applied RF field and then coupled back to the photonic waveguide by an identical output taper. A high modulation depth can be achieved due to the strong overlap between the electrical and optical fields, thus the RF 500 GHz band can be achieved.

2.7 Appendix: ITU grid

Table 2.2 tabulates the channel number, its optical frequency and corresponding wavelength. In using these channels, the frequency must be allocated to channel as the wavelength in the fiber varies as a function of the refractive index of the core material and its guiding or the waveguide effective refractive index. In RPoF the optical carrier can be assigned according to the ITU-T grid given in this table. ITU-T = International Telecommunication Union-Telecommunication, a body that sets the standard for all telecommunication carriers/operators.

References

[1] Kao K C and Hockham G A 1986 Dielectric-fibre surface waveguides for optical frequencies *IEE Proc.* **133** 191–8

[2] Maurer R D 1974 Method of producing glass for optical waveguides *US Patent* US3791714A

[3] Maurer R D and Schultz P C 1972 Fused silica optical waveguide *US Patent* US3659915A

[4] Binh L N 2011 *Guided Wave Photonics: Fundamentals and Applications with MATLAB* (Boca Raton, FL: CRC Press)

[5] Binh L N 2017 *Optical Modulation: Advanced Techniques and Applications in Transmission Systems and NetworksOptics and Photonics* (Boca Raton, FL: CRC Press)

[6] Keysight Technologies 2020 *Engineering the 5G World* (Wokingham: Keysight Technologies) https://keysight.com/de/de/assets/7119-1223/ebooks/Engineering-the-5G-World.pdf

[7] Keysight Technologies UXR1102A infinium UXR-series oscilloscope: 110 GHz, 2 channels

[8] AMO 2016 New speed record for graphene photo-detector at 65 GHz

[9] Binh L N 2013 *Digital Processing: Optical Transmission and Coherent Receiving Techniques Optics and Photonics* 1st edn (Boca Raton, FL: CRC Press)

[10] Wooten E L *et al* 2000 A review of lithium niobate modulators for fiber-optic communications systems *IEEE J. Sel. Top. Quant. Electron.* **6** 69–82

[11] Li M *et al* 2020 Lithium niobate photonic-crystal electro-optic modulator *Nat. Commun.* **11** 4123

[12] Chen L 2015 Hybrid silicon and lithium niobate integrated photonics *PhD Dissertation* Ohio University, Athens, OH

[13] Karlsson M 1995 Modulational instability in lossy optical fibers *J. Opt. Soc. Amer.* **B 12** 2071–7

[14] Binh L N 2009 Linear and nonlinear transfer functions of single mode fiber for optical transmission systems *J. Opt. Soc. Am.* A **26** 1564–75

[15] Ramos F and Marti J 2000 Frequency transfer function of dispersive and nonlinear single-mode optical fibers in microwave optical systems *IEEE Photonics Technol. Lett.* **12** 549–51

[16] NKT Photonics https://www.nktphotonics.com/lasers-fibers/

[17] Eblana Photonics https://eblanaphotonics.com/products/

[18] Lundberg L *et al* 2018 Frequency comb-based WDM transmission systems enabling joint signal processing *Appl. Sci.* **8** 718

[19] Hu H *et al* 2018 Single-source chip-based frequency comb enabling extreme parallel data transmission *Nat. Photonics* **12** 469–73

[20] Smith G H, Novak D and Ahmed Z 1997 Technique for optical SSB generation to overcome dispersion penalties in fiber-radio systems *Electron. Lett.* **33** 74–5

[21] Wei R, Yan J J, Peng Y, Yao X, Bai M and Zheng Z 2013 Optical frequency comb generation based on electro-optical modulation with high-order harmonic of a sine RF signal *Opt. Commun.* **291** 269

[22] Mishra A K, Schmogrow R, Tomkos I, Hillerkuss D, Koos C, Freude W and Leuthold J 2013 Flexible RF-based comb generator *IEEE Phot. Tech. Lett.* **25** 701–4

[23] Xiao Y-Q, Chen L, Li F and He H-Z 2013 Novel full-duplex SSB WDM-RoF system with SLM technique for decreasing PAPR *Optoelectron. Lett.* **9** 309–12

[24] Zhenbo X *et al* 2010 OFDM system implementation using compatible SSB modulation with a dual-electrode MZM *Opt. Lett.* **35** 1221–3

[25] Zhang Y *et al* 2010 Theoretical and experimental investigation of compatible SSB modulation for single channel long-distance optical OFDM transmission *Opt. Express* **18** 16751–64

[26] Nazarathy M *et al* 2009 Optical comb and filter bank (De)Mux enabling 1 Tb/s orthogonal sub-band multiplexed CO-OFDM free of ADC/DAC limits *Proc. ECOC 2009 (Vienna, Austria, 20–24 Sept.)* paper P3.12

[27] Zhang J *et al* 2012 Theoretical and experimental study on improved frequency-locked multicarrier generation by using recirculating loop based on multifrequency shifting single-sideband modulation *IEEE Photon.* **4** 2249–63

[28] Li J *et al* 2010 Analysis of the stability and optimizing operation of the single-side-band modulator based on re-circulating frequency shifter used for the T-bit/s optical communication transmission *Opt. Express* **18** 17597–609

[29] Zhang J *et al* 2011 Generation of coherent and frequency-lock multi-carrier using cascaded phase modulators and recirculating frequency shifter for Tb/s optical communication *Opt. Express* **19** 12891–902

[30] Zhang J *et al* 2012 Improved multi-carrier generation by using multi-frequency shifting recirculating loop *IEEE Photon. Technol. Lett.* **24** 1405–8
[31] Torres-Company V, Lancis J and Andrés P 2008 Lossless equalization of frequency combs *Opt. Lett.* **33** 1822–4
[32] Li J, Li X, Zhang X, Tian F and Xi L 2010 Analysis of the stability and optimizing operation of the single-side-band modulator based on re-circulating frequency shifter used for the T-bit/s optical communication transmission *Opt. Express* **18** 17597
[33] Binh L N 2008 *Digital Optical Communications* (Boca Raton, FL: CRC Press/Taylor and Francis)
[34] Binh L N 2019 *Transmission and Processing for Data Center Networking* (IOP Series in Advances in Optics, Photonics and Optoelectronics) (Bristol: IOP Publishing)
[35] Binh L N 2011 *Guided Wave Photonics: Fundamentals and Applications with MATLAB®* (Boca Raton, FL: CRC Press/Taylor and Francis)
[36] Wooten E L *et al* 2000 A review of lithium niobate modulators for fiber-optic communications systems *IEEE J. Sel. Top. Quant. Elect.* **6** 69–82
[37] EOSPACE 18 GHz optical switches
[38] Van Campenhout J *et al* 2009 Low-power, 2 × 2 silicon electro-optic switch with 110-nm bandwidth for broadband reconfigurable optical networks *Opt. Express* **17** 24020
[39] Alferness R C 1979 Polarization-independent optical directional coupler switch using weighted coupling *Appl. Physics Lett.* **35** 748–51
[40] Chung S V, Lee R A and Binh L N 1985 An exact ray representation of bound modes in a step-profile planar slab waveguide *Opt. Acta: Int. J. Optics* **32** 779–91
[41] Chung S V and Binh L N 2010 On the connection between evanescent field tracking and generalized ray representation in optics *Opt. Acta: Int. J. Optics* **31** 615–7
[42] Vujicic V, Anandarajah P M, Browning C, Zhou R, O'Duill S and Barry L P 2014 Optical multicarrier based IM/DD DWDM-SSB-OFDM access networks with SOAs for power budget extension *European Conf. Optical Communications* (Cannes, France) paper We.1.6.6. ECOC
[43] Keldysh L V 1957 Behavior of non-metallic crystals in strong electric fields *J. Exp. Theor. Phys. (USSR)* **33** 994
Franz W 1958 Einfluß eines elektrischen Feldes auf eine optische Absorptionskante *Z. Naturforsch.* A **13** 484
[44] Verbist J *et al* 2017 First real-time 100-Gb/s NRZ-OOK transmission over 2 km with a silicon photonic electro-absorption modulator *OFC 2017 (Los Angeles, CA, 19–23 March)*
[45] Binh L N 2013 *Guided Wave Photonics: Fundamentals and Applications with MATLAB Series on Optics and Photonics* 1st edn (Boca Raton, FL: CRC Press)
[46] Pallec M L *et al* 2004 42 GHz bandwidth InGaAlAs/InP electro absorption modulator with a sub-volt modulation drive capability in a 50 nm spectral range *Proc. 16th Int. Conf. on Indium Phosphide and Related Materials (Kagoshima, Japan, 31 May–4 June)*
[47] Fukano H *et al* 2006 Very-low-driving-voltage electro-absorption modulators operating at 40 Gb/s *IEEE J. Lightwave Technol.* **24** 2219–24
[48] Theurer M *et al* 2017 2 × 56 GB/s from a double side electroabsorption modulated DFB laser and application in novel optical PAM4 generation *IEEE J. Light Tech.* **35** 706–10
[49] NeoPhotonics Laser 28G and 56G EML high performance *NeoPhotonics* https://www.neophotonics.com/products/laser-28g-56g-eml-high-performance/

[50] Baier M *et al* 2020 Integrated transmitter devices on InP exploiting electro-absorption modulation *PhotoniX* **1** 4
[51] Burla M *et al* 2019 500 GHz plasmonic Mach–Zehnder modulator enabling sub-THz microwave photonics *APL Photon.* **4** 056106-0–10
[52] Ha T T 1981 *Solid State Microwave Amplifiers Design* (New York: Wiley)
[53] Cherry E 1983 Designing NDFL amps *Electron. Today Int.* 46–51
[54] Binh L N 2017 *Digital Processing: Optical Transmission and Coherent Receiving Techniques* (Boca Raton, FL: CRC Press/Taylor and Francis)
[55] Hadi M U 2018 Digital signal processing techniques applied to radio over fiber systems *UPEM (May)*
[56] Kuri K A *et al* 2012 Optical and millimeter-wave radio seamless MIMO transmission based on a radio over fiber technology *Opt. Express* **20** 29395–403
[57] Keysight Technologies UXR1004A infinium UXR-series oscilloscope: 100 GHz, 4 channels *Keysight Technologies*
[58] Pittala F *et al* 2020 220 GBaud generation enabled by a two-channel 256 GSa/s arbitrary generator and advanced DSP *ECOC2020 (Brussels, Belgium, Dec.)* PDP paper
[59] Mecozzi A *et al* 2016 Kramers–Kronig coherent receiver *Optica* **3** 1218–27
[60] Le S T *et al* 2019 5 × 510 Gbps single-polarization direct-detection WDM transmission over 80 km of SSMF *Optical Fiber Communication Conf.* Tu2B.1
[61] Buchali F *et al* 2020 1.52 Tb/s single carrier transmission supported by a 128 GSa/s SiGe DAC *Optical Fiber Communication Conf.* Th4C.2
[62] Vinod B *et al* 2020 Single-channel 1.61 Tb/s optical coherent transmission enabled by neural network based digital pre-distortion *European Conf. on Optical Communications* Tu1D.3
[63] Guoxiu H *et al* 2020 Up to 115 GBaud faster than Nyquist PDM-64QAM based on Tomlinson–Harashima precoding with single DAC *European Conf. on Optical Communications* Tu1D.6
[64] Pittalà F *et al* 2020 220 GBaud signal generation enabled by a two-channel 256 GSa/s arbitrary waveform generator and advanced DSP *European Conf. on Optical Communications (Brussels, Belgium, Dec.)* Paper PDP-24
[65] ITU 2008 Spectral grids for WDM applications: DWDM frequency grid *ITU Recommendation ITU-T G.694.1*

Chapter 3

RPoF long-distance transmission

3.1 Introductory remarks: why use single sideband (SSB) and frequency orthogonality for long-haul RPoF?

In 5G and 6G networking the delivery of data channels is embedded on RF carriers whose frequencies can be distributed over a number of spectral windows in the microwave and millimeterwave bands. Furthermore, testing of remote antennae in 5G and 6G networks is preferred with the test signals transmitted from remote central test centers which can be several kilometers away. Therefore radio photonics over fiber (RPoF) signals are required to ensure that no distortions or depletion of the RF channels occurs after long-distance transmission. This chapter thus presents the technology for such long-haul transmission systems in which the RF frequency is in the frequency band of 10 to 60 GHz and even as high as 90 GHz. Of most interest would be the free license band 56.8 GHz with the 7 GHz band. The transfer function of 100 km standard single mode fiber (SSMF) [1] and the amplitude of the modulated RF signals as a function of RF frequency are shown in figure 3.1 along with the schematic of an RPoF transmission system under test, and figure 3.2 shows the fiber transfer transmittance characteristics under RF signals modulated with the single sideband (SSB) and double sideband (DSB) formats, with optical power versus the RF shifted optical carrier.

Furthermore, the transfer functions of the SSMF under nonlinear (NL) self-phase modulation (SPM) effects are shown in figure 3.3. The normal limit of the linear range of the SSMF is about 3 dBM in the average power of the launched optical waves under RF modulation. We can observe that under such an NL limit the low frequency RF waves would suffer some distortion due the over damping of the roll-off under DSB modulation but not under SSB modulation.

The main reason for the degradation of the RF signals under DSB modulation is due to the interference of the two sideband components whose frequencies are separated by a frequency difference and thus the phase delay difference varies according to the fiber length of propagation. Thus when this phase difference equals to π radians they cancel each other, thus the dip in the transfer characteristics.

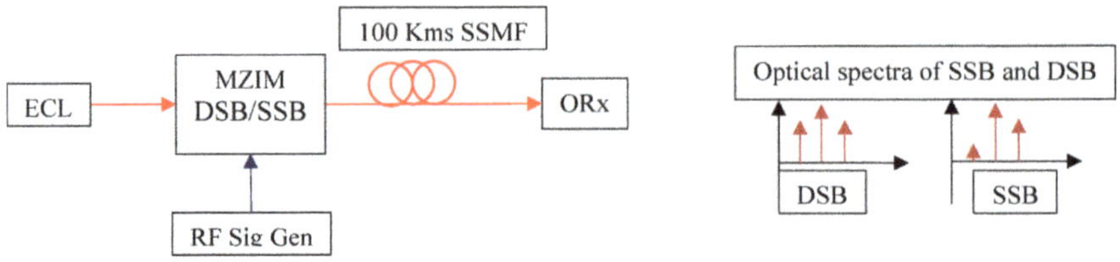

Figure 3.1. Schematic of RF modulated transmission over 1.00 km of SSMF. On the left are the spectra of DSM and SSM modulated optical carriers. SSMF = standard single mode fiber; ORx = optical receiver; MZIM = Mach–Zehnder interferometric modulator; ECL = external cavity laser; DSB = double sideband; SSB = single sideband.

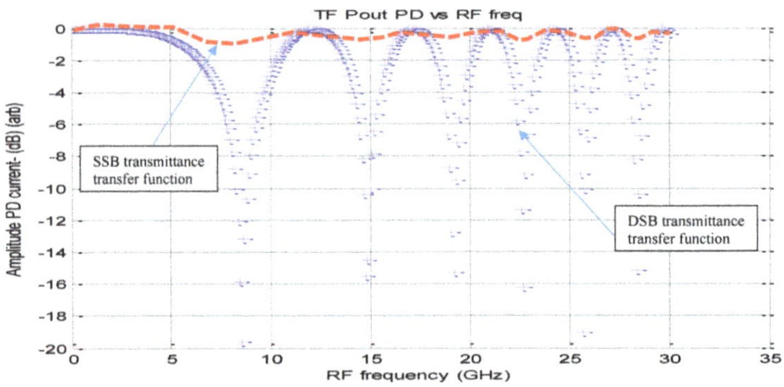

Figure 3.2. Variation of the amplitude of the detected current with respect to the frequency of the RF wave modulated lightwaves after transmission over 100 km of SSMF. + = with DSB; – = with SSB.

As described in chapter 8, on the fiber properties and transmission characteristics of signals over single mode fibers in particular the SSMFs, the transfer functions of the SSMF operating under linear and nonlinear regimes are given. In the linear region we can observe the dips at a number of RF modulating frequencies on the optical waves, thus the passbands of the signals transport over the fiber exhibit a non-fat passband and therefore severe distortion and depletion.

The main phenomenon that leads to this distortion is due to the interference of the two sideband components of the RF modulation when they propagate along the fiber transmission line. Thus, in order to transport RF carrier optical channels over very long distances (>20 km), the modulator needs to be modulated in such a way that only one sideband can be created.

On the other hand, orthogonal frequency domain multiplexing (OFDM) is also a frequency domain with orthogonality of the RF carriers in the time domain and is understood to offer resiliency to the dispersion in the single mode fiber (SMF). The transfer function of the fiber under the transmission of OFDM [2] signals is shown in figure 3.3(b). However, it requires a high speed and high sampling rate digital signal processor (DSP) at the transmitter and digital-to-analog converter (DAC) which can slow down the system operation. On the other hand, the SSB can be easier as it is simpler to generate a single sideband [3, 4]. This modulation format would improve the nonlinear intermodulation distortion of RPoF [5].

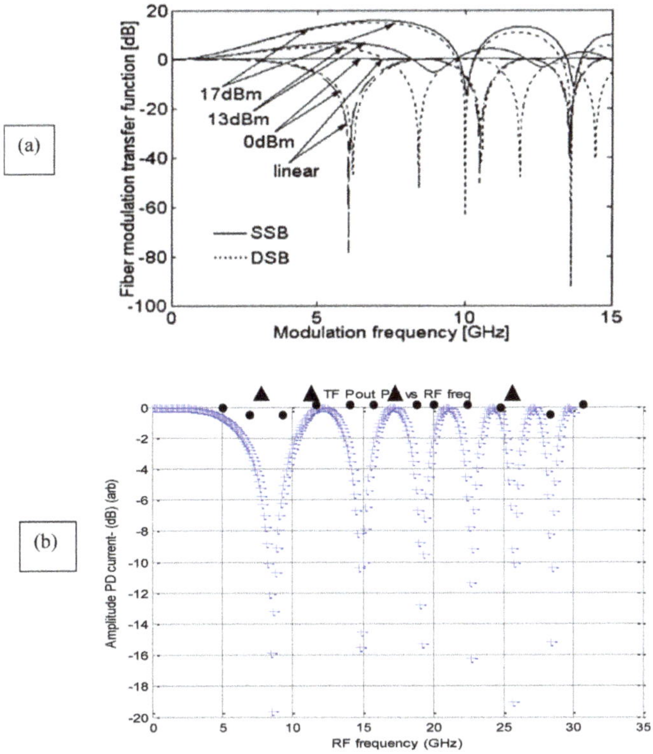

Figure 3.3. Amplitude (power) of RF waves versus the frequency of lightwaves modulated by RF waves at 100 km SSMF. The linear and nonlinear transfer characteristics of SSMF; the dotted lines are linear and the others are nonlinear with dependence on the launched power of the optical waves. SSMF transmission length = 100 km. Launched optical average power in dBm. Self-phase modulation average power limit = 3 dBm. (b) Fiber transfer function under DSB modulation (+); the black dots show SSB modulation and the black triangles show orthogonal frequency division multiplexing (OFDM).

This chapter addresses the techniques for the generation of single sideband signals and their applications in optical transmission systems for microwave and millimeter signals—microwave photonics technology.

3.2 The generation of a carrier suppression (CS)-SSB

If a DSB spectrum of a modulated optical signal is propagating through a fiber, after some reasonably long length the two sidebands are delayed differently due to the wavelength dependence of the fiber refractive index, and then they can interfere with each other. If such a delay phase reaches π (radians) then the waves become destructive. Thus there are a number of dips in the fiber transmittance transfer function, as is observable in figure 3.3. Therefore, the single sideband (SSB) modulated format is preferred, in particular for the case that RPoF systems are transmitted over a long length of SMF. Further, the nonlinear fiber distortion, the intermodulation distortion (IMD) products, can further mitigate the RF data channels and add cross-talk in other sub-RF carrier multiplexed systems.

This section gives details of the analytical expressions and experimental demonstration of the generation of SSB RF modulated optical signals. Furthermore, a suppression of the carrier can assist in the increase of the power of the sideband, hence increasing the link gain–loss of the RPoF.

3.2.1 Principles of generation of SSB

Figure 3.5 shows a single polarization in-phase/quadrature phase (IQ) modulator or QPSK (complex modulator) used to generate optical waves under the excitation of an RF oscillator signal. The IQ modulator is structured by a 'parent' Mach–Zehnder interferometric modulator (MZIM) whose two branches are also embedded with 'child' MZIMs. These two branches, the child MZIMs, are modulated by RF waves, one branch is shifted by $\pi/2$ with respect to the other, that is they are Hilbert transformed with respect to each other (see figure 3.4). Thus, there is the elimination of one sideband of the optical waves. Either the upper or lower sideband is suppressed depending on whether the branch is shifted positively with respect to the other.

Furthermore, in order to suppress the central optical carrier so that there is no interference of this central optical carrier and the sideband, we can bias the child MZIMs at their minimum transmission point, that is at the this point the phase shift of the lightwaves in the two branches are shifted by π, thus the transmission coefficient at this biasing point is minimum, or zero power out. Thus the RF signal can swing with a peak-to-peak amplitude of $2V_\pi$ which would give the maximum SSB lightwaves at the output of the IQ modulator whose integrated optic structure with electrode and excitation ports is shown in figure 3.4. The spectra of the lightwaves at different points of the parent IQ modulator shown in figure 3.5.

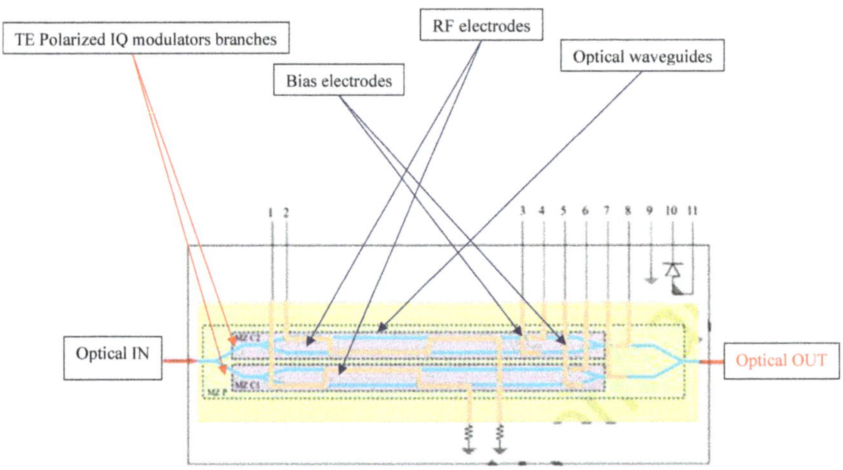

Figure 3.4. Schematic of an IQ modulator with either diffused or ridge optical waveguides with the electrode in deposited metallic thin-film. (a) Schematic diagram of the parent and child MZIMs with electrodes for RF and dc biases; dc3 = electrode for the *C* voltage applied for phase shifting between the two branches of the parent MZIM. Plane view of the IQ modulator with optical waveguide, images of RF electrodes, optical waveguide cross section, polarization beam splitter, and the combiner of the modulator.

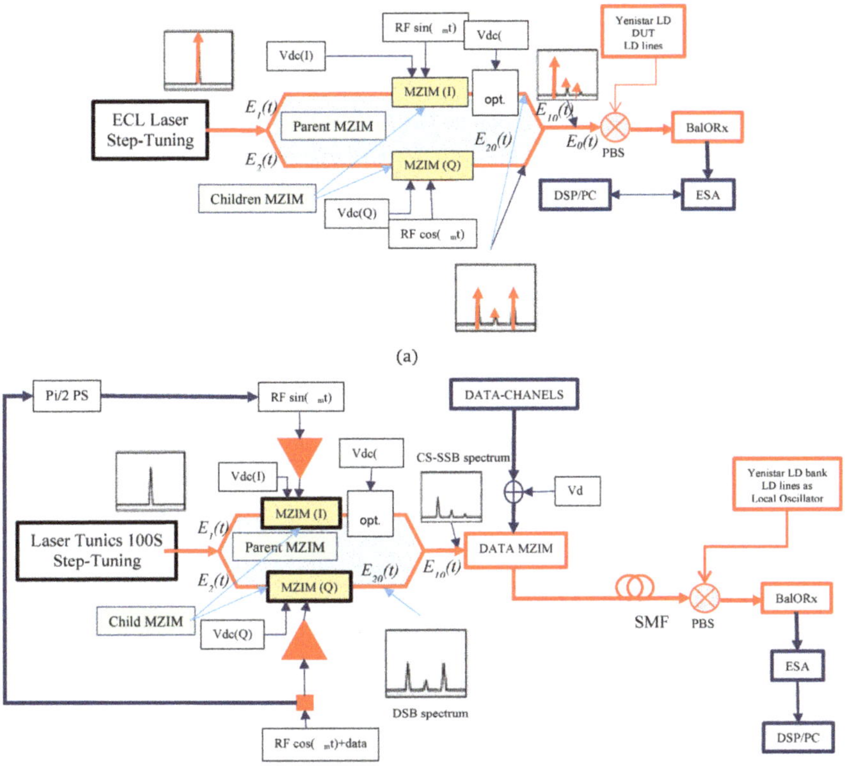

Figure 3.5. Schematic of a CS-SSB QPSK modulator and a heterodyne/homodyne mixing and detection of the RF modulated carrier with suppression of the optical carrier by an optical phase shifter (applied V_{Φ}) and suppression of an RF sideband by a Hilbert transformer via a $\pi/2$ hybrid coupler or an RF delay line of $T/4$ (quarter RF wave period). Alternatively, the suppression of the sideband can be implemented by the Hilbert transformer or a $\pi/2$ phase shift of the sinusoidal signal applied to I or Q and vice versa. (a) Schematic without the Hilbert transformer and (b) with the transformer using a $\pi/2$ phase shift to generate orthogonal RF signals. The SSB can be the upper or lower band by reversing the Hilbert transformer to apply to the upper or lower optical branch and transmission fiber.

The output optical power of the RF modulated SSB varies with the strength of the RF waves launched into the electrodes of the child MZIM.

3.2.2 Case 1: CS-SSB from the electrical applied to the optical phase

Figures 3.5(a) and (b) show schematic diagrams of the proposed SSB vector signal generation using a QPSK modulator. The QPSK or IQ modulator was introduced in chapter 2 as one of the important devices for the enabling technologies of RPoF networking and transmission. This section details the techniques for generating the SSB using a QPSK modulator with biasing and driving conditions. Analytical expressions are given and then the implementation for the SSB and suppression of the optical carrier are given in the next section.

A tunable external cavity laser (ECL) is coupled to a QPSK electro-optic effect (EO) space modulator via a polarization controller (PC) so as to match the polarization of the fiber linearly polarized mode to that of the planar channel waveguide. The QPSK modulator consists of two sub-MZIMs which are embedded in a parent MZIM. In the proposed scheme, the upper (in-phase, I) MZIM of the dual 'child' MZIM QPSK modulator in the 'parent' QPSK modulator is modulated by a single-frequency RF signal. The lower (quadrature, Q) MZIM is also modulated by a sinusoidal RF source but shifted by $\pi/2$ so as to obtain a Hilbert transform RF source and thus the summation of the two guided optical fields leads to the suppression of a sideband. The dc biases of the two 'child' MZIMs are set at quadrature and ideally with a total of V_π difference so there is a π phase difference of the optical carrier between the in-phase and quadrature branch, thus the suppression of the carrier can be obtained. Thus, finally we expect that only the RF shifted sideband exists [6].

The optical field at the output of the QPSK optical modulator (the parent MZIM) can be written as the summation of the two output fields of the two child MZIMs as

$$
\begin{aligned}
E_0 = E_I + E_Q e^{\left(j\frac{V_\varnothing}{V_\pi}\pi\right)} &= [A_x] \cos\left[\frac{\pi}{2V_\pi}\left(v_{\mathrm{RF1}(t)} + V_{\mathrm{dc}I}\right)\right] \\
e^{j\frac{V_{\mathrm{dc}I}}{2V_\pi}} e^{j\omega_0 t} &+ e^{j\pi\frac{V_\varnothing}{V_\pi}} \cos\left[\frac{\pi}{2V_\pi}\left(v_{\mathrm{RF2}(t)} + V_{\mathrm{dc}Q}\right)\right] e^{j\frac{V_{\mathrm{dc}Q}}{2V_\pi}} e^{j\omega_0 t}
\end{aligned}
\tag{3.1}
$$

where E_0, E_I, and E_Q represent the optical fields as indicated in figure 3.5, at the output of the parent QPSK and the two child MZIMs, respectively. Φ is the optical phase shift due to the applied voltage V_Φ to the optical phase shifter on the branch of the parent QPSK MZIM. ω_0 is the angular frequency of the optical carrier, A_x is the carrier amplitude, V_π is the half-wave voltage, $v_{\mathrm{RF1}}(t)$ and $v_{\mathrm{RF2}}(t)$ are the RF signals applied to the child MZIMs which can be written as $v_{RF1}(t) = V_{\mathrm{RF}}\sin(\omega_{\mathrm{RF}}t)$ and $v_{\mathrm{RF2}}(t) = V_{\mathrm{RF}}\cos(\omega_{\mathrm{RF}}t)$, where ω_{RF} and V_{RF} are the angular RF frequency and the amplitude of the RF signal, respectively. The subscripts 1 and 2 indicate the upper (in-phase) and lower (quadrature phase) branches, respectively. To generate an SSB-CS signal, the child MZIMs are biased at the quadrature point, i.e. $V_{\mathrm{dc}I} = V_{\mathrm{dc}Q} = V_{\pi/2}$ and the parent MZIM phase is also biased at the quadrature point, i.e. $V_\Phi = -V_{\pi/2}$, thus a suppression of the carrier due to this π phase difference between the two optical paths of the child MZIM, and the optical carrier signal can then be expressed as

$$
\begin{aligned}
E_0 = [A_x] \cos&\left[\frac{\pi}{2V_\pi}\left(V_m \sin(\omega_m t) + \frac{\pi}{2}\right)\right] \\
e^{j\frac{\pi}{2}} e^{j\omega_0 t} + A_x e^{-j\frac{\pi}{2}} \cos&\left[\frac{\pi}{2V_\pi}\left(V_m \cos(\omega_m t) + \frac{\pi}{2}\right)\right] \quad e^{j\frac{\pi}{2}} e^{j\omega_0 t},
\end{aligned}
\tag{3.2}
$$

$\gamma = \pi Vm/2V\pi$ denotes the modulation index of an MZIM modulator. Expanding (3.2) using the Jacobi–Anger identity [7], the total field at the output of the QPSK is given as

$$
\begin{aligned}
E_0 &= [-jA_x] \sin\left[\gamma\left(\sin(\omega_m t)\right)\right] e^{j\omega_0 t} - [jA_x] \sin\left[\gamma\left(\cos(\omega_m t)\right)\right] e^{j\omega_0 t} \\
&\approx 2A_x J_1(\gamma) e^{j(\omega_{0+\omega m})t}.
\end{aligned}
\tag{3.3}
$$

As expressed in (3.3) the CS-SSB signal can be obtained with only a +1 order sideband (the upper sideband (USB)) on the right of the optical carrier which is suppressed to a much lower level (>25 dB).

Furthermore, the upper QPSK modulator in the dual polarization (DP)-QPSK modulator is used to generate a baseband complex signal. Thus the sub-MZIMs in the QPSK modulator can be configured as in-phase (I) and quadrature phase (Q) branches modulated by two data sequences $I(t)$ and $Q(t)$, and then the vector signal can be expressed as

$$E_y(t) = A_y[I(t) + jQ(t)]e^{j\omega_0 t}, \tag{3.4}$$

where A_y is the amplitude of the optical field. A polarizer with its polarization direction adjusted by a polarization controller or by a polarization aligned fiber such as a polarization maintaining fiber (PMF) to align an angle θ with respect to the x-axis of the DP-QPSK is incorporated to combine the two orthogonally polarized optical signals, which can be expressed as

$$E_{\mathrm{POL}}(t) = \cos(\theta)A_x J_1(\gamma)e^{j(\omega_0+\omega_m)t} + \sin(\theta)[I(t) + jQ(t)]e^{j\omega_0 t}. \tag{3.5}$$

Equation (3.5) shows the composition of the generation of the optical carrier and its sidebands independently to form an equivalent SSB signal or a suppression of one sideband. The optical signals of the upper branch and polarized lower branch are combined and then coupled to the fiber for transmission through L (km) fiber. $\beta(\omega)$ is the effective wave number of the guided wave propagating in the fiber at an angular frequency ω and z is the position of the fiber at which the optical field is evaluated. Then the optical signal after the transmission can be expressed as

$$\begin{aligned} E'_{\mathrm{POL}}(t) = &-2\cos(\theta)A_x J_1(\gamma)e^{j[(\omega_0+\omega_m)t-\beta([(\omega_0+\omega_m)z])]} \\ &+ \sin(\theta)A_y\left[I\left(t - \frac{\beta(\omega o)z}{\omega_0}\right) + jQ\left(t - \frac{\beta(\omega o)z}{\omega_0}\right)\right]e^{j[\omega_0 t-\beta(\omega o)z]} \end{aligned} \tag{3.6}$$

The fiber output optical signal is then received by a PD, a square law detection that converts then the optical signal power (square of the optical fields) to electrical current signal, $i_{\mathrm{PD}}(t)$, which can be expressed as

$$\begin{aligned} i_{\mathrm{PD}}(t) = \varkappa[E'_{\mathrm{POL}}]^2(t) = &-4\cos(\theta)A_x \\ &J\sin(\theta)A_y J_1(\beta).\left[I\left(t - \frac{\beta(\omega o)z}{\omega_0}\right)\cos[\omega_0 t - \beta(\omega o)z]\right. \\ &\left.+ Q\left(t - \frac{\beta(\omega o)z}{\omega_0}\right)\sin[\omega_0 t - \beta(\omega o)z]\right]; \varkappa = \text{conversion constant.} \end{aligned} \tag{3.7}$$

Equation (3.7) shows that the generated RF vector signal is not significantly influenced by the fiber dispersion. The dispersion or the variation of the group delay with respect to the frequency components of the optical channels would induce power fading and distortion effects on RF modulated optical signals after L (km) of SSMF.

This technique can be applicable to both the transmission of RF over optical fiber [8] and the development of a coherent optical signal analyser (COSA) [9]. We can observe that for any fiber connected to the output of the QPSK CS-SSB modulator there would be a minimum distortion effect. Thus such a CS-SSB optical head can be linked to a remote optical signal analyser (OSA) by a modulated optical modulator for optical analyses.

3.2.3 Case 2: IQ modulation with phase shift in the optical domain

An experimental scheme using a CS-SSB QPSK optical modulator for RPoF is depicted in figure 3.5, which consists of a parent MZIM in which two child MZIMs are embedded on each branch of the interferometric paths [10]. The coupled optical field at the input is split equally into the two paths of a guided optic Mach–Zehnder interferometer. The optical fields in each branch can be expressed as

$$E_{01}(t) = \frac{E_0}{\sqrt{2}}\cos(\omega_c t) \tag{3.8}$$

$$E_{02}(t) = \frac{E_0}{\sqrt{2}}\cos(\omega_c t + \Delta\varphi), \tag{3.9}$$

where E_0 and ω_c are the optical field amplitude and frequency, and $\Delta\varphi$ is the relative phase shift between the carriers propagating through the two branches. Two integrated electro-optical (EO) MZIM transfer the RF signals into two amplitude-modulated lightwaves. The modulators are assumed to be driven by the same RF waves with a mutual phase difference $\Delta\Phi$.

For the sake of simplicity, the two MZIMs are assumed to be identical, lossless, and driven by identical, pure, single-tone sinusoidal RF signals, thus the optical output of the QPSK modulator is given as

$$E_1(t) = E_{01}(t)\cos\left(\frac{\pi}{2}\frac{V_{dc}}{V_\pi}\right) + \left(\frac{\pi}{2}\frac{V_m}{V_\pi}\right)\cos(\omega_m t), \tag{3.10}$$

where V_{dc} is the dc bias term, V_m and ω_m are the RF signal amplitude and angular frequency, respectively, and V_π is the MZIM half-wave voltage, which is the voltage level at which a π (radians) phase difference is exerted on the guided lightwaves passing through the MZIM optical waveguides.

Expanding equations (3.8)–(3.10) in terms of Bessel functions, we can obtain the condition of carrier component suppression as

$$\begin{aligned} E_{01}(t) = \frac{E_0}{\sqrt{2}}\cos(\omega_c t)\cos&\left\{\frac{\pi}{2}\frac{V_{dc}}{V_\pi}J\left(\frac{\pi}{2}\frac{V_m}{V_\pi}\right)\right. \\ &+ 2\cos\left(\left(\frac{\pi}{2}\frac{V_m}{V_\pi}\right)\sum(-1)^n J_n\left(\frac{\pi}{2}\frac{V_m}{V_\pi}\right)\cos(\omega_m t)\right) \\ &\left.- 2\sin\left(\left(\frac{\pi}{2}\frac{V_m}{V_\pi}\right)\sum(-1)^n J_{2n+1}\left(\frac{\pi}{2}\frac{V_m}{V_\pi}\right)\cos[(2n+1)\omega_m t]\right)\right\}, \end{aligned} \tag{3.11}$$

where $J_n(.)$ is the nth-order Bessel function. The first term of (3.11) is the carrier component. The second and third terms represent odd- and even-order sidebands.

As can be seen from (3.12), if setting $V_{dc} = V\pi/2$, the carrier and the even-order sidebands are suppressed. At this operating point we obtain the minimum transmittance of the MZIM transfer function. At the combined output signals of the two MZIMs, by using (3.11) the combined field can be given as

$$E_0(t) = \frac{E_0}{\sqrt{2}}\left[\sin\left(\frac{\pi}{2}\frac{V_m}{V_\pi}\cos(\omega_m t)\right)\right]\cos(\omega_c t) + \left[\sin\left(\frac{\pi}{2}\frac{V_m}{V_\pi}\right)\cos(\omega_m t + \Delta\Phi)\right]\cos(\omega_c t + \Delta\varphi), \tag{3.12}$$

where $\Delta\Phi$ is the phase shift between the RF signals applied to the two MZIMs.

Using (3.12) the output combined field of the QPSK or IQ modulator can then be written in terms of Bessel functions as

$$\begin{aligned} E_0(t) = \frac{E_0}{\sqrt{2}}J_1\left[\left(\frac{\pi}{2}\frac{V_m}{V_\pi}\right)\right]&\{\cos(\omega_c + \omega_m)(t) \\ &+ \cos[(\omega_c + \omega_m)(t) + \Delta\varphi + \Delta\Phi] + \cos(\omega_c - \omega_m)(t) \\ &+ \cos(\omega_c - \omega_m)(t) + \Delta\varphi - \Delta\Phi\} + [\text{higher order terms}]\,. \end{aligned} \tag{3.13}$$

In (3.13) if now setting the optical, $\Delta\varphi$, and RF signals, $\Delta\Phi$, equal to $\pm\, 90°$, one of the RF modulated optical sidebands can be suppressed. The 90° phase shift in the RF signals can be obtained by driving one MZIM with the signal and the other with its Hilbert transform or an electrical phase shift of $\pi/2$.

3.3 CS-SSB implementation

Figure 3.6(a) depicts a detailed diagram of the implementation of the generation of an SSB with CS RF modulated lightwaves with the specifications of the components. Figures 3.6(b) and (c) show a photograph of the electronic amplifiers and phase shifter using which suppression of one of the sidebands and the suppression of the lightwave carrier can be achieved.

The spectra of the output lightwaves are shown in figures 3.7(a) and (b) in which the suppression of the upper sideband and lower sidebands are obtained, respectively. An automatic control of the biasing voltages is obtained by using an automatic bias controller.

The stability of the master local oscillator (MLO)-SSB is quite stable and the maximum level to modulate the optical modulator must remain at the peak of the transmittance curve of the MZIM, which is quite simple to control as the electronic amplifier to drive the MZIM is well known for its stability.

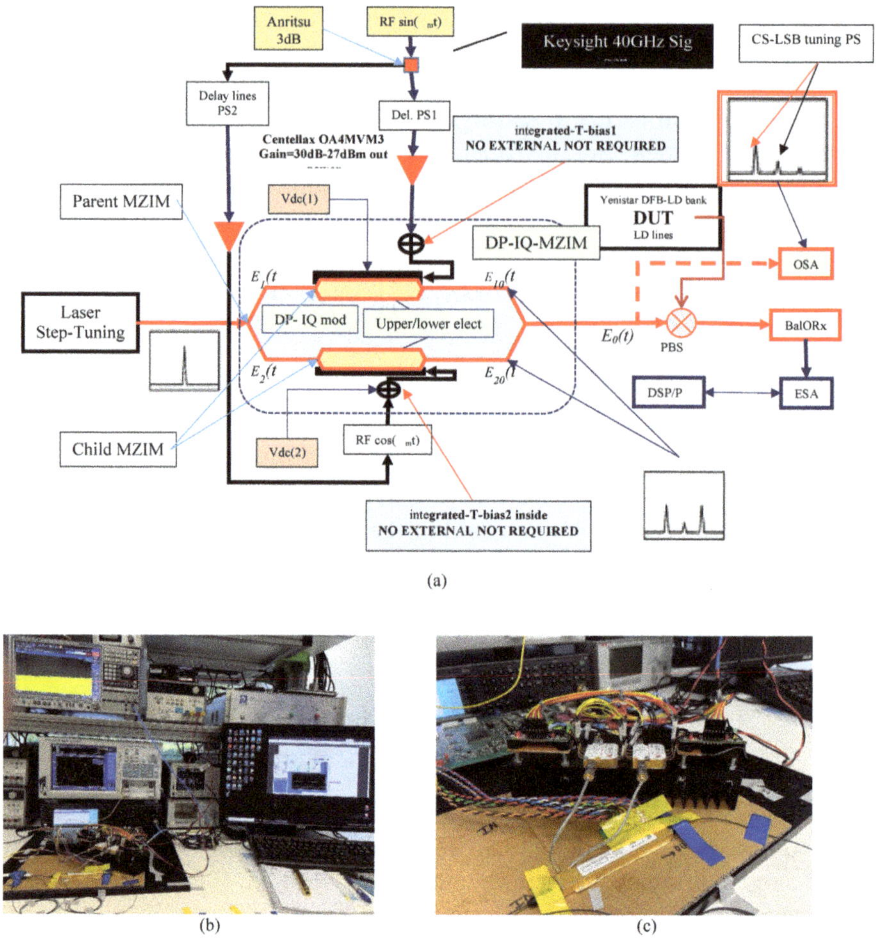

Figure 3.6. DP-IQ modulator, ABC platform (20200807), and RF amplifiers Centellax OM4MVM3 with 30 dB gain (three stage FET amplifier) and output RF amplifier at a maximum of 27 dBm. (a) The set-up platform for a CS-SSB using a DP-IQ modulator and a 30 dB gain RF amplifier. (b) Photograph of (a). (c) Photograph of Centellax amplifiers, type OA3MVM3—gain = 30 dB; output power max. = 23 dBm; bandwidth (BW) = 50 GHz.

3.4 Mitigation of fiber dispersion by frequency orthogonality modulation for RPoF

3.4.1 Introductory remarks

RoF systems enable significant simplification of optical wireless access points, that is, optical network distribution to wireless base stations. They are the key components of 5G, 6G, and 7G ultra-wideband wireless distribution networks for the Internet of Everything (IoET), smart vehicle-to-vehicle, and automatic industrial devices, particularly in the frequency region 1 (FR1), to FR2 and FR3, reaching towards the higher spectral region of the millimeterwave (mmW). Furthermore, in a

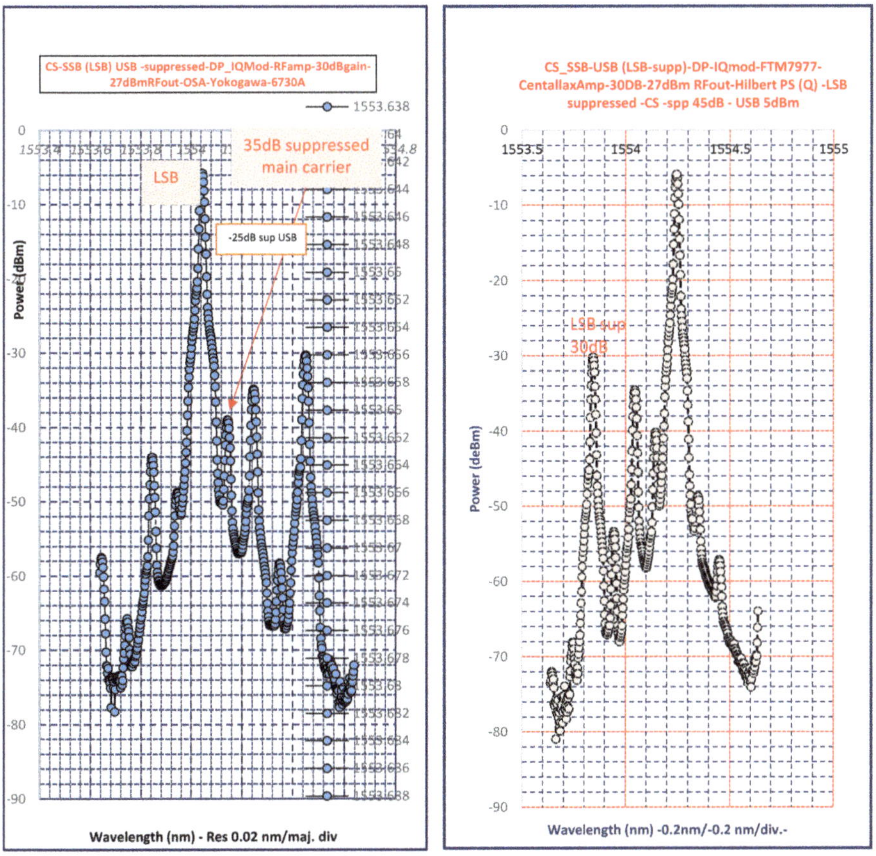

(a) Suppressed → CS-SSB (LSB) (b) Suppressed → CS-SSB (LSB)

Figure 3.7. Spectra of (a) a CS-SSB with suppression of the upper sideband (USB). (b) CS-SSB with suppression of the lower sideband (LSB).

production environment there is the need for longer distance testing systems, and in such high frequencies the transmission and receiving test signals are critical in terms of signal distortion and sensitivity as well as the safety of test personnel.

This can be achieved through the consolidation of signal processing functions in one remote centralized control platform or provisional center due to the advantages of the low loss and high bandwidth properties of optical fibers to RPoF distributing the RF channels to multiple remote radio access units (RAUs) with high fidelity. In wireless networks and test systems, simplifying RAUs can lead to installation and maintenance cost benefits, lower network expenditure, and minimizing the operation cost, thus reducing the operation expenditure (OPEX), in particular when high frequencies are employed, i.e. 5G and 6G networking.

High RF carriers provide more data information bandwidth, and also decreasing wireless cell sizes due to higher wireless signal propagation losses. Thus a large number of wireless access points are required to cover the desired areas and hence at a higher cost. RPoF links employing intensity modulation (IM) and direct detection

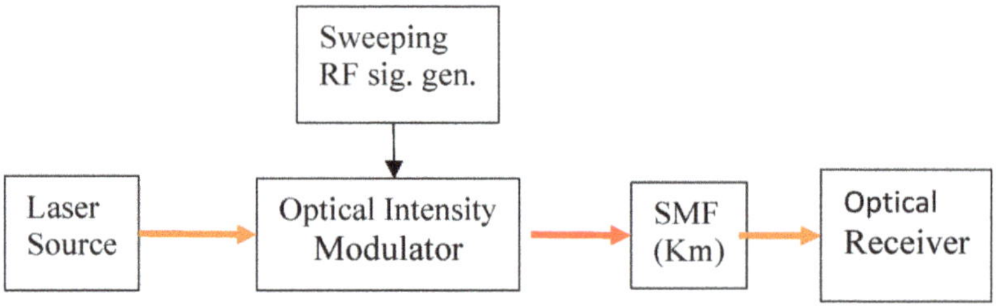

Figure 3.8. Schematic of an RPoF system test for measurement of the fiber transmittance transfer function by modulating an optical modulator with an RF sweeping signal generator.

(DD) would be preferred. IM/DD RPoF systems are attractive for low frequency band (the < 3 GHz spectral window) applications and short distances. In high RF spectral regions it requires external modulation so it is more expensive. Furthermore, the IM/DD RPoF link length can be severely limited (e.g. < 5 km for a 30 GHz 5G FR2 carrier) due to distortion effects induced by the group delays in SMF, thus producing chromatic dispersion or interference of the two sideband carriers.

To achieve longer distance links, alternative RPoF techniques such as heterodyning and homodyning or self-homodyning plus SSB transmission techniques can be used to mitigate fiber distortion since they minimize the interference of the two sidebands [11, 12], as shown in figures 3.8 and 3.9 [13]. The notches of the transfer transmittance at certain frequencies indicate the depletion due to the interference of the USB and LSB frequency shifted optical carrier after propagating over a length of SMF. The high performance of minimum distortion over the long-reach transmission can be achieved with the additional cost of a more complex coherent reception subsystem and the necessity for a narrow linewidth light source.

3.4.2 Mitigation of the fiber distortion effect

The COBRA group [14] reports the use of optical frequency modulation (OFM) to combat such distortion on RF carriers and their data channels induced by chromatic dispersion (CD). OFM is the phase modulation of the optical carrier in which the RF carriers are embedded. The tolerance of the OFM system to CD over 70 km SSMF for mmW carrier-shifted signal can be experimentally proven. The OFM system can up-convert and transport a 100 Mbps 16 QAM modulated 22.2 GHz carrier over 50 km dispersive fiber.

A testbed for evaluating the dispersion tolerance of the OFM RPoF system is shown in figure 3.10 in which two optical modulators are used, one MZIM is used to modulate the optical carrier with the RF carriers, the second in cascade with the MZIM is an optical phase modulator (OPM) on which the RF sweeping frequency is applied for the frequency modulation of the optical RF carrier. The fiber length can then be varied so as to evaluate the effects of fiber dispersion on the OFM RPoF system. The optical receiver consists of a Mach–Zehnder delay interferometer (MZDI) whose delay of one optical branch allows the self-coherent detection on the PD and an electronic amplifier. The output is then either fed into a remote antenna unit via an RF power amplifier or fed into an ESA to evaluate the carrier spreading.

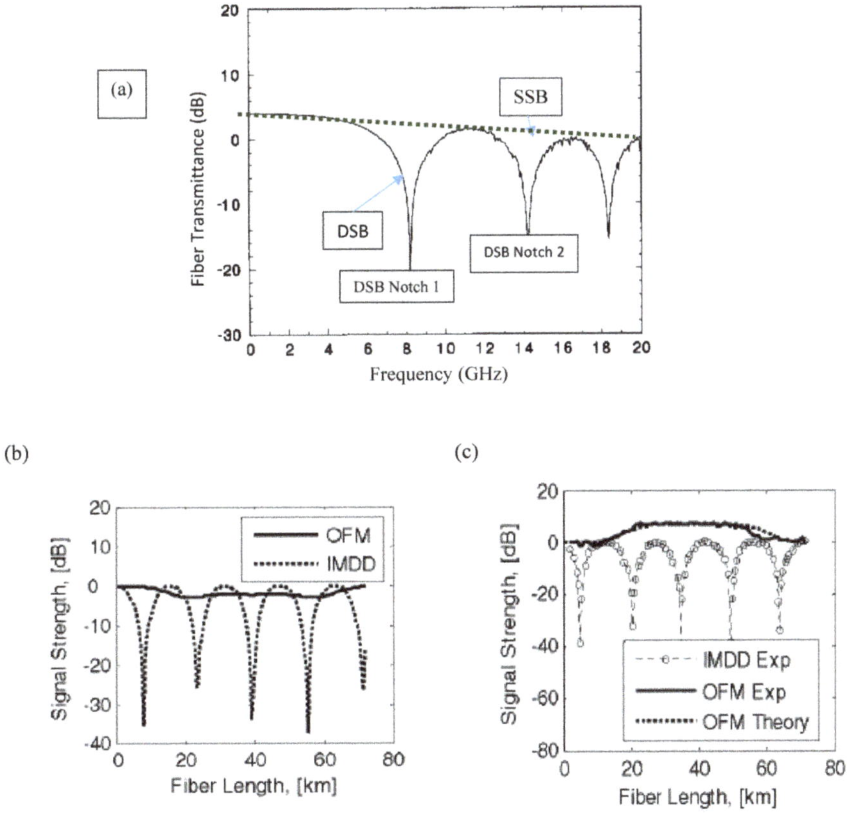

Figure 3.9. Linear fiber transfer transmittance function of a 1.0 km long SMF. (a) Plots of the transmittance as a function of the frequency of a sweeping RF sinusoidal signal generator which modulated a narrow linewidth lightwave source. Transmittance after propagation through an SMF under DSB (solid line) and SSB (dotted line) or suppression of one sideband. (b) Theoretical fiber transfer function under OFM for the experimental results.

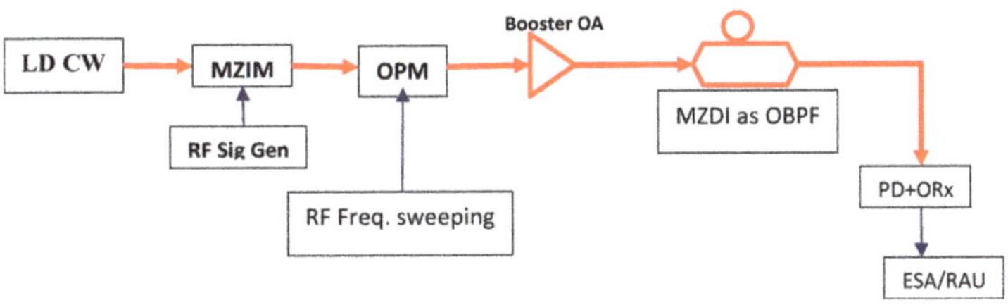

Figure 3.10. An OFM RF system for evaluating the tolerance of the system with respect to the variation of SSMF length. ESA = electrical spectrum analyzer or remote radio antenna unit; RF signal generator = sinusoidal RF carrier frequency sinusoidal waves; RF sweeping generator = generating modulation frequency of the optical RF shifted carrier MZDI (as an optical bandpass filter (OBPF)). OA = optical amplifier.

The dynamics of the dispersion of OFM RPoF can be explained as follows:

(a) When an RF modulated optical carrier is propagating through an SSMF, the optical carrier phase of the RF upper and lower bands is propagating with different delays, then they interfere with each other after some distance due the phase changes of the group velocity and the refractive index of the SMF varying as a function of frequency. These two sideband carriers are destructive at some distance when their phases have a π difference. Thus we observe the notch of the transfer function as depicted in figure 3.9 (DSB).

(b) This DSB interference due to intensity modulation can be minimized by suppression of either the upper or the lower sideband by using a dual-drive or IQ modulator with an appropriate bias and driving signal amplitude [15].

(c) The tolerance of the OFM by phase modulation of the RF shift optical carrier can effectively chirp the optical carrier in such a way that this would counter the chirping effect of the optical fiber on the optical fiber. The dispersion parameter can be expressed by

$$\beta''(\lambda) = -\frac{D(\lambda)\,\lambda_0^2}{2\pi c_0}, \tag{3.14}$$

where $D(\lambda)$ is the dispersion factor in units of ps nm^{-1} km^{-1} of the SSMF, λ_0 is the optical carrier wavelength, and c_0 is the free space velocity of light. Then the chirp parameter exerted on the lightwaves by the fiber can be varying due to the variation of the optical wavelength. If the phase modulation implemented on the lightwaves is of opposite sign with respect to that of the fiber then the two effects can be mutually compensated and thus no dispersion effects occur on the RF optical carrier with phase modulation. The parameter $\beta''(\lambda)$ represents the second-order differentiation of the propagation wave number, thus it is the differentiation of the group velocity of the optical waves, the dispersion parameters. This indicates the chirp of the lightwave carrier after propagating through a length of SMF. Hence the notch of the transfer function, as shown in figures 3.9(a)–(c). If the OFM is employed with the phase, and thus the frequency, modulation to chirp the optical carriers to mitigate the fiber chirping effects and thus the transfer function can overcome the notches, as shown in figures 3.9(b) and (c).

Alternatively, if we can suppress one sideband carrier and the central carrier then the interference effect by the DSB carriers can be eliminated. The next section will describe this technique to overcome the fiber chirping effects.

In the 1990s the common technique to compensate for the fiber dispersion was to use the dispersion compensating fiber spans. However, this technique needs an additional optical amplifier due to additional losses by the DCF which has a much smaller core diameter than that of the SSMF [16, 17]. Therefore, in the case of RPoF or optical-to-antenna (OTA) the use of DCF will add more costs to the link and thus it is preferable that modulation techniques are used. The sensitivity of the OFDM modulated data carried by an RF optical carrier can be seen in figure 3.3 under the influence of an optical bandpass filter, showing nearly 20 dB improvement under short distances <10 km and <10 dB improvement over longer transmission lines with the passband narrow or wider, respectively. The main reason is that the orthogonality of multi-level modulation OFDM on the RF carrier has allowed the recovery of these RF carriers much more effectively, or the resiliency of the frequency orthogonality modulation format. One set back is that the sampling rate of the DSP based transmitter for OFDM is high and thus there is higher power consumption and

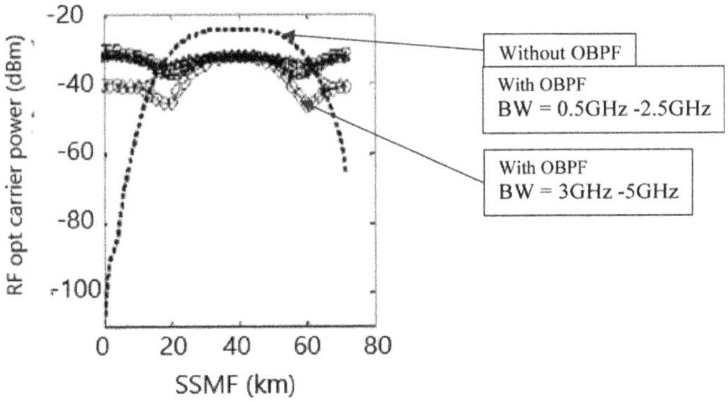

Figure 3.11. Effects of optical filtering to minimize noise due to out-of-band optical amplification on OFDM channels by using an OBPF as a function of the optical transmission lines (SSMF), and average received RF optical carrier power versus transmission distance in SSMK standard single mode fiber (SSMF) km.

higher cost. Figure 3.11 shows the effects of optical filtering using OBPF to minimize the amplified noise due to out-of-band optical amplification on OFDM channels. The filter is an OBPF filter whose pass band as a function of the length in km of SSMF. Further, this figure also shows the variation of the average received RF-shifed optical-carrier power against the SSMF transmission distance.

3.5 Carrier re-injection in SSB modulation for RPoF mitigation of IMD

Intermodulation effects and products can interfere and generate cross-modulation component noise in RPoF transmission links. These problems are normally observed in modulation formats of DSB spectra as the propagation delay, hence phase, times of the group of waves are spectrally dependent in the fiber. These waves interfere at the fiber output, in particular when the fiber transmission line is long enough to result in destructive interference. Thus the SSB modulation format of the RF carrier can extend the transmission dip to a much greater distance. It is highly critical in employing linearization techniques and incorporating RF nonlinearity due to electronic power amplification, as an example. Thus the characterization of an RPoF link under an optical SSB modulation should be carried out to determine the destructive-interference-free length so that RF modulated optical waves can be transmitted with minimum distortion over a desired transmission distance.

The recovery of the electrical RF carriers and the embedded data channels requires the original optical carrier so that its beating product under mixing with the data channels transported by the optical RF shifted carriers can be determined so that the RF carrier modulated data signals can be recovered.

Transport techniques for RPoF employing the RF carriers of mmW in the Ku-band (26.540 GHz) or W-band (75–100 GHz) are being investigated intensively for 6G and 7G networking. The optical transport of mmW signals is susceptible to the effects of fiber CD, as described in section 3.2, and thus spectral inefficiency. Optical SSB without carrier suppression (OSSB+C) would minimize the chromatic dispersion

(CD) effects in the fiber [3] and hence more is spectrally efficient. The optical spectral efficiency can be further increased by using a wavelength interleaving scheme [18–20]. In a multiple radio channel environment, the performance of an RPoF system is severely limited by the nonlinearity of the optical front-end [21] and is further degraded by fiber nonlinearities if high-gain optical amplifiers are inserted [22].

It is thus highly important to minimize the fiber nonlinear distortion effects. Previously Lim *et al* [23] have demonstrated a linearization technique to improve the carrier-to-interference ratio (CIR) of OSSB+C signals. It is noted that in RPoF the bandwidth of the data channels is much lower and more closely spaced than that for broadband optical transmission, such as intercontinental optical transmission systems. The RPoF data channel bandwidths can only be in the order of about 250–500 MHz, which are superimposed on the RF carriers or sub-carrier multiplexed (SCM) RF carriers operating in the mmW or μW spectral regions. Any nonlinearity would result in intermodulation distortion (IMD) effects or cross-talk from one to the other RF modulated data channel. This mitigation technique was based on the removal of the dominant sources of IMD in the optical domain by eliminating one sideband of the DSB modulation commonly faced in optical modulation.

In this section we describe the method to mitigate the RF nonlinearity in an RPoF link by using SSB-CS or vestigial SSB.

In figure 3.12 the optical modulator is placed in the upper arm of an optical interferometric structure and in the other arm an optical attenuator is placed, and then an optical phase shifter in cascade. Thus the interference of the optical carrier can be eliminated or suppressed at the output by shifting the phase in the lower arm of the interferometer so that a π phase difference can be obtained so as to suppress the residual optical carrier after the optical modulation in the upper arm. On the other hand, the method for generation of the SSB with suppression of the carrier given in section 3.2.2 uses the amplitude of the RF wave which is large enough for a swing of $2V_\pi$ so that a complete suppression of the carrier can be obtained. This relies on the sharp swing of the single RF oscillation so that almost no nonlinearity can be experienced. The method presented in figure 3.12 requires more components and may suffer from the phase variation of the optical paths of the interferometer. Alternatively, an optical notched filter (ONF) can be used to remove the RF carrier

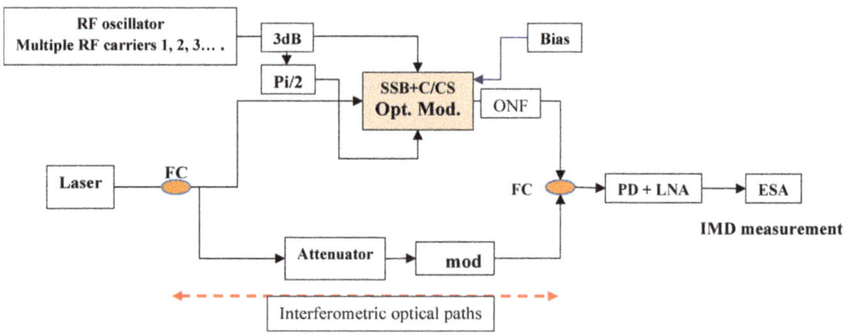

Figure 3.12. Linearization of an RPoF transmitter. LNA = linear amplifier; PD = photodetector; CS = carrier suppression; C = with carrier; FC = fiber coupler; ONF = optical notched filter.

modulated optical waves. However, in the case of multi-data channels being embedded this is almost impossible, thus the interferometric method should be used. The optical modulator can be an IQ modulator or a dual-drive MZIM modulator as shown in figures 3.27 and 3.29.

Under RPoF, to produce RF shifted data channels at the output of the PD and linear amplifier (LNA), it is essential that the original optical carrier be embedded so that the RF shifted and the non-RF shifted optical waves can beat each other in the PD so that the pseudo-coherent RF carriers plus their phase and data channel can be recovered. A $\pi/2$ phase shifter delays the RF waves and modulates the optical waves with an imaginary component of the optical waves in the lower branch of the modulator so that when both output optical waves combine at the output of the MZIM cancellation of the imaginary component occurs and hence suppression of a sideband.

Alternatively, one can use an optical filter which can have a sharp roll-off close to the RF modulated carrier to reject one of the bands. This is normally termed as vestigial sideband signal.

By this linearization method, there is no need to drive the optical modulator with a large amplitude RF Hilbert transformed into the optical modulator to achieve CS-SSB signals. However, again we note that the optical notch filtering with sharp rolling off is almost impossible for 6G networking due to the existence of several data channels associated with the RF carriers. Thus the CS-SSB seems to be the most effective method.

In order to evaluate the IMD, an RF oscillator can generate multi-RF frequency components so that the product of different RF components resulting from the nonlinearity can be measured at different frequencies of the data channels. A generic schematic for IMD determination is shown in figure 3.13 in which both RF carriers and data channels are transmitted over the fiber and create IMD products (figure 3.14).

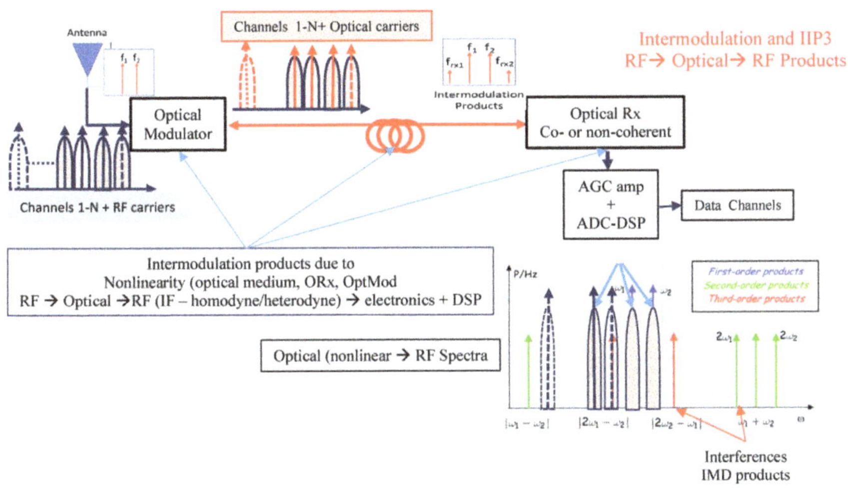

Figure 3.13. Generic schematic for the determination of IMD products. The spacing frequency between RF carriers is equal to the data channel spacing.

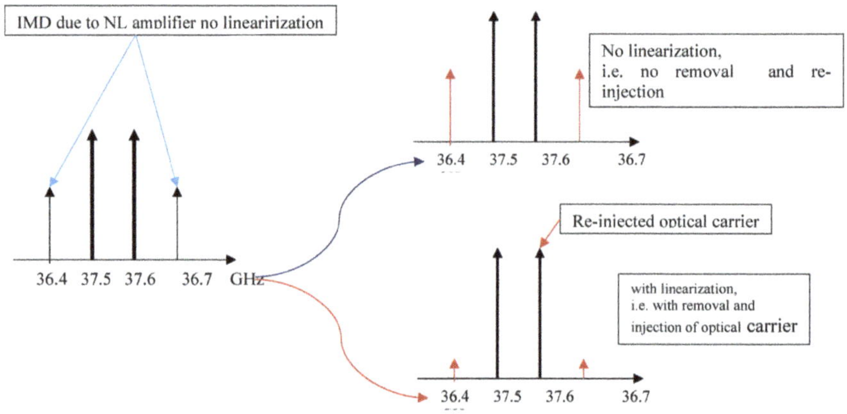

Figure 3.14. Elimination of the IMD by linearization, i.e. removal and re-injection of the original optical carrier, and reduction of intermodulation products. Only two RF carriers are launched into the optical modulator at 36.5 and 26.6 GHz, that is 100 MHz spacing.

3.6 The noise figure and SFDR of the RPoF link

An externally modulated analog RPoF based on an MZIM can operate in the μm or mmW regions. These links are composed of a modulation device, an MZIM, for converting RF optical signals from an optical fiber, normally an SSMF, then feeding an optical receiver (ORx) for optical-to-electrical conversion. A microwave RPoF link can effectively transport very high RF signals over long distances with the low loss property of the SSMF in the silica C- or L-band. While different modulation schemes and techniques can be used in RPoF links, the majority of links employed today are based on intensity modulation and direct detection (IM/DD). The previous sections of this chapter have described the IM/DD and advanced modulation formats. RPoF links can also be based on directly modulated lasers (DMLs) or external modulation via electro-optic modulators (EOMs), inline integrated lasers. A schematic of a generic RPoF based on an external modulator is shown in figure 3.15(a). From an electrical perspective, an RPoF is equivalent to a two-port microwave network, as shown in figure 3.15(b).

3.6.1 RPoF gain

The small signal link gain G_{RoF} of an RPoF represents the RF-to-RF power signal transfer gain/loss ratio in either voltage or current in terms of power, the square of the scattering power,and the scattering parameters.

We define the mid-band small signal gain, or the signal induced voltage, as being much smaller than that of the half-wave RF voltage of the RPoF link from the source power P_S to the load P_L, using a maximum power transfer theorem as =

$$G_{\mathrm{RoF}} = \frac{P_L}{P_S} = \frac{\langle i^2(t)\rangle R_L}{\langle v_S^2(t)\rangle/(4R_s)}, \tag{3.15}$$

Where the bracket indicates an ensemble average with respect to time.

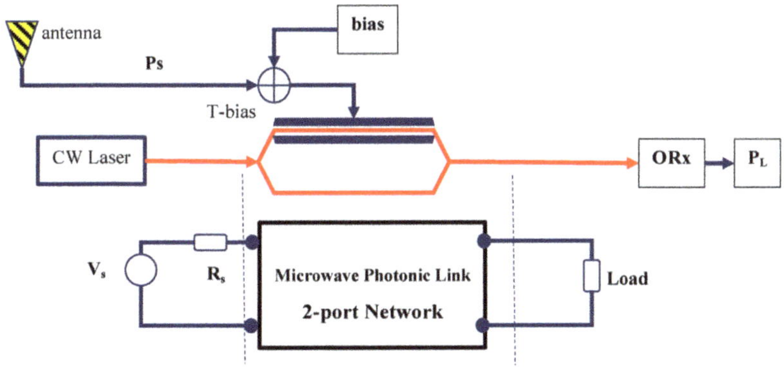

Figure 3.15. Schematic of an IM/DD RPoF microwave photonic link. (a) Physical structure. (b) Representation by a two-port network.

The average current delivered to the load impedance is produced by the PD and can be written as

$$i_L = \frac{1}{2}\Upsilon_{PD}P_m(t), \tag{3.16}$$

where Υ_{PD} is the PD responsivity and $P_m(t)$ is the instant optical power of the RF modulated optical signals.

In the case of the intensity modulation using an optical MZIM the time varying power can be written as

$$P_m(t) = \frac{P_i}{2L}\left(1 - \cos\left\{\pi\left[\frac{V_b}{V_{\pi,\,\mathrm{dc}}} + \frac{v_{\mathrm{RF}}(t)}{V_{\pi,\,\mathrm{RF}}}\right]\right\}\right); \tag{3.17}$$

where P_i is the input optical power, L is the input insertion loss, V_b is the bias voltage (dc), $V_{\pi,\,\mathrm{dc}}$ and $V_{\pi,\,\mathrm{RF}}$ are the half-wave voltages with a phase change of π in dc and RF, respectively, in one of the optically guided branches of the MZM, and $v_{\mathrm{RF}}(t) = 0.5v_S(t)$ is the RF signal exiting to the MZM.

In the small signal region, where $v_{\mathrm{RF}} \ll V_{\pi,\mathrm{RF}}$, the argument of the cosine function can be expanded using the Taylor series as the sum of a dc term, $P_{\mathrm{av,MZIM}}$, and first-order and higher order terms, and depending on amplitude of the modulating signal $v_{\mathrm{RF}}(t)$. For G_{RoF} we consider the contribution of the first-order component

$$P_{\mathrm{MZM}}(t) = \frac{P_i}{2L_{\mathrm{MZM}}}\left[\pi\frac{v_{\mathrm{RF}}(t)}{V_{\pi,\,\mathrm{RF}}}\right]\sin\left(\frac{V_b}{V_{\pi,\,\mathrm{dc}}}\right); \tag{3.18}$$

$$L_{\mathrm{MZM}} = \text{modulator insertion loss},$$

in which the latter term of (3.18) is the bias phase. From (3.18) the expression of the link gain/loss of an RPoF link can be written as

$$G_{\mathrm{RoF}} = \frac{\Upsilon_{\mathrm{PD}} R P_i \sin\left(\frac{V_b}{V_{\pi,\,\mathrm{dc}}}\right)}{4 L V_{\pi,\,\mathrm{RF}}}. \tag{3.19}$$

3.6.2 Noise figure (NF)

The NF in decibels is defined as the ratio between the signal-to-noise ratio (SNR) at the input and the SNR at the output of the link. The NF of a photonic link can then be written as

$$\mathrm{NF} = 10\log_{10}\frac{S_i/N_i}{S_o/N_o} = 10\log_{10}\frac{\mathrm{SNR}_i}{\mathrm{SNR}_o}; \tag{3.20}$$

where S_i is the power of the input signal, N_i is the noise power at the input port, S_o is the power of the output signal, N_o is the noise power at the output port, and SNR_i and SNR_O are the signal-to-noise ratios at the input and output ports, respectively. The noise model of an optical receiver can be represented as shown in figure 3.16, with the average current produced by the PD when exposed to a modulated RF signal, the dark current noise $i^2_{\mathrm{d,N}}(t)$, the quantum shot noise of the PD $i^2_{\mathrm{S,N}}(t)$, and the total equivalent noise referred to the input of the amplifier $i^2_{N,\mathrm{eq}}(t)$. This equivalent noise can be obtained by measuring the noise power at the output of the output of amplifier, that is, under input signal by shorting the input port. Thus the noisy amplifier can be represented by a parallel equivalent noise source at the input port the output can then be followed by a noiseless amplifier. If an additional amplifying stage is used then the gain given in (3.19) is to be multiplied by a gain factor of this amplifier.

3.6.3 Spurious free dynamic range

Based on the noise power spectral level one can estimate the spurious free dynamic range (SFDR) of an RPoF link. We can assume that the nonlinear distortions are approximately constant over the system mid-band spectral range. This is the in-band noise that cannot be measured explicitly. In this case, one can derive an input intercept point, IIP3 (see figure 3.17), as the intersecting point between the third-order IMD and the second-order IMD. The IMD is found by launching two RF carriers (two-tone) into the RPoF, separated by a frequency spacing, as illustrated in figure 3.18. It is normal that superimposed on the two RF carriers are the data baseband signals, but in the two-test only two pure RF tones are used.

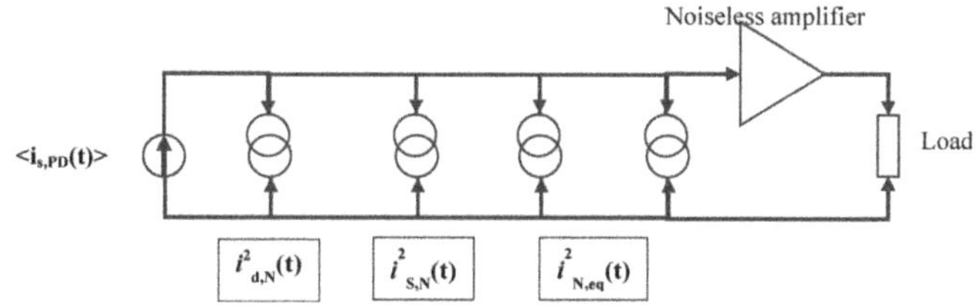

Figure 3.16. Noise model of an optical receiver in which a PD can be followed by an electronic amplifier (of transimpedance type).

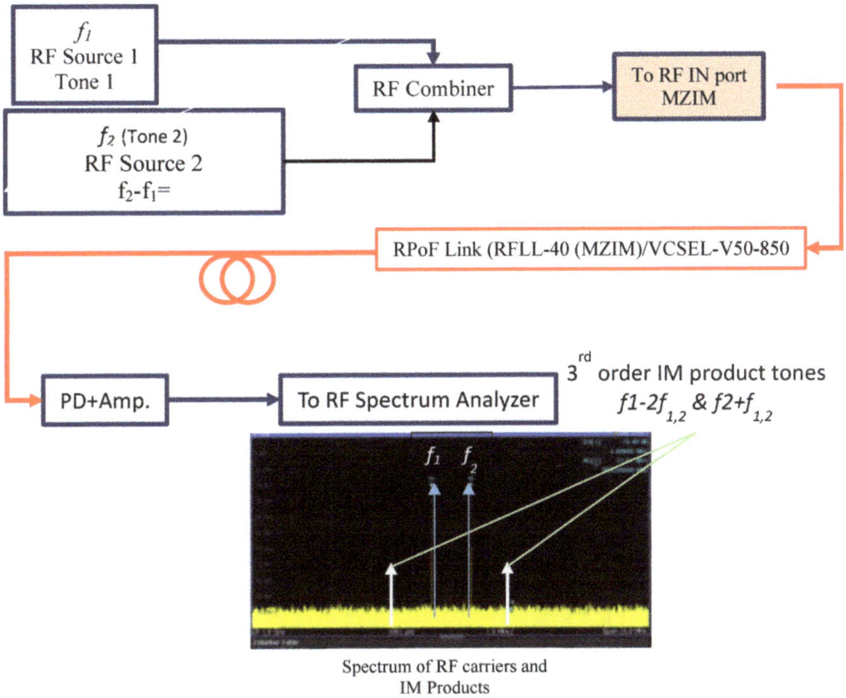

Figure 3.17. RPoF transmission: the intermodulation (IM) testbed. Testbed measurement of IIP3 generating f_1 and f_2 combined with $f_1 - f_2 = 250$ MHz, in a step of FR2 wireless channel spacing, modulating the modulator MZIM/VCSEL, increasing the signal amplifier to observe third order intermodulation products at $(2f_2 - f_1)$ and $(2f_1 - f_2)$ as shown in figure 3.18, then interpolating the IIP3 frequency.

In 5G and 6G this two-tone spacing is commonly about 250 or 500 MHz for an RF carrier frequency of about 28 to 40 GHz. Figure 3.19 illustrates an IIP3 of 18.9 dB. Thus there is a third-order $SFDR_3$ in the order of 92.7 dB $Hz^{-2/3}$ and a second-order $SFDR_2$ of approximately 94.9 dB $Hz^{-1/2}$. It is important to note that the $SFDR_2$ can be improved greatly by accurately adjusting the bias point to exact quadrature. During the tests, the bias voltage can be adjusted with a relative resolution of 0.1 V. Figures 3.17 and 3.18 show a typical testbed and schematic of the carriers and first-, second-, and third-order IM products and the method for how to measure the SFDR of an RPoF link. Furthermore, the 1 dB compression point is also defined as the 1 dB departure from the linearity due to nonlinear effects, as indicated in figure 3.19(c).

3.6.4 EAM and intermodulation in RPoF

We have employed an EAM laser (EML) in which an electro-absorption modulator is fabricated inline with a laser structure, as shown in figure 2.36 of chapter 2 and also shown in figure 3.20. A bias voltage is required to be applied to the EA modulator so that the operation of the RF peak–peak level can be applied such that linearity can be obtained, see figure 3.21. The performance is linear except at some high level

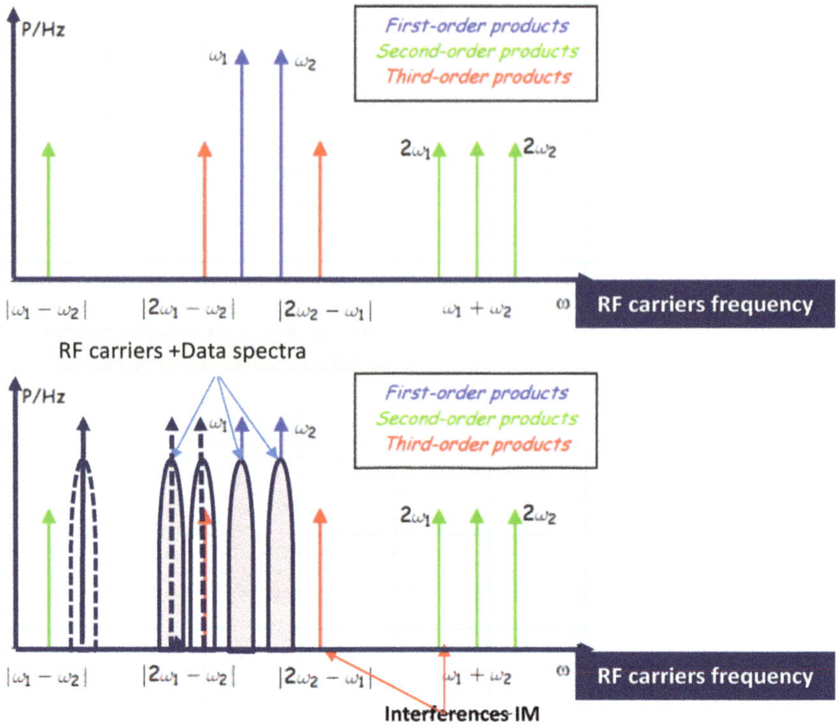

Figure 3.18. Illustration of intermodulation distortion sub-carriers generating the beating products of the two RF carriers launched into the fiber, that is the cross-talk from one RF carrier channel to the adjacent one. The power spectral density of the RF carriers and intermodulation products are plotted against the RF frequency. The order products are distinguished with different colors. Blue, green, and red show the first-, second-, and third-order intermodulation products, respectively.

launched power. At −25 dBm to −15 dBm the RF carrier signal level is in the small signaling region and high linearity is achieved. This is more suitable for moderate transmission distances. For antenna sites, a much higher power of the RF carrier is required and EML may not be suitable. The transmission capacity can be doubled by using a back-to-back EML structure, as shown in figure 2.36 of chapter 2.

3.7 Dispersion resistant chirp signaling for RPoF

3.7.1 Chirp signal generation and transmission

When the RF modulated optical carrier and its data channel are propagating through a fiber transmission line, they suffer a chirping of the carrier due to fiber chromatic dispersion. The chirping limits the transmission distance. However, if such fiber chirp effects can be countered by a modulation scheme, that is opposite to the fiber chirp, then we have a chirpless RPoF system and the distance can be expanded (see figure 3.22).

Let $\beta(\omega)$ be the effective propagation constant of the weakly guided waves in the SSMF. That is, the effective propagation constant along the fiber axis. We can

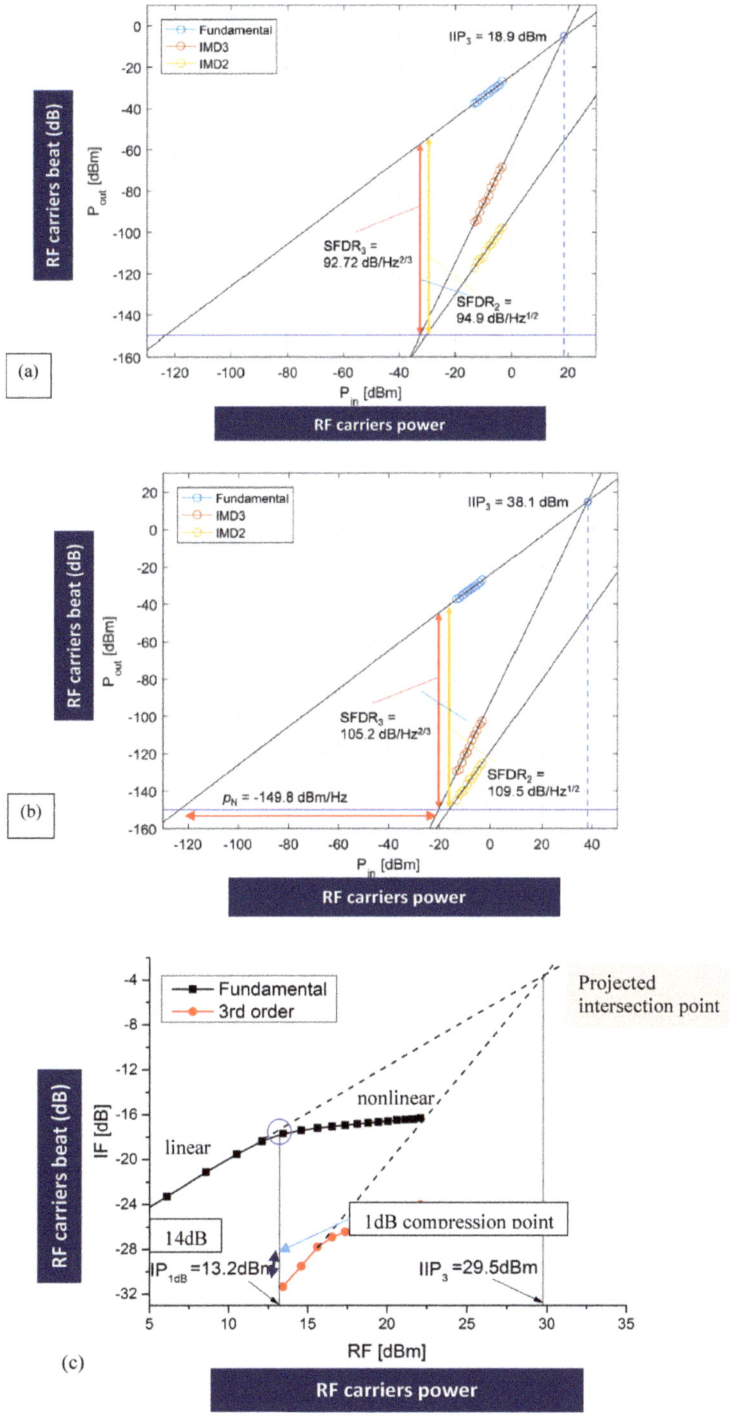

Figure 3.19. Illustration of the interception point IIP3 which is derived from the intersection point between the third-order distortion due to intermodulation and the second-order intermodulation effects as measured from

the noise power level and excited by an RF carrier with a variation of power of two narrow frequency spacings, (a) IIP3 = 18.9 dBm and (b) IIP3 = 38.1 dBm, and (c) the IIP3 of 29.5 dBm (fundamental and third-order) as excited by the RF carriers at different frequencies. IF = intermediate frequency, that is the frequency difference between RF carrier-modulated optical signal launched into the RPoF, and that transferred to the optical domain at the output.

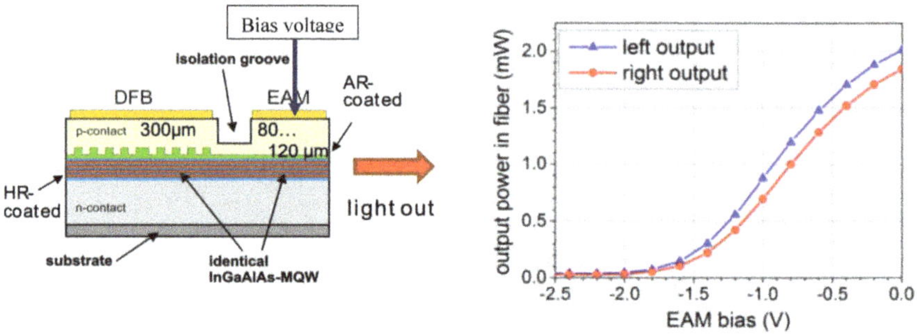

Figure 3.20. Generic structure of an EML and biasing transmittance characteristics.

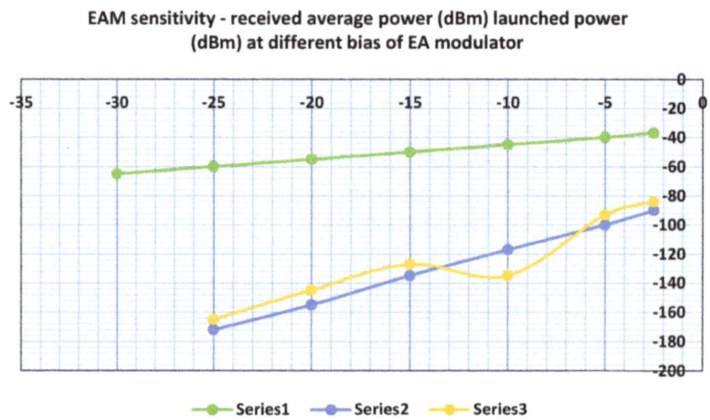

Figure 3.21. Integrated structure of an EML and sensitivity performance under various biasing voltages. Series 1: launched power and received power. Series 2: received power at biasing voltage of 0.778 V. Series 3: at a biasing voltage of 0.787 V.

expand using a Taylor series around the optical operating frequency ω_0, so the optical phases δ_0 and δ_{-1}, respectively, due to the propagation delay through the fiber length L and its first and second-order group velocity (or the group delay) and the dispersion factor, can be written as [24, 25]

$$\delta_0 = L\beta_0(\omega_0) \text{ and } \delta_{-1} = L\{\beta_0(\omega_0) - \beta_1(\omega_0)\omega_{\mathrm{RF}} + \frac{1}{2}\beta_2(\omega_0)\omega_{\mathrm{RF}}^2\}. \tag{3.21}$$

These phase factors create chirping on the carrier, the RF modulated optical carrier in the case of RPoF.

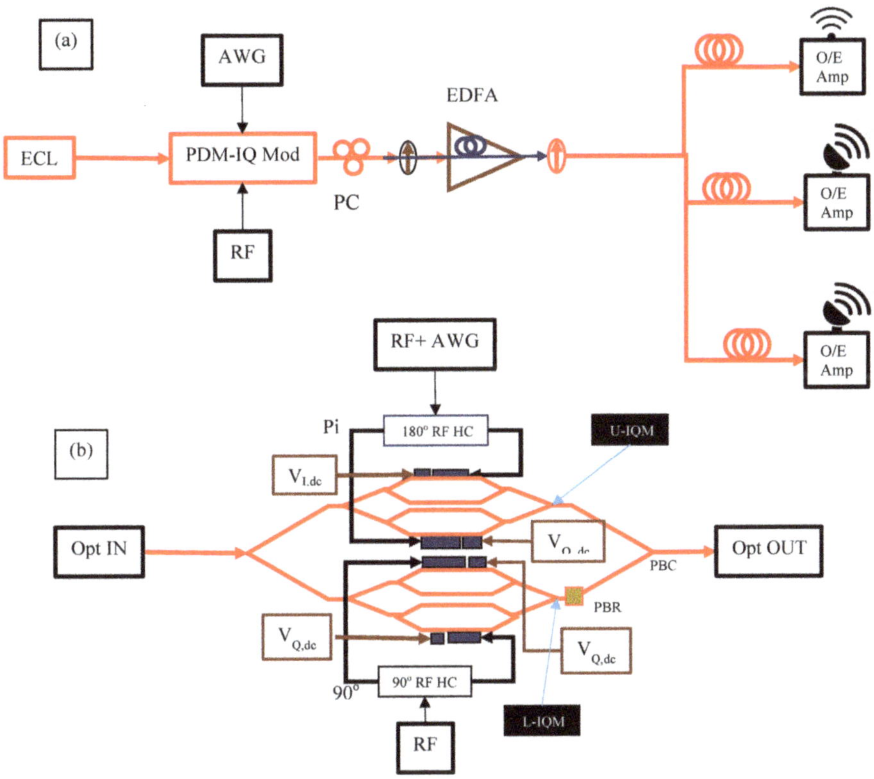

Figure 3.22. Schematic diagram of (a) the proposed generation and transmission link of a dual-chirp microwave waveform with quadruple bandwidth and anti-dispersion based on a PDM-IQ modulator and (b) a modulation and bias scheme of a DP-QPSK modulator, the specific structure and operating state of the PDM-IQ modulator. LD = laser diode; EL = external cavity laser; AWG = arbitrary waveform generator; RF = radio frequency; PDM-IQ modulator = polarization-division multiplexing Mach–Zehnder modulator; PC = polarization controller; Pol. = polarizer; EDFA = erbium doped fiber amplifier; SMF = single mode fiber; PD = photodetector; 90° PBR = polarization beam rotator; O/E = optical-to-electrical converter; HC = optical hybrid coupler; IQ = in-phase quadrature; PBC = polarization beam combiner. (b) Structure of a QPSK optical modulator and PS hybrid coupler and child MZIMs whose electrodes are properly biased. PS = phase shifter; L-IQM and U-IQM = lower and upper IQ modulators.

In 5G, 6G, and 7G wireless networking assisted by optical fiber interconnections, the RF carriers are embedded in the optical lightwaves. As the data baseband bandwidth is much lower that the frequency of the RF carrier, the dispersion of the RF carrier is so critical and must be eliminated, hence a chromatic dispersion resilient modulation format must be devised. Figure 3.22 depicts the dual polarization complex QPSK (or IQ) modulator and the arrangements of biasing and signal application to its electrodes. So that both arms of the two 'children-MZMs' of each polarized section, can be applied with RF modulated data channels. The arbitrary waveform generator (AWG) can generate the test data channel whose output can then be frequency shifted by the RF waves.

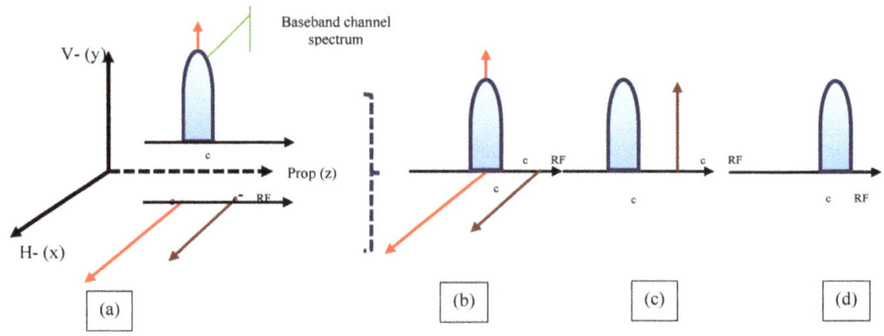

Figure 3.23. Simplified diagram to represent the evolution of optical carriers, an RF modulated carrier, and the data baseband channel.

As mentioned above, in order for the transmission of the RF modulated optical waves over the fiber to occur, a single sideband (SSB) format is preferred so that no interference depletion will occur. Furthermore, if an even longer length of the fiber is required then chirping of the carrier is even desired [26]. This can be implemented in a dual polarized QPSK modulator (or IQM) by modulating the upper IQM (U-IQM) of the DP-QPSK modulator with a double sideband and frequency doubling by biasing at the maximum transmission point.

Then the lower IQM (L-IQM) can be modulated so that a carrier suppression can be generated with only one sideband by biasing at the quadrature point, Q+ and Q− on the two MZIMs of the two paths of the parent MZIM.

This SSB components and the DSB components of the two IQMs are shown in figures 3.23 and 3.24.

Propagating along the fiber, chirping effects due to linear dispersion are imposed on the two modulated signals as well as the nonlinear phase effects. The SSB cancels out one of the bands of the DSB U-IQM and, similarly, any intermodulation products from the system nonlinearity.

It can be shown [26] that the chirp waveform as observed at the output of a single MZIM is as depicted in figure 3.25(a) and the minimum or non-chirped waveform of the DP-QPSK after 80 km of SSMF propagation is as shown in figure 3.25(b).

The electrical signal detected by a square law photodetection (O/E conversion) after transmission through a length of fiber, a dual-chirp RF current waveform, can be generated by

$$i(t) = \varkappa \sin(2\alpha) J_1(\beta_2) J_2(\beta_1) \{\cos(\omega_{\mathrm{RF}} t - 2kt^2 + \gamma) + \cos(\omega_{\mathrm{RF}} t + 2kt^2 + \gamma)\}, \qquad (3.22)$$

where $\gamma = \frac{\pi}{2} - \theta + \beta_1(\omega_0)\omega_{\mathrm{RF}} - \beta_2(\omega_0)\omega_{\mathrm{RF}}^2 - \frac{1}{2}$, $\varkappa$ is a conversion factor, β_1 and β_2 are the biased phase indices, ω_{RF} is the RF angular frequency, k is the wave number in a vacuum, ω_0 is the operating optical angular frequency, and α is the gain/loss factor in a linear scale. This shows that the phase chirp is minimum under an SSB optical signal transmission with suppression of the main optical carrier. The main dynamics is that

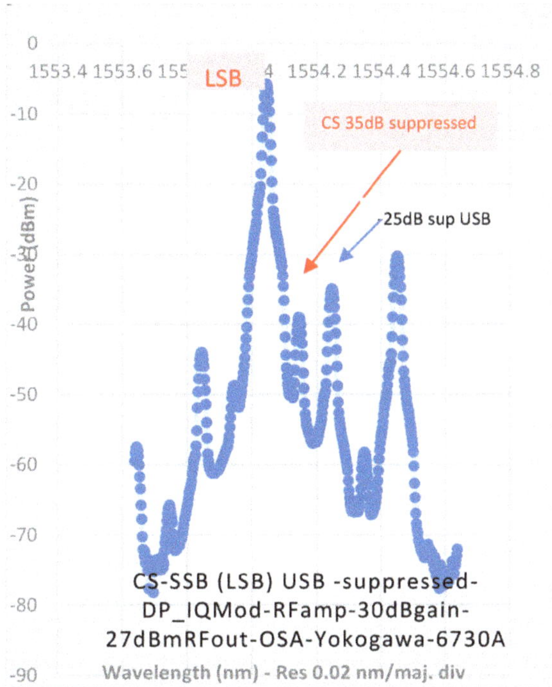

Figure 3.24. SSB spectrum at the output of the L-IQM.

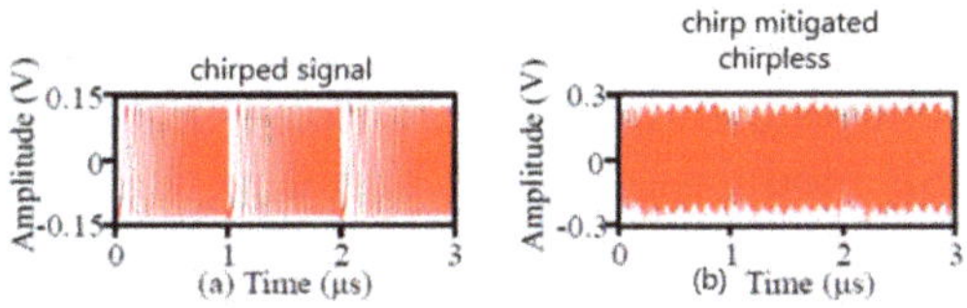

Figure 3.25. Chirped signals: time-sampled waveform as observed at a point after the O/E converter. The PD output connected to a high sampling rate digital signal analyzer, Keysight DSOX3024A oscilloscope: BW = 200 MHz, four channels at 200 GSa s^{-1}. (a) Chirped signal and (b) compensated chirpless signal.

one SSB can counter the chirp of one of the bands of the U-IQM. Both suffer fiber chirp equally due to dispersion (the phase is given in (3.21)). This allows only one sideband traveling along the fiber and thus no or minimum chirping can occur.

The chirp of the fiber and the chirp due to the modulation scheme can then be designed so that they can cancel each other to give a chirpless signal. The biasing phase θ can be adjusted so that the total chirp factor γ in (3.31) can be minimum. However, in this case we use one IQM, the L-IQM, to mitigate one of the sidebands of the U-IQM. This is particularly appropriate for the case of the RPoF due to the narrow bandwidth of the data channel spectrum.

3.7.2 Implementation

An experiment platform is shown in figure 3.22. A continuous-wave (CW) linearly polarized laser (ECL) operating at 1550 nm with narrow linewidth (less than 100 Hz), e.g. Pure Photonics Model 550PLC, of 16 dBm power is coupled into a PDM-MZIM (Fujitsu type FTM7980EDA) by a polarization maintaining fiber (PMF) or via a polarization controller (PC). The optical carrier is split 50:50 (using a 3 dB Y-splitter) into two optical paths, the U-IQM and L-IQM.

A baseband single-chirped signal from an AWG with 100 MHz bandwidth and 1 μs duration is first amplified by an electrical amplifier (EA) and then lunched into two RF ports of the U-IQM with 180° phase difference. The RF signal frequency of 12.5 GHz/25 GHz of 16 dBm power is applied to the two RF ports of the L-IQM with 90° phase shift difference.

The dc bias of the U-IQM is set so that it operates at maximum transmission point so as to generate even-order optical sidebands of the baseband single-chirped signal, which mainly consist of ±second-order sidebands and a modulated optical carrier.

The dc bias of the L-IQM is set at the Q+ point and Q− point so as to obtain SSB RF shifted optical waves and suppression of the optical carrier, thus it consists of a −first-order SSB modulated optical carrier, as shown in figure 3.24.

After PDM-MZIM the polarization multiplexed optical signal is coupled to the PC and the polarization rotation (PBR) on the output of the L-IQM is set by an applied dc voltage. The PC can be tuned under the condition of certain electrical signal amplitudes, and the main optical carrier is suppressed. The spectrogram at the output of the polarized channel is given in figures 3.23 and 3.24. The optical spectrum analyzer (OSA) with a 20 pm resolution is similar to that given in figures 3.7 and 3.24. As observed in figure 3.25, the +first-order sideband is suppressed by the pre-chirping of the modulator of higher than 24 dB. This confirms the resilience of the SSB modulation of RF signal to the fiber dispersion effect.

Therefore, DSB-based dual-chirp signals degrade significantly from CD induced distortion. An SSB modulated RF signal with optical carrier suppression can mitigate the chirping due to CD in fiber transmission. It is shown that such a 25 GHz RF carrier can be transmitted over the fiber span of a long-haul transmission line. The transmitted optical spectrum can be analyzed in an optical spectrum analyzer (OSA). The electrical signals in the real temporal domain and spectrum can be measured using the 10 MHz resolution bandwidth and 200 GSa s^{-1} Keysight DSOX3024A Oscilloscope (200 MHz with four channels).

3.7.3 Remarks

In conclusion, an RPoF system using dual-chirp microwave signals with much lower CD effects based on a PDM-IQM can be implemented. By controlling the biases of the IQM sections, dual-chirp signaling can be generated with quadruple bandwidth and CD resilience. Utilizing the correspondence between 25 Hz and the null point of power suppression of 25 km SSMF transfer characteristics, the proposed scheme can transmit over 80 km of SSMF. Thus the CD effect elimination is confirmed and is not limited by the central frequency and transmission distance. The chirping of the RF modulated optical carrier is simpler than for other SSB as described in section 3.2.

This chapter describes the principle of the transmission of radio frequency carriers and baseband data channels over long-length SMFs. Both the DSB and SSB techniques are illustrated to show the interference effects of the two sidebands evolving through the fiber chromatic dispersion and their elimination via the suppression of one sideband, thus allowing long transmission with minimum dispersion effects. The techniques described here would allow the development of a central control and transceiver system for delivering RF modulated carrier data baseband modulated channels in the mmW regions, 5G and 6G FR2 and FR3, to remote antennae as well as monitoring for wireless antennae networks via optical guided media.

3.8 Appendix 3.1: other DPM-IQM-dual polarization mux-IQ modulators

3.8.1 Single polarization IQM (or QPSK) Fujitsu type: FTM-7961 diagram

See figures 3.26 and 3.27.

3.8.2 Dual-drive MZIM: Fujitsu FTM-7937EZ/7938

See figure 3.28.

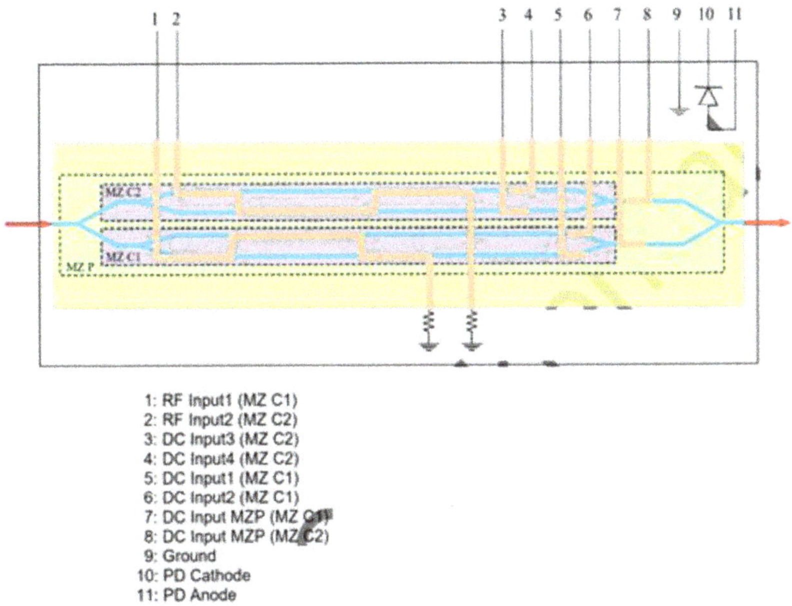

Figure 3.26. FTM-7961 structure and pin connections. The I branch and Q branch with a phase shifter and polarization rotation are incorporated in the optical paths. dc = direct current; MZ = Mach–Zehnder; C1, …, N = connection numbers; MZP = Mach–Zehnder polarization; PD = photodetector; RF = radio frequency.

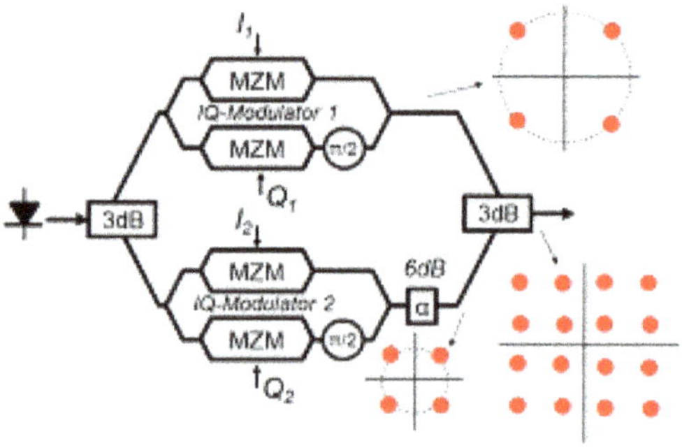

Figure 3.27. Operation and biases of a DPM-IQM; IQM1 = upper IQM; IQM2 = lower IQM. 3 dB or 50:50 splitter or combiner.

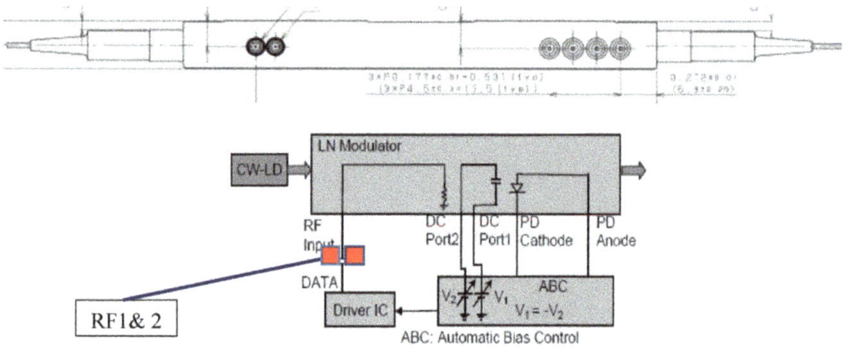

Figure 3.28. FTM-7937EZ structure and pin connection for a dual-drive MZIM. Condition for CS-SSB swinging voltage $V_{\text{p-p}} = 2V_\pi$ or in dual-drive $V_\pi(\text{RF}) = 1.8\text{ V} \rightarrow V_{\text{p-p}} \rightarrow 24$ dBm (3 dB down from 27 dBm for $V_{\text{p-p}} = 7$ V).

3.9 Appendix 3.2: RPoF link analyses and design

This section gives an introduction to the estimation of the power budget and the design of a simple RPoF link in which coherent transmission and reception are assumed with the popular modulation format Mary-QAM OFDM. Figure 3.26 shows the Fujitsu dual polarization IQ modulator in which the RF electrodes are alternatively placed on top of the optical waveguide section so as to increase the interaction length, hence lowering the driving RF signal level. This lowering inter-modulation distortion (IMD) can be achieved. Figure 3.27 displays the detected output of the IQ modulator of figure 3.26, which is the constellation of complex signal QPSK or 16QAM (or 4 × 4 two-dimensional pulse amplitude modulation) modulation format. Furthermore, a specific Fujitsu modulator with its packaged structure, the model FTM-7937EZ, can be used with appropriate circuit equivalent circuit and internal photodetector for automatic control biasing. The Mary-QAM is for the RF transport carrier embedded in the optical domain. The OFDM is for the data channel with the channel bandwidth assuming a bandwidth of 100 MHz. The optical local oscillator (LO) is designed for down-conversion to an intermediate frequency (IF) so that the processing can be conducted in a low frequency region with a much lower sampling rate.

3.9.1 Introductory remarks

This section gives the design of an RPoF link without simulation via a VPI transmission maker in order to give insights into the design procedure so that the simulation can be implemented to obtain RPoF link performance.

Given the parameters to be specified, such as the error vector magnitude (EVM), Mary-QAM OFDM format, and adjacent channel power ratio (ACPR), how do we design the components of the system presented in figure 3.29? Then there is the question of how to design the biasing and adjust the frequencies of the laser and the optical modulators.

The EVM is a parameter related to the degradation of the sampled points of the pulse sequence with respect to the origin of the constellation. This Euclidean distance and degradation, and the contribution of all the points of the constellation are critical. It is thus necessary to find out the relationship between the EVM and the SNR, and thus the optical signal-to-noise ratio (OSNR), and therefore the bit error rate (BER) of the RPoF.

Two methods are summarized for these relationships of the EVM and other system parameters, except the ACPR. One is an analytical derivation from references and the other is achieved using the Matlab function bertool.m under exact and simulation conditions as desired. From the analytical analysis the curves representing BER against SNR for $M = 16$ to 64 can be extracted from the references, while the Matlab BER curve can reach 1024 if desired, but only that for the 256QAM modulation format is given.

Once the BER with respect to the EVM can be determined, given an EVM value such as 5% or 0.05 is wanted, the BER is obtained, and hence the SNR for certain M. Thus for a maximum optical power which can be obtained from the laser source and loss through the optical components along the link, an estimate of the available optical power in front of the receiver can be obtained. Then the maximal allowable noise power as referred to the input of the electronic amplifiers and subsystem can be determined as well as the gain of the amplifiers and any noise reduction by electrical bandpass filter (eBPF). We can see the advantages of the IF down-conversion here as the noise reduction can be tremendous. This is due to the integration of the noise power or current spectral density over the device and system bandwidth. In particular the PD, but in a much narrower passband and with avoidance of a wideband optical detector such as the XPDV 3120 (U2t or Finisar type) which has a 3 dB bandwidth of 45 GHz. If an IF frequency of 3 GHz is used as the PD then a noise reduction factor of 15 or 13 dB can be achieved. Furthermore, an eBPF can be used to reduced the noise further so that a good SNR can be achieved. It is noted that for 256QAM an SNR of about 26 dB is required for a BER of 1e-4 to 1e-5. Theoretically, for practical implementation it is expected that this BER is about 1e-3, which is the minimum level for the forward error coder to perform the compensation. At this tolerance it is expected that a safety margin of 6 dB is to be used in the link power budget.

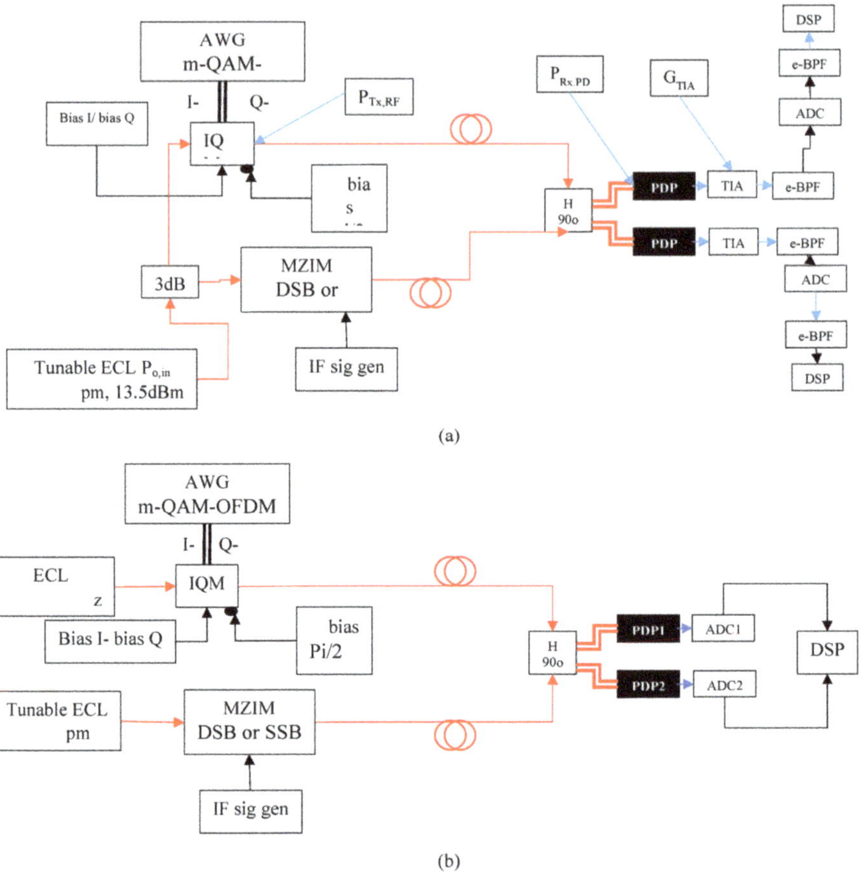

Figure 3.29. Generation of an LO IF laser by a separate tuneable ECL laser and frequency shifting so as to obtain serrodyne detection, i.e. the IF is close to or less than the baseband bandwidth of the signal. Serrodyne means that the IF is within the bandwidth of the baseband signal, that is, homodyne detection, but there is a small difference between the signal laser frequency and that of the local oscillator. The different amount is normally called the local oscillator frequency off-set (LOFO). PDP = PD pair; ADC = analog-digital converter; AWG = arbitrary waveform generator. (a) Simplified model for power budget calculation with symbols. (b) Full structure using self-heterodyne coherent reception and an RPoF transmission link. See the final design structure in figure 3.39 of the appendix.

3.9.2 Optical RPoF link structure

Figure 3.29(a) depicts an RPoF system employing coherent reception and a self-coherent technique. That is, the LO is derived from the laser source which is split into two arms, one through the signal optical modulator and one used as the LO whose frequency is shifted by an optical frequency modulation of either SSB or DSB with an appropriate filtering OBPF. The LO frequency can be shifted close to and within the passband of the data channel, in this case the coherence can be considered as serrodyne detection.

Figure 3.29(b) depicts an RPoF with a difference compared to that of figure 3.29(a) with the use of a separate laser as the LO whose frequency can be tuned and optically shifted via an optical modulator. The advantage of this LO arrangement is that the LO frequency can be tuned across the C-band to L-band as modern ECLs have advanced tremendously with a narrow linewidth to <10 kHz. The disadvantage of this separate laser is that the random variation of the phase difference between the two lasers creates some complexity in the receiver DSP. A phase matching algorithm can be used to compensate for this, as in the well-known modern coherent optical transmission.

3.9.2.1 A simplified RPoF transmission system

See figure 3.30.

The analyses can be simply considered the power budget curve shown in figure 3.31.

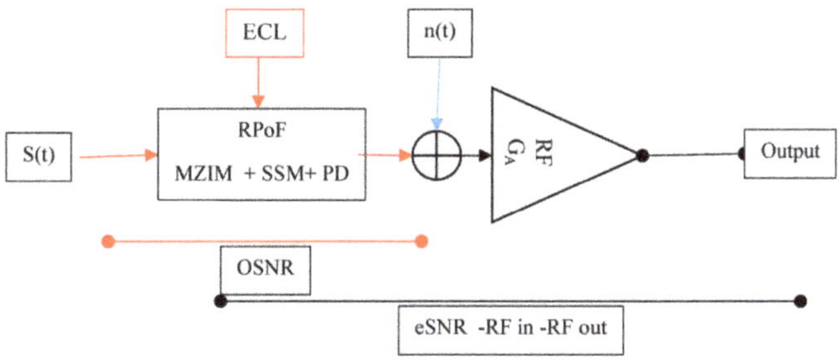

Figure 3.30. A simplified model with the OSNR and eSNR.

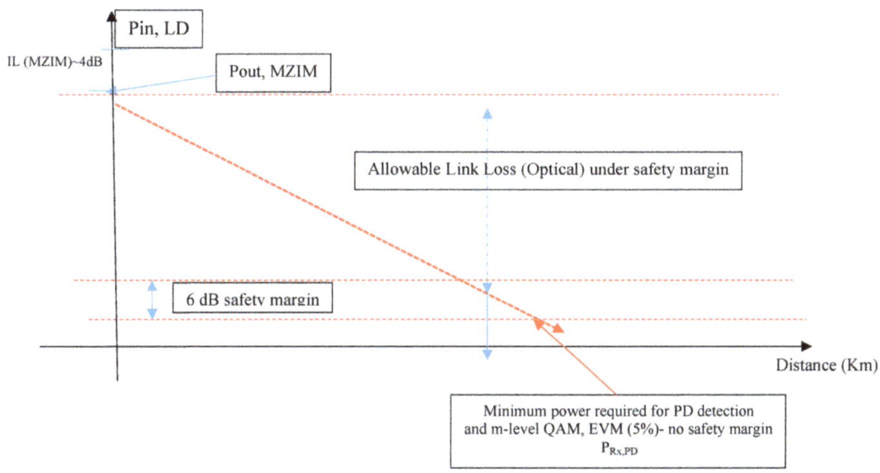

Figure 3.31. General power budget curve for communication links, in particular under m-QAM OFDM. Detected by a single PD or balanced PD pair and boosting with an LO laser. Note the safety margin level 6 dB for analog transmission.

3.9.2.2 Assumptions

The calculations with the following assumptions (real case):

- M-QAM or in-phase and quadrature.
- The transmission distance is short enough to satisfy the power budget, or the dispersion on the data channels is minimized and negligible. This is acceptable as the bandwidth of the OFDM is assumed to be less than 1.0 GHz and thus the dispersion factor in the SSMF is negligible.
- Hence the EVM is estimated by a 45° angle with an equal distance of the I- and Q-components.
- Now (see figure 3.31) given that the receiver sensitivity is at level $P_{\mathrm{Rx,PD}}$, that is the minimum optical power required for an optical receiver to accept the signal and give an output electrical signal with a specified OSNR, and hence the eSNR, under a certain condition of the EVM minimum value to be achieved under *M*-level QAM modulation.
- The bandwidth of the receiver is assumed to be under initial conditions that are very wideband, then an electrical BPF (eBPF) and amplifier can be added so as to satisfy the receiver condition as specified or desired.
- Thus, given the input power and the receiver sensitivity, a linear distance for fiber loss or link loss can be derived in a logarithmic scale, as seen in figure 3.31. The RF link loss can be considered as the RF in at the Tx port and the RF out of the receiver port. This will be explained during the analyses or calculations/estimations.

3.9.2.3 Analysis procedures

Specifications of technical parameters:

- OSNR > 10 dB.
- EVM< 5%.
- M = 256 or level in I = 8; Q-level =8 modulation for QAM; OFDM = number of sub-carrier. The number of sub-carriers can be determined by user, for example the number can be estimated by taking the optical channel spacing dividing to the bandwidth of the data channel plus that of the guard band. For example, for a 50 GHz optical channel spacing and 400 MHz data bandwidth plus 100 MHz guard band, then there can be 100 RF-shifted data channels.
- PD without TIA (electronic amplifier) and PD +TIA + eBPF, coherent with LO IF with beating signals in the IF mid-band 100 kHz.
- Power of LO IF laser ~13 dBm, laser linewidth of ~100 KHz.

The noise current and signals can be modeled as shown in figure 3.32.

The eSNR at the output of an amplifier is defined as

$$\mathrm{eSNR} = \frac{\langle i_S^2(t)\rangle}{\langle i_{NS}^2(t)\rangle}$$

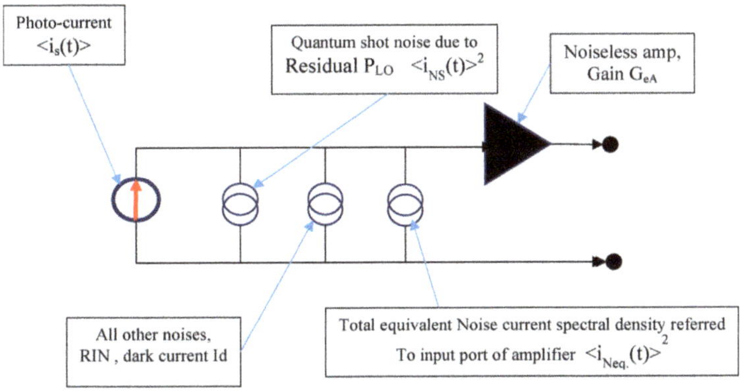

Figure 3.32. Signal and noise model in the linear small signal conditions, that is the third-order (THD) and second-order harmonic distortion (SHD) are negligible.

where $\langle i_S^2(t)\rangle$ is the expected square current as detected by the PD and $\langle i_{NS}^2(t)\rangle$ is the total noise square current of all noise sources which are (i) the quantum shot noise due to the laser power of the local oscillator and (ii) the total equivalent noise power as referred to the input of the amplifier including all electronic noise and thermal noise of the amplifier. A typical value for electronic amplifiers is $i_{N,\,\mathrm{eq}}^2(t) = [20\ \mathrm{pA}\ (\sqrt{\mathrm{Hz}})]^2$ (spectral noise current density). So the total noise current square is given as

$$i_{NS}^2(t) = \int_0^B 2qI_{\mathrm{PD}}df + \int_0^{B_{amp}} i_{N,\,\mathrm{eq}}^2(t)_{\ \mathrm{amp}}df,$$

where B_{amp} is the bandwidth of the amplifier and B is the system bandwidth. The amplifier bandwidth is not necessarily the same as the system bandwidth. This is critical as if we use a wider amplifier bandwidth the noise increases. For the RPoF case the signal channel bandwidth is much smaller than that of the carrier frequency. In this case the BW is 1.0 GHz and the PD XPIV is 45 GHz. The noise contribution by this PD is substantial.

A numerical estimation, with the assumption that the 3 dB bandwidth is uniform with 1 dB fluctuation:

- Quantum shot noise (current square) for a 45 GHz 3 dB bandwidth of the PD is $B \times 2qI_{\mathrm{PD}}$ = 45e9 × 2 × 1.6e-19 × $P_{\mathrm{LO}} \times \mathfrak{R}$, where $\mathfrak{R}$ is the responsivity of the PD ~ 0.6 A W^{-1} ~ 86.4e-10 A^2, therefore there is a 92 μA quantum shot noise current for an optical power of 13 dBm of the LO laser. This is quite substantial.
- The total equivalent electronic noise referring to the input port of the electronic transimpedance amplifier (TIA) of bandwidth 20 GHz is

$$i_{N,\,\mathrm{eq}}^2(t) = (20\mathrm{e} - 12)^2 \times 20\mathrm{e}9 \sim 400\mathrm{e} - 15 \text{ or } i_{N,\,\mathrm{eq}} \sim 6.3\mathrm{e} - 7 \text{ or } 0.63\ \mu\mathrm{A}.$$

- Thus, this current is also much smaller than that of the quantum shot noise due to the LO laser which is also proportional to the $\sqrt{B}$—the square-root of the bandwidth of the PD. This noise can be reduced if an electrical bandpass filter is used. Let use a 2 GHz bandwidth eBPF, then the total noise of the coherent receiver can be estimated as 92e-6 sqrt^{-1} (22.5) = 19.4 μA which is also much higher than the electronic noise referring to the input port.

Now assuming $m = 2$ (8 bits per symbol) per in-phase and quadrature phase components (figure 3.33). Assuming the SNR = 10 dB power, 20 dB electrical, or 100 times greater than the noise current level, a noise figure of 3 dB (or doubling in noise power), and a four times increase in noise current or voltage, i.e. an effective gain of 50% transimpedance in the signal current, then the electrical signal required to achieve this with only two levels is SNR = 10 $\rightarrow \langle i_S^2(t)\rangle = 10 \times \langle i_{NS}^2(t)\rangle$, thus the signal current to be produced at the PD output is 100/2 × 19.4 → 0.87 mA. So the optical power to be available in front of the PD is 0.87/0.6 → 1.6 mW or about 2.04 dBm in power log scale.

For $m = 8$, note that the factor 100 in electrical current can be obtained by using an electrical amplifier. If a TIA is used with a transimpedance gain of 1500 Ω then the optical power required at the PD is 0.139 mW or −8.6 dBm. If the noise is 3 dB worst or double at the output then the expected noise contribution to each constellation point is the same and the eBPF is used, i.e. connected to the output of the main amplifier so that the noise of the TIA can also be reduced.

Now, under $m = 8$ for 256QAM the current available between two levels in either the I- or Q-components is, with the noise contribution being the same, the PD current for 10% EVM which is 2 × 194 μA (3 dB NF), and 2 × 384 μA for 5%. This requires the optical power to be available at the PD. Thus the optical power to be available at the PD is 8 × 1.28 mW with a PD responsivity of 0.6 A W^{-1} per level of the M-QAM. This requires a total average power under RF modulation of 10.24 mW optical power. This is achieved only with a TIA of 1500 Ω transfer resistance and an eBPF of 2 GHz passband. This power can be achieved provided no fiber transmission is used or only a very short fiber length, i.e. a back-to-back connection and maximum power of about 13 dBm with a linewidth of 100 kHz.

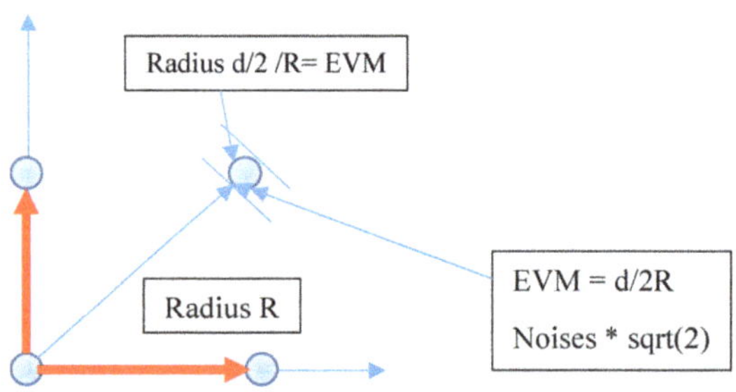

Figure 3.33. Schematic of the EVM in a QAM constellation. This is the contribution of noise with respect to the magnitude of the RF modulating the average power of the RF carrier. This can be estimated using the SNR, hence the OSNR under the m level under consideration.

3.9.2.4 Relation of BER from EVM

The EVM of a square M-QAM constellation can be written as

$$\mathrm{EVM} = \frac{\sqrt{\frac{1}{N}\sum_{n=0}^{N-1} I_{\mathrm{err}}[n]^2 + Q_{\mathrm{err}}[n]^2}}{P_{Reff}}, \tag{3.23}$$

where n is the symbol index, N is the number of symbols, $I_{\mathrm{err}} = I_{\mathrm{ref}} - I_{\mathrm{meas}}$ is the error distance of the in-phase component from the reference point (I_{ref}). Similarly, $Q_{\mathrm{err}} = Q_{\mathrm{ref}} - Q_{\mathrm{meas}}$ is the Euclidean error of the quadrature component from the reference point. The EVM has the following direct relationship to the received signal SNR [27, 28]. P_{Reff} is the received average power.

The BER can be obtained for an m-QAM from the EVM values as follows.

For the constellation of a square Mary-QAM reception EVM can be influenced by the amplified stimulated emission (ASE) or quantum shot noise, and the BER can be estimated using [29, 30]

$$\mathrm{BER} = \frac{(1 - M^{-1/2})}{\frac{1}{2}\ln(M)} erfc \left[\sqrt{\frac{3/2}{(M-1)\mathrm{EVM}_{\mathrm{rms}}^2}}\right], \tag{3.24}$$

thus for 256QAM we have

$$\begin{aligned}\mathrm{BER} &= \frac{(1 - 256^{-1/2})}{\frac{1}{2}\ln(256)} erfc \left[\sqrt{\frac{3/2}{(256-1)\; 0.05_{\mathrm{RMS}}^2}}\right] \\ &= \frac{15/16}{2.77} erfc \;\; (1.5 \times 0.6375) = 0.1243,\end{aligned} \tag{3.25}$$

where RMS is the root mean square value, M is the number of constellations, for example $M = 256$ for 8 bits per symbol modulation 256QAM, and ln is the natural logarithm,

Thus, theoretically the BER for 256QAM and 5% EVM is BER ~0.0033 or 3.3e-3. This is close to the BER that is acceptable for such modulation for an RPoF provided forward error coding (FEC) is used. The OFDM sequence would be impossible to be decoded. The most common error rate would be acceptable with an FEC of ~1e-3 or 2e-3.

3.9.2.5 Concluding remarks

Thus under more practical implications with a 3 dB margin of operation, the number of QAM constellations should be reduced by 50% or $m = 4$ or 64 QAM. If we use a balanced receiver with a pair of PDs and a 3 dB gain in optical power of the optical receiver this factor may improve a bit and this can be used as the safety margin for operation. The 6 dB margin can be used, as used commonly in analog coherent transmission systems in the 1980s when the first optical coherent transmission was developed. The QAM modulation is assumed to be 80% or an almost 30 dB extinction ratio. The tolerance for this RF modulation can be adjusted, but is allowed under the safety margin. The best way to achieve a more exact estimation is to set up a excel file for performing all these estimations.

3.9.3 Analytical power budget analysis

3.9.3.1 Signal-to-noise ratio

The signal and noise can be written in the case of an avalanche PD (APD) and PIN detector as in the following.

The square signal current can be written as $\langle i_S^2(t)\rangle = m^2 I_D \langle \hat{i_S^2}(t)\rangle$ where m is the modulation index, and I_D is the PD current converted from the average optical power on which the signal is imposed, that is, at the biasing point of the modulation device, either a laser diode or an optical modulator biased at the quadrature point. $\langle \hat{i_S^2}(t)\rangle$ is the expected value of the signal $s(t)$ which varies in the data pseudorandom random binary sequence (PRBS).

$P_{Tx}(t)$ can be written as a function of $s(t)$, the biased power level, as

$$P_{Tx}(t) = P_0(1 + ms(t)), \tag{3.26}$$

where P_0 is the mean power output of the modulator, assuming an MZIM, at the biasing point, and m is the modulation index. The modulation index can be estimated for M-QAM where the number of levels in the in-phase and quadrature is the square0root of this number M, e.g. the 256 constellation point is constructed by 8 bits per symbol in both components.

Total noise (current square) can be written as $i_{NS}^2(t) = i_{SQ}^2(t) + i_{RIN}^2(t) + i_{N,\,eq}^2(t)$ with $i_{SQ}^2(t)$ = the quantum shot noise due to the average optical power which is set at the biasing point. In case of the MZIM biased at the quadrature point the output optical power is derived from the Q-point. Superimposing on this average power is the RF signal optical power transfer at which follows the cos^2 transfer profile.

$$i_{SQ}^2(t) = 2qI_DB.\ F(M), \tag{3.27}$$

where $F(M)$ is the noise enhancement factor if the PD is of APD type, M is the avalanche gain, B is the 3 dB system bandwidth (Hz) and/or the PD device.

$i_{RIN}^2(t)$ is the noise current square due to the laser relative intensity noise and

$$i_{RIN}^2(t) = \text{RIN} * B, \tag{3.28}$$

where RIN is the relative intensity noise power in units of dBm Hz^{-1} or an equivalent linear scale mW Hz^{-1} or W Hz^{-1}, that is the power out of the band of the laser spectrum.

$i_{N,\,eq}^2(t)$ is the total electronic noise power referring to the input port of the electronic amplifier connected to the PD output. This electronic noise equivalent power can be obtained from the measured noise spectral density and integrated over the 3 dB bandwidth of the amplifier. This can normally be obtained by measuring the noise spectral density on the electrical spectrum analyzer (ESA) with no signal at the input or shorted input:

$$i_{N,\,\mathrm{eq}}^2(t) = \frac{di_{N,\,\mathrm{eq}}^2}{df}B. \tag{3.29}$$

$\frac{di_{N,\,\mathrm{eq}}^2}{df}$ is the total noise equivalent spectral density referred to the amplifier input port in units of $A^2\ Hz^{-1}$. It is commonly represented in square current per BPF *df*. Therefore, the OSNR can be written as

$$\mathrm{OSNR} = \frac{\langle i_S^2(t)\rangle}{i_{NS}^2(t)} = \frac{\langle i_S^2(t)\rangle}{i_{SQ}^2(t) + i_{\mathrm{RIN}}^2(t) + i_{N,\,\mathrm{eq}}^2(t)}. \tag{3.30}$$

Three cases may happen in which case each of the noise type can dominate the noise processes, where we can have the following scenarios.

CASE 1: Dominant quantum shot noise

$$\mathrm{OSNR}_{SQ} = \frac{\langle i_S^2(t)\rangle}{i_{NS}^2(t)} = \frac{\langle i_S^2(t)\rangle}{i_{SQ}^2(t)}. \tag{3.31}$$

CASE 2: Dominant RIN

$$\mathrm{OSNR}_{\mathrm{RIN}} = \frac{\langle i_S^2(t)\rangle}{i_{NS}^2(t)} = \frac{\langle i_S^2(t)\rangle}{i_{\mathrm{RIN}}^2(t)}. \tag{3.32}$$

CASE 3: Dominant total equivalent electronic noise

$$\mathrm{OSNR}_{N,\,\mathrm{eq}} = \frac{\langle i_S^2(t)\rangle}{i_{NS}^2(t)} = \frac{\langle i_S^2(t)\rangle}{i_{N,\,\mathrm{eq}}^2(t)}. \tag{3.33}$$

3.9.3.2 RF eSNR

The eSNR can be derived from the square-root of the OSNR due to the PD square law conversion, that is the envelop detection. The eSNR for direct detection can be written as

$$\mathrm{eSNR} = \frac{\langle i_S^2(t)\rangle R}{i_{N,\,\mathrm{eq}}^2(t)} \tag{3.34}$$

where R is the responsivity of the PD, normally < 1 for a p–i–n detector. The noise in the above equation is the same as the equivalent summation of the noise sources which are less than the signal current that can be derived for both the envelop detection and coherent detection by deriving the signal current accordingly. For direct detection the signal current is the product of the optical average power of the optical signal and the responsivity of the receiver. While the signal current of the coherent detection by a PD (for a balanced outcome a ×2 factor is used), the signal current is the product of the amplitude of the local oscillator field (the square of the average power of the CW LO) and that of the signal field, then multiplied by the PD responsivity. The signal currents for these cases, $i_{S,\,\mathrm{DD}}(t)$, $i_{S,\mathrm{Co\text{-}PD}}(t)$, and $i_{S,\mathrm{Co\text{-}PDP}}(t)$, are

$$i_{S,\,\mathrm{DD}}(t) = R * P_{S,\,\mathrm{av}} \tag{3.35}$$

$$i_{S,\mathrm{Co\text{-}PD}}(t) = R * A_S A_{\mathrm{LO}}$$

$$i_{S,\mathrm{Co\text{-}PDP}}(t) = 2R * A_S A_{\mathrm{LO}}.$$

These signal currents can be obtained from the power budget estimation. The EVM has the following direct relationship to the received SNR [28]:

$$\mathrm{SNR\ (dB)} = P_{\mathrm{OUT},\ Rx} - P_{\mathrm{noise}} = -20\log_{10}\left[\frac{\mathrm{EVM}}{100}\%\right] \tag{3.36}$$

or

$$\left[\frac{\mathrm{EVM}}{100}\%\right] = 10^{-(\mathrm{SNR\ (dB)}/20)} \tag{3.37}$$

$$\text{and } \mathrm{EVM_{RMS}} = \left[\frac{1}{\mathrm{SNR}}\right]^{1/2} \tag{3.38}$$

[31] or in log scale $20\log 10(\mathrm{EVM_{RMS}}) = -20 * \frac{1}{2}\log 10[\mathrm{SNR}] = -10\log 10(\mathrm{SNR})$.

For 5% EVM or −26 dB EVM the SNR is 26 dB.

The BER under a Gaussian probability distribution function (PDF) assumption can be written as a function of the M-ary level and the SNR for M-QAM as

$$\mathrm{BER}(P_e) = \frac{2\left(1 - \frac{1}{L}\right)}{\log 2(L)} Q\left\{\sqrt{\frac{3\log 2(L)}{L^2 - 1}}\sqrt{\frac{2E_s}{N_0\log 2(M)}}\right\}, \tag{3.39}$$

where E_s is the signal energy/power, N_o is the noise power, so $E_s/N_o = \mathrm{SNR}$, L is the number of levels (or number of bits per symbol) in each in-phase and quadrature direction, e.g. $M = 256$ and $L = 8$, and Q is the Gaussian distribution function defined as $Q(x) = \int_x^{\infty} \frac{1}{\sqrt{2\pi}} e^{-y^2/2} dy$.

The appendix in section 3.10 gives a summary of this analytical analysis. In the next section Matlab simulated curves are given from the bertool.m of the communication toolbox facility for comparison. Figures 3.34 and 3.35 show the variation of the bit-error-rate (BER) versus, respectively, the signal energy and noise ratio (SNR), the EVM (error vector magnitude) under different modulation formats. As a result, figure 3.36 shows the derived EVM against the SNR.

Now we present an example, if EVM = 0.05 then EVM = 13 dB (in the power constellation) and −26 dB in the magnitude scale.

Therefore, the design by estimation of the RPoF for a given link can follow the following procedures:

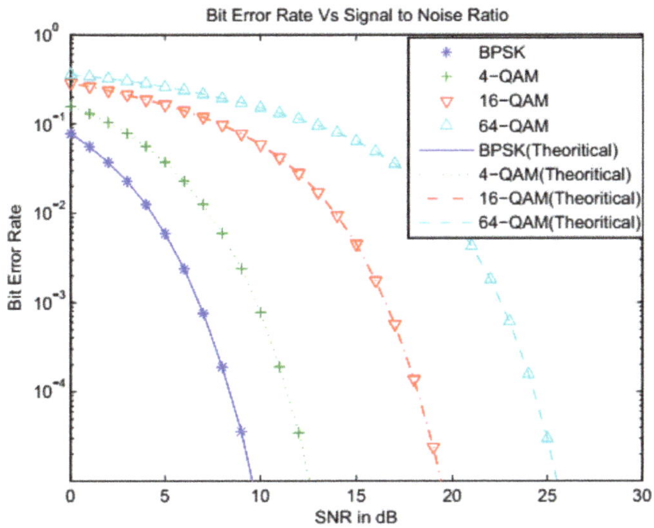

Figure 3.34. BER versus SNR in dB where M is the parameter, L is the number of bits per symbol or the number of levels in the in-phase or quadrature direction. This curve can be correlated with the EVM for the design. Copyright (2006) IEEE. Reproduced with permission from [31].

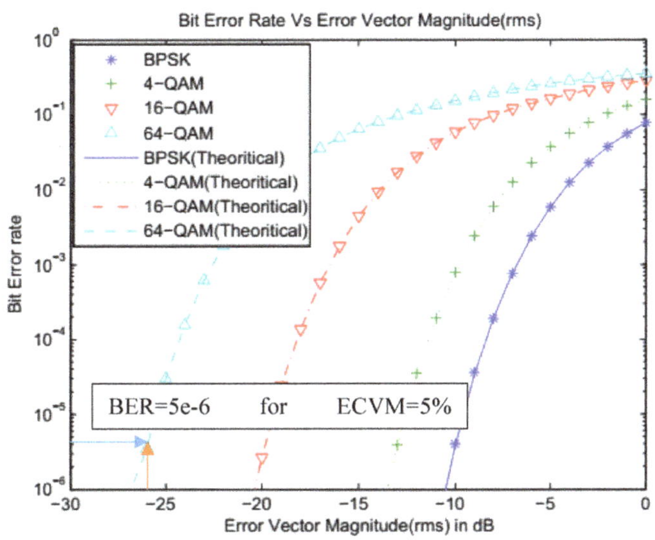

Figure 3.35. BER versus EVM (also in the dB scale) where M is the parameter, L is the number of bits per symbol or the number of levels in the in-phase or quadrature direction. 10% EVM = −10 dB; EVM 1% = −20 dB, etc… For EVM = 0.05 then EVM = −13 dB. Copyright (2006) IEEE. Reproduced with permission from [31].

1. Step 1: Given a link length modulation format M-QAM, with the OFDM of the bandwidth much smaller than that of the RF carrier to carry the QAM OFDM signals, specify the EVM.
2. Step 2: Obtain the eSNR as given by equations (3.24) and (3.34), and hence the OSNR. *Note*: The eSNR for the multi-level M-QAM.

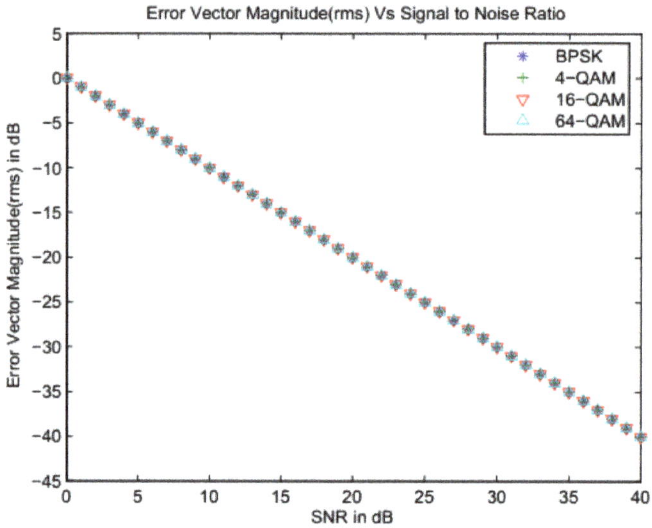

Figure 3.36. Stylized EVM versus BER performance curve.

3. Step 3: Considering the detection technique, whether it is envelop detection or coherent detection of a single PD or balanced PD pair, the noise can be considered and estimated using (3.30).
4. Step 4: Then use the power budget to obtain the power of the LO, the power of the laser at the transmitter, and the modulation index of the optical modulators.
5. Step 5: For multi-level QAM, obtain the number of levels in the in-phase and quadrature components. From step 2 the signal amplitude spacing of the levels must be considered for the maximum level or the outer-most level to satisfy the linearity condition or the IIP3 and the SFDR parameters, and hence the modulation index of the QAM signals to be applied to the bias point of the MZIM optical modulator.

3.9.3.3 MATLAB simulation

Similar to the derivations obtained in the above subsection, we can use the Matlab bertool.m from the toolbox 'communications' to plot BER versus SNR, as shown in figure 3.37 and 3.38 without and with coding (Reed–Solomon coding (R–S)) for $M = 16$ to 256. We observe some minute difference for $M = 64$ between the curves given in figures 3.34 and 3.37 from analytical and Matlab exact theoretical estimations. Thus either can be used, but the Matlab program bertool.m can be extended to much a higher order of Mary-QAM. If simulating under additional coding such as R–S codes or a 20% additional bit sequence then the BER versus SNR requires a higher SNR for the same given EVM, as shown in figure 3.38. For example, for 256QAM at least 5 dB extra is required for the SNR of the R–S code as compared to the case of no coding at all.

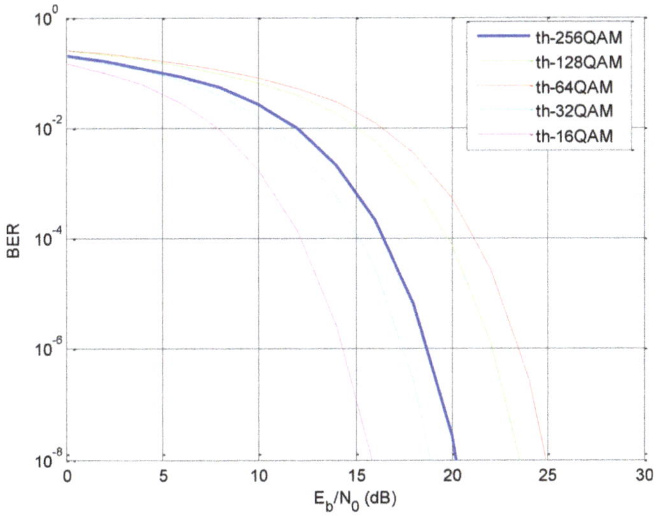

Figure 3.37. Plot of BER versus SNR with M as a parameter and without any channel coding.

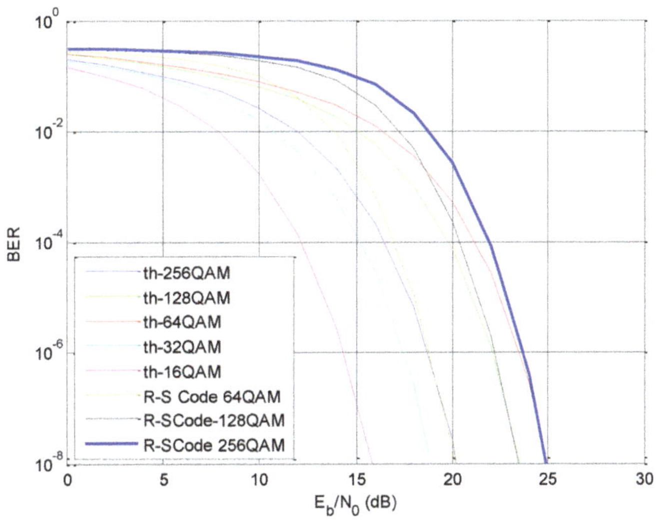

Figure 3.38. Plot of BER versus SNR with M as a parameter and with R–S channel coding.

3.9.3.4 Design example for 256QAM/OFDM

See table 3.1.

The design of the RPoF link shown in table 3.1 is associated with a number of photonic and electrical components. The estimation of the parameter of the link is based on the relationship between the EVM, SNR, and OSNR, as described in sections 3.9.2 and 3.9.3. Based on the design conclusions of the table the RPoF link is modified, as shown in figure 3.39. In particular, an eBPF is inserted after the PD or PDP balanced receiver so as to minimize the noise contribution to the EVM, which

Table 3.1. RPoF link design table.

Transmitter	Modulation format	EVM/SNR/OSNR	Notes
Modulation (RF carrier and baseband)	M-QAM/OFDM	OSNR/eSNR (dB)	
EVM	0.05 (5%)	26 dB or 398 linear scale, 20 dB if no coding or 100 times in the linear scale with respect to the noise level (BER = 1e-6) 17 dB for 1e-4 and 18 dB for 1e-5. Use 1e-5 and 18 dB SNR (or a 6 dB margin).	Equation (3.16) and figure 3.37 (no coding or FEC) using MATLAB Bertool.m for 1e-8.
Receiver	**Noise**	**Signal levels and LO power**	
Receiver method—coherent with IF to bring it down to the 5 GHz band.	Noise level at 10 dBm LO power = $2qI_DB = 2 \times 1.6\text{e-}19 \times 10\text{e-}3 \times 0.6 \times 2\text{e}9 = 6.4\text{e-}6$ or 6.4 μA. • Signal = 640 μA or equivalent optical power at the front of the PD = 0.64 mA/0.6 or 1.066 mW or about 1.25 dBm. • $\sqrt{}$(signal power × LO power) = 1.066 mW.	$\sqrt{}$(signal × LO power) → 1.06 mW. Or the beating between the signal and CW LO laser to give a coherent boosted signal at IF.All for EVM 5% or 20 dB SNR (taken as the power), if there is a voltage SNR then only 10 dB power is required. Let it operate in the worst case of 20 dB.	• B = bandwidth of eBPF ~ 2 GHz at the IF center. • PD responsivity = 0.6 A W^{-1}, BW filtered by eBPF. • Quantum shot noise dominant. • Electronic noise referred to the input of the amplifier following PD 20 pA sqrt^{-1} (Hz) → square and multiplying BW ≪ quantum shot noise.
Signal power at ORx-PD (if PDP balance → 3 dB gain with noise level × 2).	10 dBm at 3 dB splitter → IL (4 dB) IQ mod → 10 meter SMF → 6 dBm or 4 mW. LO power ~8 dBm (some loss – so the LO power is about 6.3 dBm) → signal × LO ~ 5.01 mW (square-root of product 25.2 mW^2). • This is about 6 dB above the required signal level.	OK with an 8 mW LO and 4 mW signal average power with a 3 dB power operating margin and hence 3 dB allowable further loss or 15 km SSMF.	This 6 dB above the required level may be used to meet the 3 dB operating power margin (6 dB in voltage). Thus, 3 dB is allowed for the SSMF.

Output level of optical receiver for ADC/DSP.	The minimum level required for ADC = 100–200 mV → require a TIA → 640 μA × 150 = 96 mV, in practice 1500 = Z_T transfer resistance gain. Thus the current level is OK for ADC input for conversion to digital levels for DSP.		

Modulator	Biasing/amplitude	Amplitude of IQ signals	Notes
IQ modulator	Biased at the Q-point in both the I and Q ports. Phase bias applied to obtain the $\pi/2$ shift for orthogonality.	Q+ points are bias points for both the I- and Q-branches in the MZIM.	Applying biasing, using automatic biasing control is OK as it is designed for such applications. Note only one polarization is used so chose an appropriate algorithm software GUI downloadable from the manufacturer.
	The optical average power of the signal RF carrier is the laser power. The dynamic power level varies according to the levels of M-QAM. The maximum L of the M level must be in the linear range or < 90% of the maximum level or 80% modulation, assumed to be still in the linear range or 60% level can be used.	The average power of the signal is the laser power at the output of the IQ modulator. This power gives enough sensitivity for coherent mixing as calculated above. The dynamic of the modulation index gives the RF carrier QAM signals to be demodulated. Thus for 80% modulation, index = 0.4, the dynamic power level is 0.4 × 8 mW or 3.2 mW → a 2 mA (0.6 A W^{-1} responsivity PD) swing in electrical RF or 2 mA/8 (0.25 mA) per level for 256QAM. Thus the eSNR per	• Nonlinear analysis by two-tone analyses are used here. Assuming 8% of the maximum peak-to-peak. A 20–30 dB extinction ratio optical MZIM in each branch of the IQ modulator. • To measure we need a vector signal analyser. We must use a four channel Keysight digital oscilloscope. There are two such DSAs in the O-lab. Processing can be carried out using the capturing of sequential bits and processing.

(*Continued*)

Table 3.1. (*Continued*)

Transmitter	Modulation format	EVM/SNR/OSNR	Notes
		level = 250/6.4 (noise level) ~ 40 → EVM of −16 dB. Or >5%, but with the 10 dB used in the calculation above, this EVM is improved to an acceptable level. In practice I think that there would be some acceptable fluctuation of the EVM.	• Quite a few processing algorithms are built in the DSO to obtain EVM, but not SFDR or IIP3. However, these can be requested from Keysight if available. The algorithm can be purchases or an ESA (R&S FSW) used to analyse the SFDR, provided there are two tones in the modulation signals.
System set-up details.	See the diagram in figure 3.39.		

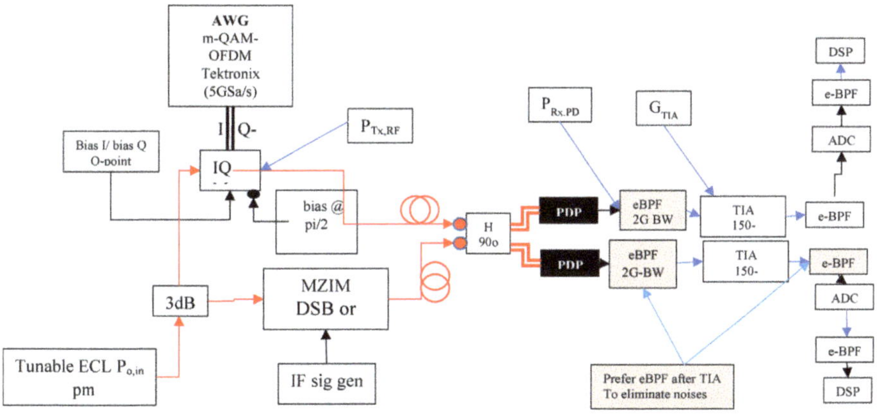

Figure 3.39 Design set-up for the RPoF. Note: $V_{\pi dc}$ is separate from the RF port → ~7–12 V for V_π of the IQ modulator. The phase bias is nearly the same.

can be derived following an inverse proportional to the square-root of the SNR in the power term. Section 3.10 provides an appendix on the relationship between the parameters EVM, SNR, and OSNR. These relationships can also be estimated using the bertool.m toolbox of Matlab.

3.10 Appendix 3: EVM, BER, and SNR relationship

As we can see from the previous sections, the parameter EVM and the BER as well as the SNR and hence the OSNR of the received signal constellation under the M-QAM constellation with OFDM as the data baseband is critical for the design of the RPoF link. That is, from a specified EVM and the constellation square points the OSNR can be determined and thus the subsystem components can be selected and specified for the link design. This section is an appendix that gives an analytical expression and some published calculated results for M up to 64 and the relationship between EVM and SNR, and EVM versus the BER.

The SNR, EVM, and BER can be related via a simple equation given in this appendix.

EVM is an alternative measure of error. It is the summation of the errors of the measured and ideal expected symbols of a constellation. EVM can be used as a performance parameter for RPoF links. The relationships between BER, SNR, and EVM can be derived as an extension from [32] and [33]. Accurate BER calculations can thus be avoided since they may require a very long sequence of bits (2^{31}–1) [34]. BER can be related to EVM, and this equivalence relationship can simplify and save processing time.

3.10.1 Signal-to-noise ratio

The SNR in dB is a ratio of the signal power and the noise power over a spectral region. The noise power is measured in the signal band, which can be difficult as the noise is embedded in such band. Assuming a Gaussian noise model for wireless channels or RPoF channels and complex signals, SNR can be defined as

$$\mathrm{SNR} = \frac{\text{signalpower}}{\text{noisepower}} = \frac{\text{squaresignal current}}{\text{squarenoise current}} = \left\{ \frac{\frac{1}{T}\sum_{t=1}^{T}\left[(I_t)^2 + (Q_t)^2\right]}{\frac{1}{T}\sum_{t=1}^{T}\left[(n_{It})^2 + (n_{Qt})^2\right]} \right\}, \quad (3.40)$$

where I_t and Q_t are the in-phase and quadrature signal amplitudes of the M-QAM modulations, and $n_{I,t}$ and $n_{Q,t}$ are the noise amplitudes of the I- and Q-components which can be quantum shot noise due to the LO power and the electronic noise of the amplifiers under coherent reception. Equation (3.40) gives a direct measure of the SNR. E_S/N_0 gives the SNR, where E_S is the symbol energy and $N_0/2$ is the noise power spectral density. It is to be noted that $E_S = \log 2(M \times E_b)$, for such systems.

3.10.2 Bit error rate

The BER is the probability of error of the transmitted and received bits over an RPoF link. Normally a long sequence of bits must be used so that a practical BER can be determined. In the case of RPoF with m-QAM modulation of the RF carrier, BER can be affected by the channel Gaussian noise [35]. For m-QAM modulation of RPoF it can be shown that [36]

$$\mathrm{BER} = P_b = \frac{2\left(1 - \frac{1}{L}\right)}{\log 2(L)} = Q\left\{\sqrt{\frac{3\log 2L}{L^2 - 1}}\sqrt{\frac{2E_b}{N_0}}\right\}, \quad (3.41)$$

where L is the number of modulation levels in each Euclidean dimension of the complex plane of the m-QAM signal, E_b is the energy per bit and $N_0/2$ is the noise power spectral density. $Q[.]$ is the Gaussian error function [37].

For raised-cosine pulses, equation (3.41) also gives the BER in terms of SNR as

$$\mathrm{BER}(P_b) = \frac{2\left(1 - \frac{1}{L}\right)}{\log 2(L)} Q\left\{\sqrt{\frac{3\log 2(L)}{L^2 - 1}}\sqrt{\frac{2E_s}{N_0\log 2(M)}}\right\}, \quad (3.42)$$

where E_S/N_0 is the SNR for the m-QAM modulation system.

3.10.3 EVM and BER relationship

EVM can be expressed by considering the contribution of the constellation point under measurement and other constellation symbols as [38]

$$\mathrm{EVM}_{\mathrm{rms}} = \frac{\frac{1}{N}\sum_{n=1}^{N}\left|S_n - S_{0,n}\right|^2}{\frac{1}{N}\sum_{n=1}^{N}\left|S_{0,n}\right|^2}, \tag{3.43}$$

where S_n is the normalized nth symbol in the sequence of measured symbols, $S_{0,n}$ is the ideal normalized constellation nth symbol, and N is the number of constellation symbols. Equation (3.43) can be normalized using a normalization factor

$$|A_0| = \sqrt{\frac{N}{\sum_{t=1}^{T}\left[V_{I,t}^2 + V_{Q,t}^2\right]}}. \tag{3.44}$$

Equation (3.44) can be extended to

$$\mathrm{EVM}_{\mathrm{rms}} = \sqrt{\frac{\frac{1}{T}\sum_{t=1}^{T}\left|I_t - I_{0,t}\right|^2 + \left|Q_t - Q\right|^2}{\frac{1}{N}\sum_{n=1}^{N}\left[I_{0,n}{}^2 + Q_{0,n}{}^2\right]}}, \tag{3.45}$$

where I_t and Q_t are the normalized I- and Q-component levels of measured symbols and $I_{0,t} = (V_{I0,t})|A_0|$ and $Q_t = (V_{Qt})|A|$ are the normalized I- and Q-levels, respectively, of 'ideal' constellation symbols. Equation (3.45) gives the interrelationship between EVM, BER, and SNR. For a Gaussian noise model, equation (3.45) can be simplified in terms of the noise of the I- and Q-components, $n_{I,t}$ and $n_{Q,t}$ as

$$\mathrm{EVM}_{\mathrm{rms}} = \sqrt{\frac{1}{T}\frac{\sum_{t=1}^{T}\left|n_{I,t}\right|^2 + \left|n_{Q,t}\right|^2}{P_0}}, \tag{3.46}$$

where P_0 is the power of the normalized ideal constellation or the transmitted constellation. For $T \gg N$, equation (3.46) can be rewritten as [39]

$$\mathrm{EVM}_{\mathrm{RMS}} = \left[\frac{1}{\mathrm{SNR}}\right]^{1/2} = \left[\frac{N_0}{E_S}\right]^{1/2}. \tag{3.47}$$

The SNR can thus be related to the EVM as

$$\mathrm{SNR} = \left[\frac{1}{\mathrm{EVM}^2}\right]. \tag{3.48}$$

Combining equations (3.47) and (3.42), the BER can be related to EVM as

$$\mathrm{BER}(P_e) = \frac{2\left(1 - \frac{1}{L}\right)}{\log 2(L)} Q\left\{\sqrt{\frac{3\log 2(L)}{L^2 - 1}}\sqrt{\frac{2E_s}{N_0 \log 2(M)}}\right\}. \tag{3.49}$$

References

[1] Binh L N 2009 Linear and nonlinear transfer functions of single mode fiber for optical transmission systems *J. Optical Soc. Am.* A **26** 1564–75

[2] Ng'oma A, Rijckenberg G-J and Koonen A M J 2007 Building extended-reach radio-over-fiber links by exploiting optical frequency multiplication's dispersion tolerance *IEEE/MTT-S Int. Microwave Symp. (Honolulu, HI, 3–8 June)*

[3] Smith G H, Novak D and Ahmed Z 1997 Technique optical SS generation to overcome fiber dispersion penalty in fiber radio-system *Electronic Lett.* **33** 73–5

[4] Laurêncio P and Medeiros M C R 2003 Dynamic range of optical links employing optical single side-band modulation *IEEE Photonics Tech. Lett.* **15** 748–50

[5] Lim C *et al* 2007 Intermodulation distortion improvement for fiber–radio applications incorporating OSSB+C modulation in an optical integrated-access environment *IEEE J. Lightw. Technol.* **25** 1602–12

[6] Huang *et al* 2016 Photonic generation of equivalent SSB vector signals *IEEE Photonics Tech. Lett.* **28** 2633–6

[7] Abramowitz and Stegun 1964 *Handbook of Mathematical Functions with Formulas, Graphs, and Mathematical Tables* (Washington, DC: United States Department of Commerce, National Bureau of Standards)

[8] Lim C *et al* 2007 Intermodulation distortion improvement for fiber-radio applications incorporating OSSB+C modulation in an optical integrated-access environment *J. Lightw. Technol.* **25** 1602–12

[9] Binh L N 2021 Coherent OSA *US patent pending*

[10] Granieri S *et al* 2003 Optical modulation with a single sideband and suppressed carrier *Proc. SPIE* **4833** 1154–8

[11] Smith G H, Novak D and Ahmed Z 1997 Technique for optical SSB generation to overcome dispersion penalties in fibre-radio systems *Electron. Lett.* **33** 74–5

[12] Capmany J, Pastor D, Munoz P, Sales S, Ortega B and Martinez A 2003 WDM-SSB generation and dispersion mitigation in radio over fiber systems with improved performance using an AWG multiplexer with flat top resonances *Microwave Photonics Conf. (September)* pp 39–42

[13] Binh L N 2009 Linear and nonlinear transfer functions of single mode fiber for optical transmission systems *J. Opt. Soc. Am.* A **26** 1564–75

[14] Ng'oma A *et al* 2007 Building extended-reach radio-over-fiber links by exploiting optical frequency multiplication's dispersion tolerance *IEEE MTT-S Int. Microwave Symp. Digest (July)* (Piscataway, NJ: IEEE)

[15] Binh L N 2017 *Optical Modulation: Advanced Techniques and Applications in Transmission Systems and Networks* (Boca Raton, FL: CRC Press/Taylor and Francis)

[16] Grüner-Nielsen L, Wandel M, Kristensen P, Jorgensen C, Jorgensen L V, Edvold B, Pálsdóttir B and Jakobsen D 2005 Dispersion-compensating fibers *J. Lightw. Technol.* **23** 3566
[17] Dar A B and Jha R K 2017 Chromatic dispersion compensation techniques and characterization of fiber Bragg grating for dispersion compensation *Opt. Quantum Electron.* **49** 108
[18] Lim C *et al* 2001 Technique for increasing optical spectral efficiency in millimeter-wave WDM fibre-radio *Electron. Lett.* **37** 1043–5
[19] Toda H *et al* 2002 A DWDM MM-wave fiber-radio system by optical frequency interleaving for high spectral efficiency *Proc. Microwave Photonics (Long Beach, CA)* pp 85–8
[20] Dat P, Rottenberg F, Kanno A, Yamamoto N and Kawanishi T 2021 3 × 3 fiber-wireless system in W-Band with WDM/PDM RoF transmission capability *J. Lightw. Technol.* at press
[21] Kurniawan T *et al* 2006 Performance analysis of optimized millimeter-wave fiber radio links *IEEE Microwave Theory Tech.* **2** 921–7
[22] Chen W H and Way W I 2004 Multichannel single-sideband SCM/DWDM transmission systems *J. Lightw. Technol.* **7** 1679–93
[23] Lim C *et al* 2007 Link characterization for optical single sideband modulation with linearization technique incorporating RF nonlinearity *IEEE/MTT-S Int. Microwave Symp. (Honolulu, HI, June)*
[24] Agrawal G P 2010 *Fiber Optic Communication System* (New York: Wiley)
[25] Binh L N 2017 *Advanced Digital Optical Communications* (Boca Raton, FL: CRC Press)
[26] Zhou P *et al* 2020 Photonic generation of tunable dual-chirp microwave waveforms using a dual-beam optically injected semiconductor laser *Opt. Lett.* **45** 1342–5
[27] Keysight Inc. What is the relationship between the Error Vector Magnitude and the Signal to Noise Ratio for a 64-QAM signal? https://edadocs.software.keysight.com/kkbopen/what-is-the-relationship-between-the-error-vector-magnitude-and-the-signal-to-noise-ratio-for-a-64-qam-signal-589738547.html (accessed Nov. 2021)
[28] Keysight Technologies 2012 What is the relationship between the error vector magnitude and the signal to noise ratio for a 64-QAM signal?
[29] Schmogrow R *et al* 2012 Error vector magnitude as a performance measure for advanced modulation formats *Photon. Technol. Lett.* **24** 61–3
[30] Sunnerud H, Skold M, Westlund M and Andrekson P A 2012 Characterization of complex optical modulation formats at 100 Gb/s and beyond by coherent optical sampling *J. Lightw. Technol.* **30** 3747–59
[31] Shafik R A *et al* 2006 On the extended relationships among EVM, BER and SNR as performance metrics *4th Int. Conf. on Electrical and Computer Engineering (Dhaka, Bangladesh, 19–21 December)*
[32] McKinley M D, Remley K A, Mylinski M, Kenney J S, Schreurs D and Nauwelaers B 2005 EVM calculation for broadband modulated signals *Technical Report* United States Government
[33] Ghairabeh K, Gard K and Steer M 2004 Accurate estimation of digital communication system metrices—SNR, EVM and ρ in a nonlinear amplifier environment *Proc. 64th ARFTG Microwave Measurements Conf. (Orlando, FL, 2–3 Dec 2004)* (Piscataway, NJ: IEEE) pp 41–4
[34] Yi S, Tsimenidis C, Hinton O and Sharif B 2004 Adaptive minimum bit error rate multiuser detection for asynchronous MC-CDMA systems frequency selective Rayleigh fading channels *14th IEEE Int. Symp. on Personal, Indoor and Mobile Radio Communications.* vol 2*(April)* (Piscataway, NJ: IEEE) pp 902–6

[35] Hanzo L, Webb W and Keller T 2000 *Single- and Multi-Carrier Quadrature Amplitude Modulation* 2nd edn (Chichester: Wiley)

[36] Forestier S, Bouysse P, Quere R, Mallet A, Nebus J and Lapierre L 2004 Joint optimization of the power-aided efficiency and error vector measurement of 20-GHz p-HEMT amplifier through a new dynamic bias control method *IEEE Trans. Microw. Theory Techol.* **52** 1132–40

[37] Goldsmith A 2005 *Wireless Communications* 1st edn (Cambridge: Cambridge University Press)

[38] Forestier S *et al* 2004 Joint optimization of the power-aided efficiency and error vector measurement of 20-GHz pHEMT amplifier through a new dynamic bias control method *IEEE Trans. Microw. Theory Technol.* **52** 1132–40

[39] Shafik R A *et al* 2006 On the extended relationships among EVM, BER and SNR as performance metrics *4th Int. Conf. on Electrical and Computer Engineering (Dhaka, Bangladesh, 19–21 December)*

Chapter 4

Radio-photonic transmission with IF processing and nonlinear suppression

This chapter outlines the following for ultra-wideband and high frequency conversion techniques and technology:

- A theoretical analysis and practical implementation of radio-photonics over fiber (RPoF) links using intensity modulation and direct detection (IM/DD) and coherent reception under the conditions without and with optical filtering. The filtering enables the down-conversion to an intermediate frequency (IF) and thus processing in a much lower spectral region. The link performance with spurious free spectral range (SFDR) dynamic ranges are obtained and all such links are compared to prove that the coherent reception system offers higher dynamic range performance. Furthermore, optical filtering further boosts the SFDR operating range.
- Introducing a further improvement of the RPoF with optical filtering techniques, the optical bandpass filter (OBPF) employing a cascade Mach–Zehnder delay interferometer (MZDI) and micro-ring resonator (MRR) as an optical filtering sub-section for optical filtering to reject one band of the RF modulated optical waves which follows the modulation of the lightwaves. The cascade MZDI sections give a transmittance transfer function with all zeros or all pass with zeros only, so peaking can occur, but with low roll-off outside the passband. The MRR structure offers a pole-zero pattern, and hence resonant peaks and a free spectral range (FSR) with sharp roll-off, and hence higher performance in filtering—the OBPF. The effects of the OBPF on linearity and distortion parameters are described.
- Illustration of the linearity of the optical modulator, the traditional MRR and phase modulation section incorporated in a branch of a Mach–Zehnder interferometric modulator (MZIM) and further improvement of the SFDR by placing the MRR in one branch and the optical phase modulator (OPM) on the other branch of the Mach–Zehnder modulator (MZM). This replacement

doi:10.1088/978-0-7503-3511-9ch4

can alter the pole-zero pattern on the z-plane of the transmittance transfer function of the interferometer MZM, thus offering a much better SFDR performance.

- A number of RPoF systems are given using both the envelop detection and coherent reception techniques. The operation principles of these systems are given in the figure captions.
- The IF frequency generation can be obtained via an optical frequency comb and array waveguide grating multi-port filtering, so that two output ports can be used as two optical beams to derive the RF frequency at the photodetector output via coherent reception.

4.1 Baseband and RF shifted band reception RPoF

4.1.1 Introductory remarks

Wireless communications systems in 5G, 6G [1], and 7G transmission and networking employ RF carriers in the microwave (μW) and millimeterwave (mmW) regions with baseband data channels superimposed on the RF carriers. Under common signal analyses, in order to recover these RF channels the sampling rate is extremely high and impossible for digital signal processors (DSPs) to handle. This section describes both theoretical analyses and reported experimental demonstrations of radio-photonic analog links in which the RF channels are down-converted to an IF region so that the baseband data channels can be sampled and digitally processed in much lower sample rate DSP systems.

RPoF systems are configured according to their locations and applications in networks, in the following scenarios:

- Case 1: the systems are at the user end in which for data processing is preferred at the baseband and the recovery of the RF carriers is not critical.
- Case 2: The recovery of the RF is critical for further transmission or identification in the case of testing or RF channel identification for further network control.

In either case spur-free intermodulation (IMD) is critical. Possible solutions for RPoF can be as follows:

a. For case 1, using optical filtering that selects only one sideband if under double sideband (DSB) conditions or a single sideband (SSB) with the main optical carrier removed or suppressed. Thus only the baseband data channel is detected and the signal band is at the baseband. The optical filter passband of the OBPF should be of the sharp roll type such that cross-talk from adjacent channels can be avoided.
b. For case 2, a DSB with an optical carrier and direct detection can be used to recover both the RF carrier and the data channel. However, the phase delay difference of the two sidebands can deplete each other, causing destructive interference and thus limiting the transmission distance.
c. Self-coherent detection of modulated RF optical channels, that is the local oscillator (LO) is derived from the original laser of the RPoF and its beating

Table 4.1. Table of abbreviations.

Abbreviation/ notation	Physical meaning	Explanation/description
Bold	Field vector	
OA	Optical amplifier	EDFAs are common for 1550 nm C-band amplification. The main factor is the spontaneous photon number n_{sp}.
EDFA	Erbium-doped fiber amplifier	
OBPF	Optical bandpass filter	Many types are used in the optical domain, including Fabry–Pérot cavity resonant filters with an FSR, and all-zero IIR type, and FIR MRR type OBPFs.
IIR	Infinite impulse response filter OBPF Single MZDI or cascade MZDI with thermal electrode tuning	OBPF—integrated optic type using SiN_4 on a SiO_2 substrate. The structure is a cascade of MZDIs, i.e. only zeros due to interferometric phenomena. Tuning of a centered passband is achieved by a thermal electrode applied voltage so that the interference is shifted.
FIR	Finite impulse response OBPFA Micro-ring or cascade of micro-rings with thermal electrode tuning of the notched or resonant point. Both zero and pole transfer function → notch and resonant.	MRR structure using an integrated optic structure of rings of radio ~10–20 μW thermal electrode deposited on top of the waveguide section → shifting of the resonant point→ OBPF, the passband bandwidth can be tuned. A longer ring length can be achieved using two N spiral rings, e.g. the double ring by Leonix (The Netherlands).
OC	Optical coupler Type 1: $N \times N$, $N \times 1$, or $N \times M$ Type 2: hybrid OC.	Splitting or combing optical waves. Hybrid type: i.e. incorporating a polarization splitter and combiner and a phase shift of $\pi/2$ for complex lightwaves, e.g. the IQ signal—normally placed in front of a balanced receiver. 4 × optical output ports for balanced reception of the LO and complex optical signals.

(Continued)

Table 4.1. (*Continued*)

Abbreviation/ notation	Physical meaning	Explanation/description
		The phase-shifter is tuned such that the $\pi/2$ component can be extracted so that the I- and Q-components can be extracted and detected by BalORx. optical internetworking forum (OIF) PD-QAM then a polarization splitter hybrid coupler can be used.
PD PDP	Photodetector Pair of PDs	Detection following square detection, that is the optical power/intensity is absorbed by the PD with a wide spectral property and produced detected current. It is negative biased so that the generated electron–hole pair can be pushed to opposite electrode signs. Thus, reverse biasing is required. PDPs can be connected back-to-back so as to obtain 3 dB gain and eliminate the common mode noise types.
MZM	Mach–Zehnder modulator. An interferometric modulator of Mach–Zehnder type which consists of two parallel paths and electrodes applied so that the phase shift of one path with respect to the other produces depletive or constructive interference at the output.	An interferometric optical structure (normal Mach–Zehnder) in $LiNBO_3$ or silicon-on-insulator (SOI) such that if there is a phase shift of $\pi \rightarrow$ depletion and so on. It is biased at different operating points. A dc and AC signal superimposed to obtain swinging optical intensity or field at the output of the modulator. Operating points = Q-point, i.e. at the $\pi/2$ phase shift of one branch with respect to (w.r.t.) the other. Q+ at the positive going transmittance point and Q– at the negative going transmittance point.
DD-MZM	Dual-drive Mach–Zehnder modulator	Two electrodes are placed on top of two optical waveguide paths of the MZM so that both biases and

		signals can be applied to the electrode input, e.g. to obtain bias at a minimum transmission point the $\pi/2$ and $-\pi/2$ biased points are to be applied to both sides of the electrode. This is to obtain a 3 dB reduction in RF signals
QPSK modulator	Two parallel MZMs (child MZMs) embedded on two branches of a parent MZM with a $\pi/2$ phase electrode shift in one child branch so as to obtain complex optical modulated optical signals.	Now common for 100G transmission systems. The biasing and signal superimposed are the same as for MZM. The biasing point can be controlled using automatic bias control.
QAM	Quadrature amplitude modulation (complex) with real and imaginary parts.	A complex modulation constellation. Normally two optical modulators are used, one for the I- and one for the Q- component. Thus PAM-n for I and PAM-m for Q give the constellation $[m \times n]$ order. If $M = m \times n$ then we have an M-QAM modulation scheme.
OFDM	Orthogonal frequency division multiplexing.	Modulation technique using frequency domain in which each sub-carrier is orthogonal with respect to other two adjacent sub-carriers. The sinc profile of these spectral components leads to no interference at sampling point. This modulation gives a 'brick wall' like spectrum of signals. So it is highly efficient for packing data channels, hence increasing the spectral efficiency.
Laser		Emission of photon particle lightwaves, commonly by a gain cavity with two end mirrors 100% and ~30% such as a Fabry–Pérot laser. If an external cavity is incorporated then a single line external cavity laser (ECL) can be generated. This gives a narrow linewidth if the reflected mirror is a high resolution grating.
DFB	Distributed feedback structure	The cavity of the laser consists of two mirrors whose structure satisfies the surface corrugated grating.

(Continued)

Table 4.1. (*Continued*)

Abbreviation/ notation	Physical meaning	Explanation/description
ECL	External cavity laser	The cavity is extended with a grating mirror which reflects only the desired wavelength with narrow resolution.
EML	External modulation laser	A laser, commonly a DFB, with an external modulator (electro-absorption type). The laser is in continuous wave (CW) operation and an RF signal is applied to the electro-absorption modulator.
Comb laser		The spectrum consists of several equally spaced lines. Commonly generated by a mode-locked laser whose pulse width is very short in the sub-femtosecond ranged and spaced by 25 ps to give 50 GHz spacing and 40 line lasers.
LO	Local oscillator (optical)	
dc	Direct current	The voltage applied at the dc level to generate a phase constant bias.
Re, Im	Real part, imaginary part	
Q-factor	The quality factor of a resonator or filter.	Due to zeros or the pole of optical circuits.
Q+ point, Q– point	Quadrature point of the optical transmittance transfer curve (optical field/power versus applied voltage) of an MZM. Q+ positive transfer curve point; Q– negative transfer curve point.	
ORx	Optical receiver	Normally structured with a PD front-end followed by an amplifier. The transimpedance amplifier (TIA) type is used if a a wider bandwidth required.
BalORx	Balanced ORx	Two PDs (PDP) connected back-to-back and then to an amplifier to give a 3 dB gain and minimizing noise due to common mode rejection.

PD	Photo detector	Normally connected back-to-back for balanced reception.
PDP/BPD	PD pair/balanced PD	
q_e	Electronic charge	1.6 e-19 coulombs
k_B	Boltzmann's constant	
T	Absolute temperature in kelvins.	
η	PD sensitivity in units of A W^{-1}	
α	Attenuation factor of the fiber link in IM/DD or coherent RoF link.	
$G_1 \ldots G_n$ $A_1 \ldots A_n$	Gain and attenuation factors in dB or linear scale of OA–SSMF–DCF spans from 1 to nth of the optical amplifier and propagation loss of fiber.	
V_π	Half-wave voltage at which the phase change of π in the intensity of $\pi/2$ on the optical field.	
N_{TIA}	Total equivalent noise power of a TIA.	Calculated by taking the given total current noise density referred to the input port × (transimpedance gain) → voltage → square and divided by output load resistance (50 Ω).
G_p	OA gain factor	Number of optical photons multiplied at the OA output.
$NF_{IM/DD}$	Noise figure for an IM/DD RPoF link in dB (log10 scale)	Noise output as compared to input noise of DUT (device under test).
F_n	Noise factor (linear).	Note that the electronic NF and optical NF are slightly different.
$NF_{F\text{-}IM/DD}$	NF of a filtered IM/DD link.	
$NF_{F\text{-}Coh}$	NF of a filtered coherent RPoF link.	Units of dB
SFDR	Spurious free distortion range	Units of dB $Hz^{2/3}$
FSR	Free spectral range	Units of Hz
OIP3	The output third-order interception point	Obtained from linear extrapolations of the fundamental and IMD3 responses.

(*Continued*)

Table 4.1. (*Continued*)

Abbreviation/ notation	Physical meaning	Explanation/description
IIP3	Input interception point corresponding to OIP3, or the maximum input power so that there is no spurious IMD.	Obtained from linear extrapolations of the fundamental and the horizontal axes representing the input power which is at the raising point, linearly w.r.t. to the input power.
IMD3	Intermodulation distortion of the third-order.	Testing by using two RF carriers of frequency spacing equal to data channel spacing in 5G and 6G data.
f_1, f_2	Frequencies of the two tones of the SFDR/IMD3 testing.	Normally in 5G and 6G the frequency spacing between the data channels embedded on an RF carrier.
$S_R(t)_{\|IM/DD}$	Signal in the time domain at the front of the optical receiver.	
$S_R(t)_{\|F\text{-}IM/DD}$	Signal in the time domain at the front of the optical receiver of the filtered IM/DD link.	
$S_R(t)_{\|F\text{-}Coh}$	Signal in the time domain at the front of the optical receiver of the filtered coherent link.	
$E_0(t)$	$E_0(t) = \sqrt{2}\,\|E_0\|\,e^{j\omega_0 t}$ the optical field generated by laser.	
$E_o(t)$	Optical field at the output of the light source (laser).	
$E_1(t)$	Optical field at the output of the modulator.	
$E_2(t)$	Optical field at the output of the OBPF.	
$\|E_0\|$	Amplitude of the laser field—square → power, average power P_0.	
E_{LO}^*	Electric field of the LO laser before modulation if any, normally carrier suppression (CS) so as to minimize the quantum shot noise.	
ω_0	Angular frequency of the laser source—external to the modulator.	
ω_{RF}	Angular frequency of the RF waves applied to the modulator.	
ω_{IF}	Angular frequency of the IF which is derived from mixing the laser input to the MZM and that of the RF shifted LO laser.	
$Z(t)$	Time domain RF signal applied to the MZM.	

$-Z(t)$	Complementary signal of $Z(t)$ which is normally to the other electrode of the DD-MZM.	The biasing point at the null point or V_π for a normal MZM (single electrode), the carrier is suppressed and two RF sidebands are generated. However, in dual drive, the DD-MZM must be biased at $V_\pi/2$ and $-V_\pi/2$ to give a total optical phase difference of π, thus suppression of the carrier. The RF signals are complementary to each other and are superimposed with the swing amplitude to the desired level of the transmittance transfer function.
P_{sig}	The average power of the RF signal applied to an optical modulator.	
$\beta = \pi/V_\pi$	The phase factor in units of rad V^{-1}.	For an applied voltage amplitude V the phase change is $\beta = \varsigma(\pi/V_\pi)$.
β_e	The effective propagation constant along the optical waveguide (channel, ridge, or strip).	The eigenvalue of the wave equation of the guided lightwaves along the optical waveguide.
$\rho(t)$ and phase $\varphi(t)$	Amplitude and phase as a function of time of the RF signal and its complementary, $Z(t)$ and $Z_C(t)$, applied to the MZM electrode.	
$J_k(.)$	kth order Bessel function of argument (.).	The order of the Bessel functions can be expanded using a Taylor series to obtain the fundamental, second-, and third-order (ω_1, ω_2, and ω_3) term coefficients. These will give the IMD, IMD2, and IMD3, and hence the SFDR.
A_1	A constant and the Bessel function J^{-1}.	
N_{out}	Total noise power as referred to the output of the optical receiver.	This is integrated over the system bandwidth or device bandwidth of the noise power spectral density, or the noise current spectral density, squared and multiplied by the load resistance.
N_{shot}	Quantum shot noise power density as referred to the ORx output port.	Due to the electronic noise generated by the PD under high power of the LO laser in coherent reception.
N_{RIN}	Laser relative intensity noise (RIN) power density as referred to the ORx output port.	RIN of lasers, phase noise due to the finite linewidth of the laser.

(Continued)

Table 4.1. (*Continued*)

Abbreviation/ notation	Physical meaning	Explanation/description
N_{OA}	Noise power density of the optical amplifier as referred to the ORx output port.	Normally the beating between the signal and amplified stimulated emission (ASE) noise of the OA, normally under the pump power and threshold level before simulated emission occurs.
N_{TIA}	Noise power density as referred to and contributed by TIA electronic noise to the ORx output port.	Calculated by taking the noise current density referred to at the input of the TIA then multiplied by transimpedance at the mid-band and then squared and divided by the load impedance (50 Ω).
$H(t)$	The temporal impulse response of a distortion compensating device so as to have either pre- or post-distortion mitigation.	Hence the frequency response to design for distortion compensation in DSP at the receiver or at the transmitter.
OBPF	Optical bandpass filter	Passband and rejection band to selectively filter desired band which includes the carrier and the data channel bands.
MZDI	Mach–Zehnder delay interferometer	All-zero photonic processor with filtering of gradual roll-off in the rejection band OBPF.
MRR	Micro-ring resonator	Integrated photonic signal processor (PSP) with a pole-zero pattern on the z-plane giving the ability to offer a high Q-factor and thus a sharp rejection band OBPF.
PSP	Photonic signal processor/processing	Processing of signals in the optical plane.
RPAL	Radio-photonic analog link	Analog optical link using a radio signal modulated with data channels superimposed on the RF shifted optical carrier, mostly under coherent reception.
RPoF	Radio-photonics over fiber	The term 'photonics' is used due to the manipulation of photons in the external modulator.
RoF	Radio over fiber	The general term for the transmission of an RF modulated optical beam over the fiber.

with the transmitted RF modulated data channel such that the detection can recover both the RF carrier and the data channels. The advantage of this technique is that there is no frequency off-set (FO) between the LO and the modulated channel. The disadvantage is that the LO power is at a near equal level to the data channel and no gain in receiver sensitivity can be achieved.

d. Coherent reception with either DSB or SSB modulation, of either homodyne or heterodyne type. However, SP processing must be used so that the local oscillator off-set (LOFO) can be eliminated. This is the most effective and stable technique but is more complex.

All reception techniques are dependent on where the RPoF is located in the network for economical operation expenditure (OPEX) or capital expenditure (CAPEX) as required for network deployment. In 6G and 7G under massive deployment of RPoF methods, b, c, and d above are the preferred configurations. In this section we look at the reception sensitivity of the configurations of the above cases, as given in figure 4.1.

In RF systems with the RF carrier frequency in the mmW region the demodulation of such high frequency oscillation poses difficulties in the sampling front-end

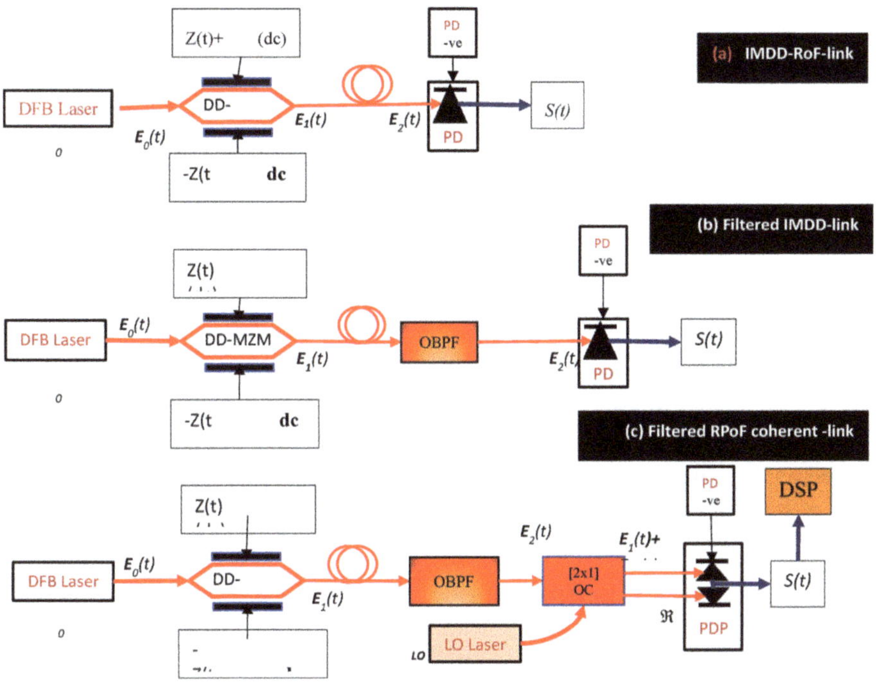

Figure 4.1. RPoF links for (a) an IM/DD RPoF link with envelop detection, (b) an optically filtered (OBPF) IM/DD RPoF link, and (c) an optically filtered coherent RPoF link. OBPF = optical bandpass filter; LO = local oscillator (optical); OC = optical coupler; PD = photodetector; DD-MZM = dual-drive Mach–Zehnder modulator; DFB = distributed feedback laser; dc = direct current of voltage applied at the dc level to generate a phase bias; bold = field vector. (Adapted with permission from [7].)

before further processing in the receiver. In the case of 6G and 7G systems, remote radio astronomy systems, etc, the antennae can be operating in the high frequency RF multi-band region with the data channel occupying only a fraction of the high frequency. Thus the electronic sampling at the front-end is extremely high but the data rate is small. This would not be economical and would be more difficult.

A photonics pre-processing system (PSP) can be useful [2, 3] to provide significant advantages such as tunability, wideband operation, flexibility, and size, weight, and power benefits. Reconfigurable integrated optical filter circuits can have the potential to assist and are useful in such RF front-end processing and other systems [4–6]. Furthermore, current 5G and 6G networking is moving the RF carriers to the μW and mmW regions, that is, in the frequency of 10–100 GHz, in particular in the sub-6G band and upper 6G band and the 7G band. Under such ultra-high frequency and ultra-wide broadband (UWB) wireless transmission, the transport of such carriers must be in the optical domain close to the wireless radiation node. This chapter focuses on the RoF techniques and analyses for such links under different transmission and reception methods, including direct detection and coherent reception with and without optical filtering to extract the selected RF shifted optical channels.

Inherent to most photonics, analog RPoF links require E/O and O/E converters, hence these are critical for the links to exhibit a wide SFDR, which is dependent on the modulation and reception sub-systems. The RPoF link budget depends on the received optical power, modulator π-phase level, and link loss/gain. The link gain suffers an additional noise spectral density floor, and the transmission and reception influence the SFDR factor.

PSP devices such as tunable filters can be inserted in-line in the link and can introduce nonlinear distortion and link losses, thus worsening the system's NF and hence reducing its SFDR. In-line optical amplifiers (OAs) can be used to compensate for this loss but impose additional ASE noise.

4.1.2 SFDR

Linearity in terms of the intermodulation (IMD) distortion products of an RPoF was introduced in chapters 1 and 2. We distinguish two definitions of the SFDR in terms of the linear scale and $\log_{10}$ scale. The additional factor $10\log_{10}(B)$ can be considered as the contribution of the bandwidth to the noise level, in particular when setting the resolution bandwidth in an electrical spectrum analyzer. Thus again we have

$$\mathrm{SFDR} = \left[\frac{\mathrm{OIP3}}{N_{\mathrm{out}}}\right]^{2/3} \text{in units of dBm Hz}^{-2/3} \tag{4.24a}$$

or

$$\mathrm{SFDR} = \frac{2}{3}\left[\mathrm{OIP3} - P_{\mathrm{ONF}} - 10\log_{10}(B)\right].$$

4.1.3 Signal reception in RoF systems

Various RPoF links for RF filtering and coherent techniques are given figure 4.1. Their transfer functions are formulated and the sensitivity and performance are given.

- Figure 4.1(a): External modulation of an RPoF link with direct direction (IM/DD). The intensity modulation is conducted via an external MZM modulator of either single drive or dual dive (DD-MZIM) features.
- Figure 4.1(b): An in-line OBPF RPoF with an IM/DD structure of the DD-MZIM.
- Figure 4.1(c): In-line OBPF intensity modulation with coherent detection using an LO laser.

4.1.3.1 Non-filtered direct detection IM/DD RPoF

A non-filtered RPoF link can consist of:

(i) An intensity modulator, for example a dual-drive interferometric modulator (DD-MZIM) biased at the Q+ (quadrature) point. RF signals are applied in a complementary pair to the two electrodes of the MZIM, that is with a π phase shift to each other with the $Z_c(t)$ of$Z(t)$ denoting the RF signal and its complementary version.

(ii) Direct detection (DD) is used and a PD detects the envelop of the RF modulated signals after transmission over a fiber length. The schematic of a typical IM/DD RPoF is shown in figure 4.1 (a), the laser is an external cavity laser (ECL).

Under such PD square-law detection, the received signal is given as

$$s_R(t)_{|\mathrm{IM/DD}} = \frac{\mathscr{R}}{2} \mid E_0 \mid^2 \{1 - \sin[2Z(t)]\}, \tag{4.1}$$

where $E_0(t) = \sqrt{2} \mid E_0 \mid e^{j\omega_0 t}$ is the optical electric field and R is the photodiode responsivity (in A W^{-1}) or a conversion to the number of electrons by one received photon. Figure 4.1(b) illustrates the insertion of an optical bandpass filter after the modulator. Under a bandpass input RF signal centered at ω_{RF}, with amplitude $\rho(t)$ and phase $\varphi(t)$, β is the scaling factor as desired, operating in the linear or nonlinear region of the optical modulator:

$$Z(t) = \beta\rho(t)\sin[\omega_{\mathrm{RF}}t + \varphi(t)]. \tag{4.2}$$

Using a Fourier series approximation and Bessel functions, this equation can be expressed as the ratio of the output optical field and that of its input as

$$\begin{aligned}\frac{E_1(t)}{E_0(t)} = &\cos\left[\frac{\varphi}{2}\right] \sum_{\text{even } k=-\infty}^{+\infty} J_k\left\{\beta\rho(t)e^{jk[\omega_{\mathrm{RF}}t+\varphi(t)]}\right\} \\ &+ j\sin\left[\frac{\varphi}{2}\right] \sum_{\text{odd } k=-\infty}^{+\infty} J_k\{\beta\rho(t)e^{jk[\omega_{\mathrm{RF}}t+\varphi(t)]}\},\end{aligned} \tag{4.3}$$

where φ is the dc phase due to the MZIM bias at the quadrature point (= $\pi/2$ at the Q+ point), $\beta = \pi/V_\pi$ is phase factor with respect to the half-wave voltage ($\beta = \pi/2$ at Q+), and $J_k(.)$ is the kth order Bessel function of the first kind.

Equation (4.3) indicates (i) two optical sidebands and (ii) a central optical carrier. If the MZIM is operating under carrier suppression or the biasing point is at the minimum transmission of the optical–RF transfer characteristics, then the received field only displays either odd- or even-order sidebands with frequency doubling.

4.1.3.2 A passband filtered reception IM/DD RPoF

Consider an RPoF system using an MZM biased at the quadrature point Q+ ($\varphi = \pi/2$) and DD. With an in-line OBPF which is tuned to pass only one signal sideband, the kth sideband, and rejecting the carrier and other sidebands, the optical field at the output of the OBPF can be obtained using equation (4.3) as

$$E_2(t) = j\,\sin(\varphi/2)E_0(t)J_k\left\{\beta\rho(t)e^{jk[\omega_{RF}t+\varphi(t)]}\right\} \tag{4.4}$$

and the electrical signal after PD detection is

$$\mathrm{s}(t)_{|\text{F-IM/DD}} = \eta\,|\,E_0\,|^2\,J_k^2\{\beta\rho(t)\}. \tag{4.5}$$

Equation (4.5) is the RPoF link field transfer function (TF) in which an OBPF filters the signal with an extraction of the original optical carrier. The subscript 'F-IM/DD' stands for filtered IM/DD. If the OBPF covers the original optical carrier then the RPoF is a vestigial filtered link or RPoF link of the IM/DD plus the vestigial OBPF. In the former case there would be some linear distortion if the passband roll-off of the OBPF is not sharp enough. Thus this type of filter can only be used in the filtering of several RF sub-carrier multiplexing (SCM) channels because each RF data channel bandwidth is much smaller than that of the RF carrier frequency and the optical filter would be difficult to design. Although a micro-ring resonator (MRR) can be designed, there remains much difficulty in its tuning and stability.

For the first sideband $k = 1$ and operates under small signal conditions, that is the peak-to-peak amplitude of the RF signals falls in the linear range of the transfer characteristics. By expanding the Bessel function $J_1[\beta\rho(t)]$ by a first-order Taylor series, we have

$$s_R(t)_{|\text{F-IM/DD}} = \eta\,|\,E_0\,|^2\left(\frac{\beta}{2}\right)^2\rho(t)^2. \tag{4.6}$$

Obviously, under DD the envelop of the RF signal, or the data channel, is recovered and any phase information $\varphi(t)$ is lost.

4.1.3.3 A self-coherent or coherent reception RPoF link

The RPoF link using both coherent reception (via an LO laser) and an in-line OBPF is shown in figure 4.1(c). Under this structure reception, RF phase information can be preserved when coherent reception is used. The coherent reception is essential when the RF carrier is modulated by a multilevel format with two-dimensional or complex modulation, such as QPSK or m-QAM, so that an increasing capacity of the data channel can be obtained.

Coherent reception can be either self-coherent or LO coherent beating of the modulated channel with the laser oscillation term, which can be derived from its own original optical carrier or from an LO laser. Hence a down-conversion of the detected signals can result from the frequency difference of the original optical carrier and that of the LO. This is the intermediate frequency (IF) and the data recovery is processed at this IF region. The beating RF signal is then filtered by the OBPF and, under a single PD detection or non-balanced reception, is given by

$$\frac{1}{\sqrt{2}}E_0(t)\Big[J_0\big\{\beta\rho(t)\big\} + J_1\big\{\beta\rho(t)e^{j[\omega_{\mathrm{RF}}t+\varphi(t)]}\big\}\Big]. \tag{4.7}$$

The beating term between the carrier and the first-order sideband contains the RF phase reference. Under a CS-RPoF self-coherent reception, as shown in figure 4.2, the carrier suppression (CS) maximizes the power in the modulated sidebands and minimizes the carrier contribution to detector shot noise, this improves the link gain, the NF, and SFDR compared to the case of non-filtered links [8].

If the filter it set to select only the first upper sideband (USB) then the filtered optical field is $E_2(t) = E_0(t)[J_1\{\beta\rho(t)e^{j[\omega_{\mathrm{RF}}t+\varphi(t)]}\}]$ by setting k = 1 and $\varphi = \pi$ in equation (4.4).

The signal after a coherent down-conversion and under a balanced reception of a PDP connected back-to-back is given by the beat term between the signal and the LO as

$$\begin{aligned} s_R(t)_{|\text{F-IM/DD}} &= \alpha 2\mathscr{R}\,\mathrm{Re}\,\{E_2(t)E_{\mathrm{LO}}^*\} \\ &= \alpha 2\mathscr{R}.\ \mathrm{Re}\,\{E_0(t)E_{\mathrm{LO}}J_1\{\beta\rho(t)\sin(\omega_{\mathrm{RF}}t + \varphi(t))\}\}, \end{aligned} \tag{4.8}$$

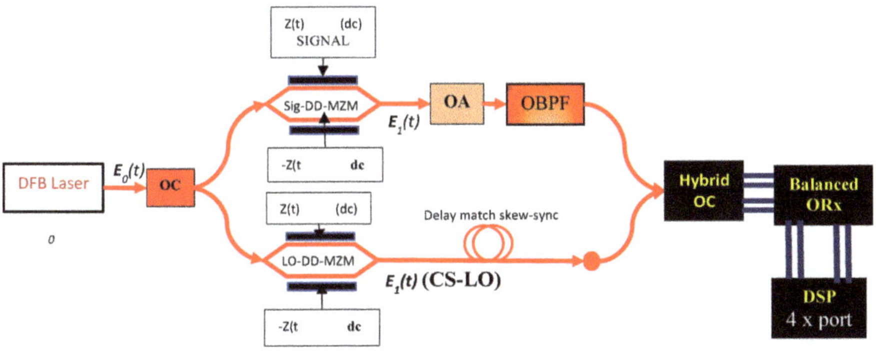

Figure 4.2. A filtered coherent link using a common laser for the signal and LO. This is also the experimental configuration. OBPF = optical bandpass filter; LO = local oscillator (optical); CS = carrier suppression; OC = optical coupler; PD = photodetector; Bal. ORx = balanced optical receiver (PDP connected back-to-back); DD-MZM = dual-drive Mach–Zehnder modulator; DFB = distributed feedback laser; dc = direct current of voltage applied at the dc level to generate a phase bias; bold = field vector; DSP = digital signal processor (on-line or off-line); Sig = signal. (Adapted with permission from [7].)

where $E_{LO} = \sqrt{2} \mid E_{LO} \mid e^{j[\omega_{LO}t]}$ is the electric field of the LO laser, is the difference in the centre frequencies of the signal and LO lasers and for balanced detection, otherwise. $\mathscr{R} = \frac{\eta q_e}{h\nu}$ is the PD responsivity, where η is the quantum efficiency of the PD, q_e is the electronic charge, $\alpha = 2$ for a PDP balanced gain.

4.1.4 NF and SNR of RPoF links

This section outlines briefly the link analyses of all configurations as given in section 4.1.3.

4.1.4.1 The NF of RPoF system configurations

The RF NF after reception for an RPoF IM/DD link can be written as

$$\mathrm{NF_{RF}} = 1 + \frac{N_{\mathrm{out}}}{G_p k_B T}, \tag{4.9}$$

where N_{out} is the total output RF noise power spectral density (in units of W Hz^{-1}), G_p is the link RF power gain/loss (dimensionless), k_B is the Boltzmann constant, T is the absolute temperature (in kelvins), and N_{out} is the sum of all noise power spectral densities referred to the output port of the receiver, given as

$$N_{\mathrm{out}} = N_{\mathrm{shot}} + N_{\mathrm{RIN}} + N_{\mathrm{OA}} + N_{\mathrm{TIA}}, \tag{4.10}$$

where N_{out} is the total output RF noise power spectral density, and N_{shot}, N_{RIN}, N_{OA}, and N_{TIA} are the quantum shot noise [9] due to the LO power producing a dc current at the output of the PD, the laser RIN, the OA signal–ASE beat noise, and the detector transimpedance amplifier (TIA) noise powers, respectively.

The TIA noise can be estimated by taking the total equivalent noise current referred to the input of the TIA multiplying by the transimpedance gain and the 50 Ω matching load resistance, respectively. Note that the signal–ASE beat term is larger than the ASE–ASE beat term and thus is taken to be dominated by the signal–ASE beating. These noise power spectral density terms of a bandwidth of 1.0 Hz are [18]

$$N_{\mathrm{shot}} = \alpha 2 q_e I_{\mathrm{pd}} R_L \text{ in units of W Hz}^{-1} \text{ to be integrated over the system BW}, \tag{4.11}$$

$$N_{\mathrm{RIN}} = \varepsilon \mathrm{RIN}\, I_{\mathrm{pd}}^2 R_L \text{ in units of W Hz}^{-1} \text{ to be integrated over the system BW}, \tag{4.12}$$

$$\begin{aligned} N_{\mathrm{OA}} = \alpha 2\mathscr{R}.\ n_{\mathrm{sp}} h\nu (G_{\mathrm{OA}} - 1) I_{\mathrm{pd}} R_L &\text{ in units of} \\ \text{W Hz}^{-1} \text{ to be integrated over the system BW}, \end{aligned} \tag{4.13}$$

$$N_{\mathrm{TIA}} = I_N^2 R_L = \int_0^{\mathrm{TIA\ 3dB\ BW}} \left[\frac{di_{\mathrm{eq},\,N}^2}{df} \right]^2 df \text{ in units of W or mW (dBm log scale)}, \tag{4.14}$$

where q_e is the electronic charge, R_L is the load impedance, RIN is the relative intensity noise of the laser, ε is the RIN suppression factor under a balanced detection, n_{sp} is the OA spontaneous emission photon number, h is the Planck constant, ν is the optical frequency, G_{OA} is the amplifier gain, $\alpha = 2$ for balanced detection, and I_{pd} is the dc photocurrent at a detector after absorption of the LO power and changes based on the link configuration of either IM/DD or filtered coherent reception.

The total noise spectral density term can be integrated over the appropriate bandwidth of the system and sub-systems, from all noise sources, to obtained a total equivalent noise current produced at the output of the receiver. Thus we can derive the optical power equivalent to this total noise and thus for a desired SNR we can estimate the required original optical power to overcome this noise. The link power budget can then be obtained as described in chapter 3.

Advanced low RIN lasers with high power and narrow linewidth are commercially available and are the most suitable for coherent reception. Thus the LO residual power quantum shot noise in a coherent RPoF dominates. For example, lasers with low RIN = −165 dBc Hz^{-1}, the quantum shot noise dominates over the RIN. The receiver TIA noise is included in the total noise current power at the output of the amplifier, determined by taking the noise current referred to the input $\frac{di^2_{eq,N}}{df}$ in units of $A^2\mathrm{Hz}^{-1}$ as given by the manufacturer and multiplying by the amplifier gain and then the load impedance, then integrating over the amplifier 3 dB bandwidth as given in equation (4.14); unlike other noise powers, it is independent of I_{pd}.

When an OA is used, it should be placed in front of the OBPF so as minimize the ASE noise. The noise power density for cascade OA units with gain and loss stages can be written as

$$N_{S\text{-}ASE} = \alpha 2Nh\nu n_{sp}R_L I_{pd}L_E[(G_1 - 1)G_2G_n.\ A_1A_2.\ ..A_n + (G_2 - 1)G_3.\ ..G_n.\ A_1A_2...A_n]. \quad (4.15)$$

G_n and A_n are the gain and input magnitude of the nth stage of the amplification and fiber, respectively, of the nth stage and is the optical loss following the last filter stage. The amplifier gain, $G_n = 1/A_n$ and $G_n \gg 1$; N is the number of amplifier stages:

$$N_{S\text{-}ASE} = \alpha 2\mathscr{R}h\nu n_{sp}R_L I_{pd}L_E[n]. \quad (4.16)$$

The NF of the various link configurations shown in figures 4.1(a)–(c) can also be estimated. The link RF power is the gain defined as the ratio of the output RF power to the input RF power (the RPoF is considered as a black box) and is required to formulate the RPoF transfer function.

4.1.4.1.1 Case 1: non-filtered IM/DD link without OA

As shown in figure 4.1(a), $I_{pd}\left(\frac{\mathrm{IM}}{\mathrm{DD}}\right) = \mathscr{R}\alpha P_0/2$ under small signal approximation. This link is denoted by a subscript 'IM/DD'.

$$G_{S\text{-}ASE} = \frac{1}{4}\mathscr{R}^2\beta^2(\alpha P_0)^2 R_L R_S. \quad (4.17)$$

Using equation (4.9) the NF for an IM/DD RoF link can be expressed as

$$\mathrm{NF}_{\mathrm{IM/DD}} = 1 + \frac{4q_e}{k_B T R_S \mathscr{R} P_{\mathrm{sig}}(\pi / V_\pi)^2} + \frac{N_{\mathrm{TIA}}}{k_B T G_p}, \tag{4.18}$$

where α is the link optical loss in the linear scale, E^2 is the laser average optical power, $| E_0 |^2 = \frac{P_0}{2}$, $P_{\mathrm{sig}} = \alpha P_0$, R_S is the electronic input impedance, and NF includes all noise contributions.

4.1.4.1.2 OA cascade RPoF

For a cascade chain of in-line OAs, the link is categorized as an optical amplification repeater link in which the typical SSMF-DCF-OA configuration [10] is normally used as a length of SSMF followed by a shortened length of dispersion compensating fiber (DCF). The DCF is to compensate for the dispersion of the SSMF length and the OA gain is to compensate for the loss of the DCF and SSMF so there is no loss in such span (figure 4.3). This is because the loss for the DCF is high due to the smaller core region. The coupling loss and the propagation loss of the DCF are a bit higher (~11 dB for 20 km to compensate for 100 km of SSMF at the 1550 nm region) and can be obtained as

$$I_{\mathrm{pd,\ F\text{-}IMDD}} = \frac{\mathscr{R}(\alpha P_0)}{2}\left(\frac{\beta}{2}\right)^2 \rho(t)^2, \tag{4.19}$$

$$G_{\mathrm{F\text{-}IM/DD}} = \left(\left| I_{\mathrm{pd}} \right|^2_{\mathrm{F\text{-}IM/DD}}\right)\frac{2R_S R_L}{\rho(t)^2} = \frac{\beta^4 \mathscr{R}^2(\alpha P_0)^2 R_L R_S}{32}\rho(t)^2. \tag{4.20}$$

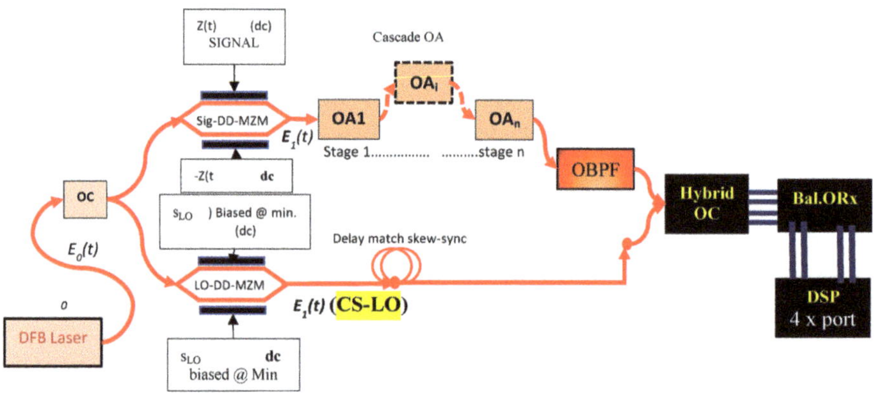

Figure 4.3. A filtered coherent RPoF with distributed optical amplification for a lossless link and multiple filtering stage. OBPF = optical bandpass filter; LO = local oscillator (optical); CS = carrier suppression; OC = optical coupler; PD = photodetector; Bal. ORx = balanced optical receiver (PDP connected back-to-back); DD-MZM = dual-drive Mach–Zehnder modulator; DFB = distributed feedback laser; dc = direct current of voltage applied at the dc level to generate a phase bias; bold = field vector; DSP = digital signal processor (on-line or off-line); sig = signal. (Adapted with permission from [11].)

Under small signal operation, and considering shot noise, OA noise, and detector noise, the NF for the filtered IM/DD RPoF link is

$$\mathrm{NF}_{\text{F-IM/DD}} = 1 + \frac{8q_e\{1 + n_{\text{sp}}[n]L_E\}}{k_B T R_S \mathscr{R} P_{\text{sig}}(\pi / V_\pi)^2} + \frac{N_{\text{TIA}}}{k_B T G_p}, \tag{4.21}$$

where N_{TIA} is the transimpedance noise, R_s is the load resistance, G_p is the gain of the TIA, P_{sig} is the signal average power, $P_{\text{sig}} = \alpha P_0$, and α is the total link loss and gain in the linear scale. As in traditional unfiltered links, increasing P_0 increases I_{pd} and G_p; $n_{\text{sp}} \sim 2$.

4.1.4.1.3 An OBPF coherent RPoF

For the optically filtered coherent link (denoted by the subscript F-Coh), shown in figures 4.1(c) and 4.2, using equation (4.8) we have

$$G_{\text{F-Coh}} = 4\alpha \mathscr{R} E_0 E_{\text{LO}} \left(\frac{\beta}{2}\right)^2 2R_L R_S = 8\beta^2 \mathscr{R}^2 P_{\text{sig}}(t) P_{\text{LO}} R_L R_S, \tag{4.22}$$

with $\alpha = 2$ for balanced detection. Balanced detection enhances not only the RF power gain but also reduces the laser RIN. The total NF for this coherent filtered RPoF is

$$\mathrm{NF}_{\text{F-Coh}} = 1 + \frac{q_e\{1 + n_{\text{sp}}[n]L_E\}}{2k_B T R_S \mathscr{R} P_{\text{sig}}(\pi / V_\pi)^2} + \frac{N_{\text{TIA}}}{k_B T G_p}. \tag{4.23}$$

4.1.4.2 SFDR of RPoF links

For a nonlinear RPoF system which operates under the excitation of two purely sinusoidal signals of frequencies f_1 and f_2, IMD components are generated and located at frequencies $(2f_1 - f_2)$ and $(2f_2 - f_1)$ which in turn create cross-talk into other existing signal channel bands. By biasing the modulator at the quadrature Q+ or Q– points and controlling the amplitudes (peak-to-peak) of the RF signals one can minimize the nonlinear IMD effects, hence ensuring the minimization of even-order IMD terms. Likewise, balanced detection can also suppress the third-order term, IMD3.

The SFDR is defined (given in chapter 2 and section 4.1.2) as

$$\mathrm{SFDR} = \left[\frac{\mathrm{OIP3}}{N_{\text{out}}}\right]^{2/3}, \tag{4.24a}$$

where OIP3 is the output power at the third intercept point and is obtained by linear extrapolations of the fundamental and IMD3 responses. Also,

$$\mathrm{OIP3} = \mathrm{IIP3} * G_p, \tag{4.24b}$$

where IIP3 is the input power corresponding to the OIP3.

The two-tone tests with an appropriate frequency spacing in the filtered links results in the fundamental and IMD3 given by

$$\text{fundamental} = J_0\{\beta\rho(t)\}\,.\,J_1\{\beta\rho(t)\} \quad \text{and} \quad \text{IMD3} = J_0\{\beta\rho(t)\}\,.\,J_2\{\beta\rho(t)\}\,. \tag{4.24c}$$

Equating these two terms and keeping only the first-order terms gives

$$\text{IIP3} = \frac{8}{2\beta^2 R_S}. \tag{4.24d}$$

The SFDRs of all three RPoF links given in figures 4.1(a)–(c) can then be estimated.

4.1.4.2.1 The SFDR of a non-filtered IM/DD link

Referring to figure 4.1(a) the SFDR is given by

$$\text{SFDR}_{\text{IM/DD}} = \left[\frac{\mathcal{R}P_{\text{sig}}}{q_e + \dfrac{N_{\text{TIA}}}{\mathcal{R}(\alpha P_0)R_L}}\right]^{2/3}. \tag{4.25a}$$

4.1.4.2.2 The SFDR of an OBPF filtered IM/DD link

Referring to figure 4.1(b), the SFDR is given by

$$\text{SFDR}_{\text{F-IM/DD}} = \left[\frac{\mathcal{R}P_{\text{sig}}}{2q_e\left\{1 + n_{\text{sp}}[n]L_E\right\} + \dfrac{N_{\text{TIA}}\beta^2 R_S \mathcal{R}P_{\text{sig}}}{\mathcal{R}(\alpha P_0)R_L}}\right]^{2/3}. \tag{4.25b}$$

4.1.4.2.3 SFDR of F-Coh filtered coherent link

Referring to figure 4.1(c), the SFDR is given by

$$\text{OIP3}_{\text{F-Coh}} = 32\mathcal{R}^2 P_{\text{LO}} P_{\text{sig}} R_L, \tag{4.26}$$

$$\text{SFDR}_{\text{F-Coh}} = \left[\frac{8\mathcal{R}P_{\text{sig}}}{q_e(1 + n_{\text{sp}}[n]L_F) + \dfrac{N_{\text{TIA}}}{4\mathcal{R}P_{\text{LO}}R_L}}\right]^{2/3}. \tag{4.27}$$

It is observed that a linear relationship of the terms or the NF and SFDR vary in a nonlinear exponential relationship and are inversely proportional to each other. Thus there is an optimum operating condition for the reduction of the NF and the increase of the spur-free range.

4.2 Optical filters for RPoF link pre-processing

As presented above, optical filtering can bring the processing of the signal bands to either the baseband or an IF region. Thus the role of OBPF is critical, in particular its passband and roll-off characteristics. In an optical domain OBPF, similarly to the case of electrical filters, the design can follow an all-zero network (all pass) pole-zero network (resonance and pass). In the photonic domain the pole of a circuit type is equivalent to the resonance phenomenon (e.g. micro-ring resonator) at which the optical circuit produces a very high amplitude optical field at its output. On the other hand, a zero of an optical circuit is equivalent to an interference phenomenon such as the MZDI or destructive interference. Thus we can have two types of OBPF [12]: type 1 is the cascaded MZDI and (ii) type 2 is the MRR.

4.2.1 An integrated optic OBPF using cascading MZDIs

In-line OBPF can be implemented and fabricated using an MZDI. An integrated optic narrow-passband OBPF based on cascaded Mach–Zehnder interferometers (MZIs) can be fabricated using a Si-on-insulator (SOI) platform, as shown in figure 4.4(a). Its transfer matrix is given in figure 4.4(b) by cascading a number of stages of an MZDI. Such an OBPF can be implemented in an RPoF link to significantly improve the SFDR.

The roll-off of the passband of a number of pass filters can be sharpened by cascading the MZDI stages. This improves the SFDR of the RPoF. A filter using a three-stage cascaded MZI structure can be realized with a 3 dB bandwidth of about 1536 MHz and an FSR of about 13.5 GHz. The performance of a down-converting RPoF employing such an OBOF can offer an SFDR of 104.1 dB $Hz^{2/3}$. It is noted that the MZDI is a passive all pass type photonic circuit; thus it is quite stable and no oscillation will occur due to the fact that the zeros of the transmittance transfer function of the circuit cannot create any oscillation, unlike a pole-zero circuit such as the MRRs described in section 4.2.

4.2.2 An integrated optic MRR

An OBPF can be realized using basic or advanced MRR structures which consist of a ring of optical waveguides, and coupling of lightwaves into and out of such a ring

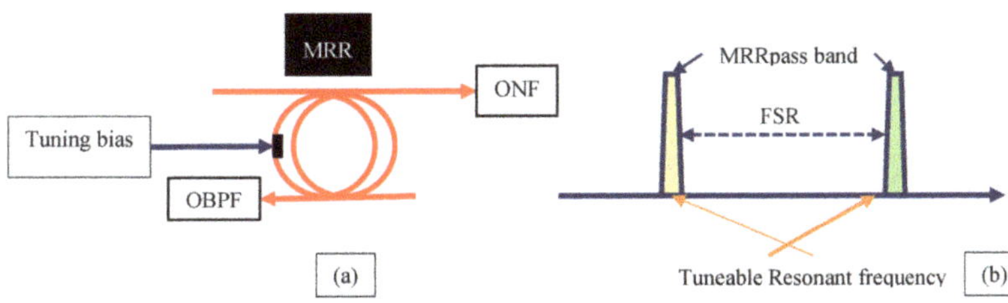

Figure 4.4. (a) Schematic of an *N*-stage cascaded MZDI tuneable filter, in which each stage of the filter can be changed independently. (b) Optical transmittance transfer function between input ports 1 and 2 and output ports 1 and 2. (b) The transmittance transfer matrix representing the transfer function of the cascade MZDI filter.

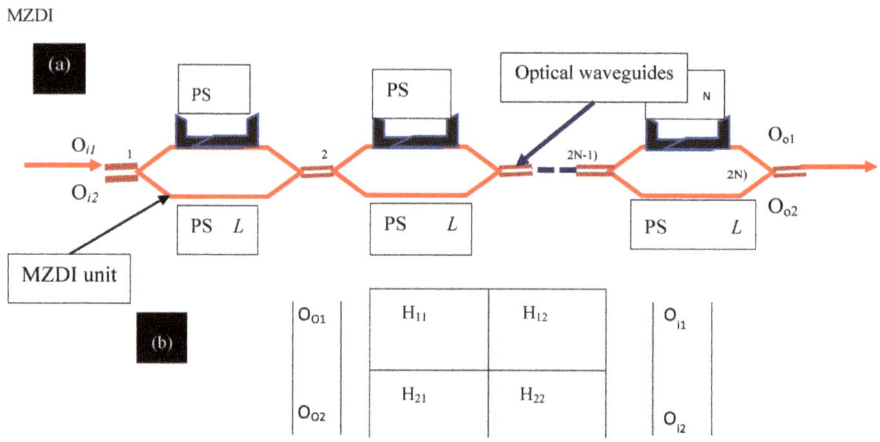

Figure 4.5. Device measurement. (a) An OBPF of an MRR double ring structure. (b) Transfer matrix response—resonant band.

by a directional coupler. A more advanced MRR can be a cascade or rings, etc. The optical passband is critical and can be designed so that it can accommodate the data band.

A typical MRR structure is shown in figure 4.4(a), an MRR using a double ring MRR so as to extend the length of the resonance ring. The ring radius is 125 μm, the waveguide ridge width can be 1.2 μm with 2 μm gaps. It is noted that the refractive index of Si_3N_4 is about 1.73 and for silica it is 1.484 at 1550 nm. This high difference in refractive index leads to a high energy concentration in the waveguide core. The coupling ratio of the coupler is about 0.01 (10%). The Q-factor of the MRR filter is about 1e6 when tuned to resonance. The ring propagation loss is 0.1 dB cm^{-1}.

Figure 4.4(b) shows the passband and the tunability of the MRR, with an electrode deposited on top of a section of the ring. Pyroelectric effects occur via the thermal heating of the electrode. Thus the micro-ring length can be changed minutely and thus also the resonance frequency. The lightwaves can be coupled in and out via the coupler. The two outputs can be an OBPF and notched filter ONF as shown in figure 4.5(a).

The cascade structure of a number of MZDIs is shown in figure 4.5(a) to form an all-pass optical transfer device, that is, the z-transform transfer matrix of figure 4.5(b) is consisting of only the zeros in the numerator of the transmittance function.

4.3 Optical filtering in self-coherent RPoF links

4.3.1 Remarks

With the explosive growth in 5G, 6G, and 7G wireless [13] and radar communications [14], the RF operating frequency is approaching several tens of GHz to the μW and mmW regions or even hundreds of GHz into the lower x-ray region. The data channel baseband can be extended to 1.0–2.0 GHz per channel which is superimposed on the RF carriers.

Ultra-high frequency receivers for RPoFs are being developed and deployed quickly for adoption in dynamic carrier interchangeable frequencies in 5G, 6G, and 7G global wireless Internet information networks. In planning a manufacturing product testing environment, RPoF links should be configured with reconfigurable and tunable features so as to can significantly simplify multi-frequency region testing. Furthermore, such a multiple spectral region switchable RPoF test platform can reduce the network operational expenditure (OPEX) if a centralized control and test platform can be universally conducting, monitoring, and testing massive remote wireless micro-cells operating at different spectral bands from the μmW to mmW. However, the frequency range of electronic devices and systems that can be flexibly processed is limited to only some tens of GHz [15] and noise effects [16].

In addition, in radio astronomy telescopes, the requirements for receiving and processing bands demand that receivers cover the L-band to Ku-band [17]. For these high frequency multi-band (HFMB) applications, spectrally tunable broadband receivers are essential. This capability seems difficult for the use of DSP-based reception systems in which an ultra-fast sampling rate must satisfy the Nyquist amplifying theorem to recover the RF carrier as commonly employed in broadband optical receivers of digital optical communication systems [18].

Thus for UHMB signals it would be advantageous that the processing signals can be down-converted to an intermediate spectral region or even to the baseband in which a high effective number of bits (ENOB) ADC and DSP can be used for a highly sensitive receiver.

The photonic domain processing of RF photonic-assisted receivers/front-ends could assist in overcoming such UHMB receiver-bandwidth limitations, that would offer the potential for full-band (from the L-band to the W-band) μW and mmW photonic front-end receiver [19, 20]. Furthermore, a photonic pre-processor offers the advantages of low/loss, wide bandwidth, tunability, and high immunity to RF radiation.

A schematic diagram of a full-band photonics pre-processing microwave reception system is proposed, as shown in figure 4.6. Broadband signals received by the

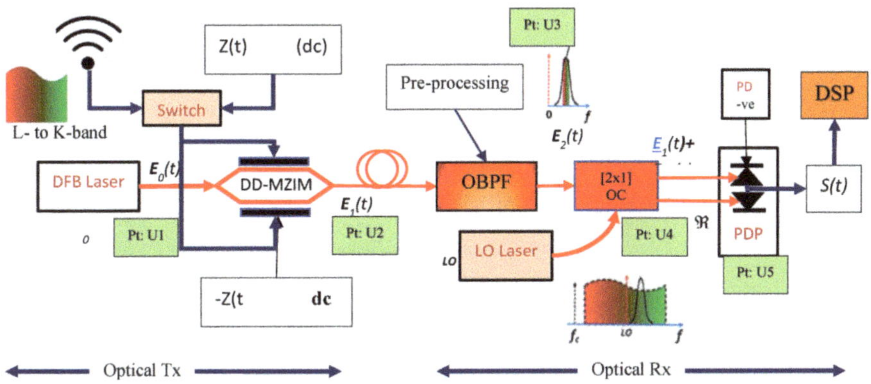

Figure 4.6. Schematic diagram of the RF photonic transceiver employing coherent reception. The transceiver contains two parts, the transmitter and receiver with the full-band signal. Tx = transmitter; Rx = receiver.

antenna are up-converted to the optical domain by the EO modulator, and the desired information band is extracted by the high performance (UHMB) radio-photonic filter (RPF). The operational principles of the receiver are as follows: (i) the incoming signals are filtered by an OBPF, an MRR, or cascade MZDIs; (ii) they are mixed with an LO laser; (iii) the summing field of these fields is beating in a photodetector (PD) so that the signal products of a dc term and the signal term product of the signal amplitude and LO amplitude and frequency difference of the two optical carriers, this brings the signal bands to be processed to an intermediate region whose frequency can be tuned by tuning the LO laser frequency; and (iv) a DSP can process the phase difference between the lasers and signal carrier without any difficulty [21].

The signal is then mixed with the LO in the optical domain to down-convert to an IF for sampling by the analog-digital converter (ADC) and is post-processed by the DSP. This is proposed in the next section.

4.3.2 OBPF for a down-conversion RPoF link

A self-coherent down-converting RPoF system is shown in section 4.1.3.3 with an OBPF of multi-stage MZDIs filter, as shown in figure 4.4. A continuous wave (CW) laser is launched to an MZM biased at the null point or at V_π so as to suppress the carrier and is then modulated by $f_{\mathrm{RF,LO}}$. The two sideband carriers of the output of this CS-MZIM is then separated via a frequency interleaver. Thus, two separate two optical carriers are generated with frequency doubling due to the nonlinear switching by $2f_{\mathrm{RF,LO}}$. These two sideband carriers are split via an optical interleaver, as shown in figure 4.7. The optical interleaver is also an optical MZDI with the delay line matched to the frequency spacing of the two channels, e.g. 50 GHz. The system link is shown in figure 4.8, which is an alternative set-up of the RPoF link of figure 4.9.

The upper sideband $\omega_o + \omega_{\mathrm{RF,LO}}$ is used as an optical LO for down-conversion and the lower sideband $\omega_o - \omega_{\mathrm{RF,LO}}$ serves as the optical carrier and is phase-modulated by the input broadband RF signal [22]. The optical signal path is then fed through the cascade MZDI OBPF, the narrow-passband SOI filter. The narrow-

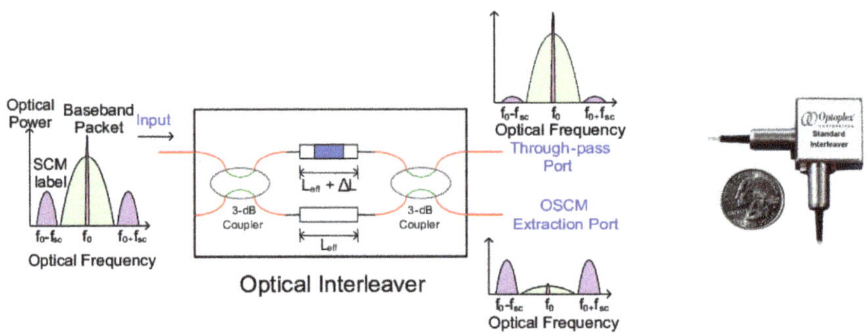

Figure 4.7. Experiment set-up of the down-conversion RPoF employing self-coherent reception with an OPM to tune the frequency difference or IF spectral range.

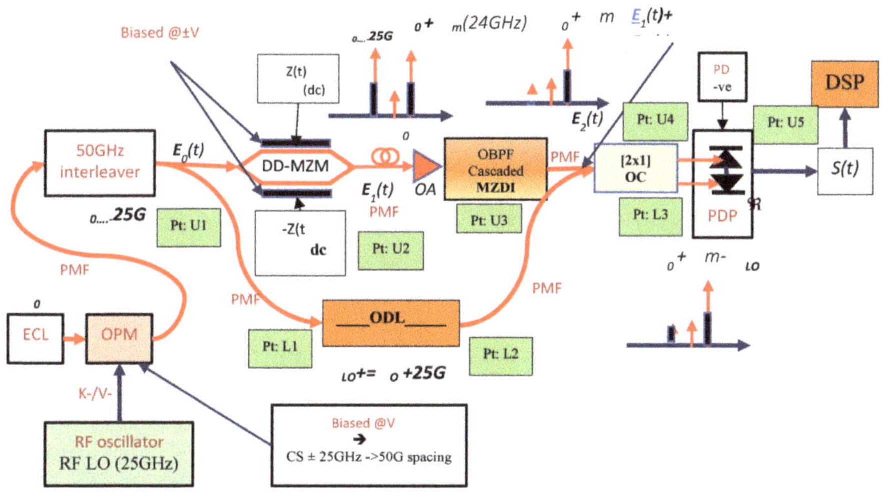

Figure 4.8. Alternative set-up of an RPoF by an interleaver and modulator before splitting. ODL = optical delay line used for synchronization of the modulated path and the LO optical path.

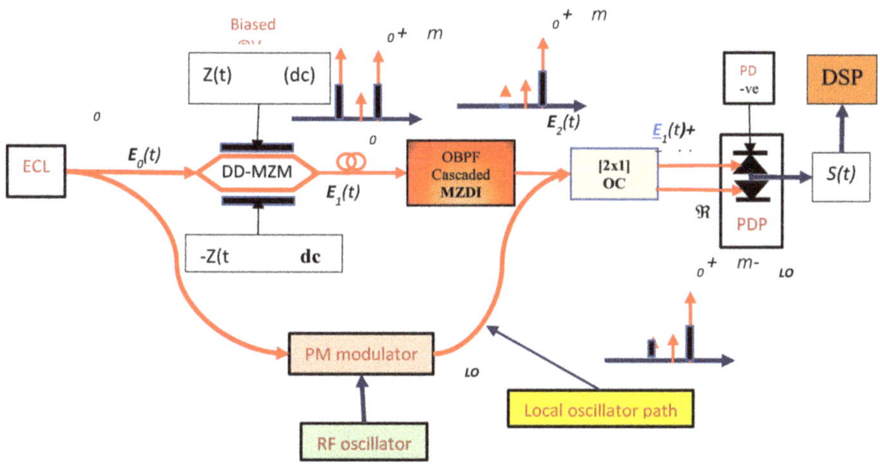

Figure 4.9. An optical interleaver by Optoplex using an MZDI, with a delay optical path in two arms of an interferometer.

passband and large FSR of the filter can filter out the signal band of interest over a wide RF frequency band from the X-band to the Ka-band. To maintain phase correlation of the upper and lower paths, their path lengths and polarization states are carefully matched using an optical delay line and PCs. The filtered signal and the optical LO are then combined in an optical coupler and detected by a balanced photodetector (BPD) or PDP connected back-to-back so that a 3 dB gain can be obtained and also high noise rejection of the common mode operation.

The parameters of the system are as follows. (A) For the transmitter section: (i) CW laser power = 12.8 dBm, with a center wavelength of 1550 nm and a 1.0 kHz linewidth; (ii) tunable LO power = 16 dBm; (iii) MZM bandwidth = 20 GHz and phase modulator (PM) = 40 GHz; V_{π} ~ 5 and 7 V for the MZIM and PM, respectively. (B) In-line amplifier and transmission line: (i) OA gain inserted before the MZM modulator and before the splitter, insertion loss ~ 9 dB, gain of the optical amplifier (if desired) inserted in front of the OBPF ~ 22 dB to compensate the complete link loss. (C) General system parameters: (i) channel spacing 25 GHz, similar for the interleaver spectral domain (see figure 4.9); (ii) PDP responsivity of ~0.62 A W^{-1}, BW = 45 GHz; and (iii) the electrical spectrum is measured by an FSW R&S electrical spectrum analyzer (ESA).

An RPoF operating RF region is in the K-band. An RF (LO) oscillator centered at 12 GHz drives the null-biased CS-MZM to obtain the carrier suppression of ~ 30 dB. For SFDR and IMD characterization a two-tone signal spaced by 1.0 MHz, with frequencies of 24.0085 and 24.0095 GHz, is excited to the optical phase modulator (OPM). By tuning the micro-heaters on the phase-shifter, the desired response can be achieved and by heating the chip one can tune the resonance frequency and select the first modulated band, and then the filtered signal is coupled with the upper sideband optical LO.

Figure 4.10(a) shows both the optical carrier and even-order optical sidebands. Carrier suppression occurs due to biasing at the minimum transmission point and RF driving signals in the complementary, in addition to the optical filtering effect of the OBPF, a carrier suppression higher than 30 dB can be achieved. Signals are detected by a balanced PDP and the beating of the LO and the signal results in down-conversion to 8.5 and 9.5 MHz as well as the third-order IMD (IMD3) at 7.5 and 10.5 MHz. The nonlinear IMD products shown in figure 4.10(b), of the transfer response of the RPoF link, are due to the phase modulation and the photodetection. An RPoF link SFDR of 104 dB $Hz^{2/3}$ is obtained with a phase noise floor of ~ −128 dBm Hz^{-1}, which is higher than a normal basic phase noise floor and is due to the ASE noise induced by the OAs.

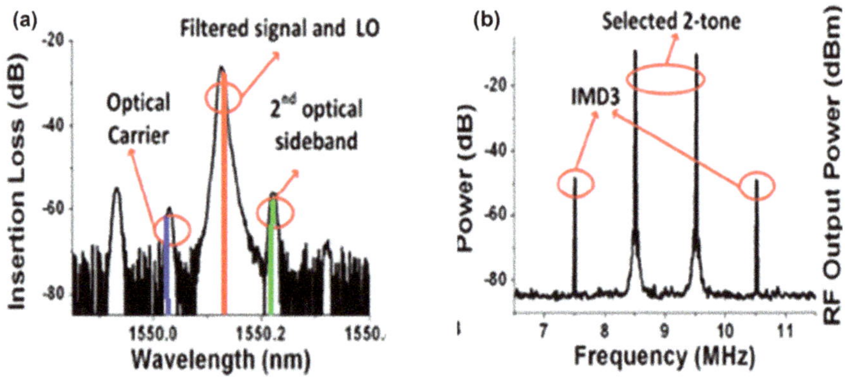

Figure 4.10. Experimental results of the APL application. (a) The optical LO coupling with the signal filtered out by the proposed filter. (b) The electrical spectrum of a down-converted IF signal, with a measured SFDR or 104 dB $Hz^{2/3}$.

In addition to the OBPF using a cascade MZDI photonic circuit as a PSP for rejecting unwanted sidebands, as described in section 4.2.1 [23], Yu *et al* [23] further demonstrated, instead of the all-zero photonic networks given in section 4.2.1, using a resonant photonic circuit, the micro-ring resonator (MRR) which is a zero-pole pattern photonic circuit. The MRR would give an ultra-sharp passband optical filtering property. It can be realized by integrated photonic technology based on silicon nitrate (Si_3N_4) together with its potential for tuning by thermo-optic effects. Such an MRR-OBPF can be used in a coherent RPoF link to significantly improve the SFDR of the link. Furthermore, the free spectral range (FSR) of >200 GHz allows extending the operation of the RF signal band and carrier from the L-band to the Ku-band.

Ultra-broadband RF receivers and those for 5G–7G wireless networks are highly important for higher frequency band wireless communications, radar communications, and multi-band applications in radio astronomy telescopes. Such ultra-broadband receivers are inherently difficult to establish with advanced electronics due to limitations in the linewidth resolution in microelectronic integrated circuits, unless distributed amplification circuitry is employed, but with significant power consumption. On the other hand, photonics technology has advanced notably over the last few decades, therefore optical receivers/front-ends can overcome the bandwidth limitations in electrical receivers via the intermediate photonic domain. It is noted that the higher the frequency is the better the photonic processing can be. The opposite would be true for electronics.

Thus, it is believed that photonic domain processing can be mixed with electronic digital and analog processing and should be employed wherever possible to take advantage of either domain. In the processing of RF signals in a full-band (from the L-band to the W-band) it is possible to employ the coherent reception with the beating of the mixed local oscillator laser and that of the signal RF shifted optical carriers to obtain down-conversion to an appropriate frequency (IF) band in which the high resolution and effective number of bits (ENOB) of the analog-to-digital converter (ADC) and digital to analog converter (DAC), in particular the sampler, can be significantly reduced and DSP can be used. Hence a photonic-assisted optical RF receiver can offer such effective and efficient processing. Furthermore, optical filtering can be used based on the Si_3N_4 MRR filter so that noise can be minimized and hence reduction of the noise loop level and increasing SFDR can be achieved.

In the last section 4.2.1 the lattice all-zero photonic filter was described, and in this section a lattice filter employing an MRR or an all pole-zero lattice photonic circuit is presented, whose transfer function exhibits a high Q-factor to extract and filter the desired signal band, thus significantly reducing the optical shot noise in the receiver and at the same time eliminating the interference due to intermodulation RF waves. However, the IMD3 exists due to the close RF carriers of the closely packed data channels which are superimposed on the RF carriers. This section thus considers the impacts of these intermodulation distortions in such filtered coherent RPoF links.

The resolution and processing range of the filter are lower than 420 MHz and larger than 112.9 GHz (MRR-FSR > 225.78 GHz). The roll-off and out-of-band rejection of the MRR-OBPF is > 40 dB. The center frequency of the filter can be tuned wider than one FSR by tuning the MRR phase-shifter section. The performance of the full-band all-optical RF receiver is described, and the SFDR of the receiver from the L-band to the Ka-band (limited by the bandwidth of the modulator in our experiment) is larger than 111.6 dB $Hz^{2/3}$.

A schematic diagram of the spectral tuning pre-processing OBPF photonic receiver is revisited and depicted in figure 4.6. An MRR is used as the OBPF. An ECL operating in continuous mode with a 10 kHz linewidth[1], centered in the silica fiber low-loss spectral window at 1550 nm, is used as the lightwave carrier.

The lightwaves are modulated via a DD-MZIM fed by a pair of RF signals which are $\pi/2$ shifted with respect to each other. This generates a single sideband (SSB), the suppression of either the upper sideband (USB) or lower sideband (LSB) depends on the positive phase shift of one to the other.

The suppression of the lightwaves can be done by biasing the electrodes of the two branches of the DD-MZIM so that a π phase difference between them can be created. This RF can be changed to tune the LO frequency.

The modulated signal would be amplified by an optical amplifier to compensate for the optical loss of the system, then passing the OBPF as described in this section, for signal extraction ands fed to a coherent balanced opto-electronic receiver. The field of the optical signal extracted by the OBPF is combined with that of the optical LO in the USB in a polarized 3 dB coupler and detected by the balanced photodetector (BPD). This mixing products down-converts the optical signal to an intermediate spectral region in electrical domain. The down-converted signal is monitored in an ESA or a digital sampling signal analyzer (DSA). In this system the RF LO is an CS-LSB or CS-USB carrier. This SSB is used to maximize the power in the modulated sidebands and to minimize the carrier detector shot noise. However, a setback of this technique is that the sideband power can be about 10–20 dB below that of the unsuppressed optical carrier. The method given here is most suitable for very narrowband RF carrier applications. With the increasing passband of 5G, 6G, and 7G systems, a tunable passband optical filter is preferred.

Due to the high performance of the signal processing in the optical domain, the desired signal would be extracted clearly without any other interference sidebands, which would reduce the nonlinear sideband power and increase the SFDR. The BPD employed in this system is used to provide additional gain and suppress the common mode noise.

RPoF system parameters can be listed as shown in table 4.2.

4.3.3 Filtered self-coherent reception RPoF link

The optical receiver of the RPoF link can be of self-coherent type. That is, the LO optical beam can be derived from the laser source of the optical transmitter, as shown in figure 4.11(b). The ECL output is split by an optical 3 dB coupler. One output is through the modulation DD-MZIM and the other via an optical phase modulator in which the optical carrier can be shifted by an RF oscillation input.

[1] Modern lasers such as the Pure Photonics 550PCL model, Encore Lasers, Toptica, can have narrow linewidths and are tunable.

Table 4.2. MRR-RPoF link typical parameters.

The lightwave ECL source	λ at *1*549.407 nm and optical power P_L = 16.0 dBm	Linewidth = 10 KHz
MZIM modulator	BW = 40 GHz, 3 dB point	Insertion loss ~ 4.5 dB
OA optical amplifier Saturated power at the optical amplifier output = 17 dBm, or loss/gain = 16.7 dBm – 17 dBm = 0.3 dB gain → compensating the OBPF MRR waveguide insertion loss (I_L) = ~19.7 dB	Gain =20 dB	Higher power OS can be used and available to 23 dBm saturated power

This optical path acts as an LO beam to the OBPF pre-processor coherent receiver, similar to that of figure 4.6.

The split optical path can be modulated by an RF shifted optical modulator such as that shown in figure 4.11. The self-coherent receiver with an OBPF can covert the signals to the baseband or an intermediate frequency band which can cover the frequency bands of the 5G/6G or 7G RF-carrier. There is no frequency shift of the LO with respect to the signal carrier as they are derived from the same source. An OBPF passband of ~430 MHz and an FSR of > 200 GHz is obtained with a sharp roll-off. This allows multiplexing of many RF subcarriers in RPoF channels, for example, with a 225 GHz FSR four optical carriers can be used with RF shifted carriers in which a number of RF subcarriers can be multiplexed, in particular when OFDM data channels are used.

Two-tone test signals are used to measure the intermodulation products of the RPoF link in the C- to K-band. The signals filtered by the OBPF (MRR type) are detected by a balanced photodetector pair (BPD) and down-converted to 6.5 and 8.5 MHz together with the third-order intermodulation distortion (IMD3) at 4.5 and 10.5 MHz. The IMD products are induced by the system's nonlinear transfer response, including nonlinearities of all the sub-system components. The SFDRs of the receiver in the order of 114–116 dB $Hz^{2/3}$ can be obtained for the RF spectral regions of the C- to K-bands under the set-up conditions for the ESA of a resolution bandwidth (RBW) of 1 kHz and a view bandwidth (VBW) of 1 kHz at 30 dBm optical input power. The synchronization of the system is shown in figure 4.11(a).

The phase noise floor is ~ –151 dBm Hz^{-1}. This level is mainly contributed by the optical amplifier ASE noise and LO coherent residual power quantum shot noise. The TIA electronic noise is substantially smaller under such a narrow passband of an RF modulated data channel after filtering. The SFDR of the C-band (signal at 4.8065 and 4.8085 GHz, IF at 6.5 and 8.5 MHz or higher at 1.3–1.7 GHz) for 6G and 7G systems and the X-band (signal at 10.0065 and 10.0085 GHz), with the IF frequency set at 6.5 and 8.5 MHz for the case of an RF carrier only and no data, can be achieved by tuning the RF signal applied to the optical phase modulator (OPM) for self-coherent reception.

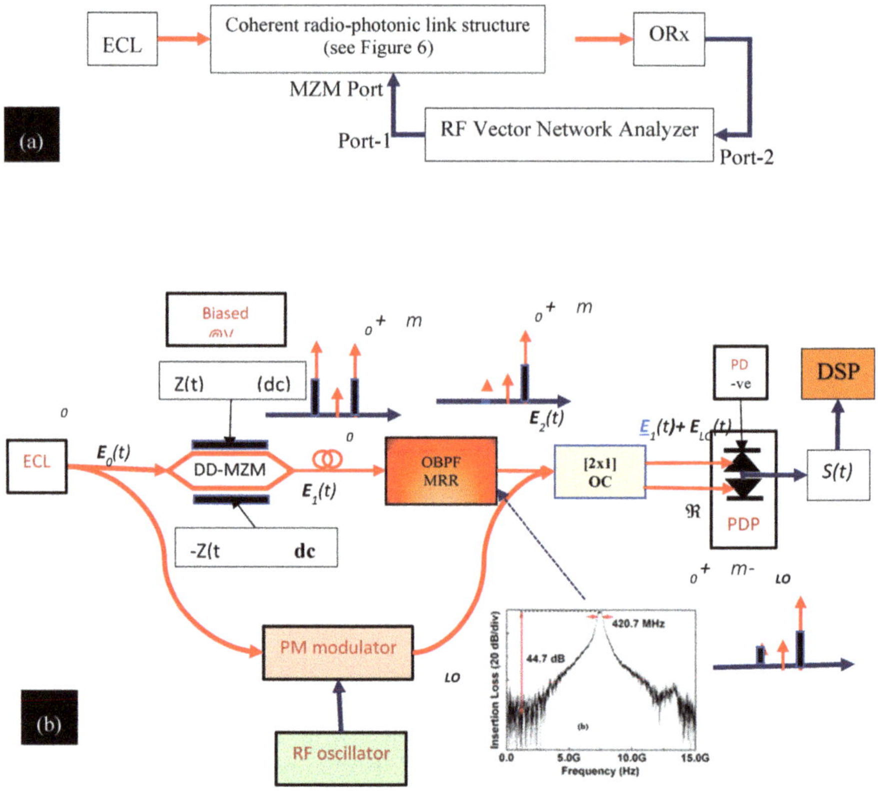

Figure 4.11. Radio-photonic link measurement with a vector network analyzer. (a) Schematic diagram of the measuring system for measuring the transfer response of the filter, as the S_{21} parameter by network analyzer sweeping. (b) RPAL with MRR in the signal path to filter one sideband with an MRR sharp passband and tunability. IF = $\omega_0 + \omega m - \omega_{LO}$. The inset shows the optical passband. Copyright (2016) IEEE. Adapted with permission from [23].

In the Ku-band the conditions are a signal at 13.5065 and 13.5085 GHz, IF at 6.5 and 8.5 MHz, and for the K-band the conditions are a signal at 19.0065 and 19.0085 GHz, a down-converted IF at 6.5 and 8.5 MHz for RF carrier only and no data channels, and the SFDR parameters can be obtained as 118.6 dB Hz$^{2/3}$ (for 13.5065 GHz), 116 dB H$^{2/3}$ (for 13.5085 GHz), 114.8 dB Hz$^{2/3}$ (for 19.0065 GHz), 119.1 dB Hz$^{2/3}$ (for 19.0085 GHz), respectively, for the Ku- and K-bands.

The SFDRs of the system with a two-tone test signal of frequency from 1.5 to 39.5 GHz with a spacing of 2 GHz are used. In such a frequency range the largest range that could be realized with SFDR ranges from 111.6 dB Hz$^{2/3}$ to 118.6 dB Hz$^{2/3}$. That is only a 2 dB fluctuation in the entire spectral region.

The RF operating regions for the OBPF and self-coherent reception system can be tuned to several nanometers in the wavelength of the optical domain. This is enabled by tuning the center of the passband by thermal effects of the MRR. The tunability of such an OBPF in the RPoF can be tuned across 40 GHz.

4.3.4 Remarks

In this section an RPoF system based on an MRR-OBPF signal pre-processing receiver is given with its advantages and disadvantages in either direct detection or coherent reception with an OBPF of resonant type, the MRR. The tunability of the MRR over a wide optical window allows potential multiplexing of several radio frequency components on the same optical carrier.

The only limitation of such photonic pre-processing coherent receivers is the high resolution signal processor with a large signal processing range and nonlinear intermodulation products which can create cross-talk to other multiplexed channels.

4.4 Linearization of the RPoF link

4.4.1 Linearization by modulator linearity

As illustrated and described in section 4.2, modulators with high linearity and a broadband are critical components for RPoF analog communication links or ultra-broadband access 5G and 6G networks. Previous works have shown that photonic signal processors can assist in the processing of signals in the photonic domain and analog circuits [24]. In section 4.2 these PSP techniques were employed to design an OBPF for an RPoF.

Furthermore, an ultra-linear optical modulator composed of a phase modulator and ring resonator in a Mach–Zehnder (MZ) configuration exhibits an SFDR value of 130 dB $Hz^{2/3}$. In this section, we introduce a new configuration of MZM and MRR on either branch of the interferometer and a phase modulation (PM) section to offer high linearity in the transfer transmittance function, that is, optical power in and optical power out. It is noted that the electrode of the PM is a lumped type, thus the modulation frequency is limited to the low frequency < 10 GHz region. This structure offers under simulation a superior feature SFDR of ~ 133 dB $Hz^{2/3}$, modulation bandwidth extension, and high link gain.

In RPoF transmission technology, external modulators are preferred over direct modulators so as to minimize the relative intensity noise (RIN) of the directly modulated lasers due to the widening of the laser linewidth. Extensive works on linear modulator designs have been published over the years. They have been extensively exploited in ultra-high capacity digital optical transmission [25].

Under RPoF, the key performance measures of these external modulators include the linearity or SFDR, RF bandwidth capability, low half-wave (V_π) voltage in the dc and operating spectral window, and reasonable low insertion loss. The SFDR seems to be the least understood dynamic parameter. An ultra-linear optical modulator with SFDR capability in the range of 130–140 dB can be possible. The design was based on a modified version of a Michelson Gires-Tournois Interferometer by Dingel *et al* [26]. An SFDR > 130 dB (with a 1 MHz resolution bandwidth electrical spectrum analyser) could be obtained. Such a low noise and high SFDR modulator can find many applications in RPoF link design and systems applications. Furthermore, another key advantage is the high bandwidth capability (>10 GHz) and large link RF-in to RF-out gain.

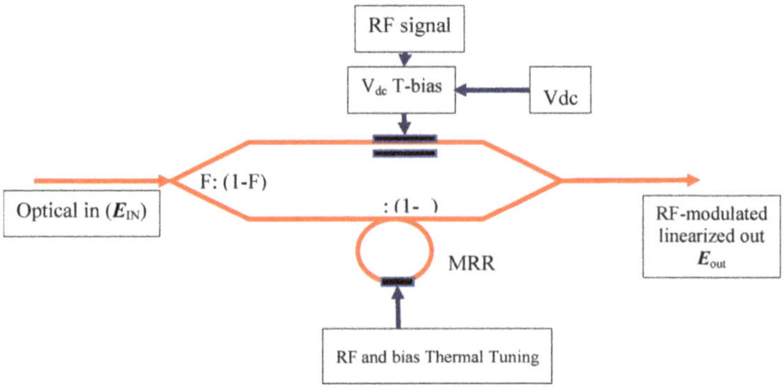

Figure 4.12. Schematic of a high linear optical modulator, MRR-MZM, in an integrated optic structure. MRR = micro-ring resonator with a cross coupling coefficient κ.

Furthermore, in chapter 6 there are a number of simulation platforms for the linearization strategies described in this section.

4.4.1.1 Linearization of MRR-MZIM transmittance

Figure 4.12 shows the asymmetric MZM structure consisting of an MZM incorporated with a micro-ring resonator (MRR-MZM) as adopted from the structure reported in [30], that offers much higher nonlinearity operation for RF signals. An incident electric field (E_{in}) is split via a 3 dB coupler. One of the beams propagates in the upper arm of length L_1, with a dc bias control. The other arm (L_2) of the Mach–Zehnder interferometer (MZI) configuration contains an MRR and an optical phase modulator (OPM) in cascade. The external RF input is coupled to the modulator electrodes of the PM and MRR via an RF splitter or directional coupler with a power split ratio κ:$(1 - \kappa)$, where κ is the intensity coupling/splitting ratio, as shown in figure 4.12. Here, the relative phases of the RF inputs into the phase modulator are shown.

The SFDR is evaluated in units of dB $Hz^{2/3}$ [27, 28, 29] or in dB, which is to be converted from the measured value in dB over a bandwidth set in the spectrum analyzer, typically about 1.0–3.0 MHz. Thus the logarithmic value of base 10 will add a factor of abut 60 dB in dB $Hz^{2/3}$ in the SFDR which takes into account the bandwidth of the spectrum analyzer under measurement by taking the $\log_{10}$ of such a bandwidth and the factor 2/3 of the slope of the second and third harmonics. A graphical approach can be used to evaluate the SFDR for MRR-MZIM as shown in [26], it is calculated using the transmittance in the logarithmic scale. It can be shown that the estimated SFDRs of the MRR-MZM and MZIM with $(F, \tau) = (0.009, 0.195)$, contrasted with respect to that of a standard MZM at relatively low normalized modulation frequencies (normalized to the FSR of the MRR). For these values of F and τ, the MRR-MZIM provides a much improved SFDR (> 132.6 dB) or > 24 dB higher than MZIM. Furthermore, the link gain of the MRR-MZIM performs reasonably well at relatively low frequencies. It is note that the splitting ratio of < 1%

makes it very difficult in the manufacturing of integrated optical circuits, not to mention the phase effect in the splitting section.

Similar plots for a wide range of values of F and τ can be achieved easily, resulting in a two-dimensional contour map of SFDR values [30]. Further analysis of this contour map can help to identify regions of high SFDR values. Thus the ability to have control over both F and τ results in greater design flexibility. Specifically, τ is typically fairly fixed after completion of the manufacturing process, thus having the extra degree of freedom in F allows for better performance.

This linearization technique of an MZIM with integration of an MRR and phase modulation sections can improve the SFDR from 111 dB $H^{2/3}$ to 132 dB $H^{2/3}$ or 22 dB gain in the linearity range [30].

4.4.1.2 Remarks on the MRR-MZIM structure

The operation of the MRR-MZM works with the following principles:

1. The MRR acts as a ring resonator with a passband of about 500–1000 MHz, enough to pass any a single broadband data channel embedded on the RF carrier.
2. The MRR provides an additional dynamic delay line or phase extension under an applied RF signal, then the optical phase which allows the extension of the phase under RF signal on the phase modulation path (the other side of the MZM) until it can balance the phase of a typical symmetric MZIM. Hence the linearization of the MRR-MZIM.
3. If the MRR is not under an applied RF, it is a passive coupling and delay line. Under an applied RF signal it acts as a resonant device and returns to its passive delay line when the RF signal returns to zero level (hence no phase effect). Thus there must be a dc phase bias applied to the electrode incorporated in the electrode in the MRR, this must be just at the onset of a resonant point of the MRR so that when a RF signal is applied the MRR is pushed into a resonant mode. This is what is required for the MRR-MZIM to function as a linearized modulator for RPoF.
4. The passband of the MRR must be wide enough so that the baseband data channels of an RPoF link can be accommodated when the RF carrier is embedded with such data channels. The design described here can handle only about the 500 MHz to 1.0 GHz band.

The next sections outline a few techniques in the linearization operation of the RPoF by suppression of the nonlinear harmonics and intermodulation effects by parallel MZMs.

In summary, the MRR-MZIM offers high linear modulation (SFDR ~ 136 dB $Hz^{2/3}$) for an RPoF link, and 24 dB improvement from the common MZM type with additional improvement by employing external tunable RF phase delay line(s).

The incorporation of an MRR is to add an additional delay or dynamic asymmetric MZM structure which is flexible or tunable with phase control via the pyro-optic effect with an applied voltage on the electrode embedded on the ring. The RF signal as applied to the OPM branch can enable a higher signal level, or alternatively the structure follows the structure of an asymmetric MZIM. The wide FSR ensures that no interference to any other signals operate in the FSR spectral window, thus the high number of sub-carrier multiplexings in the RPOF.

Sections 4.2 and 4.3 have adopted a number of published techniques and advanced reports on optical to wireless transmission systems, in particular the conversion of UWB-HF carriers down to the IF domain, enabling ease of processing in the digital domain and thus cost effectiveness in operation expenditure (OPEX) and capital expenditure (CAPEX) for ultra-wideband (UWB) networking which are expected to be deployed in terms of micro-cells in 5G and 6G networking. It is expected that several thousand such micro-cells are deployed. The photonic-assisted RF channel transport systems reported here offer advantages in such transmission systems to points where they are very close to the antenna sites.

Analytical techniques are given offering estimations of link performance, in particular the SFDR. The linearity of the optical modulators and its effects on SFDR are illustrated with an additional MRR, the OBPF pre-processor, an OPM on the paths of the MZIM. However, this may not offer UWB performance due the complexity in the MRR operational principles in the upper mmW region.

The MRR can only be stable, that is not in the oscillation mode, if the gain and loss of the lightwaves circulating in the ring is less than unity. Passive MRR can satisfy this condition but suffers propagation loss. On the other hand the OBPF can be made of a number of MZDIs.

An alternative structure for OBPF is the structure of cascade MZDI. Each MZDI is a first-order filter. MZDI is an all-zero network and thus it is very stable to cascade a number of MZDI stages for an OBPF which produces an all-zero optical network, unlike an MRR. An MRR is a pole-zero network, thus there is a high possibility of oscillation. The sharp roll-off resonant peak passband with a selection band approaching a band pass of less than 500 MHz. This narrow and sharp roll-off passband enables improvement of the signal-to-noise ratio of the signal at the output. The high Q-factor of the MRR performs much better than the cascade MDI filter structure but not the passband. This may not be preferred in a wideband RF carrier multiplexed RPoF link.

4.4.2 Linearization by parallel modulation

Suppression of the higher order harmonics but not the fundamental harmonic ((FH) or first tone) is another technique, indeed this is the preferred way to linearized the modulation process for the RPoF. The main feature of this technique is to have two parallel modulators which are operating in a way that the unwanted nonlinear harmonics generated from such modulators are opposite and thus mutually canceling, but the FH remains unchanged. For the case of MZM modulation the biasing can then be performed at the quadrature points but in different regions where the transmittance curve is in the positive and negative direction, respectively, for the two modulators, respectively. This section presents a number of such parallel structures.

4.4.2.1 Parallel modulators

The schematic diagram of the proposed linearized analog photonic link based on an PM-DPMZM is shown in figure 4.1. A lightwave from an LD is introduced into an PM-DPMZM. The IQ modulator is an integrated modulator which consists of two child MZMs and integrated into one parent MZM in two branches. The lightwave is split into two parts and modulated at the two child MZMs. In the lower branch the input lightwave is split and coupled. Similarly, the other branch goes into the other child MZM. The lower branch is also integrated with a phase delay such that a quadrature component can be generated with respect to the in-phase signals of the upper child MZM (figure 4.13).

The IQ modulator consists of:

- Two parallel MZMs, the MZM child 1 and MZM child 2, here named the I-child MZM and Q-child MZM.
- The parent MZM is the main MZM in which two child MZMs are placed on the two interference arms of the parent MZM.
- Two bias electrodes for biasing the child MZMs with assigned pins (5,6) and (7,8).

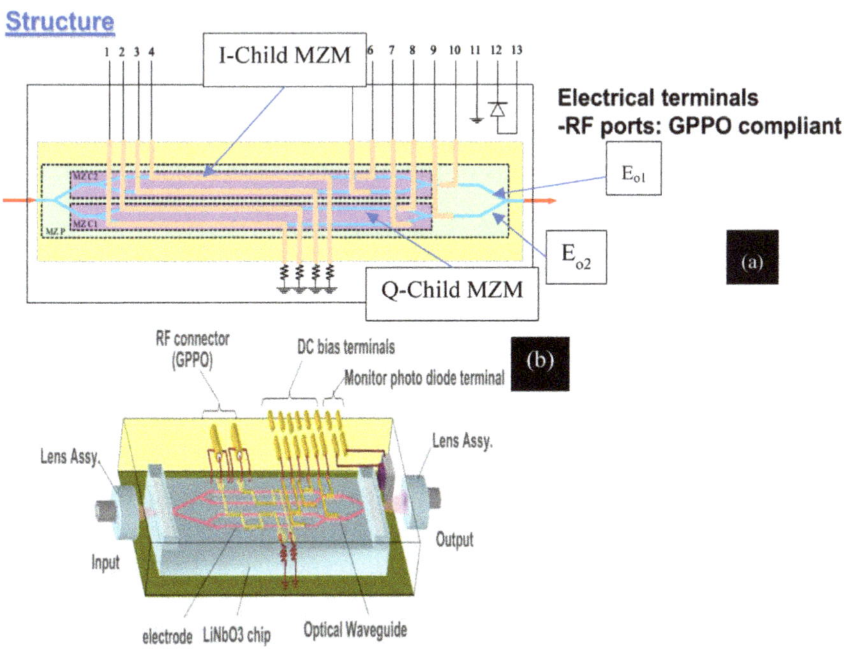

Figure 4.13. Schematic of an IQ modulator and the electrode layout of optical waveguides forming two child MZMs and the parent MZM over which are the electrodes for the dc bias and RF signals. (a) Schematic of the structure of the electrode and pin arrangement. (b) Space dimension of the modulator and electrodes for wideband operation.

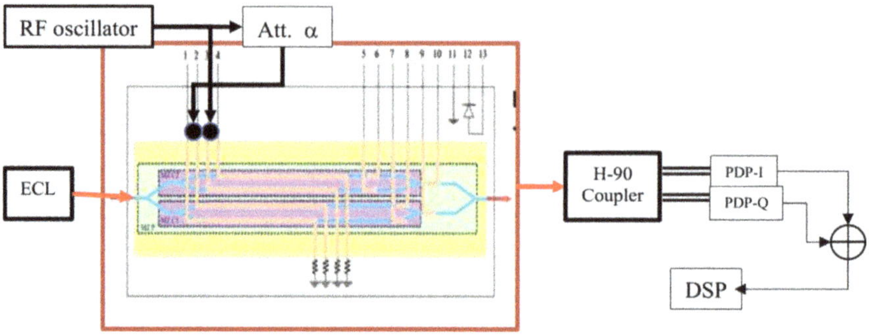

Figure 4.14. Schematic of an IQ modulator and the electrode layout of optical waveguides forming two child MZMs and the parent MZM over which there are electrodes for the dc bias and RF signals.

- A differential phase pair of electrode pins (9,10) for biasing the phase with respect to each other so that a $\pi/2$ phase shift can be exerted for in-phase and quadrature formation.
- If such an IQM modulator is polarized and multiplexed with another IQ modulator, then the device is a DP-IQ modulator or DP-QPSK modulator. Under this arrangement a polarization rotator using electrodes is formed so that they can be aligned in one direction or have orthogonal alignment for transmission over an optical fiber. The IQ modulator is arranged in association with a hybrid coupler and balanced receiver is shown in figure 4.14.

4.4.2.2 Suppression of higher order harmonics

Considering the optical fields at the output of the MZM1 (in-phase child MZM) and MZM2 (Q-phase MZM), they can be represented by the following equations, which are the sum of the two optical fields of the branches of the child MZM:

$$\tilde{E}_{0I} = \frac{1}{2} E_0 e^{j\omega_c t} \left\{ \left[1 + \cos\left(\frac{\phi_{I,\,\mathrm{dc}} + \phi_{I,\,m}(t)}{2} \right) \right] \right\} \tag{4.28}$$

$$\tilde{E}_{0,\,Q} = \frac{1}{2} E_0 e^{j\omega_c t} \left\{ \left[1 + \cos\phi_{Q,\,P} \cos\left(\frac{\phi_{I,\,\mathrm{dc}} + \phi_{I,\,m(t)}}{2} \right) \right] \right\}, \tag{4.29}$$

where:

- E_0 and ω_c are the amplitude and angular frequency of the input optical carrier, respectively, and $\phi_{I,\,\mathrm{dc}}$ and $\phi_{I,m}(t)$ are the phase of the dc biasing voltage and the RF modulated phase, respectively, exerted on the optical waves in the I-phase child MZM arm.
- $\phi_{Q,\,\mathrm{dc}}$ and $\phi_{Q,m}$ are the phase of the dc biasing voltage and the RF modulated phase, respectively, exerted on the optical waves in the Q-phase child MZM arm.
- It is assumed that there is no application of RF waves or dc bias on the second arms of both child MZMs. The dc biases are the differential bias voltages developed across the electrodes of the two arms of the MZM.

Furthermore, one branch of the child MZM is not applied and the bias voltage as the differential bias with respect to this waveguide electrode is the lightwave path with no phase shift.

- The time-variant phases exerted on the lightwaves guided through the upper and lower arms of the child MZMs can be expressed as

$$\phi_{I,\,m}(t) = \frac{\pi}{V_{\pi,\,\mathrm{AC}}} v_{\mathrm{RF},\,I}(t) = \frac{\pi}{V_{\pi,\,\mathrm{AC}}} V_{m,\,I} \cos \omega_{\mathrm{RF}}(t), \tag{4.30}$$

$$\phi_{Q,\,m}(t) = \frac{\pi}{V_{\pi,\,\mathrm{AC}}} V_{m,\,Q}\, \alpha \cos \omega_{\mathrm{RF}}(t), \tag{4.31}$$

$$\phi_{I,\,\mathrm{dc}} = \pi \frac{V_{b,\,I}}{V_\pi} = \beta_I \text{ and } \phi_{Q,\,\mathrm{dc}} = \pi \frac{V_{b,\,Q}}{V_\pi} = \beta_Q;\ \phi_{Q,\,P} = \pi \frac{V_{b,\,P}}{V_\pi}; \tag{4.32}$$

$$\beta_{I,\,m} = \frac{\pi}{V_{\pi,\,\mathrm{AC}}} V_{I,\,m},\ \beta_{Q,\,m} = \frac{\pi}{V_{\pi,\,\mathrm{AC}}} V_{Q,\,m} \tag{4.33}$$

$$\phi_{Q,\,P} = \text{bias phase, the phase shift between I and Q branches, and } V_{b,\,P},$$

where $V_{b,I}$ and $V_{b,Q}$ are the dc voltages applied to the dc bias electrodes in the I- and Q-branches, respectively, is the applied dc voltage to this phase shift electrode. In a QPSK modulator there is one phase shift section embedded in the quadrature branch so that one can generate two orthogonal components at the output of the modulator by setting this phase shift to $\pi/2$ with an appropriate applied voltage.

The amplitude of the RF signal is set to the default values of $V_{m,I}$ and $V_{m,\,Q}$ from which we define the amplitude ratio between the RF wave amplitudes of the RF signals applied to the I- and Q-ports of the QPSK modulator, as

$$\alpha = \frac{V_{I,\,m}}{V_{Q,\,m}}, \tag{4.34}$$

where the AC half-wave voltages are different from the dc half-wave voltages as the bias electrodes are completely independent with the traveling wave electrodes for RF signals, ω_{RF} is the angular frequency of the RF oscillator signal applied to the I- and Q-branches of the MZM, and α is the attenuation factor of the RF signal applied to the Q-MZM.

These two output fields can be written in vectorial form as

$$\begin{bmatrix} E_{0I} \\ E_{0,\,Q} \end{bmatrix} = \begin{bmatrix} E_O e^{j\omega_c t}\left[1 + \cos\left(\dfrac{\phi_{I,\,\mathrm{dc}} + \phi_{I,\,m}(t)}{2}\right)\right] \\ E_O e^{j\omega_c t}\left[1 + \cos \phi_{Q,\,P} \cos\left(\dfrac{\phi_{Q,\,\mathrm{dc}} + \phi_{Q,\,m}(t)}{2}\right)\right] \end{bmatrix}, \tag{4.35}$$

$$\tilde{E}_{T,O} = \tilde{E}_{0I} + \tilde{E}_{0,Q} = E_O e^{j\omega_c t}\left\{\left[1 + \cos\left(\frac{\phi_{I,\mathrm{dc}} + \phi_{I,m}(t)}{2}\right)\right] + \left[1 + \cos(\phi_{Q,P}) * \cos\left(\frac{\phi_{Q,\mathrm{dc}} + \phi_{Q,m}(t)}{2}\right)\right]\right\}. \tag{4.36}$$

These two vectorial components interfere with each other at the output of the parent MZM as given in equation (4.36). The modulation induced phase can be written as given in equations (4.30) and (4.31). It is noted that the half-wave voltages V_π for dc and AC are different due to their inter/independent properties.

The output field of the I-branch can also be superimposed by a data channel on top of the RF carrier modulating the I-child MZM. The Q-MZM is considered as the local oscillator arm which will beat with the data RF carrier modulated lightwaves of the I-phase child MZM. The bandwidth of such a data channel is much smaller than that of the RF carrier frequency so is not included here for simplicity.

Now that the total field is propagating through a length of fiber, that is, no $\pi/2$ phase difference and detected by only one PD, then the produced current at the output of the PD can be written as the square of the total field and then the beating in such a PD as

$$I = |\tilde{E}_{T,0}| = \left|\tilde{E}_{0I} + \tilde{E}_{0,Q}\right|^2 = |\tilde{E}_{0I}|^2 + \left|\tilde{E}_{0,Q}\right|^2 + 2\left|\tilde{E}_{0I}\tilde{E}_{0,Q}\right| \rightarrow I(t)$$

$$= \left|(E_O{}^2 e^{j2\omega_c t})\left\{\begin{array}{c}\left[1 + \cos\left(\frac{\phi_{I,\mathrm{dc}} + \phi_{I,m}(t)}{2}\right)\right]^2 \\ + \left[1 + \cos^2(\phi_{Q,P}) * \cos + \right. \\ \left.\left(\frac{\phi_{Q,\mathrm{dc}} + \phi_{Q,m}(t)}{2}\right)\right]^2 \\ 2|E_0|^2\left(1 + \cos\left(\frac{\phi_{I,\mathrm{dc}} + \phi_{I,m}(t)}{2}\right)\right) \\ \left(1 + \cos(\phi_{Q,P})\cos\left(\frac{\phi_{Q,\mathrm{dc}} + \phi_{Q,m(t)}}{2}\right)\right)\end{array}\right\}\right|^2. \tag{4.37}$$

However, if the fields of the two child MZMs are orthogonal by a $\pi/2$ phase shift in the Q-child MZM, then when they are separated and detected by two balanced pair of PDs (one pair for the I-component and one pair for the Q-component) and added to obtain the total current, then the total current can be written as

$$
\begin{aligned}
I(t) = |\tilde{E}_{0I}|^2 + \left|\tilde{E}_{0,Q}\right|^2 = E_O e^{j\omega_c t}\left|1 + \cos\left(\frac{\phi_{I,\mathrm{dc}} + \phi_{I,m}(t)}{2}\right)\right|^2 \\
+ E_O e^{j\omega_c t} e^{j\pi/2}\left|1 + \cos\phi_{Q,P} * \cos\left(\frac{\phi_{Q,\mathrm{dc}} + \phi_{Q,m}(t)}{2}\right)\right|^2,
\end{aligned}
\tag{4.38}
$$

$$
\begin{aligned}
I(t) = |\tilde{E}_{0I}|^2 + \left|\tilde{E}_{0,Q}\right|^2 = E_0^2 e^{j2\omega_c t}\left|1 + \cos\left(\frac{\phi_{I,\mathrm{dc}} + \phi_{I,m}(t)}{2}\right)\right|^2 \\
+ E_0^2 e^{j2\omega_c t}\left|1 + \cos\phi_{Q,P} * \cos\left(\frac{\phi_{Q,\mathrm{dc}} + \phi_{Q,m}(t)}{2}\right)\right|^2,
\end{aligned}
\tag{4.39}
$$

$$
\begin{aligned}
I(t) = 2\mathscr{R}^2 E_{0*}^2 e^{2j\omega_c t}\Bigg[2 + \cos^2\left(\frac{\phi_{I,\mathrm{dc}} + \phi_{I,m}(t)}{2}\right) \\
+ 2\cos\left(\frac{\phi_{I,\mathrm{dc}} + \phi_{I,m}(t)}{2}\right) + \cos^2\left(\phi_{Q,P}\right)\cos^2\left(\frac{\phi_{Q,\mathrm{dc}} + \phi_{Q,m}(t)}{2}\right)\Bigg] \\
+ 2\cos\phi_{Q,P} * \cos\left(\frac{\phi_{Q,\mathrm{dc}} + \phi_{Q,m}(t)}{2}\right).
\end{aligned}
\tag{4.40}
$$

The factor $\pi/2$ added to consider the $\pi/2$ phase shift in the Q-branch (equation (4.38)) represents the current added after the hybrid coupler H-90 balanced detection with the factor of 2 added. The factor $e^{j\pi/2}$ is considered as disappearing due to the $\pi/2$ phase shift at the hybrid coupler. R is the responsivity of the PD. Note that the two fields are also added via the H-coupler. The H-coupler must be biased with $\pi/2$ so as to split the in-phase and quadrature phase so they are independently detected by the balanced receiver. Under this condition the detected currents from both the in-phase and quadrature phase are added to the input of a summer amplifier to give:

$$
I(t) = |\tilde{E}_{0I}|^2 + \left|\tilde{E}_{0,Q}\right|^2 \tag{4.41}
$$

$$
\begin{aligned}
I(t) = \mathscr{R}E_0^2\Bigg[\cos^2\left(\frac{\phi_{\mathrm{dc},I} + \phi_{I,m}(t)}{2}\right) + \cos^2\left(\frac{\phi_{\mathrm{dc},Q} + \phi_{Q,m}(t)}{2}\right)\cos^2(\phi_{Q,P}) \\
+ 2\cos\frac{\phi_{\mathrm{dc},I} + \phi_{I,m}(t)}{2}\cos\phi_{Q,P}\cos\left(\frac{\phi_{\mathrm{dc},Q} + \phi_{Q,m}(t)}{2}\right)\Bigg],
\end{aligned}
\tag{4.42}
$$

where R is the responsivity of the PD. Ignoring the dc term (quantum shot noise contribution) and setting $\phi_{Q,P} = k\pi + \pi/2$ and $\phi_{Q,\mathrm{dc}} = (2n+1)\pi - \phi_{I,\mathrm{dc}}$, where k and n are integers (figure 4.14),

$$I(t) = -\mathscr{R}E_0^2[\sin\phi_{\text{dc},I}\sin\phi_{I,m}(t) - \sin\phi_{\text{dc},Q}\sin\phi_{Q,m}(t)]. \tag{4.43}$$

To evaluate the dynamic range performance of the link, a two-tone analysis should be applied, in which the RF signals with frequencies of ω_1 and ω_2 are applied. Setting $V_m(t) = \cos(\omega_1 t) + \cos(\omega_2 t)$ as the two-tone input RF signal and setting the $\pi/2$ phase shift of the quadrature phase MZM or $V_{Q,P} = \pi/2$, then we obtain

$$\begin{aligned} I(t) = -\mathscr{R}E_0^2\{\sin[\beta_{I,\text{dc}}(\cos\omega_1 t) + \beta_{I,\text{dc}}(\cos\omega_1 t)]\} - \mathscr{R}E_0^2\{\sin[\beta_{\text{dc},Q}(\cos\omega_1 t) \\ + \beta_{Q,\text{dc}}(\cos\omega_2 t)]\}, \end{aligned} \tag{4.44}$$

$$\begin{aligned} I(t) = -\mathscr{R}E_0^2\sin\phi_{I,\text{dc}}\left\{\sum_{p=-\infty}^{p=+\infty}\sum_{q==-\infty}^{q=+\infty} J_p(\beta_{I,\text{dc}})J_q(\beta_{Q,\text{dc}}) * \sin\right. \\ \left.\left[p_1\omega_1 + q\omega_2 + (p+q)\frac{\pi}{2}\right]\right\} - \mathscr{R}E_0^2\sin\phi_{Q,\text{dc}} \\ \left\{\sum_{p=-\infty}^{p=+\infty}\sum_{q==-\infty}^{q=+\infty} J_p(\beta_{Q,\text{dc}})J_q(\beta_{Q,\text{dc}}) * \sin\left[p_1\omega_1 + q\omega_2 + (p+q)\frac{\pi}{2}\right]\right\}, \end{aligned} \tag{4.45}$$

$$\begin{aligned} I(t) = -\Gamma_{I,0} + \Gamma_{I,1}(\cos\omega_1 t + \cos\omega_2 t) \\ + \Gamma_{I,3}[\cos(2\omega_1 - \omega_2)t + \cos(2\omega_2 - \omega_1)t] + \dots, \end{aligned} \tag{4.46}$$

where $J_n(\beta_i)$ provides the nth-order Bessel function of the first kind, $\Gamma_{I,1} = -\Re E_0{}^2 * \sin(\phi_{\text{dc},I}) * J_0(\beta_{\text{dc},I})J_1(\beta_{\text{dc},I})$, and $\Gamma_{I,3} = \Re E_I^2 * \sin(\phi_{\text{dc},I}) * J_1(\beta_{\text{dc},I})J_2(\beta_{\text{dc},I})$. As can be seen, the IMD2 and SHD are eliminated. So equation (4.46) can be rewritten as

$$\begin{aligned} I(t) = -\mathscr{R}E_0^2\Big\{\Big[-\sin\phi_{I,\text{dc}}J_0(\beta_{I,\text{dc}})J_1(\beta_{I,\text{dc}}) \\ - \sin\phi_{Q,\text{dc}}J_0(\beta_{Q,\text{dc}})J_1(\beta_{Q,\text{dc}})\Big] * \\ (\cos\omega_1 t + \cos\omega_2 t) + \Big[\sin\phi_{I,\text{dc}}J_0(\beta_{I,\text{dc}})J_1(\beta_{I,\text{dc}}) + \sin\phi_{Q,\text{dc}} \\ J_0(\beta_{Q,\text{dc}})J_1(\beta_{Q,\text{dc}})\Big] * [\cos(2\omega_1 - \omega_2)t + \cos(2\omega_2 - \omega_1)t] + \dots\Big\}. \end{aligned} \tag{4.47}$$

From equation (4.46) the IMD3 can also be canceled when the following condition is satisfied:

$$\begin{aligned} \left[\sin\phi_{I,\text{dc}}J_0(\beta_{I,\text{dc}})J_1(\beta_{I,\text{dc}}) + \sin\phi_{Q,\text{dc}}J_0(\beta_{Q,\text{dc}})J_1(\beta_{Q,\text{dc}})\right] = 0 \text{ and } - \sin\phi_{I,\text{dc}} \\ J_0(\beta_{I,\text{dc}})J_1(\beta_{I,\text{dc}}) - \sin\phi_{Q,\text{dc}}J_0(\beta_{Q,\text{dc}})J_1(\beta_{Q,\text{dc}}) \\ \neq 0 \end{aligned} \tag{4.48}$$

Therefore, from the sufficient condition (equation (4.48)), by carefully adjusting the biases of the IQ modulator to make $\varphi i_3 = k\pi + \pi/2$ and $\varphi i2 = (2n + 1)\pi - \varphi i_1$, which are expressed above (k and n are integers), and by adjusting the input RF power and the microwave attenuator to tune β_1 and β_2 to satisfy the condition in equation (4.48), all the IMD2, SHD, and IMD3 can be eliminated Furthermore, for the small signal modulation condition, by adopting $J_n(m) \approx mn/(2nn!)$, equation (4.48) can then be further simplified as

$$-\frac{\sin \phi_{I,\,\mathrm{dc}}}{\sin \phi_{Q,\,\mathrm{dc}}} = \frac{\beta_{Q,\,m}^{3}}{\beta_{I,\,m}^{3}}$$

$$-\frac{\sin \phi_{I,\,\mathrm{dc}}}{\sin \phi_{Q,\,\mathrm{dc}}} \neq \frac{\beta_{Q,\,m}^{1}}{\beta_{I,\,m}^{1}}. \qquad (4.49)$$

Thus, by adjusting the biases of the SP-QPSK and the amplitudes of the modulated RF power in both branches, a multi-octave linearized analog photonic link should be realized, with all the IMD_2, SHD, and IMD_3 eliminated. It is also noted that the ratios of the RF amplitude applied to the two MZMs are proportional to the inverse of the cubic optical power entering into the two optical modulators. For example, if the ratio is 1:3 for the RF then the coupling ratio of the optical waves into the two modulators is 27:1, or a cubic inverse factor. This would give an approximate tuning ratio which one would use in an experimental platform. Due to the experimental fabrication tolerance such tuning is required to obtain the optimized cancellation of nonlinear distortion (figure 4.15).

4.4.2.3 Conditions for suppression in practice

As observed in section 4.4.2.2 the conditions for suppression of NLD can be given as

1. *Condition of detection*: the two MZMs must be detected by independent PDs and added in balanced mode to be optimized.
2. *Condition on the RF power signals driving the MZM (I) and MZM(Q)*: these must be unequal by a ratio α, e.g 1:3 for RF [MZM(I)]/RF[MZM(Q)] = α.
3. *Condition on the optical power launched into the MZM(I) and MZM(Q)*: $\beta \to \beta$ must be the inverse cubic factor of that of α, that is $\beta = (\alpha)^{-3}$; for example, for a 1:3 RF ratio α, then β = 27:1 or one optical signal is attenuated 10.54 dB w.r.t. to the other.
4. *Push–pull and biasing conditions for MZM(I) and MZM(Q)*: For maximum swing of the RF, the I and Q MZMs must be biased at the most linear position and driven to a maximum value of the V_π RF peak at $V_{\pi\mathrm{dc}/2}$ (±dBias) or driving into the nonlinear region. The other MZM must be linearly biased but with a π phase difference or $-V_{\pi\mathrm{dc}/2}$ (±dBias) and driven with π shift (or the inverse signal of the RF so the push–pull effect can happen). dBias is defined as the differential variation to superimpose on the biasing voltage.

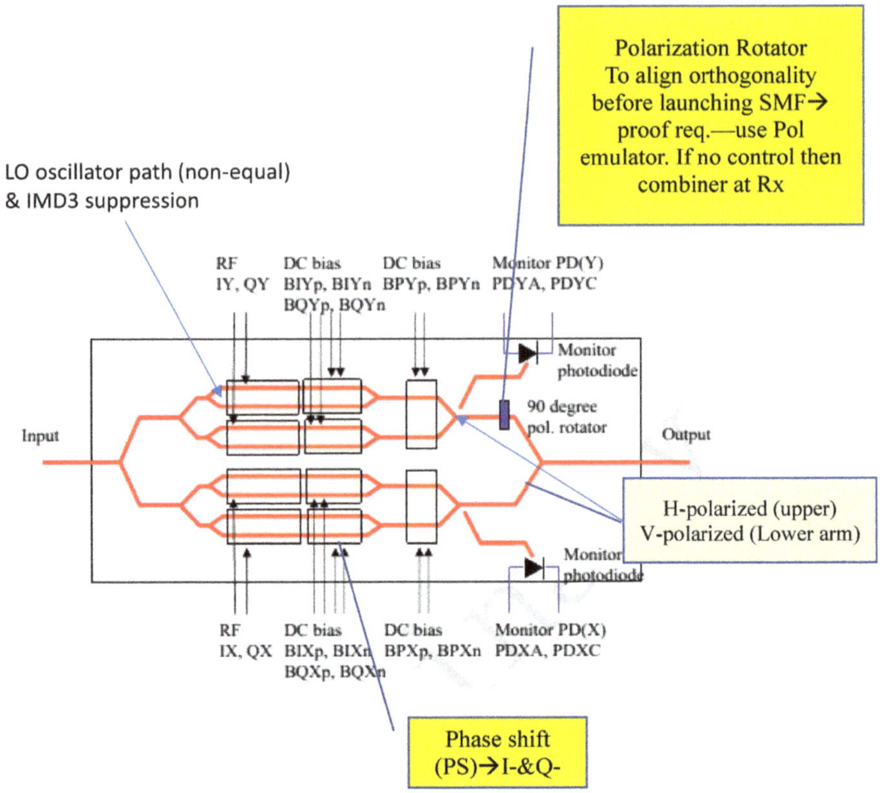

Figure 4.15. Linearization by suppression of nonlinear distortion harmonics and intermodulation components, using dual polarization QPSK modulators incorporating a polarization rotator. In comparison with the single polarized QPSK modulator of figure 4.14 in which there is no polarization rotator.

Under these conditions, it can be much simpler to use two parallel MZMs as shown in the simulation model in figure 4.16.

4.4.2.4 Parallel MZMs with polarization multiplexing

If two parallel MZMs are multiplexed in orthogonality then the polarized channels are detected and converted to current separately as they do not interfere with each other.

4.4.2.5 Suppression of nonlinear distortion using parallel modulators

In order to increase the linearity of the operation of an RPoF link the suppression of the nonlinear distortion is necessary. This suppression can be obtained if two MZMs can be connected in parallel such that one can be arranged so that the nonlinear distortion is opposite to the other when they are detected in a photodetector. Thus the optically modulated signals at the outputs of these parallel modulators must be orthogonal with each other so that they do not interfere before being received by the PD. The converted currents and their nonlinear components in the electrical domain would cancel each other out. Figure 4.17 shows the simulated performance of a dual

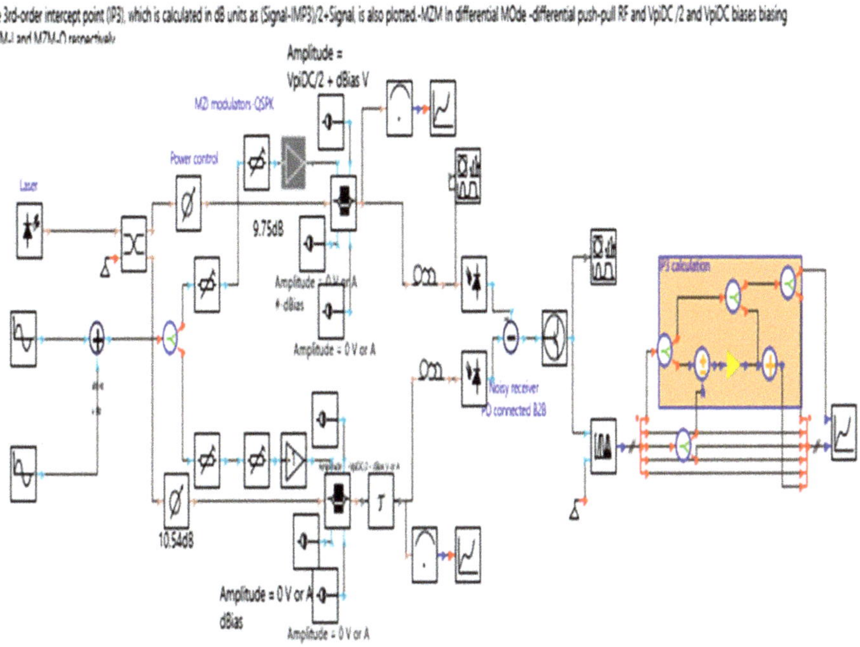

Figure 4.16. Model for NLD suppression using two parallel MZMs.

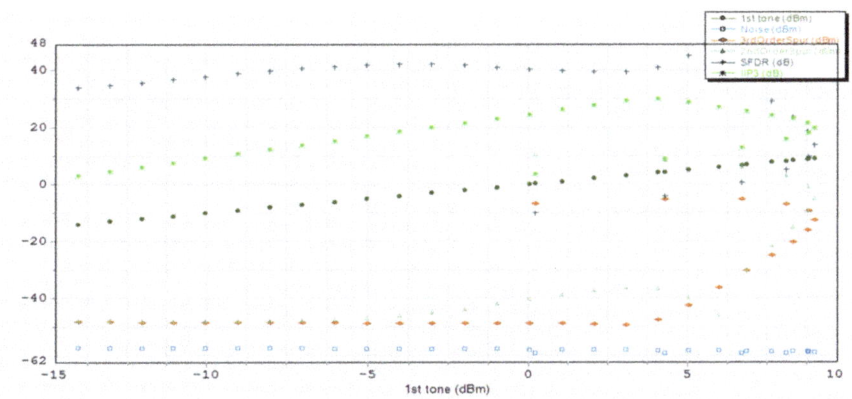

Figure 4.17. First carrier power, SHD, THD (IMD3), IIP3, and SFDR with an optical power ratio of 10.54 dB (27 times) and an RF signal ratio of 8.75 dB, with RF input 20 dBm biases at $V_{\pi/2}$ and $3V_{\pi/2}$. Note that $V_{\pi dc}$ (the half-wave voltage of the dc electrode) and $V_{\pi RF}$ (the half-wave voltage under AC operation) are different.

polarization for suppression of the distortion using VPI Photonics (a company in Berlin, Germany) platform. The biases of the MZIM of figure 4.18 are set at biases at $V\pi/2$ and $3V\pi/2$, respectively. The RF signal input power is 20 dBm to the input port of the two MZIM, with one attenuated with respect to the other. The power of

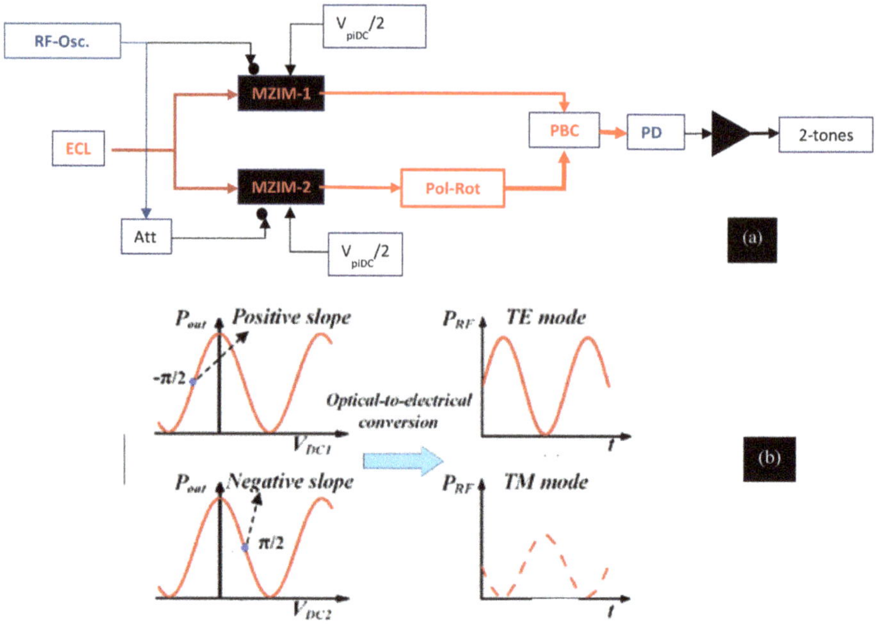

Figure 4.18. (a) Parallel MZM modulators and polarization rotation with an RF attenuator to mitigate nonlinear harmonics and intermodulation distortion (IMD) products. (b) Biasing schemes for two MZMs and polarization rotation to generate orthogonality.

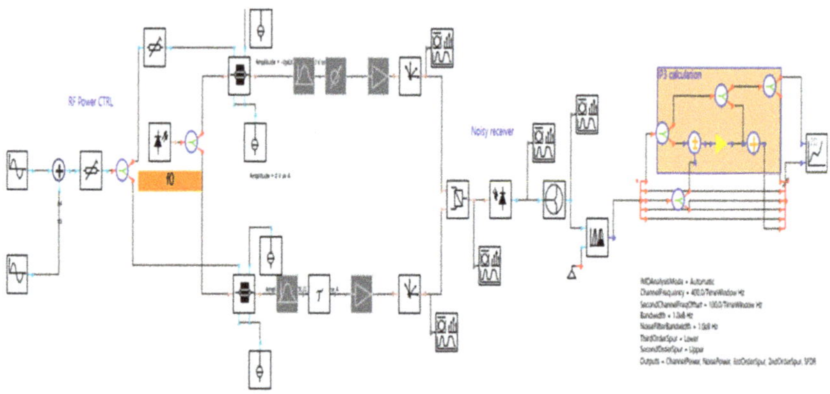

Figure 4.19. Model using VPI TransmissionMaker modeling.

the first order, the second order distortion (SHD), the third order inter-modulation (IMD3) the third order intersection point IIP3, and the SFDR factor are obtained and display against the first tone power. The simulation model for the structure given in figure 4.18 is given in figure 4.19. Furthermore, figure 4.20 gives the performances of the noises, IMD3 and SFDR are also given against the first tone power.

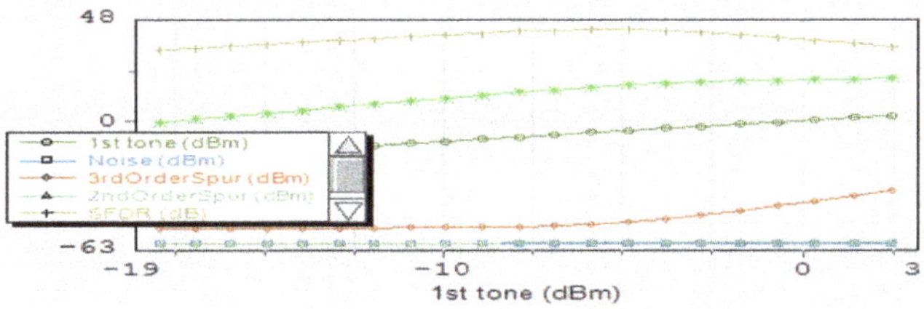

Figure 4.20. Noise, IMD3, SFDR, and second harmonics versus the power input of the RF fundamental first tone.

4.5 Coherent reception and frequency shifting in RPoF

4.5.1 Coherent reception types

4.5.1.1 The demand for lower frequency band and signal processing

The significant demand for 5G and 6G global communication networks has reached the stage where the interaction and interfacing between wireless and optical networking are getting closer and closer. Under the expected expansion of the RF carrier frequency towards the spectral regions of the sub-6G to upper 6G spectral windows, the μW and mmW spectral windows, so as to enable transport and interconnections of the wireless cells in back-haul via free space, but via the optical guided media—optical fibers.

The main difference between the wideband signaling in long- and short-haul optical fiber systems (OFSs) and RF transport optical systems (RFTOSs) is that the later requires demodulating the RF carriers whose frequencies are extremely high, but the bandwidth of the data channels needs to be narrow enough, that is, much smaller than that of the RF carrier. Thus the main difficulty is the extremely high sampling of the RF signals to satisfy the second Nyquist sampling theorem to ensure the error-free recovery of embedded data channels.

Such a difficulty was very well known in the early years of communications theory. The most common technique is the down-conversion of the processing plane to an intermediate (IF) spectral region.

As the RF carrier channels are transported in the optical domain, that is, the optical carrier is shifted by the RF center frequency, such down conversion in the optical domain would allow the photodetection to be much simpler. Therefore, the bandwidth of the pass-band of the receiver can be much narrower and thus the sampling rate is lower.

Advanced optical communications technology coupled with electronic processing techniques has in recent years offered many technologically advanced methods for high speed signal processing in both analog optical and digital electronic systems, in particular in terms of DSP capability.

However, it is quite expensive and less sensitive if the sampling rate is too high, as the resolution of the ADC becomes less effective or the number of bits must be reduced to offer the wideband width required by the ADC. Thus, if processing in the IF, low sampling rate ADC and DSP systems can have a higher number of bits, thus higher sensitivity, e.g. 6 bit ADC at 5–10 GSa s^{-1}.

This section briefly summarizes the optical down-conversion technique for optical RF transport systems or radio frequency over fiber (RFoF). The heterodyne techniques are the most common methods, that is, the mixing of the high frequency signals and the lower ones coherently to give beating signals in terms of the summation and subtraction of frequency terms. The summation terms are too far away from the detectability of the optical receiver and thus are depleted and only the lower product terms are recovered.

Then, a number of simple RFoF transport systems are proposed for experimental implementation of heterodyne reception

4.5.1.2 Heterodyne and self-heterodyne methods

The term heterodyne indicates the mixing of two different frequency sources, while self-heterodyne means that the two sources are derived from the same source. This has the important implication that the phase of the two sources are the same and hence their phase fluctuations as well. Thus there is no need for any phase compensation in the processors. This phase compensation is necessary for the heterodyne case. The advanced high sampling rate of ADCs and DSP has no problem providing these features.

The operation of each structure is given and explained briefly in the captions of figures 4.21–4.28.

4.5.1.3 Heterodyne RFoF 1 (H-RFoF #1)

Figure 4.21 shows the generic structure of a heterodyne RFoF in which the LO is derived from a separate laser, the tuneable ECL2 which can be shifted in its frequency by an IF amount determined by the frequency of the RF synthesizer. The input laser passes through an optical IQ modulator. The optical carrier is modulated by an RF carrier and a data channel which is normally achieved using an OFDM. The RF carrier can be modulated with an M-QAM, format such as 256QAM. This why an IQ modulator is used.

Figure 4.21(b) provide the detailed structure with a separate laser for the IF LO laser. The beating of the two lightwaves occurs in a coherent receiver. The optical noise source is an optical amplifier (EDFA) if no input is excited.

4.5.1.4 Self-heterodyne RFoF 1 (SH-RFoF #1)

Self-heterodyne reception systems are provided in figures 4.22 and 4.23, with coherent beating plus balanced detection and envelop detection, respectively. As indicated, the laser source is split into two optical paths. One is modulated by an RF modulated data channel and the other is shifted by a desired RF oscillation. This RF frequency determines the IF region in which the data channel can be detected and processed in the digital domain with a much lower sampling rate.

4.5.2 LO frequency shifting techniques and RoF system

4.5.2.1 Tuning of a tuneable laser and IF modulation

Figure 4.24 shows the generation of an LO IF laser by a separate tunable ECL laser with frequency shifting to obtain serrodyne detection, i.e. the IF is very close to or less than the baseband bandwidth of the signal. Serrodyne means the IF frequency is within the bandwidth of the baseband signal, that is homodyne detection, but there is a small difference

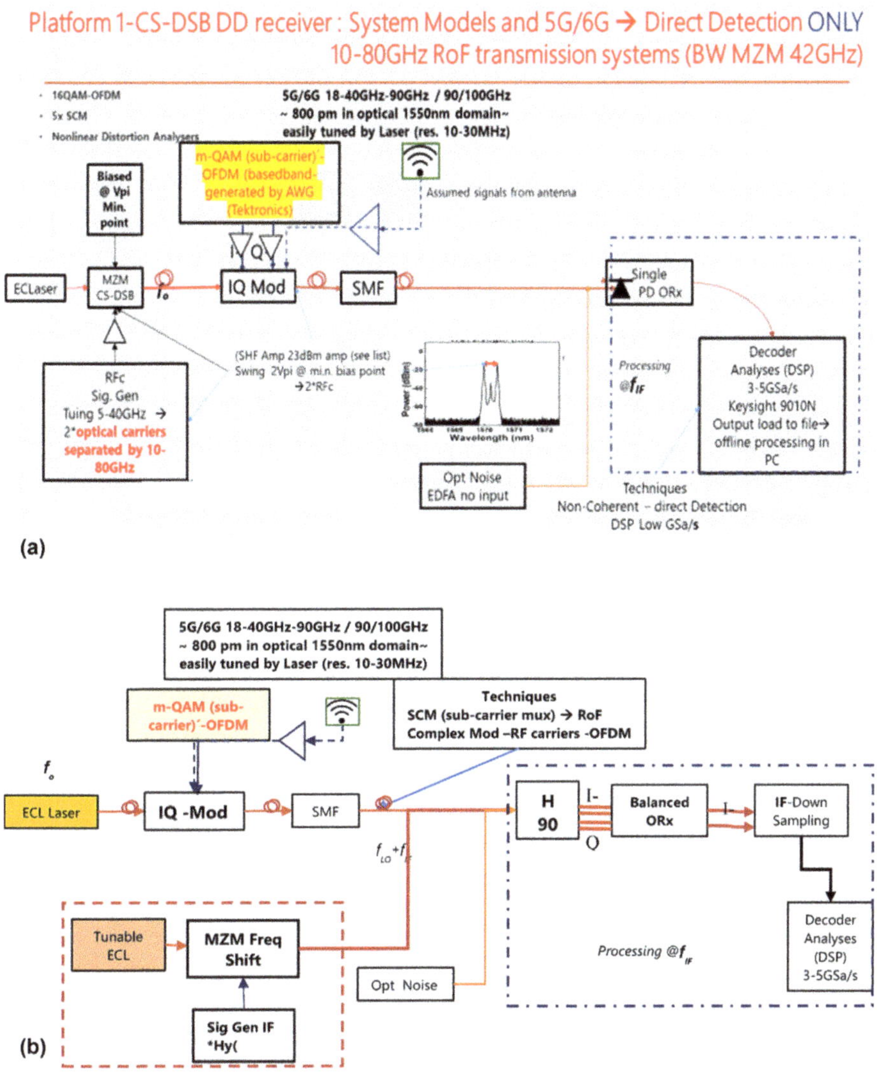

Figure 4.21. Generic structure of a heterodyne RFoF in which a separate laser provides the LO, a tuneable ECL2 which can be shifted in its frequency by an IF amount. (a) PPT image. (b) Detailed structures with a self-heterodyne method and separate lasers for the IF LO laser and beating in the coherent receiver photodetector.

between the signal laser frequency and that of the local oscillator. This is practically true, as the frequency of the two lasers can never be matched to be exactly the same. The difference in their frequency values is normally called the local oscillator frequency off-set (LOFO) and can generally be corrected in the DSP phase matching algorithm.

4.5.2.2 Frequency shifting by optical injection

Optical injection is a technique to control the stability and precision of the generation of a laser. A stable external laser is used for beating by injection into

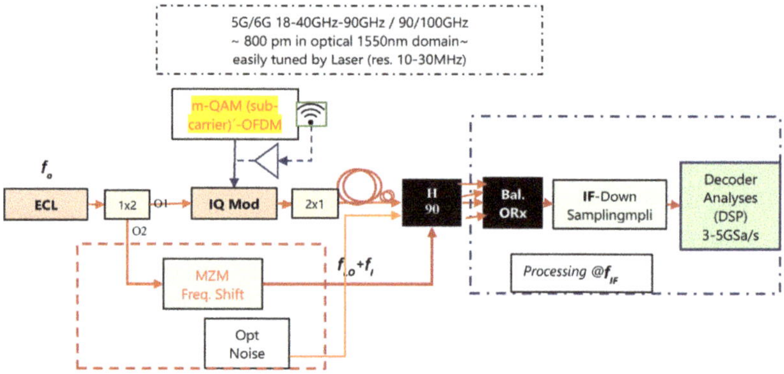

Figure 4.22. Generic structure of a self-heterodyne RFoF. The IF local oscillator is applied to the coherent receiver via coupling into an optical hybrid coupler (H-90) and demodulated by an opto-electronic balanced receiver in which the beating between the optical signals and the IF local oscillator generates the low frequency product of their amplitude and phases, then is down-sampled by an ADC and DSP processing. IQ = in-phase–quadrature phase; H-90 = hybrid coupler with 90° phase shift to separate the I- and Q-components with four optical output ports. MZM = Mach–Zehnder interferometric modulator. *Note*: Attention should be paid to the polarization of the fiber coupling at the input port of the modulator and the hybrid coupler H-90.

the slave laser. Figures 4.25(a), (b), and (c) show such an injection laser scheme in an RFoF system with heterodyne detection and IF processing. An IF tuneable source is used via optical injection to obtain a heterodyne RFoF using an IF LO derived from optical injection and an OFC using two optical modulators and injection to generate the tuneable IF LO laser. The structure of optical injection for controlling the IF shifted from the comb optical lines from the OFC is also shown, which is generated from the modulation of the amplitude (MZM) and phase (OPM).

The optical injection technique involves a heterodyne RFoF with an IF local oscillator coming from optical injection into a DFB, the slave laser. The selected LO frequency is $f_{\mathrm{DFB}} + n\Delta.f_C$, where n is the mode line of the optical frequency comb generator, and $\Delta.f_C$ is the comb spacing, which is normally generated by a cascaded MZM and PM driven by a clock source whose rate equals to the PRBS rate of the RF channel. *Note*: no optical isolator is to be used, thus a packed DFB laser is not to be used but a DFB chip with a fiber coupled platform or a DFB without an isolator. The biasing current will generate tuning of the frequency to be locked into one of the longitudinal modes of the comb spectrum. The oscillator frequency is derived from the PRBS source which is used in the modulation of the formats m-QAM and OFDM.

4.5.2.3 Serrodyne frequency shifting

Figure 4.26 shows a coherent scheme in which the local beam can be derived from a fiber laser and external cavity made up of two optical circulators and an optical phase modulator. The frequency of the derived secondary beam is very close to that of the generated lightwaves. Both of these beams can be used as the RF modulated beam and the LO beam. This can be a very stable way to generate the lightwaves for RPoF links.

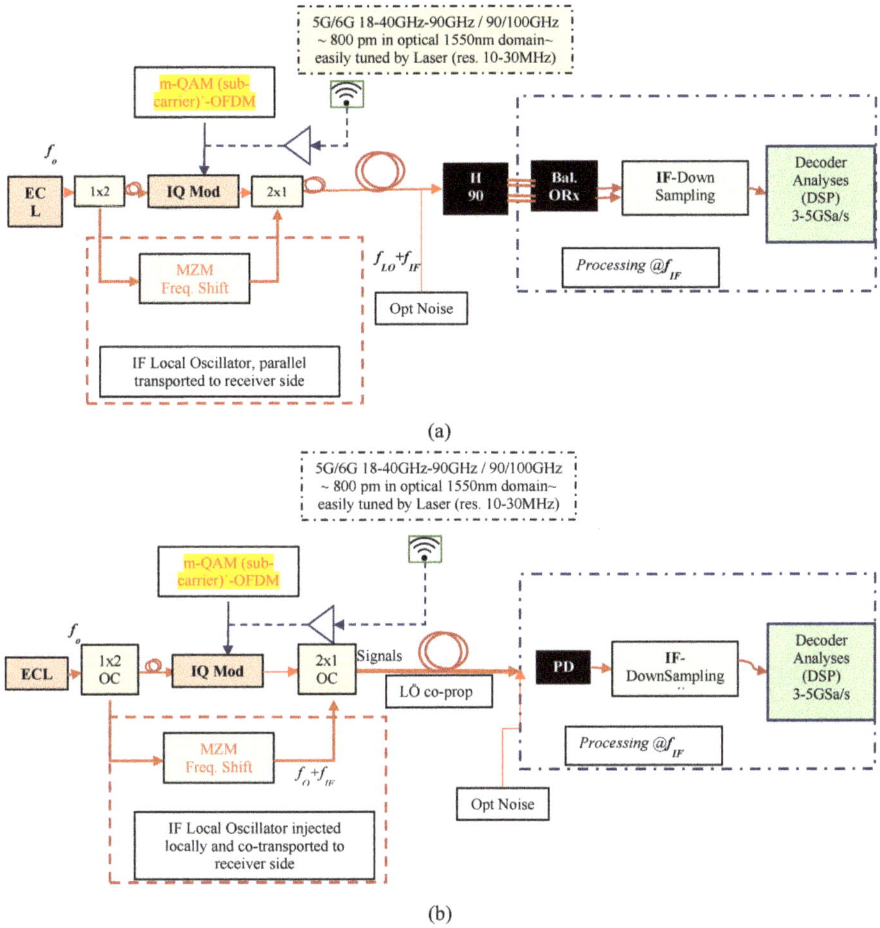

Figure 4.23. A self-heterodyne RFoF with an IF local oscillator co-propagating with optical signals and (a) coherent balanced reception and (b) envelope detection. A noise source is used to increase the injected noise so as to measure the SNR quantity of the sensitivity of the receiver.

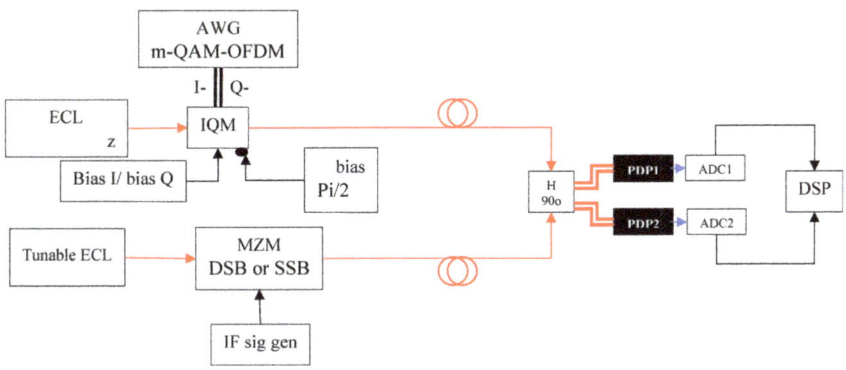

Figure 4.24. Generation of an LO IF laser by a separate tuneable ECL laser and frequency shifting so as to obtain serrodyne detection, i.e. the IF is close to or less than the baseband bandwidth of the signal. PDP = PD pair; ADC =A/D converter; AWG = arbitrary waveform generator.

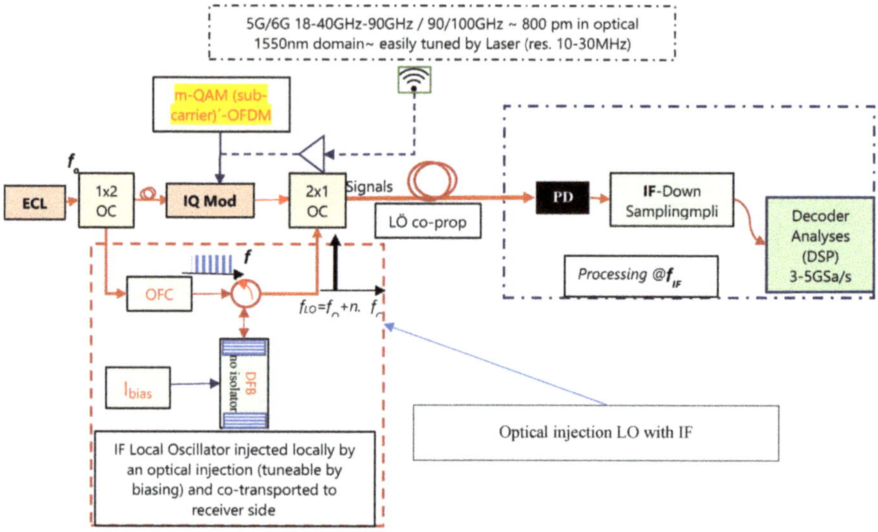

(a) Heterodyne RFoF using IF LO derived from optical injection. Legends: OFC Optical Frequency Comb)

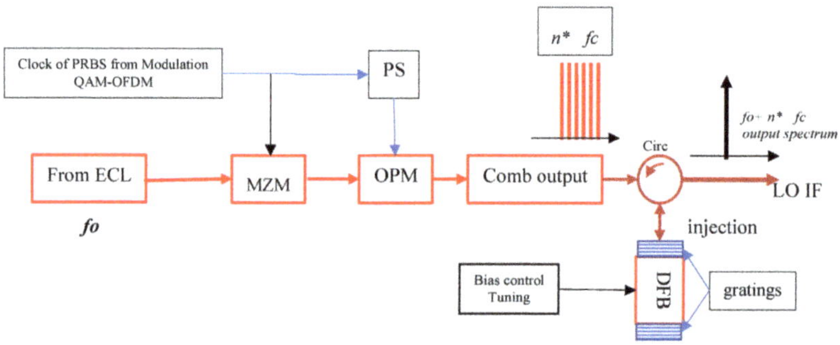

(b) OFC by two optical modulators and injection to generate tuneable IF LO laser; OPM = optical Phase modulator; MZM = Mach Zehnder Modulator (interferometric, field or intensity)

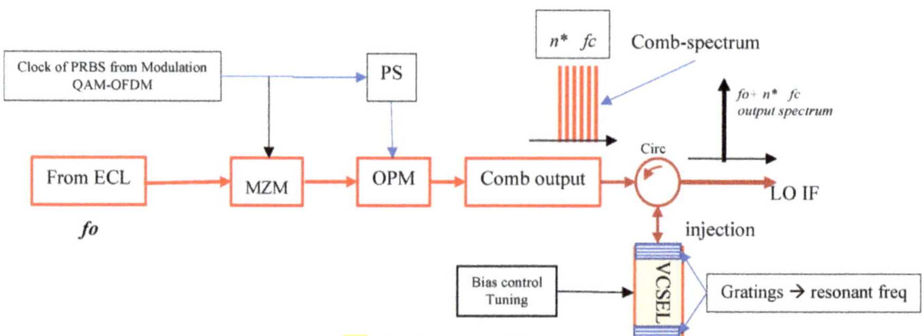

(c) Injection using VCSEL

Figure 4.25. An RFoF system employing an IF tuneable source via optical injection. (a) A heterodyne RFoF using an IF LO derived from optical injection. OFC = optical frequency comb. (b) An OFC from using two optical modulators and injection to generate a tuneable IF LO laser. (c) The structure of optical injection for controlling the IF shifted from the comb optical lines from the OFC laser, which is generated from the modulation of amplitude (MZM) and phase (OPM). (Adapted from [31, 32].)

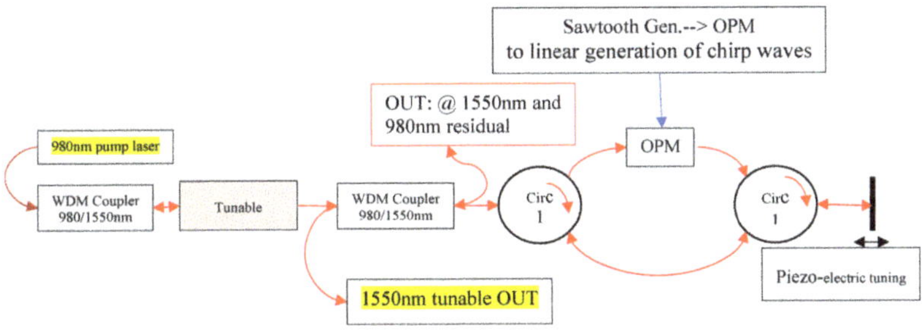

Figure 4.26. Frequency shifting of a reference laser using a tuneable cavity via a DFB type fiber laser which is resonant at 1550 nm and shifting close to this central spectrum of the fiber laser. Circ = circulator at 1550 nm (three ports); OPM = optical phase modulator; DFB = distributed feedback fiber laser. The gain material is an erbium-doped silica fiber pumped with 980 nm of power at about 250 mW. (Reproduced from [33].)

4.5.2.4 Generation of a signal laser and LO IF laser with both signals and the LO IF parallel to the coherent receiver side

The generation of a signal laser and IF LO laser using an OFC with double sideband (DSB) spectra can be obtained with the use of a comb laser [34]. The comb lines can be generated using two optical modulators, one is an intensity modulator and the other is an optical phase modulator. The comb lines are selected by a multiple wavelength demultiplexer (an array waveguide grating) and fed into an optical modulator which is modulated by an RF carrier modulated data channel, for example OFDM-256QAM. The other beam as an LO laser can be derived from another output port of the array waveguide grating (AWG). The frequency differences between the two optical output ports are normally determined by the AWG structure. This determines the heterodyne frequency of the detection scheme. Alternatively, this can be used to generate ultra-high frequency RF waves. The frequency difference can reach several hundreds of GHz.

In coherent reception for RPoF, both output beams are transported parallel to the receiver, as shown in figure 4.26(a). Figure 4.26(b) shows the optical frequency comb (OFC) generated with single sideband (SSB) modulation and shifted spectra can be generated. The AWG acts as a demultiplexing device with ports in and ports out as $1 \times N$, where N is the port number separating the two optical lightwave ports. The optical signal and the IF optical port can be fed into a hybrid coupler, H-90, in which the in-phase (I-) and quadrature (Q-)components are separated by a 90° shift in the optical domain and two output ports per in-phase and quadrature phase as detected by two pairs of balanced PDPs connected back-to-back, and can be detected in a push–pull mode. These channels are processed in an ADC-DSP system.

4.5.2.5 The signal laser and LO IF laser for coherent RPoF systems

The generation of a signal laser and IF local oscillator laser using an OFC with double sideband spectra is shown in figure 4.27 [34]. An RPoF system using combined signal and LO lasers via a mode spacing AWG (central frequency input and edge input as described above in section 4.5.2.4). The AWG is fed with the comb lines and filters out the equally spaced comb lines. Two desired comb lines can be selected and both transported to a Tx modulator and the other used as a local

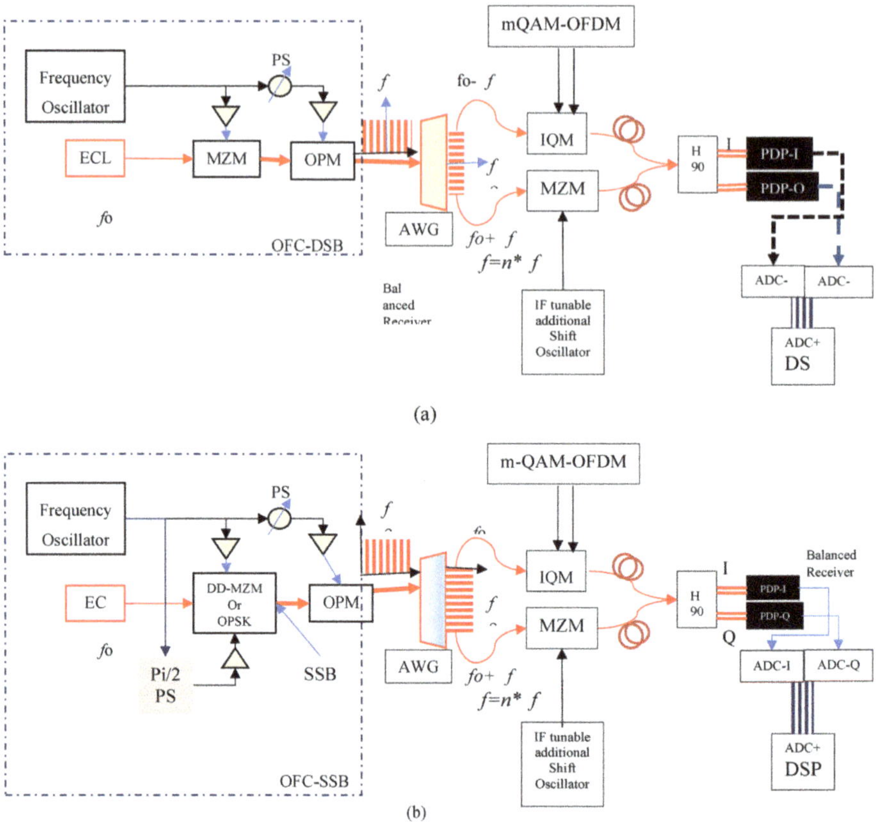

Figure 4.27. (a) The generation of a signal laser and IF local oscillator laser using an OFC with double sideband spectra. Both beams are transported parallel to the receiver. (b) An OFC with single sideband (SSB) spectra. AWG = array waveguide grating (demultiplexer); ports in and ports out $1 \times N$, N = port number separating two optical lightwave ports, the optical signal and the IF optical port; H-90 = hybrid coupler in which the I- and Q-components are separated by a 90° shift in the optical domain and two output ports per in-phase and quadrature, thus balanced receiver ports can be detected in the push–pull mode. (Reproduced from [34].)

oscillator beam. The fiber output can be detected by envelop detection of the combined field, that is single PD coherent detection. The electrical signals at the output of the receiver can be sampled and processed in the IF which is determined by the frequency spacing between the two AWG ports. Figure 4.27(c) shows the possible algorithm of the DSP following an ADC. Thus, the structure is mostly advantageous in the case that the detected signals are to be amplified and transmitting to a free-air antenna (figure 4.28).

4.6 Appendix: basic photonic active and passive components for an RFoF using an IF local oscillator

4.6.1 Passive components: 2 × 2 fiber coupler (wideband)

See figure 4.29.

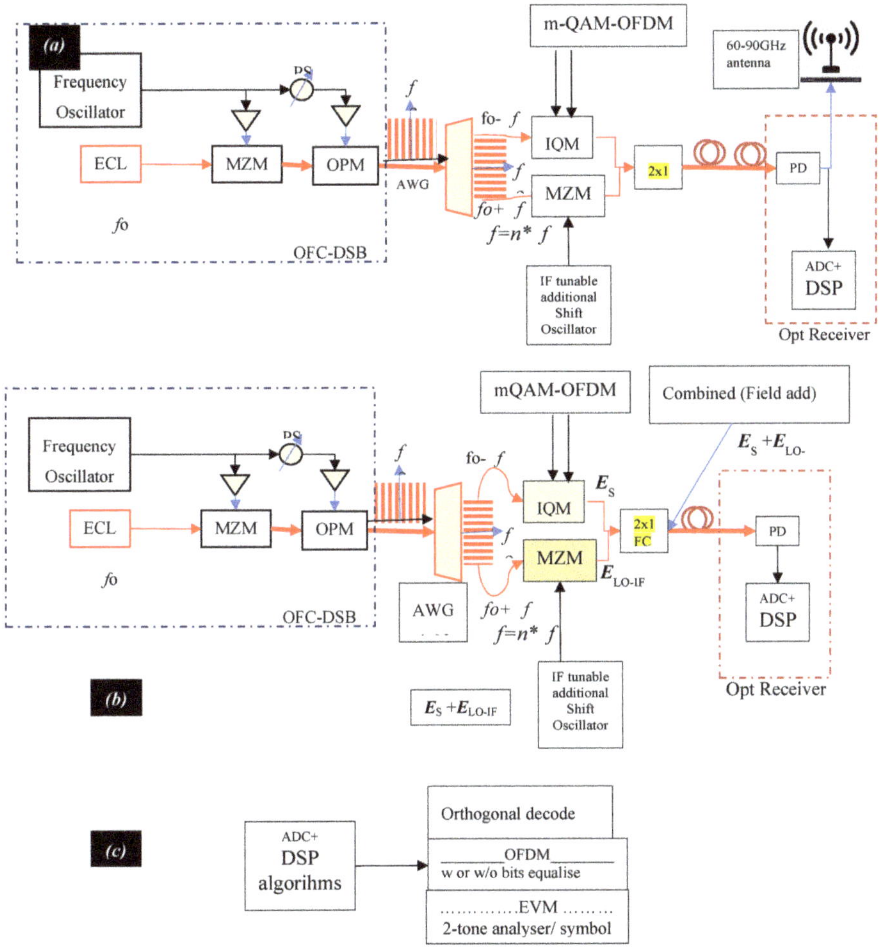

Figure 4.28. Generation of a signal laser and IF LO laser using an OFC with double sideband spectra [34]. (a) An RoF structure using combined signals and LO lasers via a mode spacing AWG (central frequency input and edge input). (b) An OFC at the Tx output and envelop detection of the combined field. (c) Algorithm of the DSP after ADC. Ports in and ports out $1 \times N$, N = the port number separating two optical lightwave ports, the optical signal and the IF optical port. Both are detected by a single PD in which both signals and IF LO lasers are beating to give the optical field product term. This coherent detection is 3 dB power, less sensitive than that of the balanced detection given in figure 4.25, or 6 dB in electrical voltage at the output. The inset shows the DSP algorithm. *Note*: This structure is mostly advantageous under the case that the detected signals are to be amplified and transmitting to a free-air antenna. OFC = optical frequency comb; ADC = analog-to-digital conversion; EVM = error vector magnitude (in %); IF = intermediate frequency; LO = local oscillator; δf = frequency spacing of the output ports of the AWG. AWG = array waveguide grating (demultiplexer). Reproduced with permission from [32].

4.6.2 Active components

4.6.2.1 Single polarization QPSK

See figure 4.30.

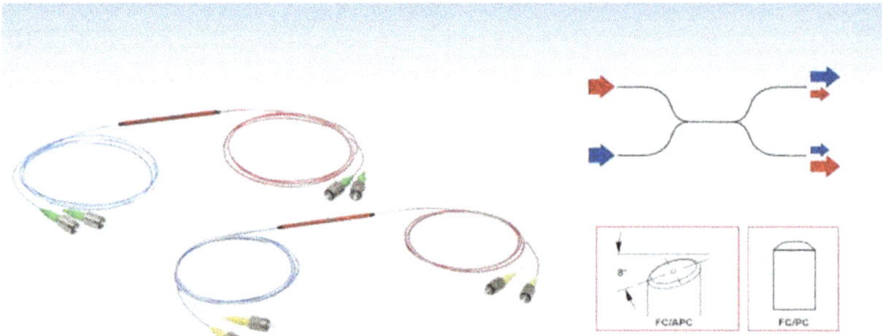

Figure 4.29. A photograph of a 2 × 2 coupler with colored input/output ports and an angle fiber (in green). Note that there are also PMF couplers in which the polarization at the input is coupled with a defined polarized mode at the output ports. Reproduced with permission from [33].

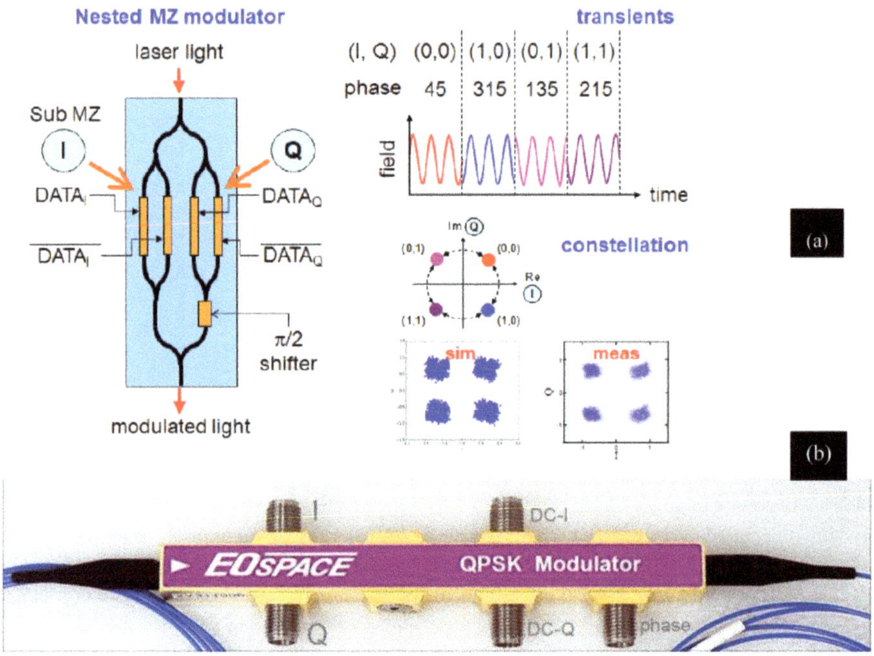

Figure 4.30. An EO space QPSK modulator with K- or SMA connectors for dc bias and RF signal with in-phase and quadrature phase ports. (a) Schematic structure and modulation. (b) Image of the EO space QPSK modulator device. The $\pi/2$ phase shift section is controlled by a dc applied voltage, this $\pi/2$ shift creates the quadrature feature with respect to the other arm, thus two optical paths, one in-phase and the other quadrature phase, are created and then merged or added up at the output of the QPSK or IQ modulator. Nested MZM = parent MZM; sub-MZM = child MZM. Copyright (2014) IEEE. Reprinted with permission from [34].

4.6.3 QPSK modulator and dual polarization QPSK modulator

4.6.3.1 Dual polarization QPSK

See figure 4.31.

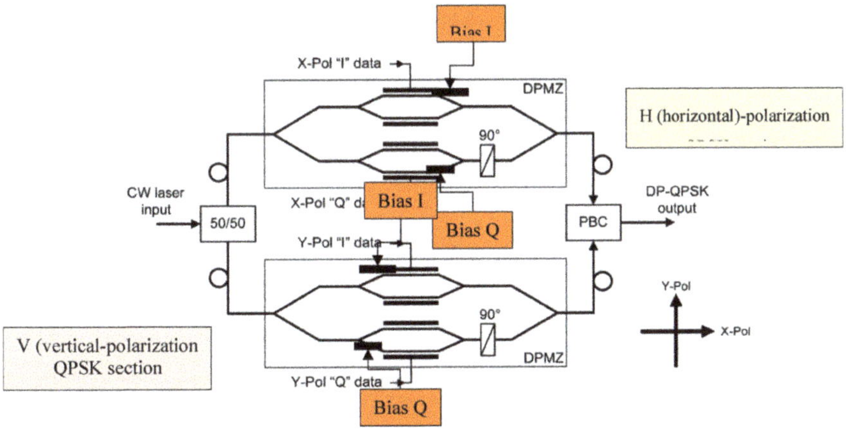

Figure 4.31. A dual polarization in-phase/quadrature phase modulator (IQM) with two QPSK modulators coupled and decoupled via a polarization beam splitter and combiner (PBS and PBC). Automatic bias control (ABC) is normally employed. *Note*: all dc biases to the electrodes of the I- and Q-component phases are separate from the RF ports. Thus the V_π of the dc biases are about 7–12 V. If they are in the same RF port then the voltages are about 3–5 V dc. The electrodes are of the single traveling wave type and not in a dual arrangement as in this figure. The positions of the electrodes are not exactly as shown above. See the technical data of the suppliers. See figure 4.30 for an EO space QPSK modulator arrangement of RF and dc biases. Copyright (2014) IEEE. Reprinted with permission from [34].

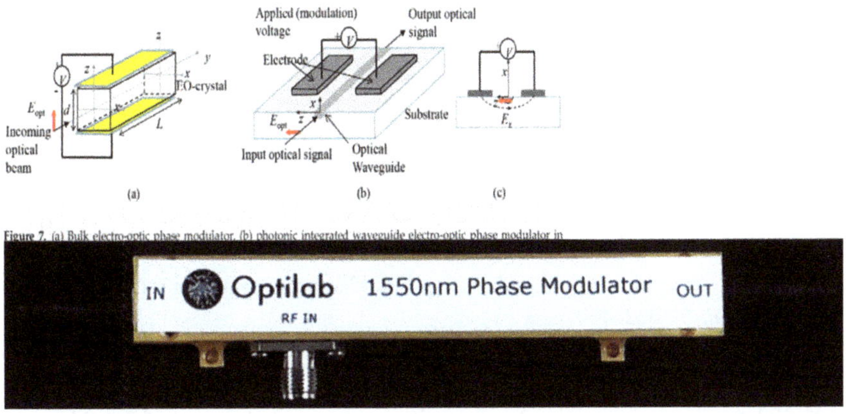

Figure 4.32. Optical phase modulator, integrated structure, and products. *Note*: There is only one RF port.

4.6.3.2 OPM: optical phase modulator

See figure 4.32.

4.6.3.3 AWG: array waveguide grating

See figure 4.33.

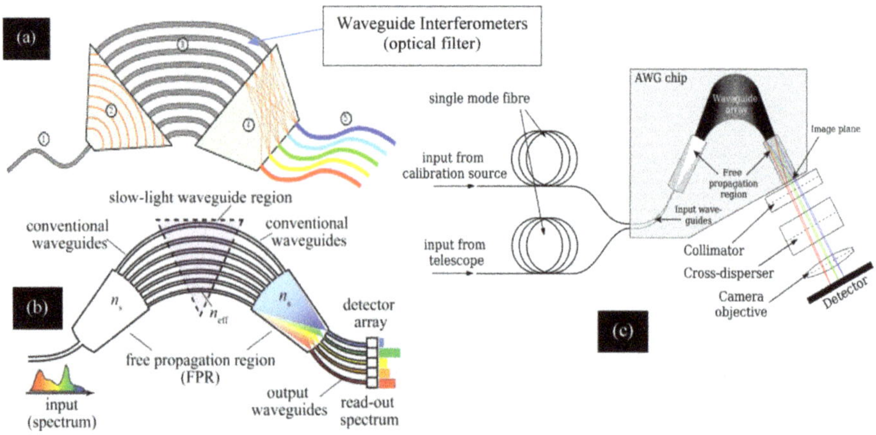

Figure 4.33. An AWG optical device, input port 1 or higher, and output ports. (a) Single input multiple output filter ports. (b) Single input to planar optical waveguide acting as a splitter and distributed to optical waveguides and a pair of optical waveguides form interferometers of different lengths to give filters of different frequencies. (c) The arrangement of two port fiber inputs to the AWG and output of multiple filtered optical ports.

References

[1] Agarwal A, Banwell T and Woodward T K 2011 Optically filtered microwave photonic links for RF signal processing *IEEE J. Lightw. Technol.* **29** 2394–401

[2] Chang W S C 2002 *RF Photonic Technology in Optical Fiber Links* (Cambridge: Cambridge University Press)

[3] Seeds A J and Williams K J 2006 Microwave photonics *IEEE J. Lightw. Technol.* **24** 4628–41

[4] Woodward T K, Banwell T C, Agarwal A, Toliver P and Menendez R 2009 Signal processing in analog optical links *Proc. IEEE Avionics, Fiber-Optics and Photonics Technology Conf.* paper TuC2

[5] Capmany J, Ortega B, Pastor D and Sales S 2005 Discrete-time optical processing of microwave signals *IEEE J. Lightw. Technol.* **23** 702–23

[6] Minasian R A 2006 Photonic signal processing of microwave signals *IEEE Trans. Microw. Theory Technol.* **54** 832–46

[7] Agarwal A *et al* 2011 Optically filtered microwave photonic links for RF signal processing applications *IEEE J. Lightw. Technol.* **29** 2394–401

[8] Xu M and Karim A 2006 Microwave photonic link with carrier suppression for increased dynamic range *Fiber Integr. Opt.* **25** 161–74

[9] Binh L N 2018 *Noises in Optical Communications and Photonic Systems* (Series on Optics and Photonics) 1st edn (Boca Raton, FL: CRC Press, Taylor and Francis)

[10] Binh L N 2008 *Digital Optical Communications Digital Optical Communications* (Series in Optics and Photonics) 1st edn (Boca Raton, FL: CRC Press)

[11] Agarwal A, Banwell T and Woodward T K 2011 Optically filtered microwave photonic links for RF signal processing applications *IEEE J. Lightw. Technol.* **29** 2394–401

[12] Binh L N 2007 *Photonic Signal Processing* 1st edn (Boca Raton, FL: CRC Press, Taylor and Francis)

[13] Kanno A *et al* 2011 40 Gb/s W-band (75–110 GHz) 16-QAM radio-over-fiber signal generation and its wireless transmission *Opt. Express* B **19** 56–63
[14] Khan M H *et al* 2010 Ultrabroad-bandwidth arbitrary radiofrequency waveform generation with a silicon photonic chip-based spectral shaper *Nature Photon.* **4** 117–22
[15] Bagheri R *et al* 2006 Software-defined radio receiver: dream to reality *IEEE Commun. Mag.* **44** 111–8
[16] Naglich J, Maxey C and Guyette A C 2014 Putting the radio in 'software-defined radio': hardware developments for adaptable RF systems *Proc. IEEE* **102** 307–20
[17] Ipatov A *et al* 2010 Radio telescope focal container for the Russian VLBI network of new generation *Sixth Int. VLBI Service for Geodesy Astronomy*
[18] Binh L N 2008 *Digital Optical Communications Systems* (Boca Raton, FL: CRC Press)
[19] Ghelfi P *et al* 2014 A fully photonics-based coherent radar system *Nature* **507** 341–5
[20] McKinney I D 2014 Technology: photonics illuminates the future of radar *Nature* **507** 310–2
[21] Binh L N 2017 *Signal Processing: Optical Transmission and Coherent Receiving Techniques* (Boca Raton, FL: CRC Press, Taylor and Francis)
[22] Li P *et al* 2013 Down-conversion and linearization of X- and K-band analog photonic links using digital post-compensation *Optical Fiber Communication Conf. 2013* 59 paper JW2A
[23] Yu H *et al* 2016 All-optical full-band RF receiver based on an integrated ultra-high-Q bandpass filter *IEEE J. Lightw. Technol.* **34** 701
[24] Binh L N 2019 *Photonic Signal Processing* 2nd edn (Boca Raton, FL: CRC Press)
[25] Binh L N 2008 *Digital Optical Communications Systems* (Boca Raton, FL: CRC Press)
[26] Dingel B and Aruga T 1999 Properties of a novel non-cascaded type, easy-to-design, ripple-free optical bandpass filter *IEEE J. Lightw. Technol.* **17** 1461
[27] Bridges W and Shaffner J 1995 Distortion in linearized electrooptic modulators *IEEE Trans. Microwave Theory Technol.* **43** 9
[28] Tazawa H and Steier W 2005 Bandwidth of linearized ring resonator assisted Mach–Zehnder modulator *IEEE Photonics Tech. Letters* **17** 1851
[29] Tazawa H, Kuo Y-H, Dunayevskiy I, Luo J, Jen A, Fetterman H and Steier W 2006 Ring resonator-based electro optic polymer traveling-wave modulator *IEEE J. Lightw. Technol.* **24** 3514
[30] Madamopoulos N and Dingel B 2006 Performance analysis of the ultra-linear optical intensity modulator *Proc. SPIE* **6390** 63900D
[31] Zhu K F *et al* 2019 Improved heterodyne scheme for RoF system based on optical carrier suppression and optical injection locking techniques *IEEE J. Opt. Commun.* **42** 133–9
[32] Lu M *et al* 2012 Highly integrated optical heterodyne phase-locked loop with phase/frequency detection *Opt. Express* **20** 9736
[33] Laroche M *et al* 2008 Serrodyne optical frequency shifting for heterodyne self-mixing in a distributed-feedback fiber laser *Opt. Lett.* **33** 2746–8
[34] Dat P T *et al* 2014 High-capacity wireless backhaul network using seamless convergence of radio-over-fiber and 90-GHz millimeter-wave *IEEE J. Lightw. Technol.* **32** 3910–23

Chapter 5

Digital RFoF access transmission technology

5.1 Introduction

5.1.1 Overview

Over the last decade, fixed lined optical transport has become the predominant front-haul technology for 4G, 5G, 6G, and 7G mobile access networks (MANs). Furthermore, transmission and networking have also evolved into new functional splits and related data center based networks, thus the evolution of the distribution of networking into cloud-based radio and wireless access networking.

This chapter presents, first, the capacity demands and metro and access optical networking employing complex modulation formats and coherent reception or higher order modulation formats in one dimension, the pulse amplitude modulation (PAM). Such ultra-high capacity transmission is most applicable in the networking evolution from long-haul, metro, and access into all metro-like networking. Furthermore, these are also much in demand for client access to data centers (DCs), both internally within the dc or dc interconnections (DCIs), and interconnections between DCs, cloud radio access networks (RANs), and fog RANs. The target of this chapter is economical 180 Gbps complex signaling and direct detection for ultra-high bit rates, and in coherent transmission technology this aggregate bit rate can reach more than one Tbps with 200 GBd as the symbol rate.

Second, the modulation formats and complex signaling by double sideband (DSB), duo-binary (DB), and PAM, or DSB-DB-PAM-4, generation for data transmission, under either a pure optical carrier or RF modulated optical carrier. Third, the experimental platform is reported and the performance of the transmission systems is provided.

The intensive transport of data capacity to the access nodes would be due mainly to the information hungry 6G and 7G networks in which wireless transmission using multiple-input and multiple-output (MIMO) antennae is used to obtain gain to compensate for losses at the mmW range and beam steering. Thus a proposed

doi:10.1088/978-0-7503-3511-9ch5

integrated circuit scheme is made using RF-MIMO-radio-photonic technology (RAPT). This is given in section 5.7.3.

5.1.2 Network evolution

Over the first two decades of the twenty-first century, the explosion of wireless networks and individual user devices has put tremendous pressure on Internet data networking. The evolution from traditional seven-layer telecoms networks to global data center cloud-based networks is illustrated in figure 5.1. Figure 5.2 shows the trends of market demands on data delivery and the transmission rates for such intercontinental and local access networking. The Internet of Everything (IoET) requires minimum delays for data interconnections and thus the long delays of traditional telecoms networks cannot meet such demands. Thus we can see the flattening of telecoms and data cloud networking so as to provide fast, high capacity networks with minimum delay times. Figure 5.3 shows the limitations of such approaches and the ways to overcome them.

5.2 Transmitter: complex signaling by the phasor

This section briefly presents the basics of digital coherent transmission techniques using complex signaling and orthogonality for access and long-haul systems, including interconnection in DCs with extremely high capacity.

Orthogonal frequency division multiplexing (OFDM) offers compact channel packing due to the orthogonal properties of the spectral components [1]. OFDM can cope with a virtually unlimited amount of inter-symbol interference (ISI) provided there is fast sampling and digital formation of the data channels. In 6G and 7G RPof systems, the data channel bandwidth is much lower than of the RF carrier in the mmW region. ISI can be caused by fiber chromatic dispersion (CD) and fiber

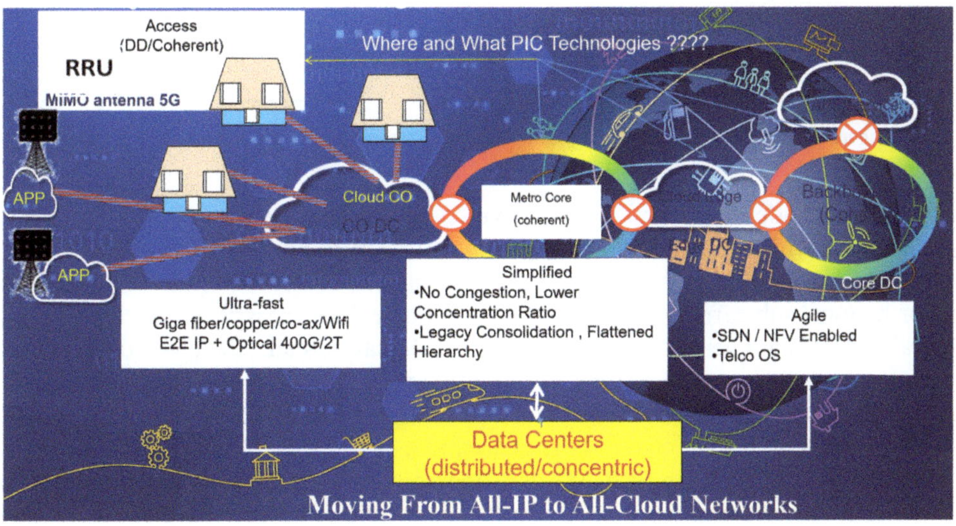

Figure 5.1. Evolutionary networks: flattening telecoms cloud networks and dc centric networks.

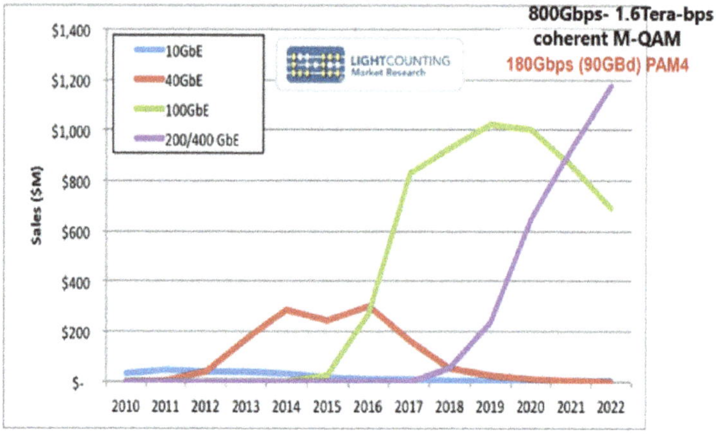

Figure 5.2. Market demands for transmission links with Ethernet bit rates much higher than 100 Gbps and RF sub-carrier multiplexing, employing direct detection (IM/DD) or coherent reception techniques for higher Ethernet bit rates.

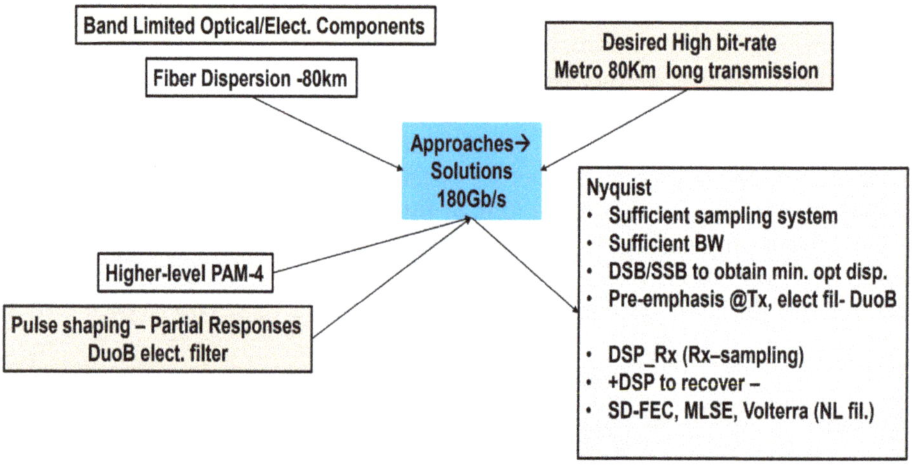

Figure 5.3. Limitations and principles to overcome band limited transmission techniques.

polarization mode dispersion (PMD), which are serious issues in long- and short-haul systems with RF carriers in the μW and mmW region on the optical carrier.

In principle, OFDM is the division of a high bit rate data stream into several low bit rate orthogonal sub-carrier (SC) streams which can be the RF carriers in multiplexing. In general, the SCs are generated in the electronic digital domain, therefore these systems typically consist of many SCs (typically more than 2^N with $N = 16$ or higher). Channel estimation is realized by training symbols. In optical fiber communication systems (OFCSs) with an OFDM channel the SCs can be generated in the optical domain.

OFDM SC multiplexing (SCM) can operate under coherent reception. Several wavelength carriers can also be multiplexed to give coherent wavelength division multiplexing (WDM) systems.

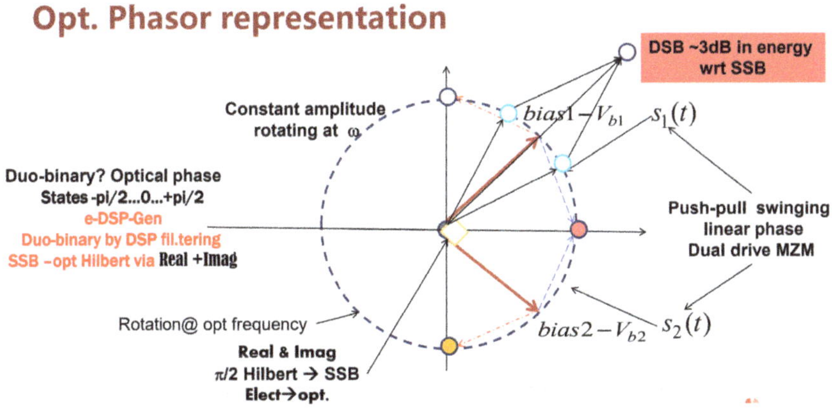

Figure 5.4. Optical phasor representation of complex modulation for duo-binary and single sideband (SSB).

For optical OFDM systems, the transmitter consists of an external optical modulator so as to keep the laser linewidth stable and narrow. An OFDM signal is generated from an arbitrary waveform generator (AWG) applied to the RF multiplexer and fed to the modulator, hence modulating the optical carrier.

After transmission through an optical fiber, commonly a length of SSMF, the signals are received by an opto-electronic receiver which can be of either coherent (Co-) or a direct detection (DD-) type. A DD-OFDM can be realized by sending the optical carrier along with the OFDM band so that DD with a single photodiode (PD) can be used at the receiver.

In a Co-OFDM system, the optical carrier can be suppressed at the transmitter and a local oscillator (LO) is used to mix both polarized modes of the signals via an optical hybrid coupler, as shown in figure 5.2, so that both polarized data channels can be recovered in a separate system. This would increase the channel capacity by a factor of 2.

The superior performance of a Co-OFDM with respect to the optical signal-to-noise ratio (OSNR) requirements, PMD tolerance, and spectral efficiency makes it an excellent candidate for long-haul transmission systems, whereas DD-OFDM, which requires fewer components at the receiver than Co-OFDM, is more suitable for cost-effective short reach applications. The spectra of both these OFDM schemes are shown in figure 5.4. The DSP processing of the DD-OFDM is much simpler as only the real part of the OFDM signals is required to be generated. The alternative name for DD-OFDM is discrete multi-tone (DMT).

An optical wave under modulation with in-phase and quadrature phase components or complex modulation is illustrated in figure 5.5. A structure schematic of a coherent optical receiver for Co-OFDM is given in figure 5.6. In the transmitter there are two modulation sections, one for the real part or the in-phase component and the other for the imaginary part or the quadrature phase component. Both modulators can be combined to provide multi-level modulated levels with constellations of 2D and their phases can be detected via coherent reception. For example M-ary or N-ary quadrature amplitude modulation (QAM) so the complex plane can be illustrated by $M \times N$ constellation points, as is illustrated later in figure 5.7.

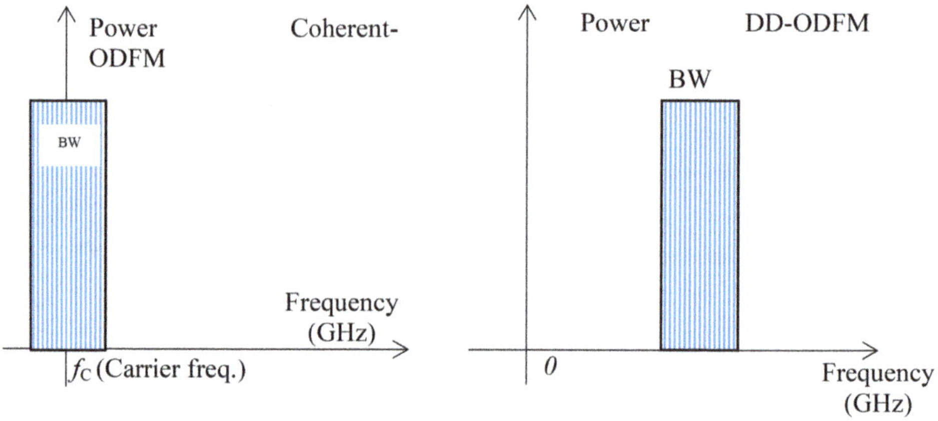

Figure 5.5. Spectra of optical OFDM—Co-OFDM and DD-OFDM. There is a sharp roll-off at the spectrum boundary and hence compact channel multiplexing.

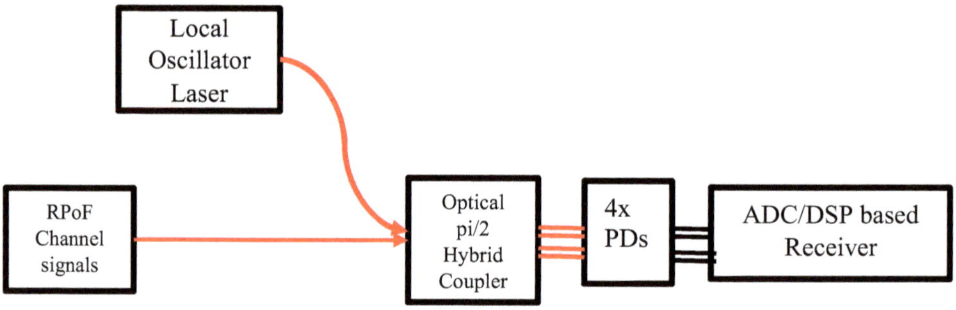

Figure 5.6. Schematic of an optical receiver for Co-OFDM. PD= photodetector; ADC = analog-to-digital converter; DSP = digital signal processor; RPoF = radio-photonics over fiber.

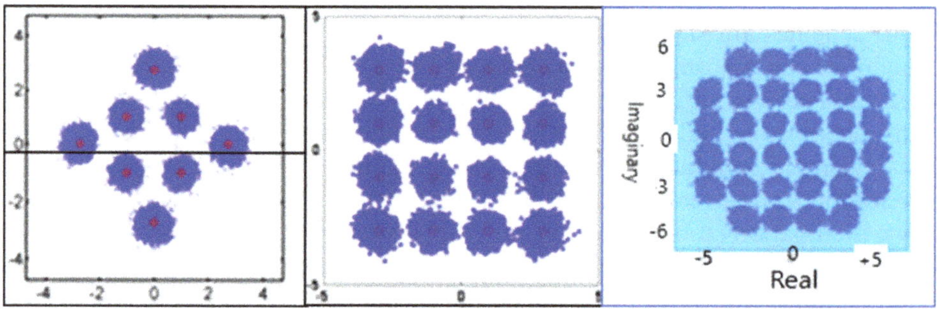

Figure 5.7. The optical spectrum of a 122 Gbps OFDM data channel, intensity versus RF shift wavelength of four channels with a total bandwidth (BW) of about 28 GHz or 28 × 0.8/100 = 0.224 nm.

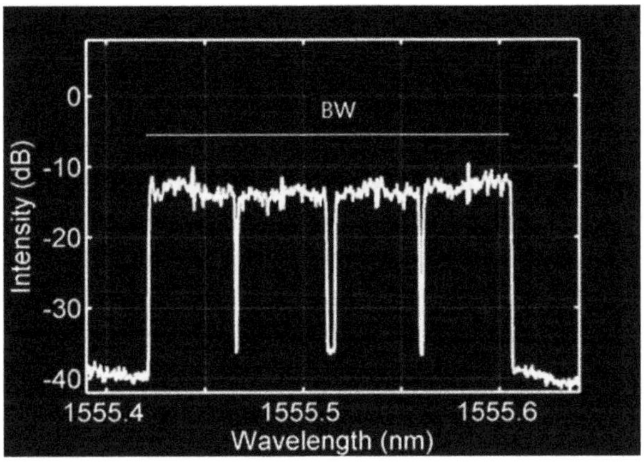

Figure 5.8. Multi-level modulated RF carrier constellations under (a) 8QAM, (b) 16QAM, and (c) 32QAM. BW = 3dB-bandwidth.

Both Co-OFDM [1] and DD-OFDM [2, 3] systems with 200 Gbps transmission experiments have been reported. To increase the bit rate to 100 Gbps, polarization division multiplexing (PDM) can assist to double the 50 Gbps and thus reduce the signal bandwidth, and hence the component operating speed.

Under digital processing ADC a DSP can be used after the analog front-end, converting the optical signals to the electronic domain. In such a PDM-OFDM, the polarized channels can be separated, processed, and correlated via MIMO algorithms [4].

The $\pi/2$ optical hybrid coupler at the front of the opto-electronic converter/receiver can be used in association with a training sequence to optimize the receiver sensitivity. Co-OFDM transmission of 10 $\times$ 122 Gbps over a 1000 km SSMF without any dispersion compensation could be achieved. In this experiment a 122 Gbps OFDM signal, consisting of two PDM channels with four SCM OFDM bands, is used.

In association with non-rectangular 8QAM constellations and symbol coding, as depicted in figure 5.8(a)), a 122 Gbps OFDM signal can be packed over only a 23 GHz spectral window, as shown in figure 5.7.

At the opto-electronic (OE) receiver, the signal is coherently detected, first with the mixing of the LO and polarized signal channels via a $\pi/2$ optical hybrid coupler and an external cavity LO laser. A real-time digital signal analyzer (DSA) is used to sample the electrical outputs of the balanced detector pair coupled to the optical output of the optical $\pi/2$ hybrid coupler port and the acquired data are post-processed. The samples are processed off-line. Polarization reversed rotating can be realized through MIMO correlation processing. RF-aided phase noise compensation can be used. After 1000 km SSMF transmission, the bit error rates (BERs) for all channels are below the threshold with a 7% overhead forward error corrector (FEC). The OFDM signal can be continuously recovered with a CD tolerance higher than 18 500 ps nm^{-1}. Such a large CD dispersion tolerance is attractive for high speed transmission systems and RPoF systems because an inline dispersion compensator may not be needed at all, thus simplify the optical amplifications and extending the repeatered distance (figure 5.8).

5.3 Transmission using band limited signaling of DSB-DB-PAM-4

This section presents briefly the basics of digital coherent transmission techniques using the real part of the complex signaling with orthogonality, the DMT modulation, for access and long-haul systems, including interconnections in data centers with extremely high capacity.

We take here a typical example of a transmission system for short-haul and high bit rate with low cost.

5.3.1 Parameters

The operational parameters of the system would be the following for low-cost transmission systems for wireless–optical transmission systems:

- Baud rate: 90 GBd; bit rate = 180 Gbps; DSP + DAC at Tx.
- Modulation formats: PAM-4 DB and DSB/SSB (SSB ◄ gain in dispersion); duo-binary in the electrical domain or by filtering with a bandwidth half that of the full channel bandwidth.
- Modulator: External dual drive MZIM; bandwidth (BW) = 30 GHz; DAC = 92 GSa s^{-1}; biased at the $\pi/4$ quadrature point; swing over the linear range.
- Pre-emphasis (digital) + tight filtering (raised-cosine to fit the signal BW).
- Fiber: standard single mode fiber (SSMF) or Corning SMF-28, SMF-28e.
- DSP-receiver: direct detection (DD) with a receiver (Rx) sensitivity = 150 V A^{-1}; BW= 50 GHz. DSP: sampling rate 160 GSa s^{-1}; BW = 63 GHz; processing algorithm = Volterra filtering (linear and nonlinear distortion) plus the maximum-likelihood sequence estimation (MLSE) (residual distortions).

In order to overcome the limited bandwidth of the components, both the optical and electronic ones, a duo-binary approach in the electrical filtering domain is considered and implemented. That is, the filter is only half of the 3 dB bandwidth. Thus a four-level PAM (PAM-4) can produce a seven-level eye diagram at the output of the transmitter and received at the reception end. DSP can be used to implement and recover the data channel.

5.3.2 The DSB-DB-PAM-4 transmission system

See figure 5.9.

5.3.3 Performance

Figures 5.10(a)–(e) depict, respectively, the received eye diagrams of the transmission system employing the following modulation formats (a) DSB in back-to-back (B2B), (b) SSB in B2B, (c) DSB over 40 km of SSMF, (d) DSB+CDC over 40 km SSMF, and (e) SSB in B2B without $-\pi/4$ rotation. Obviously, the DSP is

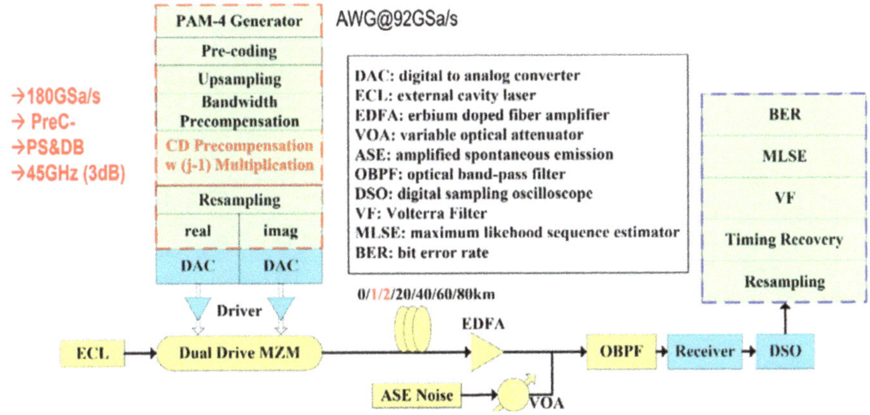

Figure 5.9. Complex signaling leading to 180 Gbps by direct detection and digital signal processing. Processing algorithms in a stack or sequence of execution for generating the modulation format at the transmitter and recovery processing at the receiver side. A complex system equivalent to two Euclidean real and imaginary orthogonal systems. Reprinted from [5], copyright (2020) IOP Publishing. Reproduced with permission. All rights reserved.

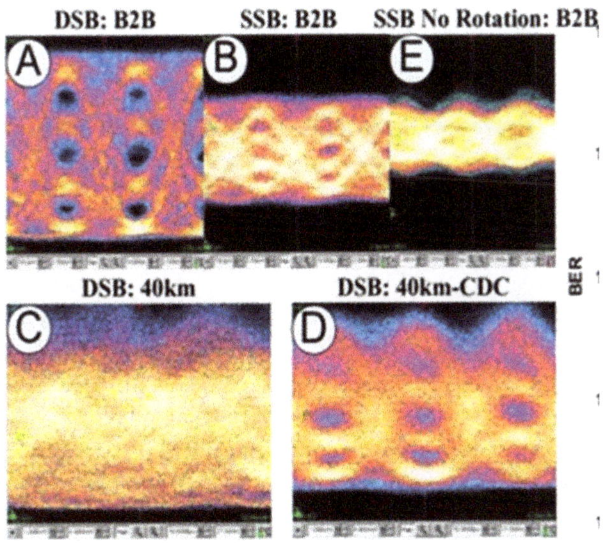

Figure 5.10. Received eye diagrams of systems under modulation (a) DSB in B2B, (b) SSB in B2B, (c) DSB after 40 km fiber, (d) DSB+CDC after 40 km fiber, and (e) SSB in B2B without $-\pi/4$ rotation. B2B = back-to-back transmission. Reprinted from [5], copyright (2020) IOP Publishing. Reproduced with permission. All rights reserved.

required to 'open' the eye for errorless recovery of the data bits. The BERs of these transmission systems are illustrated in figure 5.10.

Figure 5.11 depicts the BERs corresponding to the transmission systems (a)–(b) of figure 5.10. The SSB shows much better performance due to the fact the only a single

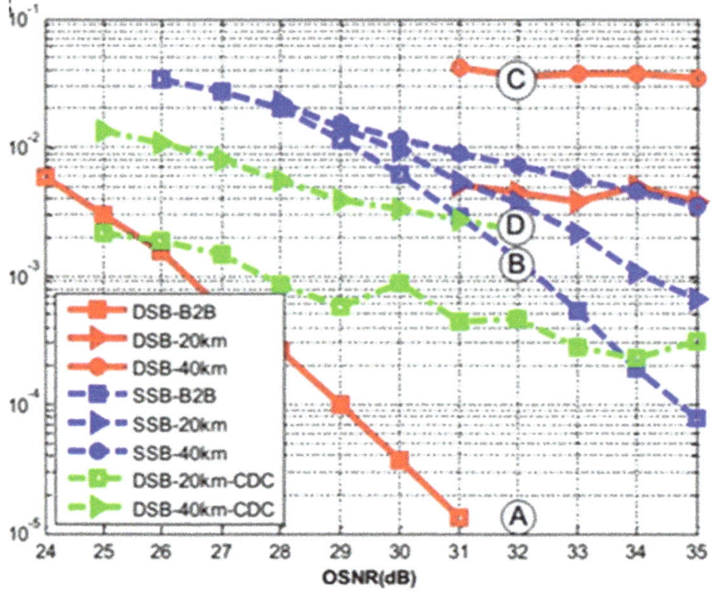

Figure 5.11. BERs corresponding to the transmission systems (a)–(e) of figure 5.10. Reprinted from [5], copyright (2020) IOP Publishing. Reproduced with permission. All rights reserved.

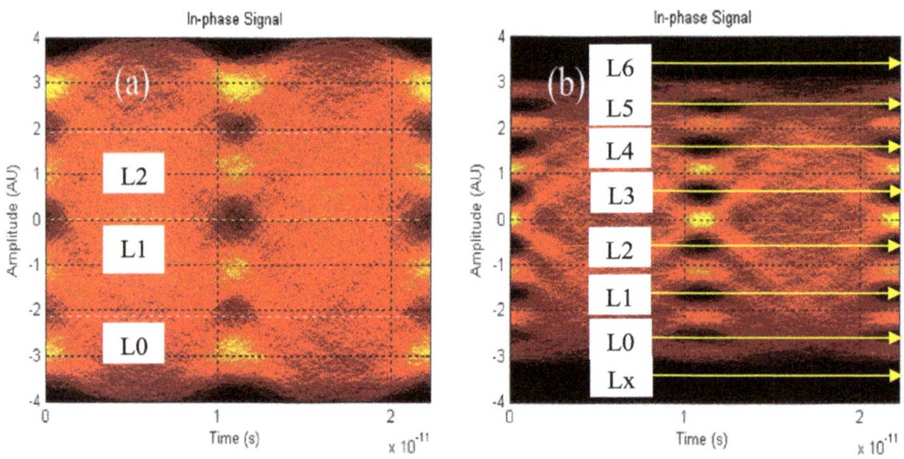

Figure 5.12. Received eye diagram for modulation format (a) PAM-4; and (b) DB-PAM-4 in-phase signals or the real part of the detected signals. Note the seven-level received signals due to the half band filtering. Receiving digital processing is implemented to translate such a doubling multi-level to PAM-4 signaling at the transmitter. L = level for decision.

sideband channel is transmitted and thus no interference effects of the other bands will occur as in the DSB modulated signals. Figure 5.12 shows the recovered eye diagrams of PAM-4 and DB-PAM-4. The in-phase signals or the real part of the detected signals indicate that the digital processing of duo-binary PAM-4 is like

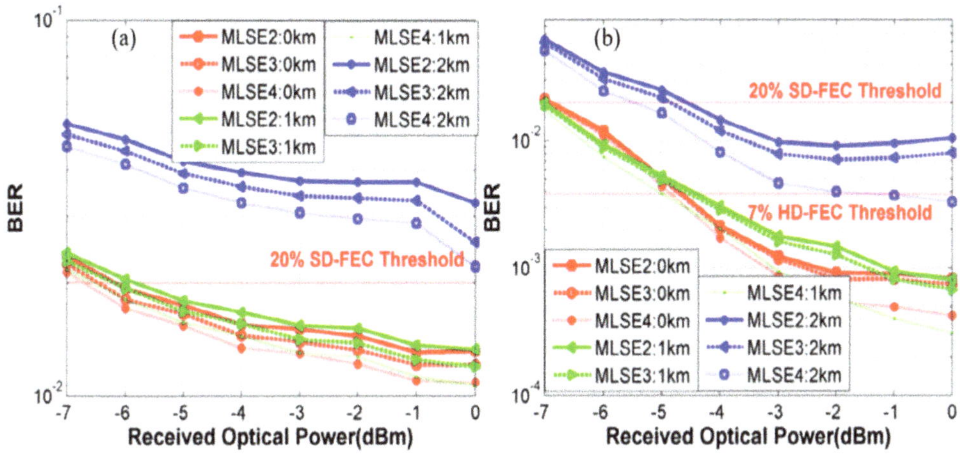

Figure 5.13. Performance of (a) PAM-4 and (b) duo-binary (electrical) PAM-4 systems, using MLSE equalizers [5].

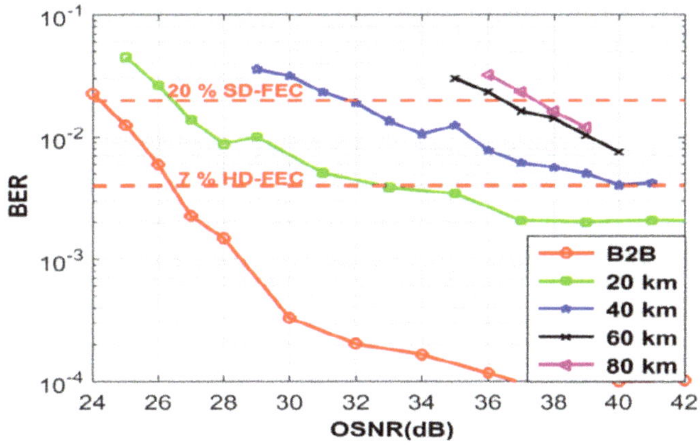

Figure 5.14. Transmission system sensitivity employing DB-PAM-4 modulation format CD pre-compensation of different fiber lengths as a parameter [5].

the processing of an equivalent PAM-8 and does not have difficulty using a high sampling rate DSP. Note the seven-level received signals are due to the half band filtering. Receiving digital processing is implemented to translate such a doubling multi-level to PAM-4 signaling at the transmitter

Figure 5.13 gives the performance of a transmission system employing the PAM-4 and PAM-4 duo-binary modulation formats with electrical filtering systems. Figure 5.14 depicts the transmission performance of a system employing DB-PAM-4 with dispersion pre-compensation over different lengths of fibers when the extension of the transmission distance is required, such as in the case of access distance in remote areas.

5.4 The single sideband duo-binary PAM-4 (SSB-DB-PAM-4) system

Figure 5.15 depicts schematics of the DAC based transmitter and ADC based DD optical receiver (ORx). The modulation format is SSB-DB in the electrical domain, that is, the electrical signals are modulated at a half-rate or filtered at a 50% rate. This electrical signal sequence is applied to a dual drive Mach–Zehnder interferometric modulator (DD-MZIM). The DD-MZM is biased in such a way that the suppression of the optical carrier occurs, that is, under D-phase biases of π difference on two optical waves passing through the two optical paths of the MZIM. This modulation format can offer minimum interference which is serious under the double sideband. In addition, the DB in the electrical domain reduces the required bandwidth to half, thus narrower bandwidth components can be used.

A length of SSMF is used for signal transmission and its loss is compensated by an optical amplifier (EDFA type). An inline optical filter is inserted which can be tunable to extract only the signal band. Under RPoF the RF carrier shifts such a band by an spectral amount of f_{RF}. In this case the RF shifted optical carrier still exists and such filtering provides a vestigial sideband-modulated spectrum. Hence the beating of such an RF modulated optical carrier and the optical signal band in the PD result in an RF carrier under the data channel envelop.

At the receiver, an RF broadband amplifier is used to recovered the number of SCM channels. A fast sampling unit is used to obtain digital samples of the RF carriers and data channel envelop. The processing under advanced DSP can compensate for samples of data and RF carriers with such lower bandwidth components. For example, the seven-level PAM-4 under a duo-binary approach can be processed easily. Figure 5.16 shows the optical spectra of the 90 GBd signals

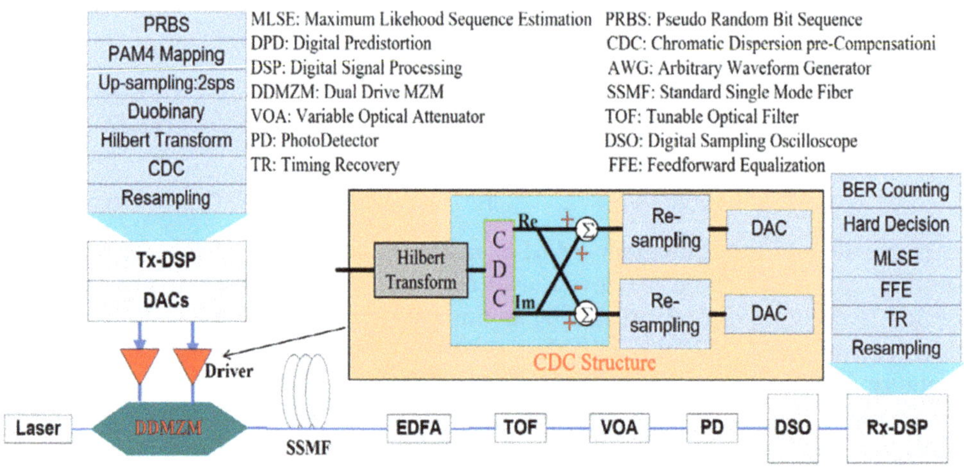

Figure 5.15. The experimental set-up and off-line DSP of an SSB-DB-PAM-4 system including software processing steps at the transmitter and reception. The Hilbert transform is effectively a $\pi/2$ transfer function. The insets show the detailed processing algorithms [5].

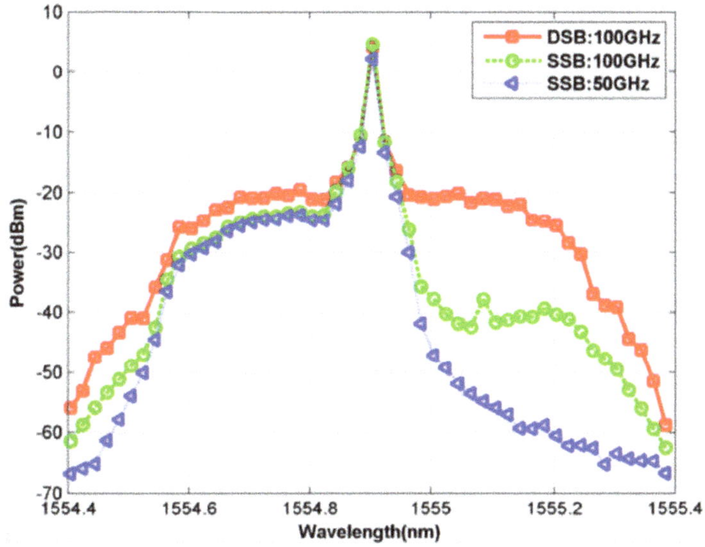

Figure 5.16. Optical spectra of 90 GBd signals of a DSB/SSB-DB-PAM-4 [5].

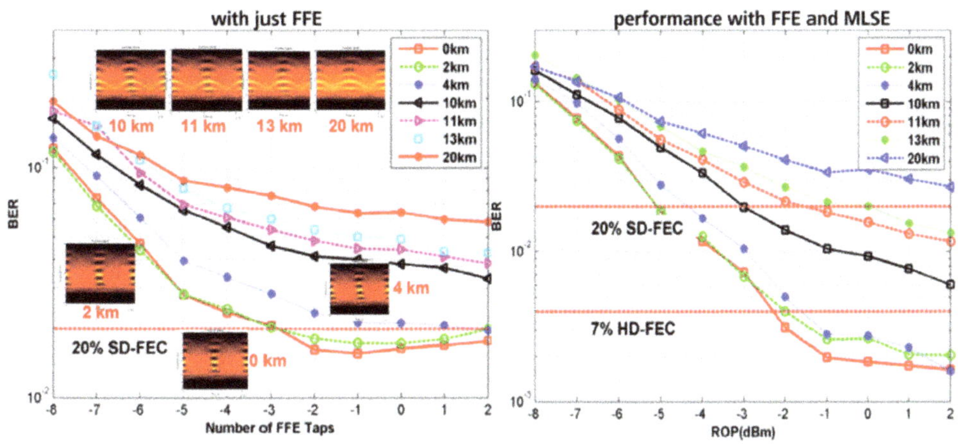

Figure 5.17. Optical spectrum of 90 GBd DB-PAM-4 signals and performance under processing (a) with just FFE and (b) with FFE and additional MLSE equalizers [5].

of a DSB/SSB-DB-PAM-4. It shows the effectiveness of optical filtering. The main difficulty is the sharp roll-off of the filter band edges. It requires at least 30–40 dB roll-off so as to obtain a signal band distinct from the noise level. We note that in photonics, the wider the bandwidth the easier the optical filtering. The optical carrier is spaced by 100 GHz with guardbands or 0.8 nm in wavelength space.

Figure 5.17 shows the optical spectrum of 90 GBd DB-PAM-4 signals and its corresponding performance under processing with just feed-forward equalization (FFE) and with FFE and additional MLSE.

5.5 Optical time interleaving for >200 GBd

An increasing symbol rate is important to minimize system components, in particular in the environment of DCIs or data center internal connection networking (DCN) using a sub-carrier multiplexed RPoF.

The optical spectrum of 90 GBd DB-PAM-4 signals and the corresponding performance under processing with only FFE and with FFE and additional MLSE. This is possible only if a high speed digital-to-analog converter (HS-DAC) is available to generate a high symbol rate signal baud rate which can approach 200 GBd and beyond.

The sampling rate of DACs can reach 128 GSa s^{-1} and even ~200 GSa s^{-1} by transmission line interleaving and propagate over distributed electrical Faraday-boxed lines. Integrated SiGe circuit technologies [6, 7] can offer broadband amplifiers using distributed amplification circuits in which the transition frequency–gain product can be achieved. Alternatively, 120 GSa s^{-1} can be achieved using complementary metal-oxide semiconductor (CMOS) technology [8]. This can be used for RF carriers of up to 60 GHz. The processing of the data channel, commonly using an OFDM, can be processed in the lower sampling rate with a higher effective number of bits and higher-level ADC, e.g. a 16 bit ADC. Figure 5.18 shows recent digital coherent experimental results with symbol rates over 110 GBd with optical time multiplexing based on an integrated photonic delay interferometer.

The modulated data channels are generated via two AWGs and multiplexed through electrical time domain multiplexing (ETDM) digital multiplexers to form 212 GBd signals.

The electronic speed in SiGe can offer a bandwidth to >110 GHz with a transition frequency of 475 GHz. Time domain multiplexing in a time multiplexer can be used to increase the baud rate by time division multiplexing. Figure 5.19 shows such an ETDM scheme for ultra-high bit rate access transmission. The ETDM signals are then driving the GaAs based IQ modulator.

Alternatively, in the second technique doubling of the baud rate can be time multiplexed in the analog domain by time interleaving in the optical domain.

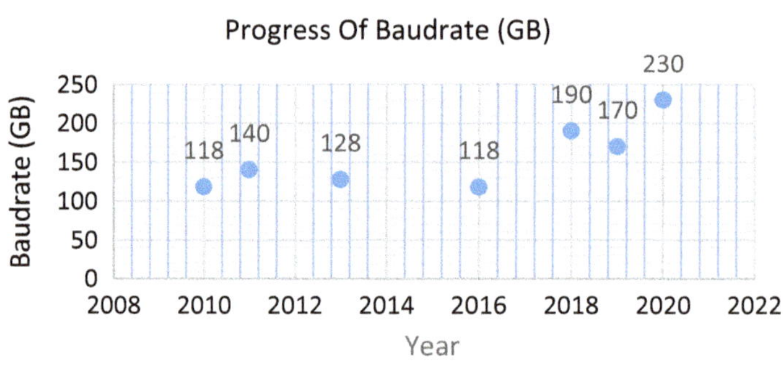

Figure 5.18. Annual progress > 110 GBd for systems employing digital complex modulation and coherent reception with IQ optical modulators. An optical transmitter structured to reach such ultra-high symbol rates can be seen in figure 5.21.

A photonic pre-processing front-end using an integrated optic Si–Ge circuit, shown in figure 5.17(b), can be used with time division multiplexing of the two optical guided paths of modulators using electro-absorption modulators (EAMs) with integration of electronic driving amplifiers. The generic scheme of the modulation system is shown in figure 5.17(b). A germanium film is deposited on the SiGe waveguide to structure an EAM due to the highly efficient absorption of such films and its crystallographic properties matching those of SiGe.

In increasing the baud rate two approaches, through either electronic or photonic circuits, are possible. First, the electronic approach, in which the sampled channels are time domain multiplexed via a high speed digital switch, which is now possible with SiGe technology. This first approach is a DAC based strong approach. Alternatively, the photonic approach is structured such that the optical channels

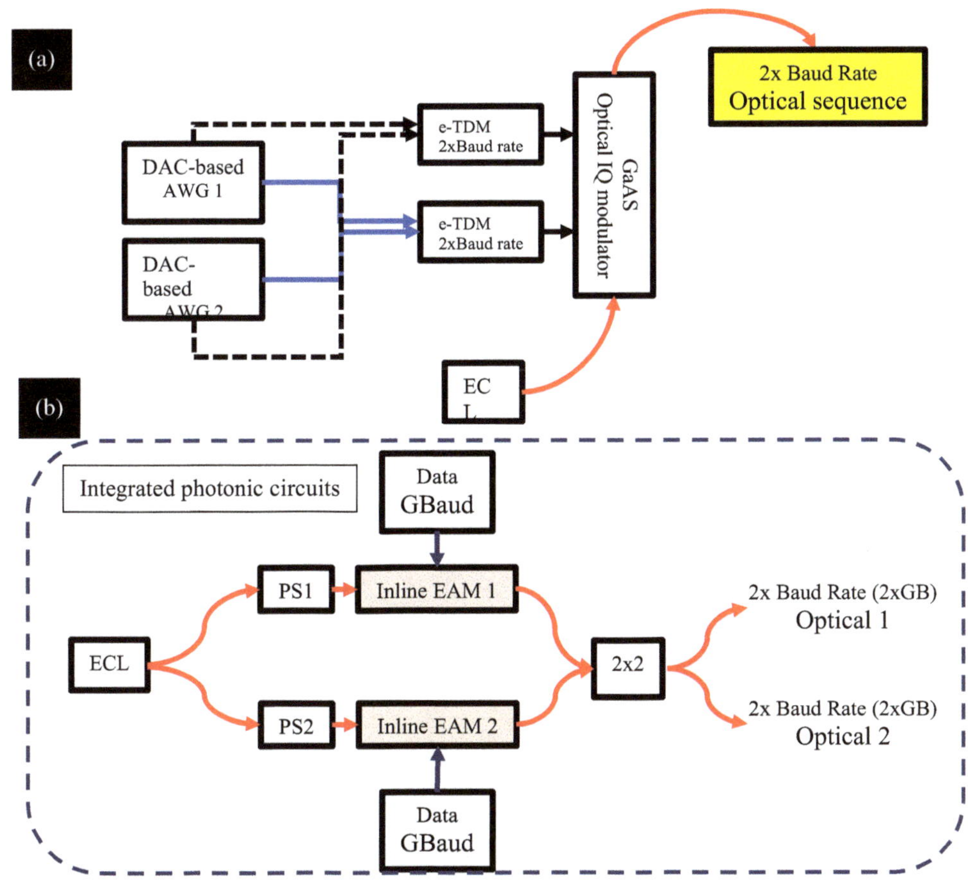

Figure 5.19. (a) A DAC based and ETDM approach to double the symbol rate via integrated photonic circuits on a SiGe platform and Ge-film for EAM. (b) Schematic of a DAC-less doubling symbol rate via optical multiplexing. DAC = digital-to-analog converter; ETDM = electronic time domain multiplexer (switch); ECL = external cavity laser; EAM = electro-absorption modulator; PS = phase shifter; GB = gigabaud (symbol); AWG = arbitrary waveform generator.

are modulated in two parallel optical modulators and then multiplexed at the combined output [9]. This is a DAC-less approach. The schematics of these two approaches are shown in figures 5.18(a) and (b), respectively.

A value of 192 GBd is possible with ETDM of a single-carrier under QPSK modulation and an integrated electro-optical (EO) front-end. A symbol rate of 220 GBd for QPSK signaling is achieved by using a 256 GSa s^{-1} AWG, a >50 GHz 6 dB bandwidth GaAs QPSK modulator, and advanced nonlinear DSP [14, 15]. The highest symbol rate transmission system is of great interest for distribution to massive data transport and storage in 6G and 7G networking in access networks. Eye diagrams at 28 and 56 GBb are given in figures 5.18 and 5.20. The time interleaving is implemented by an optical delay line on one of the guided wave paths of the MZM structure.

The binary eye diagrams of the upper side branch (USB) and lower side branch (LSB) as detected, monitored, and tested as a single branch indicate the good integrated optic modulation of the EAM (figure 5.19). Then the combined two binary non-return-to-zero sequences to form the PAM-4 sequence is detected for 28 and 56 GBd. This proven high-performance PAM-4 could be transmitted with a high baud rate using photonic based ETDM.

For RPoF a composite baud rate can be achieved by a number of sub-carrier multiplexing methods. Each of the RF sub-carriers carries a data channel at a bit rate of 1 Gbps. The frequency spacing between the optical carriers determines the total number of RF sub-carriers, the data band, and the guardband. The OFDM format is the selected modulation format for the data channels.

The progress over the years for the short-haul baud rates of OFCSs for access networking is shown in figure 5.20. It is expected that the rate has reached the level where the DSP power is at maximum capacity under the current Si–Ge micro-electronic technology.

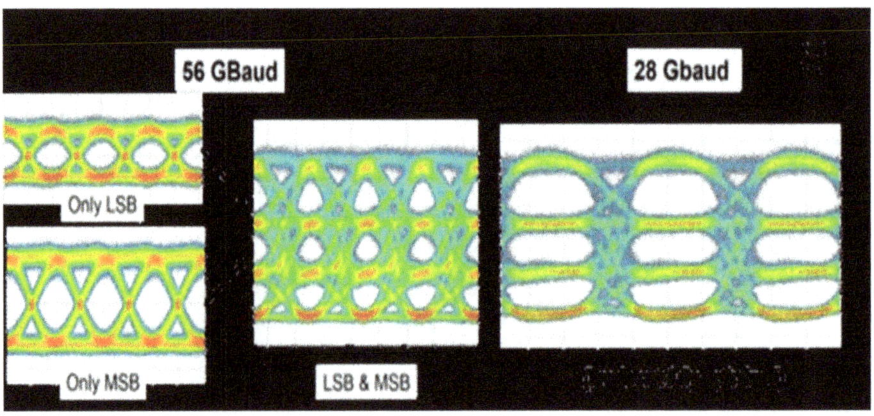

Figure 5.20. Eye diagrams as received at 28 and 56 GBb using the transmitter given in figure 5.18. Using single 128 GSa s^{-1} SiGe DACs, 128 GBd is possible [6, 7], and a higher symbol rate of 120 GSa s^{-1} for CMOS DACs is possible [8]. To realize higher symbol rates, ETDM [10, 11] or spectral multiplexing techniques [12, 13] can be used. L = level for decision [5].

Figure 5.21 depicts the schematic diagram of an optical transmission system arranged back-to-back with a DSP based transmitter and a DSP based coherent receiver.

The photonic multiplexing platform would enable the baud rate and the composite bit rate to reach higher than $N \times 192$ Gbps with $N > 4$ by using four photonic paths which can be fabricated precisely, as shown in figure 5.21.

Furthermore, under polarization multiplexing the aggregate bit rate can reach >1.0 Tbps using only one wavelength.

5.6 Tbps super-channel transmission by DSP based coherent QAM

PDM-QPSK has been exploited extensively in commercial 100 Gbps long-haul transmission systems, in particular the wavelength multiplexing system technologies in 400 GE/1 TE back-haul transport. Furthermore, single wavelength channel Tbps transmission systems are attractive. Hardware platforms for 1–N Tbps can be critical. Tbps can be considered as a super-channel which is defined as an optical channel comprising a number of sub-rate sub-channels whose spectra would be in

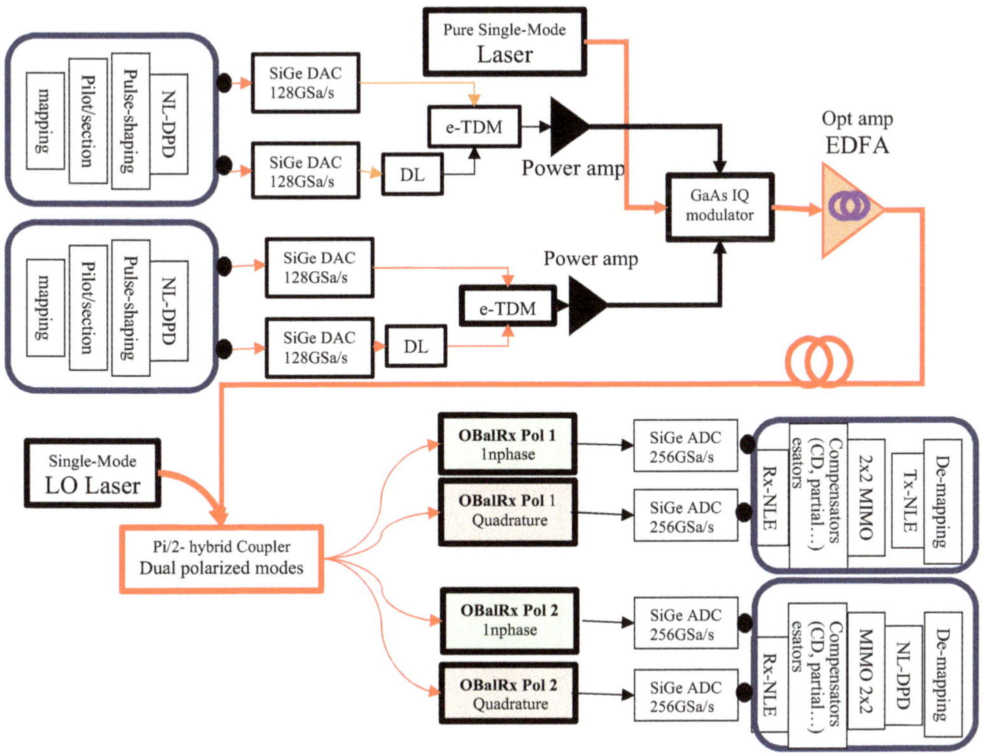

Figure 5.21. Schematic of a B2B testbed of a dual polarization complex QAM transmission system with 230 GBd Tx and reception sub-systems. DPD = data probabilistic distribution; DAC = digital-to-analog converter; ADC = analog-to-digital converter; LO = local oscillator; OBalRx = optical balanced receiver; CD = chromatic dispersion; TDM = time division multiplexer; NLE = nonlinear equalization; DL = optical delay line.

the narrowest allowable spectrum. In order to achieve efficient spectral efficiency, phase shaping and Nyquist pulse shaping are required [16].

Thus in this section we describe a detailed design and experimental platform for delivery of Tbps using Nyquist-QPSK at symbol rate of 28 to 32 GSa s^{-1} and ten sub-carriers. The generation of sub-carriers has been demonstrated using either re-circulating frequency shifting (RFS) or nonlinear driving of an IQ modulator to create five sub-carriers per main carrier, thus two main carriers are required. Nyquist pulse shaping is used to effectively pack multiplexed channels whose carriers are generated by comb generation techniques. A DAC with sampling rate varied from 56 to 112 GSa s^{-1} is used for generating the Nyquist pulse shape including the equalization of the transfer functions of the DAC and optical modulators.

Figure 5.22 shows a schematic of possible transmission for Tbps, using a comb laser and for either broadband channels or sub-carrier multiplexing of several 6G and 7G channels over a number of optical wavelength carriers. Figure 5.23 shows the detailed structure of comb laser based Tbps transport with multiplexers and demultiplexers for bidirectional transmission of several 6G and 7G RF modulated optical channels.

A comb laser provides several optical carriers of equal spectral spacing, e.g. 50 GHz for 56 GBd or 100 GHz for 112 GBd, which are demultiplexed and fed into an array of optical modulators. The data of either broadband or wireless data channels are generated from ultra-high speed DACs (UHS-DACs) and amplified by RF amplifiers to the linear region limit to drive the MZMs.

In the case of RPoF the data channels and the RF carriers are also generated using a UHS-DAC or RF coupler with an electro-photonic oscillator (see chapter 7). The RF modulated optical channels are multiplexed and transmitted over a single optical transmission line, the SSMF.

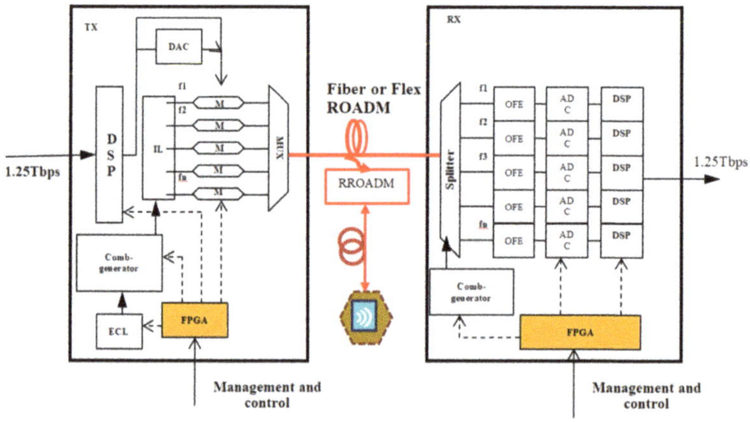

Figure 5.22. A possible structure for a Tbps super-channel or SCM multi-carrier RPoF transmission system, with an SSMF transmission fiber and flexible reconfigurable optical add drop multiplexer (ROADM). OFE = forward error; ADC = analog-to-digital converter; Mux = multiplexer; ECL = external cavity laser; FPGA = ultra-high speed field programmable gate array; M = MZM; IL = insert loss; DAC = digital-to-analog converter. Copyright (2017), reprinted from [16] by permission of Taylor & Francis Ltd.

The multi-RF modulated optical carriers are transmitted over a long fiber transmission line which can pass through several reconfigurable optical add/drop multiplexer (ROADM) inline units in which the carrier can be dropped from and/or injected into the transmission line. This kind of ROADM can be termed as a reconfigured optical-RF add/drop multiplexers (RO-RF-ADM) which can be used to distribute and accept sensing signals from several micro-cell antennae.

At the receiver site, the transmitted multi-channels are demultiplexed and received by either non-coherent or coherent balanced receivers and are then processed via a DSP based system.

An FPGA system is used to provide control signals to various elements of the transmitter sections.

Various compensation algorithms can be used such as feed-forward equalization (FFE) provided from the DSP. The compensated channels are then providing signals for either further transmission or to be used in user terminals.

5.6.1 Nyquist pulse and packing spectra

Packing of channel spectra in a Tbps transport system is critical to increase the capacity over a limited bandwidth allocated for such terabit systems. Pulse shaping is a technique applicable to this requirement. A raised-cosine pulse shape with a certain roll-off factor can satisfy such requirements. The raised-cosine filter is an implementation of a low-pass Nyquist filter.

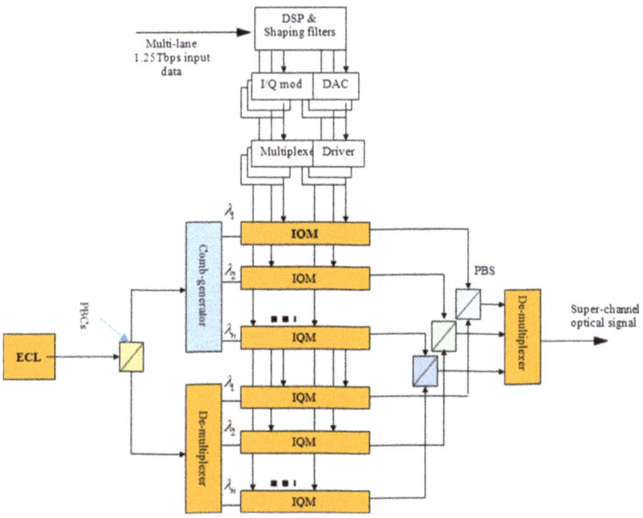

Figure 5.23. Generic detailed schematic of a super-channel or multi-carrier radio-photonic SCM sub-carrier multi-wavelength carrier transmitter. PBC = polarization beam combiner; PBS = polarization beam splitter; I/Q = in-phase/quadrature phase; DAC = digital-to-analog convertor; ECL = external cavity laser; Mod = modulator; IQM = in-phase quadrature modulator or complex modulator. Copyright (2017), reprinted from [16] by permission of Taylor & Francis Ltd.

5.6.2 Super-channel system requirements

For the next generation of backbone transport, the transmission distance should be comparable to the previous generation, namely a 100 Gbps transmission system. The most important requirement is that 1 Tbps transmission for long-haul systems should be 1500~2000 km, and for metro applications ~300 km. Several channels can be packed together using comb lasers, whose spectra are shown in figure 5.24. Modulated spectra can be observed, as shown in figure 5.25. With a baud rate of 200 Gbd, the total bit rate can reach higher than 1.0 Tbps.

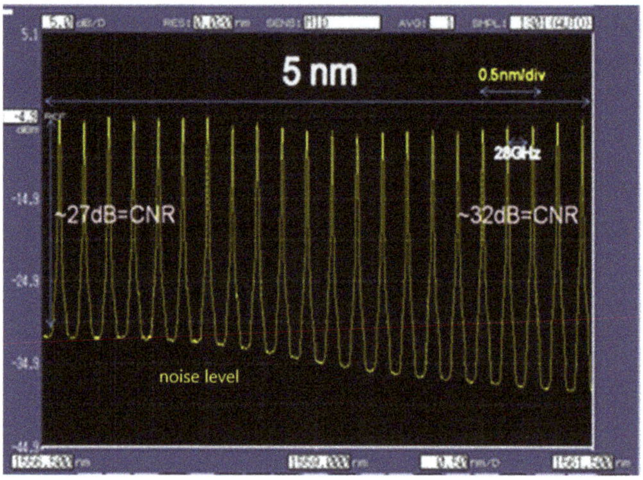

Figure 5.24. Comb spectrum of a 28 GHz frequency shifting comb-generator. Note: inline optical amplification embedded in the comb resonant loop [16].

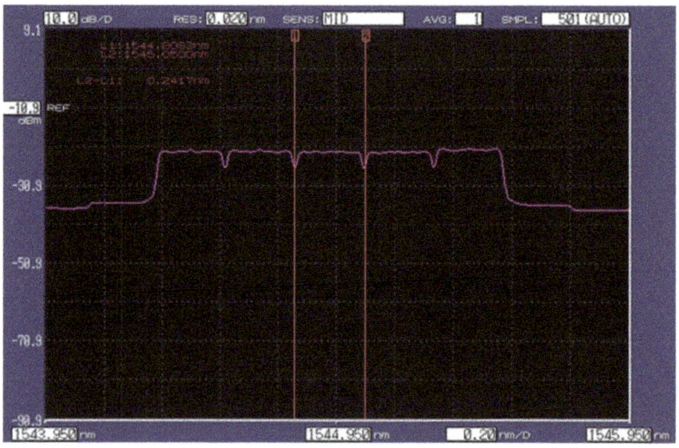

Figure 5.25. Packed optical spectra of five selected sub-carriers of a modulated comb laser lines [16].

5.7 RPoF front-haul networking to antennae sites

5.7.1 Techniques

Figure 5.26 depicts a proposed distribution of 6G and 7G channels which include RF modulated optical carriers and data channels. The transmission and distribution of data channels go to and from a macro-cell in which several 6G and 7G micro-cells of antennae sites are implemented in the optical domain for a short access distance to the antenna or arrays of MIMO antennae elements. The sensing of RF electromagnetic waves is as given in chapter 6 employing plasmonic electro-optic modulators.

The digital RPoF with several multiplexed sub-carriers of the main RF carriers and associated shifted spacing frequency is transmitted (the up-link) to the broadband unit (BBU) and then demultiplexed and grouped into sets of channels to be distributed to the antennae of micro-cells via a number of RO-RF-ADMs. The reverse direction, the down-link, can be configured accordingly.

In parallel with the up- and down-links is the monitoring and service provisional channel whose wavelength is outside the RPoF channel spectral window. These provisional lines are used for monitoring the operational conditions of antenna sites from central remote control and network stabilization.

The digital RPoF channels are coming from the metro- or terrestrial transport back-haul network.

5.7.2 Multi-carrier RPoF in cloud networks

5.7.2.1 Motivations of the proposed system/networks

The motivations for the proposed system/networks are as follows:

- Increasing the convergence of traditional telecoms and data center based networks (DCNs) and 6G and 7G networks.
- The convergence of wireless and photonic networks is essential to take advantage of the infrastructure and capacity demands of 5G, IoT, and cloud-based access networks.

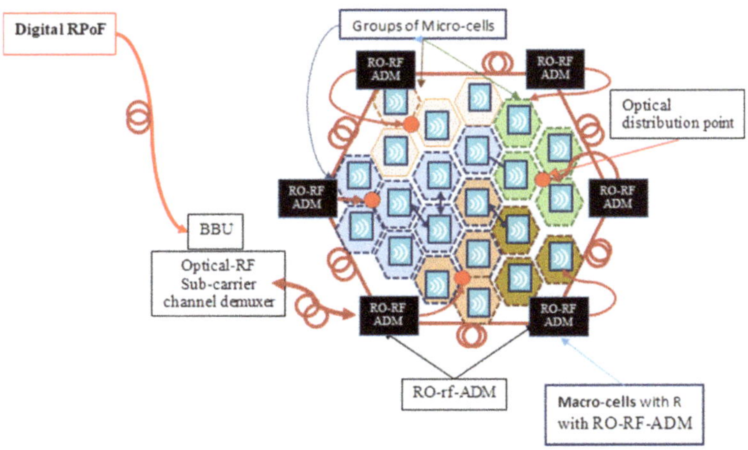

Figure 5.26. Transmission and distribution of data channels to and from a macro-cell in which there are several 6G and 7G micro-cells of antennae sites. RO-RF-ADM = reconfigurable optical-RF add/drop multiplexer; BBU = broadband unit; RPoF = radio-photonics over fiber system.

- Continued demand for low cost, e.g. through implementation of WDM passive optical networks (PONs) and wireless access.
- The requirement of a mmW spectrum for 5G, 4K video will be driven by rapidly growing wireless data traffic demand, of the order of a 1000-fold increase in 2020, or even a 10 000-fold increase by 2030 when extensive artificial intelligent (AI) systems will be deployed extensively.

It is likely that a 'hybrid' electro-photonic solution will be developed, in which both a PON and wireless access must be offered.

The key to low cost must be a common platform that supports heterogeneous integration for both photonic and mmW components (in particular for the 30–60 GHz and also the 90 GHz mmW band).

Although massive strides have been made with attempts to integrate all components on a single platform (i.e. using the system on a chip (SoC) approach) on either an optical (InP)/electronic (CMOS) or other high permittivity substrate, such as GaAs or low temperature co-fired ceramic (LTCC), these will typically fail at the mmW due to performance limitations, such as narrow bandwidths and low radiation efficiencies caused by surface wave losses.

This section aims to present a brief account of a transport system of massive capacity with multi-Tbps applications for a wireless environment using beam formation and control of such a beam formation for delivery to a targeted location. Furthermore, the use of a photonic printed circuit board (PCB) on a substrate and polymeric waveguides created on a printed circuit board with a hybrid integration of electronic and antenna fabrication is proposed. The main target is to overcome the phase non-uniformity arising from the difficulties due to electrical cable interconnections to the MIMO antenna element, as shown in figures 5.27 and 5.28. The massive data transport for multi-site MMO antennae systems could be in cloud-based networks and network edges. Distributed DCNs are expected to carry massive data channels (e.g. OFDM-256QAM), that is, the RF carrier is modulated with 256QAM formats and embedded with OFDM data channels. The bit rate per RF carrier reaches >10 Gbps.

For an MIMO antenna operating in the high mmW region, an integrated photonic plane must be used so as to minimize the radiation of such ultra-high frequency, namely the integrated photonic PCB which can be fabricated using polymeric waveguides and integrated RF electronics. The technology is proposed in the next section.

5.7.3 RF-MIMO-RAPT and integrated photonics

The back-haul transport of mmW channels over fibers is critical for 6G and 7G networks [17–21]. In order to minimize and compensate for the free-space transmission loss of mmW and under a target transmission distance of 300–500 m, for example in the scenario of a sports stadium, multiple antenna ports are required. MIMO antennae with 2D array elements of 128×512 can be used to offer an antenna gain of more than 25–30 dB and beam steering capability.

The subsystem must be installed in an outside environment and operating temperature from −40 °C to +85 °C. A compact radio-photonic PCB with integrated circuits of electronics and photonics is desired.

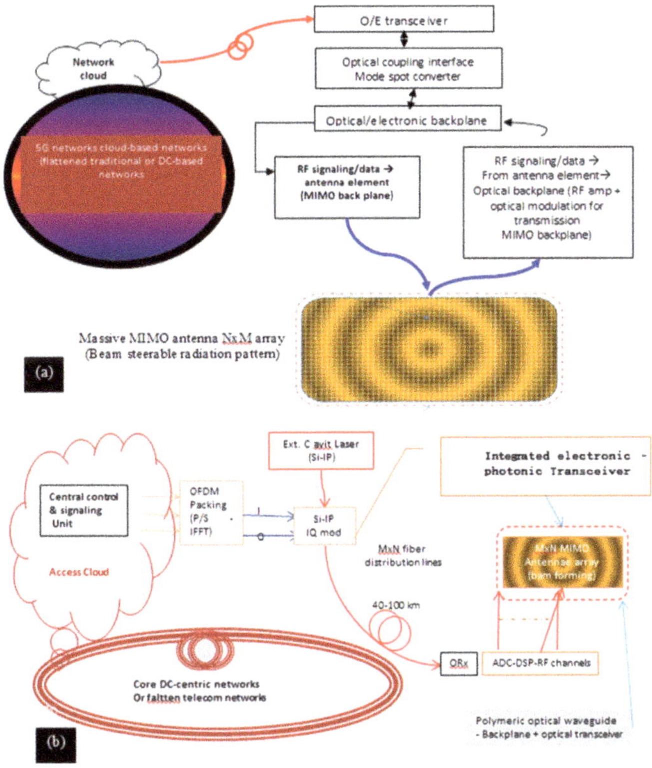

Figure 5.27. Generic RF modulated optical carrier plus data transfer to and from MIMO antenna in cloud-based networks. (a) Functional diagram and (b) technical functions with integrated photonic circuits (see figure 5.30).

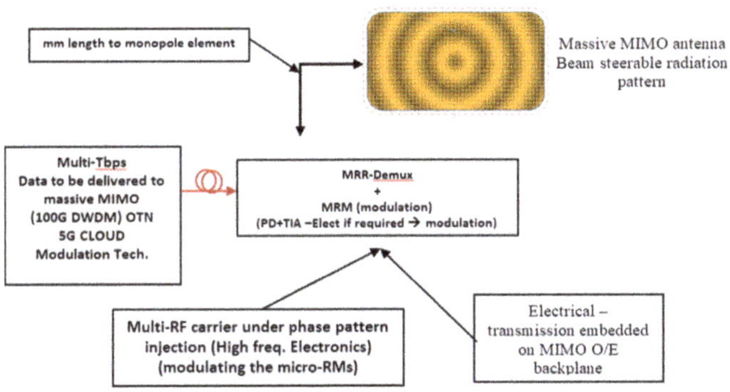

Figure 5.28. 6G and 7G electro-photonic system functionalities. MRR = micro-ring resonator; MRM = micro-ring modulator.

Under such a large area of PCB and for ease of fabrication and production, polymer thin film and waveguide structures would offer a solution. The stability of this technology has reached a good level. Polymer waveguide technology [22] is now mature enough to provide such a back and front plane on a PCB for radio-photonic

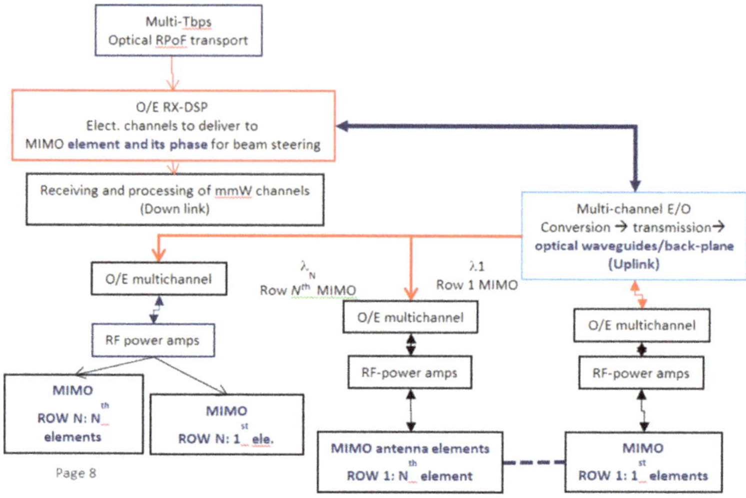

Figure 5.29. A 6G and 7G RPoF multi-carrier functional flow diagram, up-link and down-link. Conversion from the optical to electrical (O/E) and then from the electrical to optical (E/O) domain.

circuits (RPCs) on board and integration with RF broadband integrated circuits. The principles and technological conceptual schematics are shown in figures 5.29 and 5.30. In addition, multi-RF modulated carriers are used. Such a multi-carrier RPoF is proposed in section 5.6.2.

Referring to figure 5.30, the integrated photonics and RF mmW carriers can be explained as follows:

- The frequency difference between the sub-carrier lines of the comb laser and that of the carrier of a DWDM channel is the mmW frequency of the radio waves to be generated with its phase, which is created by an integrated photonic phase modulator. This modulator can be fabricated using polymeric waveguides over a PCB layered by another polymeric waveguide.
- The polymeric PM has a cross section that is sufficiently narrow, < 20.0 mm × 50 mm, so loss and coupling between the waveguides of different materials will have to be considered. So, if the PM can tolerate the phase changing speed then a slower technique can be used to avoid this complexity.
- The photodetector can be flip-chip bonded on the surface to receiving optical signals and the comb line to generate the mmW RF waves to feed to antenna element. Thus the optical waveguides in the substrate can be coupled to the vertical via an etched 450 mirror—very good etching can be achieved using dry etching techniques.
- Other integration techniques can be configured. An SiP platform would be the most likely technology.

Electronic integrated circuits (ICs) and opto-electronic devices are integrated discreetly. Antenna horn elements in the mmW, horn waveguides, can be achieved by grouting the metallic layer. Optical lightwaves are guided along the polymer

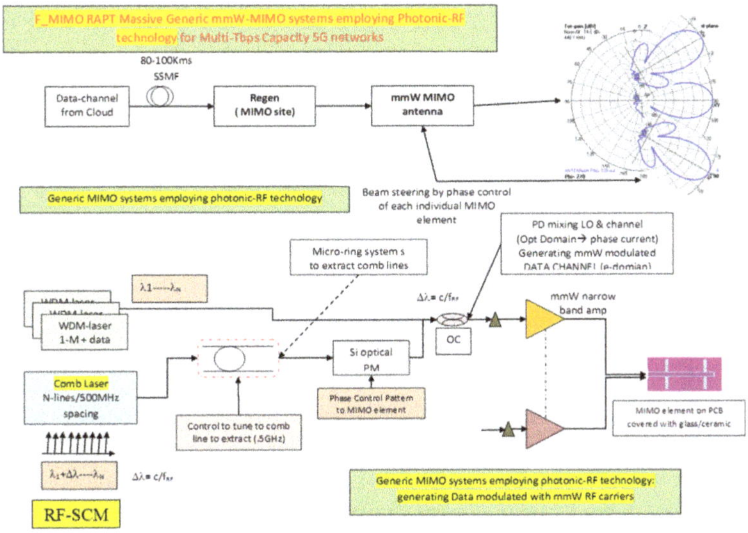

Figure 5.30. Layout and structural functionalities of an integrated photonic chip for an RPoF to antenna MIMO radiation site. RF-MIMO-RAPT = RF multi-input multi-output radio-photonic technology; SCM = sub-carrier multiplexing.

waveguides. RF transmission lines can be made with appropriate selection of the metal linewidth and if required the Faraday line box can be fabricated.

Thus, under the current advanced technologies of Si on insulator (SOI), Si-integrated photonics, and SiGe electronics, the radio-photonics SoC is highly implementable.

5.8 Remarks

This chapter has presented the following: (i) a review of our recent progress in high baud rate DB-PAM-4 systems to meet capacity; (ii) the requirements in metro and access; (iii) how a DB-PAM-4 signal as high as 90 GBd can be generated by a 92 GSa s^{-1} DAC; (iv) that the DB-PAM-4 system shows better performance than the conventional PAM-4 system; (v) how, if dispersion pre-compensation can be performed with a DD-MZM, the reach-distance can be extendable to 80 km to cover the metro point-to-point scenario; and (vi) how SSB-DB-PAM-4 is also feasible to greatly increase the spectral efficiency and capacity of the system.

A number of critical issues to be recognized are as follows: (i) realization in an integrated module; (ii) integrated electro-photonic form factor modules, such as CFP2, CFP4, and photonic IC plus electronic IC plus master control unit (MCU; for bias stability control); (iii) four lanes leading to a total of 500 Gbps which can be realized using silicon–integrated photonic (Si–IP) hybrid technology; and (iv) comb lasers plus Si–IP 180 Gbps × 12 lanes to produce an ~2 Tbps module multiple source agreement (MSA) form [23]. Furthermore, peaking of a combined driver + DSP + optical modulator (analog + digital domain) allows a wider BW (35–40 GHz) to higher baud rate (>230 Gbps).

Finally, we outline the two basic techniques to double the symbol rate so that reaching higher than 200 GSa s^{-1} can be achieved, thus the aggregate bit rate for complex signal modulation and polarization multiplexing to offer the possibility of higher than 1.0 Tbps, which can occupy only the 75 GHz bandwidth. Furthermore, Tbps transmission systems for short-haul with super-channels are briefly given for the interconnection of data centers or the delivery of several 5G or 6G wireless networks.

References

[1] Shieh W and Djordjevic I 2009 *OFDM for Optical Communications* 1st edn (New York: Academic)

[2] Jansen S *et al* 2009 121.9-Gb/s PDM-OFDM transmission with 2-b/s/Hz spectral efficiency over 1000 km of SSMF *J. Lightw. Technol.* **27** 177–88

[3] Schmidt B *et al* 2009 100 Gbit/s transmission using single-band direct-detection optical OFDM *National Fiber Optic Engineers Conf. (San Diego, CA, 22–26 March)* (Washington, DC: Optica) paper PDPC3

[4] Qian D *et al* 2009 108 Gb/s OFDMA-PON with polarization multiplexing and direct-detection *Optical Fiber Communication Conf. and National Fiber Optic Engineers Conf.* OSA Technical Digest, paper PDPD5

[5] Binh L N 2020 *Transmission and Processing for Data Center Networking: Ultra-High Capacity Data Center Networking* (IOP Series in Advances in Optics, Photonics and Optoelectronics) (Bristol: IOP Publishing)

[6] Buchali F *et al* 2020 1.52 Tb/s single carrier transmission supported by a 128 GSa/s SiGe DAC *Optical Fiber Communications Conf. and Exhibition* (Piscataway, NJ: IEEE) paper Th4C.2

[7] Vinod B *et al* 2020 Single-channel 1.61 Tb/s optical coherent transmission enabled by neural network based digital pre-distortion *European Conf. on Optical Communications* (Piscataway, NJ: IEEE) paper Tu1D.3

[8] Guoxiu H *et al* 2020 Up to 115 GBaud faster than Nyquist PDM-64QAM based on Tomlinson–Harashima precoding with single DAC *European Conf. on Optical Communications* (Piscataway, NJ: IEEE) paper Tu1D.6

[9] Verbist J *et al* 2017 100 Gb/s DAC-less and DSP-free transmitters using GeSi EAMs for short-reach optical interconnects *Optical Fiber Communication Conf. (Düsseldorf, Germany, 11–15 March)* (Washington, DC: Optica)

[10] Zhang J *et al* 2016 WDM transmission of twelve 960 Gb/s channels based on 120-GBaud ETDM PDM 16QAM over 1200-km TeraWave TM fiber link *Optical Fiber Communications Conf. and Exhibition* (Piscataway, NJ: IEEE) paper Tu3A.2

[11] Zhang J *et al* 2016 Generation and transmission of high symbol rate single carrier electronically time division multiplexing signals *IEEE Photon. J.* **8** 7902506

[12] Matsushita A *et al* 2016 Super-Nyquist 9-WDM 126-GBaud PDM-QPSK transmission over 7878 km using digital-preprocessed analog-multiplexed DAC for long-haul applications *42nd European Conf. on Optical Communication* (Piscataway, NJ: IEEE) paper W.3.D.2

[13] Nakamura M *et al* 2020 Entropy and symbol-rate optimized 120 GBaud PS-36QAM signal transmission over 2400 km at net-rate of 800 Gbps/λ *Optical Fiber Communications Conf. and Exhibition* (Piscataway, NJ: IEEE) paper M4K.3

[14] Pittalà F *et al* 2020 800ZR+ DWDM demonstration over 600km G.654d fiber enabled by adaptive nonlinear TripleX equalization *Optical Fiber Communications Conf. and Exhibition* (Piscataway, NJ: IEEE) paper M4K-5

[15] Schädler M *et al* 2020 Recurrent neural network soft de-mapping for nonlinear ISI in 800 Gbit/s DWDM coherent optical transmissions *J. Lightw. Technol.* **39** 5278–86

[16] Binh L N 2017 *Advanced Digital Optical Communications* (Boca Raton, FL: CRC Press)

[17] Dat P T, Kanno A, Inagaki K and Kawanishi T 2014 High-capacity wireless backhaul network using seamless convergence of radio-over-fiber and 90-GHz millimeter-wave *J. Lightwave Technol.* **32** 3910–23

[18] Kanno A, Dat P T, Kuri T, Hosako I, Kawanishi T, Yoshida Y and Yasumura Y 2012 Coherent radio-over-fiber and millimeter-wave radio seamless transmission system for resilient access networks *IEEE Photonics J.* **4** 2196–204

[19] Heinrich Hertz Institute Hybrid PIC (photonic integrated circuits) https://hhi.fraunhofer.de/en/departments/pc/research-groups/hybrid-pics.html

[20] Zawadzki C and Schirmer M 2019 The waveguide chip—photonic components made of polymers *laser+photonics*, January 2019

[21] Dasgupta S, Shih M-Y, Gorczyca T, Erben C, Tolliver T, Kapusta C and Breitung E 2004 A polymer-based platform technology for integrated photonics *Proc. SPIE 5517, Linear and Nonlinear Optics of Organic Materials IV (15 October 2004)* https://doi.org/10.1117/12.563835

[22] Zakariyah S S 2014 Integrated polymer waveguides technology: present and future, *Technical Report* University of Derby

[23] Howard 2020 How much do you know about MSA standard? *FS Community*

Chapter 6

5G, 6G, and 7G radio-photonic networking: multiple-band radio-photonic antenna testing and networking

6.1 The demand for testing of antennae operating in multiple-frequency regions

In radio-photonic networking the delivery of multiple signals and data to multiple sets of antennae would offer a much simpler solution for flexible and low cost operating expenditure (OPEX) for network operators, hence lowering the costs for users and clients. At the same time, during production it is necessary to send test signals from a remote central test center to several antennae operating at different frequencies for analysis of their performance. Similarly, a test center located at a distance can transmit test signals to multiple antenna sites located remotely and possibly mounted on transmitting towers. This is impossible to implement using coaxial cables, but instead requires optical guided media, standard single mode optical fibers (SSMFs). Single mode optical fibers (SMFs), normally the commonly used SSMFs, can certainly be the guiding medium to carry optical signals in which the RF carrier frequencies can be tuned to appropriate operating regions of a specific group of antennae, in the microwave or millimeterwave (μW or mmW) spectral regions.

Figure 6.1 depicts a schematic of a test system/network for a radio-photonic (or optical to wireless) network test environment in multi-frequency bands, multi-macro-cells, and multi-wavelength spectral regions. The signals generated and transmitted from a broadband business unit–central test unit (BBU–CTU) to optical line transmission (OLT) and optical 5G–6G wireless network units (O-W NUs)—several wavelength channels can each carry signals at 100 Gbps per wavelength or even higher at 1.6 Tbps per wavelength employing multi-level 16QAM and

doi:10.1088/978-0-7503-3511-9ch6

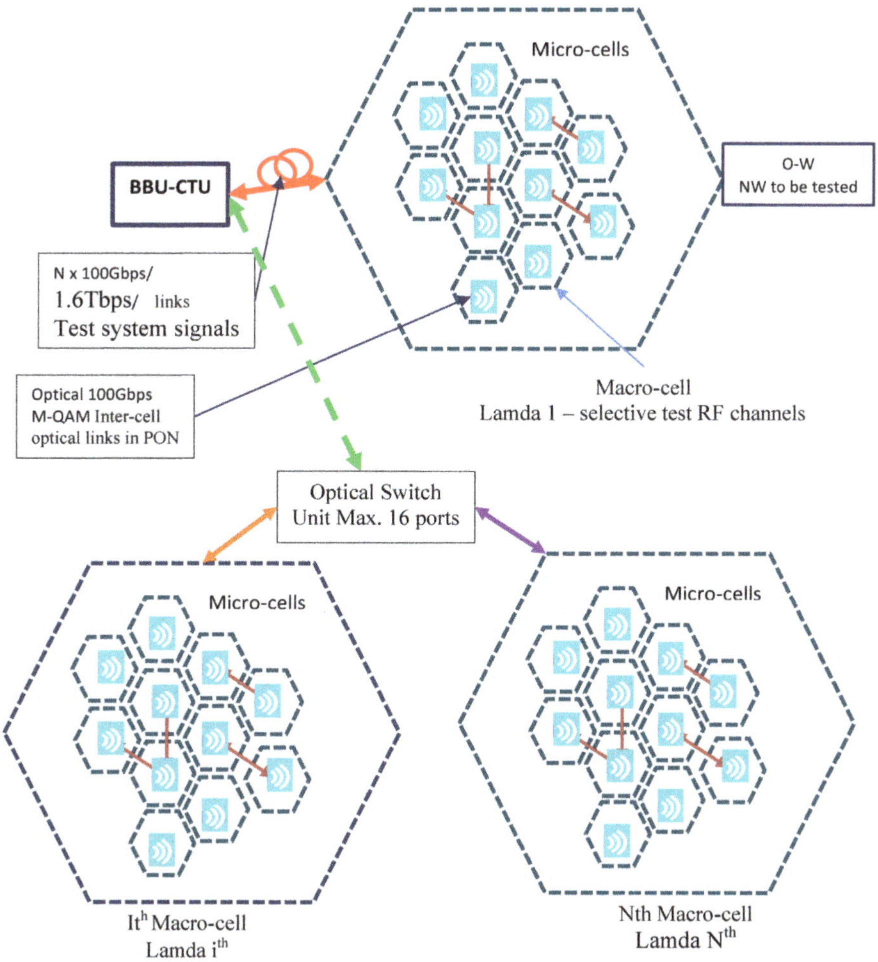

Figure 6.1. Schematics of a test system/network for an optical to wireless networking test environment in multi-frequency bands, multi-macro-cells, and multiple wavelengths. The structure can be altered in a test building to match the test area. The CTU is normally remote and far from the high RF for safety reasons, thus the term radio-photonics over fiber (RPoF) is used. BBU–CTU = broadband business unit–central test unit; OLT = optical line transmission; O-W NU = optical 5G–6G wireless network unit; NW = network; RF = radio frequency; PON = passive optical network; M-QAM = multi-level quadrature amplitude modulation.

polarization multiplexing at 100 GBd. A wavelength channel can be switched sequentially or in parallel for simultaneous testing.

For simplicity and to satisfy economic considerations it is preferred that one optical source is used which can carry radio waves at different frequency bands. We thus provide a brief description of the structure of a multi-band RPoF and then a passive optical network (PON) for such a multi-band distribution optical source carrying radio waves for either test systems or the distribution of transmission wireless waves via antennae, as shown in figures 6.1 and 6.2.

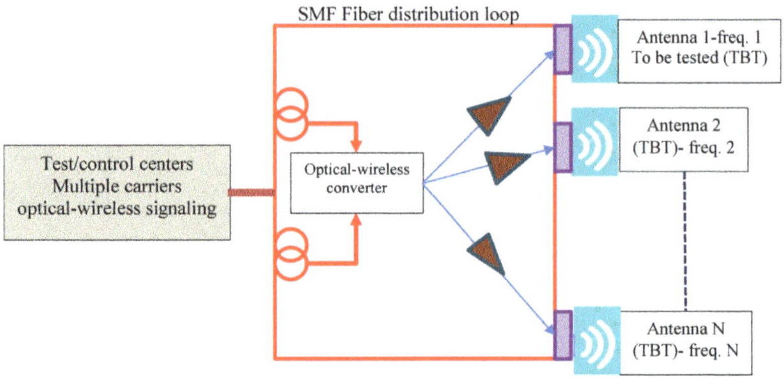

Figure 6.2. A schematic of the signaling for tests of multiple-band antenna systems at remote distances. The optical network can be a passive optical network (PON).

6.2 Generation of a CS-SSB to carry the test signals

The carrier-suppressed single sideband (CS-SSB) is described in this chapter and is proven to be highly resilient to distortion and nonlinear intermodulation (IMD) effects. Coherent reception incorporating an advanced DSP algorithm can improve the reception sensitivity significantly and is suited for RPoF systems in which the data channel bands are much lower than the carrier frequencies. For RPoF employing coherent reception and transmission techniques, multiple local oscillation lasers need to be generated, the master local oscillator (MLO).

6.2.1 Implementation of a multi-RF-band master laser oscillator (MLO)

This section provides details of the experimental set-up and analyses of the carrier suppression and the suppression of either an upper or lower sideband. This acts as the sweeping local oscillator for the proposed coherent optical signal analyser (COSA).

6.2.1.1 Implementation of SSB and carrier suppression

The schematic of the set-up and the driving conditions to achieve such a sweeping LO offering the highest resolution for optical spectral analyses are provided in figure 6.3.

6.2.1.2 QPSK optical modulator

Schematics of a quadrature phase shift keying (QPSK) optical modulator are shown in figure 6.4, which consists of two child MZMs and form a parent MZM. In one of these two branches an optical phase shifter is incorporated so that a $\pi/2$ phase can be generated via the electro-optic (EO) effect by an applied voltage if the substrate is a crystal such as $LiNbO_3$. The thermo-electric effect is used if the substrate is an insulator such as the silicon-on insulator (SOI) type. Figure 6.4(a) shows a schematic of the structure, while figure 6.4(b) gives more details of the surface view of the QPSK modulator, including the biases of all the MZMs and the monitoring of the optical

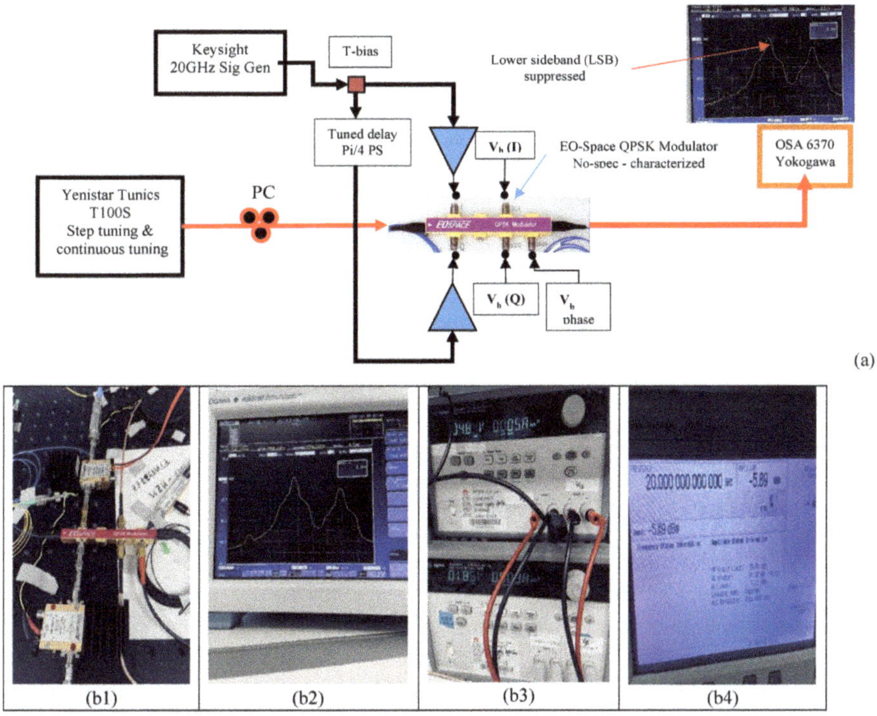

Figure 6.3. Schematics of the set-up testbed for a CS-SSB as the master LO for a COSA. (a) Schematic of the set-up. (b) Photographs of (b1) the IQ modulator and large signal amplifier SHF-807 to achieve sufficient RF swing, (b2) spectrum suppression of one band but not yet the carrier (middle), (b3) power supplies to Q biases (b4) mmW signal generator.

power by partially coupling of the output optical waveguide by an integrated coupler. Figure 6.4(c) shows the transfer characteristics of an MZM and the output optical power is plotted against the applied voltage and the biasing point at which minimum transmission of power is achieved. Figure 6.4(d) shows an image of a commercial QPSK modulator with electrodes and fiber ports at the input and output.

The ratio of the maximum and minimum transmission points is defined as the extinction ratio (ER). For $LiNbO_3$ it is best at about 30 dB while for the other types of modulators it can typically reach about 10 dB. With such an ER the linear range is reasonably large and thus most suitable for optical–wireless linear modulation or to carry the wireless carrier over an optical fiber.

The input fiber can be a polarization maintaining type whose polarization axis can be rotated to about $\pi/4$ with respect to the optical waveguide horizontal direction so that both polarized modes can be coupled equally to the modulator for the best power matching to the two polarized IQ modulator branches.

Two such QPSK modulators can be duplicated and placed in parallel with a polarization splitter and combiner at the input and output port, respectively, to create a dual polarization in-phase quadrature (DP-IQ) modulator (or polarization multiplexed (PM)-IQ) for multi-level complex optical modulation or higher order modulation to

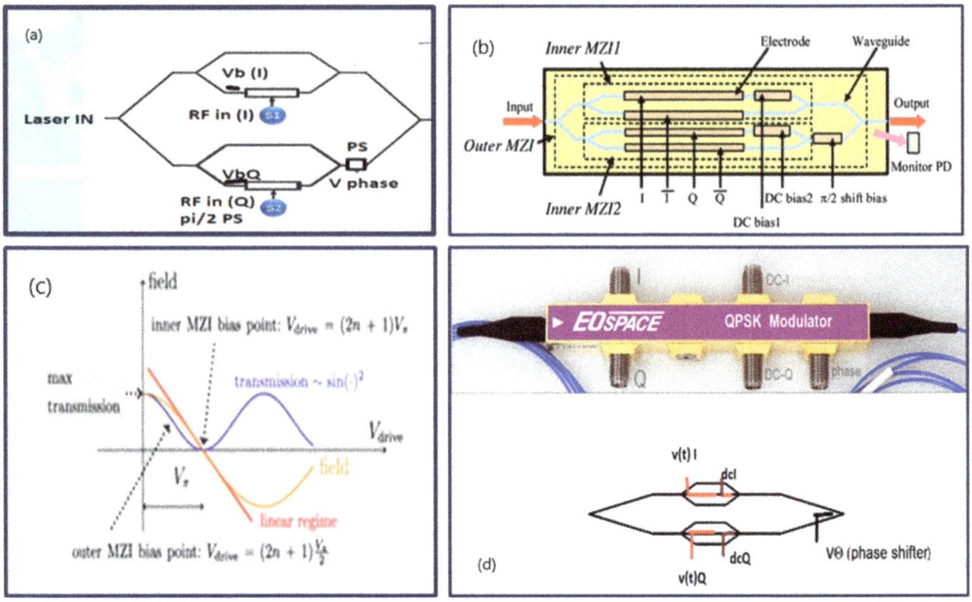

Figure 6.4 Complex modulator. (a) The generic structure with RF and dc biases of the QPSK modulator, by EO-Space Inc. Biasing voltages V_b (I), V_b (Q), and V_{phase} (PS) are the applied dc voltages. A bias T is required. Types of IQ modulator (single polarization): (b) Fujitsu IQ modulator structure (single polarization) with a phase shifter on one branch of the parent MZM; (c) transfer transmittance characteristics of each MZM of the QPSK/IQ modulator with a phase shifter on one branch of the parent MZM (outer); and (d) an EO-Space QPSK modulator structure following the structure in (c).

increase the total modulation capacity. The bandwidth for such multi-level modulation requires a basic baud rate. These DP-QPSK (IQ) modulators can carry two wireless carriers in the optical domain. This is very beneficial when multi-level modulations of the wireless carrier are required. Figure 6.5 shows the structure of such a DP-IQ modulator which is normally employed in coherent optical transmission systems for long-haul and is currently used intensively in local access or data center networking for ultra-high capacity transmission networks.

Typical QPSK modulators manufactured by EO-Space Inc and Fujitsu Inc are shown in figure 6.4. Exemplar modulator parameters of a QPSK modulator are as follows: V_p (I) dc < 10 V; V_π (Q) dc < 10 V; V_π (phase) < 15 V; V_π (1 GHz) < 3 V (3.3 V max); X-cut chirpless 3 dB at 25 GHz, 6 dB projected to 45 GHz for each MZM (I) and MZM(Q); insertion loss (IL) < 5 dB; S_{11} < 10 dB (dc 20 GHz); and ER > 20 dB. The dc bias electrodes are separated from the high speed RF electrodes and the length is not long so the voltage level is higher for the V_π of the RF. Normally they are denoted as $V_{\pi,\text{dc}}$ and $v_{\pi,\text{RF}}$.

6.2.1.3 Driving conditions and analytics of a CS-SSB

6.2.1.3.1 Case 1: A CS-SSB from an electrical signal applied to the optical phase of a DD-MZM

The schematic diagram of SSB vector signal generation by a QPSK modulator is shown in figure 6.4. An external cavity laser (ECL) is coupled to a DD-MZM or a

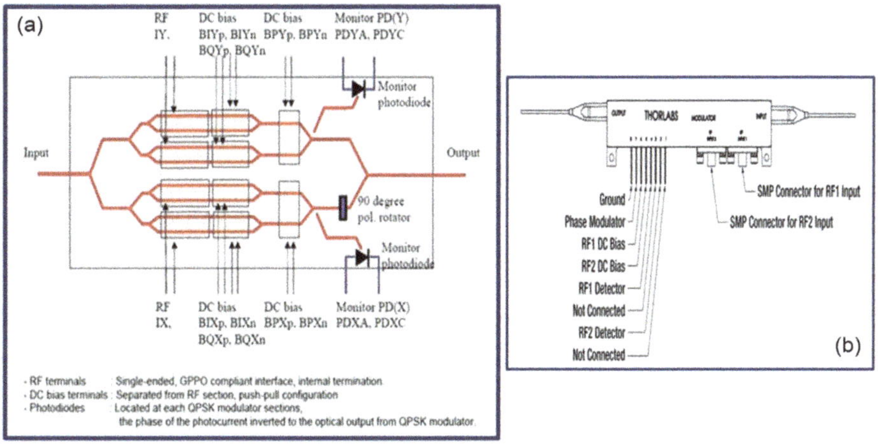

Figure 6.5. (a) A Fujitsu PM-QPSK optical modulator, FTM-7992HQ. (b) A Thorlab DD-MZM.

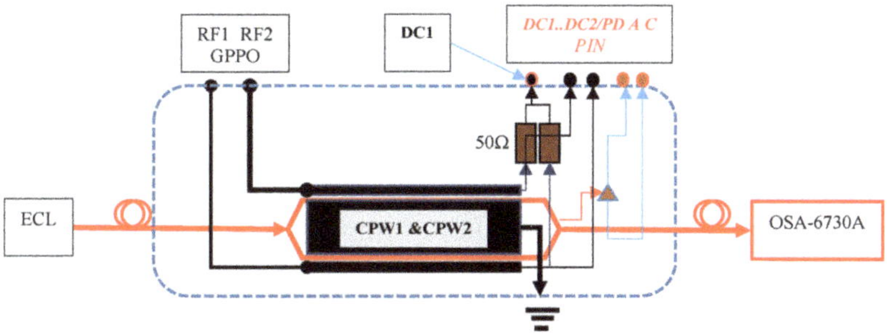

Figure 6.6. An SSB RPoF system.

QPSK modulator via a polarization controller (PC). As described in chapter 3, a QPSK modulator is fabricated with two sub-MZMs and embedded in a parent MZM. In the upper in-phase (I) MZM of the dual child MZM QPSK the modulators embedded in the parent QPSK modulator are modulated by a single-frequency RF signal. The lower quadrature (Q) MZM is also modulated by a sinusoidal RF source but shifted by $\pi/2$, or equivalently a Hilbert transform of the RF source and so the summation of the two orthogonal optical fields leads to the suppression of a sideband. The dc biases of the two child MZMs are set at quadrature and ideally with a total of V_π difference so there is a π phase difference of the optical carrier between the I- and Q-branches to suppress the optical original carrier. Finally, we expect that only one RF shifted sideband can exist (figure 6.6).

The output field of the QPSK optical modulator (the parent MZM) can be written as the summation of the two output fields of the two child branch MZMs as

$$E_0 = E_I + E_Q \exp\left(j\frac{V_\varnothing}{V_\pi}\pi\right) = [A_x]\cos\left[\frac{\pi}{2V_\pi}\left(v_{\mathrm{RF1}(t)} + V_{\mathrm{dc}I}\right)\right] e^{j\frac{V_{\mathrm{dc}I}}{2V_\pi}} e^{j\omega_0 t} + e^{j\pi\frac{V_\varnothing}{V_\pi}} \cos\left[\frac{\pi}{2V_\pi}\left(v_{\mathrm{RF2}(t)} + V_{\mathrm{dc}Q}\right)\right] e^{j\frac{V_{\mathrm{dc}Q}}{2V_\pi}} e^{j\omega_0 t}, \tag{6.1}$$

where E_0, E_I, andE_Q represent the optical fields at the output of the parent QPSK and two child MZMs, respectively, Φ is the optical phase shift due to the voltage V_Φ applied to the optical phase shifter on the Q-branch of the parent QPSK MZM, ω_0 is the angular frequency of the optical carrier, A_x is the carrier amplitude, V_π is the half-wave voltage, $v_{RF1}(t)$ and $v_{RF2}(t)$ are the RF signals applied to the child MZMs which can be written as $v_{RF1}(t) = V_{RF}\sin(\omega_{RF}t)$ and $v_{RF2}(t) = V_{RF}\cos(\omega_{RF}t)$, where ω_{RF} and V_{RF} are the angular frequency and the amplitude of the RF signal, respectively.

The subscripts 1 and 2 indicate the upper (in-phase) and lower (quadrature) branches, respectively. To generate a CS-SSB signal, the child MZMs are biased at the quadrature point, i.e. $V_{dcI} = V_{dcQ} = V_{\pi/2}$, and the parent MZM is biased at the Q+ point of the transmittance transfer characteristics, i.e. $V_\Phi = -V_{\pi/2}$, hence a suppression of the carrier due to this π phase difference, and thus the optical carrier signal can be given as

$$\begin{aligned} E_0 = [A_x] \cos\left[\frac{\pi}{2V_\pi}\left(V_m \sin(\omega_m t) + \frac{\pi}{2}\right)\right] e^{j\frac{\pi}{2}} e^{j\omega_0 t} \\ + A_x e^{-j\frac{\pi}{2}} \cos\left[\frac{\pi}{2V_\pi}\left(V_m \cos(\omega_m t) + \frac{\pi}{2}\right)\right] e^{j\frac{\pi}{2}} e^{j\omega_0 t}, \end{aligned} \tag{6.2}$$

where $\gamma = \pi V_m/2V_\pi$ is the phase modulation index. Expanding (6.2) using the Jacobi–Anger identity, the total field at the QPSK output is

$$\begin{aligned} E_0 &= [-jA_x] \sin[\gamma(\sin(\omega_m t))] e^{j\omega_0 t} - [jA_x] \sin[\gamma(\cos(\omega_m t))] e^{j\omega_0 t} \\ &\approx 2A_x J_1(\gamma) e^{j(\omega_0 + \omega_m)t}, \end{aligned} \tag{6.3}$$

which indicates a CS-SSB signal with only +1 order sideband with suppression of the original optical carrier.

Further, a polarization multiplexed QPSK (PM-QPSK) modulator can be considered in which there are two QPSK modulators in parallel associated with a polarization beam splitter (PBS) and combiner (PBC), as shown in figure 6.7.

The upper QPSK modulator of a PM-QPSK modulator can be used to generate the baseband vector signal optically. In this case, the optical field at the output of a child MZM of a QPSK modulator can be expressed in terms of the I- and Q-binary data sequences $I(t)$ and $Q(t)$, given as

$$E_y(t) = A_y[I(t) + jQ(t)]e^{j\omega_0 t}, \tag{6.4}$$

where A_y is the field amplitude at the input.

The polarization direction of the two fields can be adjusted via a polarization controller (PC) or matched by a polarized fiber. PC2 can rotate by an angle θ with respect to the x-axis of the DP-QPSK. The output branch PBC combines the two orthogonally polarized optical signals of the two QPSK sections, given as

$$E_{POL}(t) = \cos(\theta) A_x J_1(\gamma) e^{j(\omega_0 + \omega_m)t} + \sin(\theta)[I(t) + jQ(t)]e^{j\omega_0 t}. \tag{6.5}$$

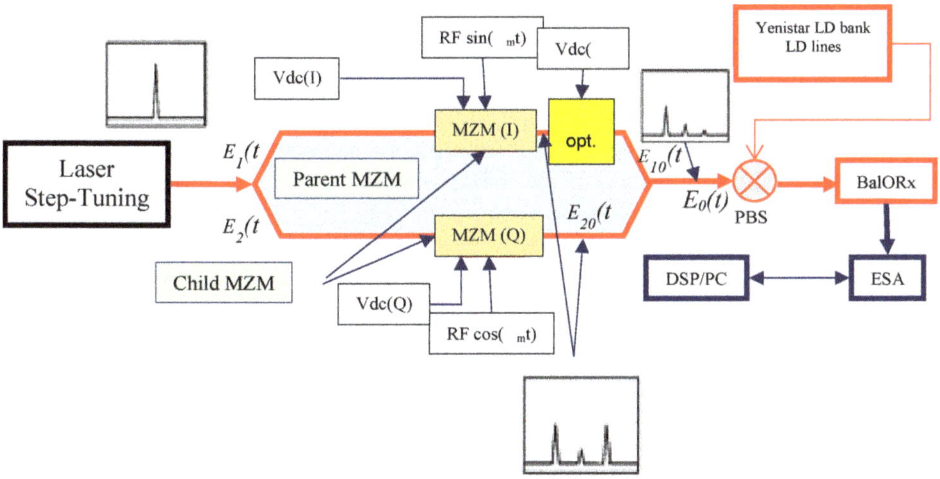

Figure 6.7. Schematic of a CS-SSB QPSK modulator and heterodyne/homodyne mixing and reception of the RF modulated carrier with suppression of the optical carrier by an optical phase shifter (applied V_Φ) and suppression of an RF sideband by a Hilbert transformer via a $\pi/2$ hybrid coupler or an RF delay line of $T/4$ (a quarter of the RF wave period). Alternatively, the suppression of the sideband can be implemented by the Hilbert transformer or a $\pi/2$ phase shift of the sinusoidal signal applied to I or Q and vice versa.

Hence the optical carrier and two SSB channel signals can be generated independently. The optical channels are transmitted through a length of SSMF. If $\beta(\omega)$ denotes the propagation constant in the fiber guided mode at an angular frequency ω and z is the fiber location, then the optical signal after a distance z of the SSMF transmission line is given as

$$E'_{\mathrm{POL}}(t) = -2\cos(\theta)A_x J_1(\gamma)e^{j[(\omega_0+\omega_m)t-\beta([(\omega_0+\omega_m)z])]} + \sin(\theta)A_y\left[I\left(t-\frac{\beta(\omega_0)z}{\omega_0}\right)+jQ\left(t-\frac{\beta(\omega_0)z}{\omega_0}\right)\right]e^{j[\omega_0 t-\beta(\omega_0)z]}. \tag{6.6}$$

After the signal travels this transmission distance and is received by a single PD, the OE receiver converts it to an electrical signal $i_{\mathrm{PD}}(t)$, given as

$$i_{\mathrm{PD}}(t) \propto [E'_{\mathrm{POL}}]^2(t) = -4\cos(\theta)A_x J\sin(\theta)A_y J_1(\beta).\left[I\left(t-\frac{\beta(\omega_0)z}{\omega_0}\right)\cos[\omega_0 t-\beta(\omega_0)z]+Q\left(t-\frac{\beta(\omega_0)z}{\omega_0}\right)\sin[\omega_0 t-\beta(\omega_0)z]\right]. \tag{6.7}$$

This equation indicates that the generated RF vector signal can be immune to nonlinear-induced dispersion in the fiber, hence less IMD. For the case of coherent optical spectrum analyses, we can see that for any SMF connected to the output of the QPSK CS-SSB modulator there would be no effect whatsoever from the length of the SMF. Thus, such a CS-SSB optical head can be linked to a remote OSA by such an modulated optical modulator for optical analyses.

Indeed, the polarization of the two optical fields of the two QPSKs can be rotated such that the PBC can act as an attenuator if such polarization is not aligned to its direction. This can assist in the suppression of the IMD, as demonstrated by a simulation in a later section of this chapter. This is possible because the polarized field can be resolved partially into the other polarized direction, which can be in the opposite direction, and thus mitigates the nonlinear IMD generated by the other polarized channel signal.

6.2.1.4 IQ modulation with a phase shift

An RPoF system using a CS-SSB QPSK modulator is shown in figure 6.7, which is composed of a parent MZM with two child MZMs embedded on each branch. An optical phase shifter is incorporated on the upper child MZM. The optical fields at the output of the MZM before entering the PBC are given as

$$E_{01}(t) = \frac{E_0}{\sqrt{2}} \cos(\omega_c t) \tag{6.8}$$

$$E_{02}(t) = \frac{E_0}{\sqrt{2}} \cos(\omega_c t + \Delta\varphi), \tag{6.9}$$

where E_0 and ω_c are the field amplitude and optical carrier frequency, respectively, and $\Delta\varphi$ is the relative phase shift propagating through the two branches. This phase is mainly due to $\Delta\Phi$ of the phase shifter. Two integrated EO-MZMs produce the amplitude modulation of the optical carrier as they are excited with the same RF signal and mutually with a phase difference $\Delta\Phi$. Assuming only a single RF oscillation, that is a single-tone sinusoidal RF signal, the optical output of the QPSK modulator is given by

$$E_1(t) = E_{01}(t) \cos\left(\frac{\pi}{2}\frac{V_{\mathrm{dc}}}{V_\pi}\right) + \left(\frac{\pi}{2}\frac{V_m}{V_\pi}\right) \cos(\omega_m t) \tag{6.10}$$

assuming a lossless modulator, where V_{dc} is the dc bias voltage, V_m and ω_m are the amplitude and angular frequency of the RF signal, and V_π is the half-wave voltage of the parent MZM. The condition of CS is achieved by expanding equations (6.8)–(6.10) in terms of the Bessel function as

$$\begin{aligned} E_{01}(t) = \frac{E_0}{\sqrt{2}} \cos(\omega_c t) \cos\Bigg\{ &\frac{\pi}{2}\frac{V_{\mathrm{dc}}}{V_\pi} J\left(\frac{\pi}{2}\frac{V_m}{V_\pi}\right) + \\ &2\cos\left(\left(\frac{\pi}{2}\frac{V_m}{V_\pi}\right)\sum(-1)^n J_n\left(\frac{\pi}{2}\frac{V_m}{V_\pi}\right)\cos(\omega_m t)\right) - \\ &2\sin\left(\left(\frac{\pi}{2}\frac{V_m}{V_\pi}\right)\sum(-1)^n J_{2n+1}\left(\frac{\pi}{2}\frac{V_m}{V_\pi}\right)\cos[(2n+1)\omega_m t]\right)\Bigg\}, \end{aligned} \tag{6.11}$$

where $J_n(.)$ denotes the nth-order Bessel function of the first kind. The first term is the carrier, and the second and third terms are the odd- and even-order sidebands, respectively. Equation (6.12) can be obtained if the dc biases are set at $V_{dc} = V_\pi/2$, and the suppression of the original carrier and even-order sidebands can be obtained.

This condition corresponds to the lowest intensity transmission point in the MZM transfer transmittance characteristics. The output signals of both MZMs can be recombined by means of a polarized beam combiner. Using equation (6.11) the output field can be written as

$$E_0(t) = \frac{E_0}{\sqrt{2}}\left[\sin\left(\frac{\pi}{2}\frac{V_m}{V_\pi}\cos(\omega_m t)\right)\right]\cos(\omega_c t) + \left[\sin\left(\frac{\pi}{2}\frac{V_m}{V_\pi}\right)\cos(\omega_m t + \Delta\Phi)\right]\cos(\omega_c t + \Delta\varphi), \tag{6.12}$$

where the dc bias is chosen for CS, and $\Delta\Phi$ is the phase shift between the electrical signals applied to both MZMs. The output field of the IQ modulator operating under these biasing conditions can be written in terms of Bessel functions using equation (6.12) as

$$E_0(t) = \frac{E_0}{\sqrt{2}}J_1\left[\left(\frac{\pi}{2}\frac{V_m}{V_\pi}\right)\right]\{\cos(\omega_c + \omega_m)(t) + \cos[(\omega_c + \omega_m)(t) + \Delta\varphi + \Delta\Phi] + \cos(\omega_c - \omega_m)(t) + \cos(\omega_c - \omega_m)(t) + \Delta\varphi - \Delta\Phi\} + [\text{higherorder terms}]. \tag{6.13}$$

It is noted from equation (6.13) that by setting the optical phase difference $\Delta\varphi$ between the two branch optical fields, and between the phase shifter of the RF signals, $\Delta\Phi$, as ±90°, one of the sidebands can be nullified. A $\pi/2$ phase shift (Hilbert transformer) or delay line can be inserted in one branch and, driving the other MZM with the signal with its Hilbert transform, suppression of one sideband can be obtained. Thus either suppression of the lower or upper sideband can be achieved by shifting the delay in an alternative branch.

6.2.1.4.1 Details of the connections of modulator driving conditions

6.2.1.4.1.1 Fujitsu FTM-7937EZ dual-drive MZM

- First, no V_{dc} (differential $V_{dc1} - V_{dc2}$) → adjust the PC to obtain max optical power $P_{0\max}$. Then one may adjust $V_{dc,diff(1,2)}$ to obtain the maximum due to non-zero $P_{0\max}$ → working out the V_π of the DD-MZM at dc (figure 6.8).
- Then apply a voltage to $V_{dc,diff(1,2)}$ to obtain the minimum P_0. We can adjust $V_{dc,diff(1,2)}$ to obtain the Q-point between the two arms, or middle linear voltage points.
- Apply $v_{RF(t)}$ to only one arm to estimate the RF. By observing the spectrum of the fundamental and second harmonics → $V_{\pi/2}$ (AC) → swinging the V_{RF}

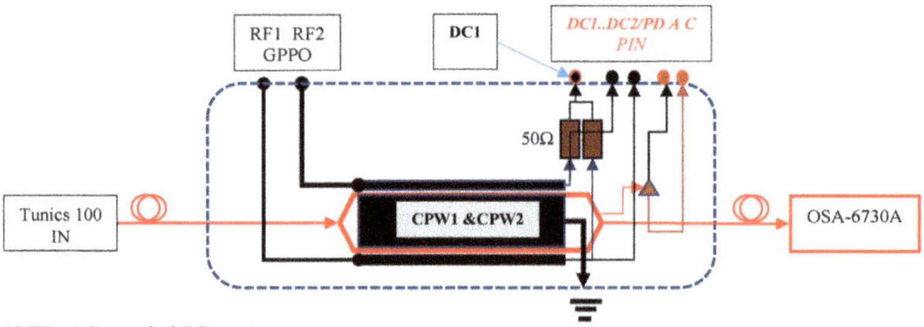

(a) NOTE: AC coupled RF ports

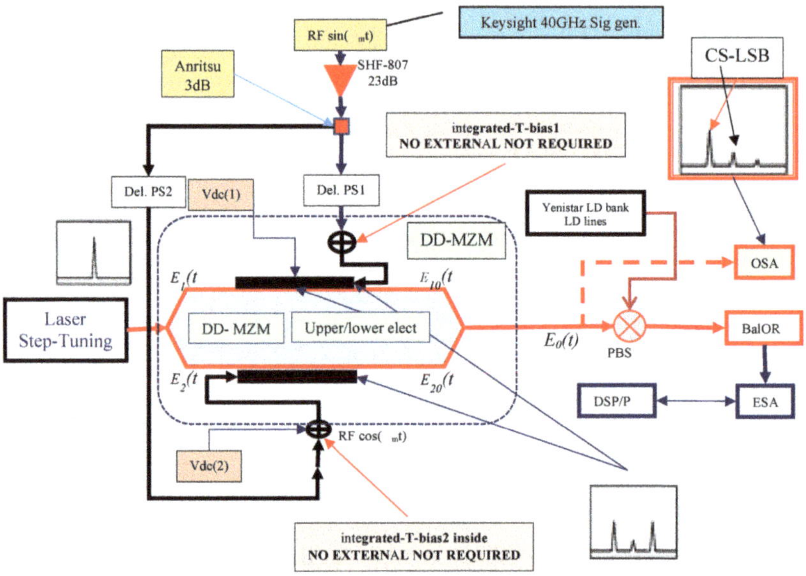

(b) NOTE: AC coupled RF Ports → impossible to use bias-T – differential Vdc (1,2) to achieve Min, Max & Q-points.

Figure 6.8. A CS-SSB from using a dual-drive Mach–Zehnder modulator. (a) The driving DD-MZM FTM-7937EZ, schematic, and connection pins (RF GPPO) + dc1 and dc2 + PD monitor pins; (b) detailed driving connections. *Note*: for AC coupled RF1 and RF2 no T bias is required, only setting a differential V_{dc} to obtain the max, min, or Q points. OSA = optical spectrum analyzer; DSP = digital signal processing; ESA = electrical spectrum analyzer; BalORx = balanced receiver; DD-MZM = dual-drive MZM.

to obtain the DSB → applying both $V_{RF(1)}$ and $V_{RF(2)}$ with a phase shift (PS) → $\pi/2$ → leading to a CS-SSB.

- *Note*: no T bias is required for both RF ports as the AC coupling is applied to both RF ports and differential V_{dc} is applied to dc1 and dc2.

6.2.1.4.1.2 EO-Space QPSK optical modulator

The procedures and conditions for a CS-SSB using a QPSK EO-Space modulator.

How to determine v_π(AC)?

- First turn both $V_{dc(I)}$ and $V_{dc(Q)}$ → the zero transmission point to give a maximum output P_0 out of the parent MZM (figure 6.9).

- Turn the MZM child 2 (Q−) off or set $V_{dc(Q-)} \sim 0$. Adjust $V_{dc(I-)}$ to obtain the max, min, and Q points of MZM(I).
- Similarly to the above immediate step, for MZM (Q−) to setting MZM(I−) to Q−. Setting MZM(Q−) to Q+ → the difference between MZM(I−) and MZM (Q−) = π (rad.) → carrier suppression.

Now apply $v_{dc(I)}$ to obtain the min, max, and Q points by observing the spectrum of the sidebands. Doing the same for the child MZM(Q−):

- Setting V_{dc} → Q− and Q+ for MZM(I−) and MZM(Q+), respectively → carrier suppression.
- Applying an RF signal so that P_{RF} = peak value = $v_{\pi/2}$ ($V_{RF(AC)}$) to both the I- and Q-ports with a PS of $\pi/2$ to either with respect to other.

6.3 CS-SSB generation by modulation in a DD-MZM, QPSK, or PM-IQ modulator

6.3.1 DD-MZM

6.3.1.1 Analytical details

Referring to figure 6.10, by applying a bias of $V_{\pi/2}$ to the lower arm with respect to the upper arm, we can obtain a $\pi/2$ PS with respect to that of the upper arm. Applying a PS to the RF path of the upper arm with respect to the RF electrode in the lower arm leads to a total phase shift of π, hence a suppression of the optical carrier. However, when the RF sinusoidal oscillating voltage is superimposed on the dc bias then there is a variation of the phase and the carrier is not suppressed, thus we have the modulation of the optical carrier by the RF signal in both the upper and lower sidebands. One of these bands can be further suppressed by an optical Hilbert transform by inserting a $\pi/2$ phase shift in one branch of the RF input ports.

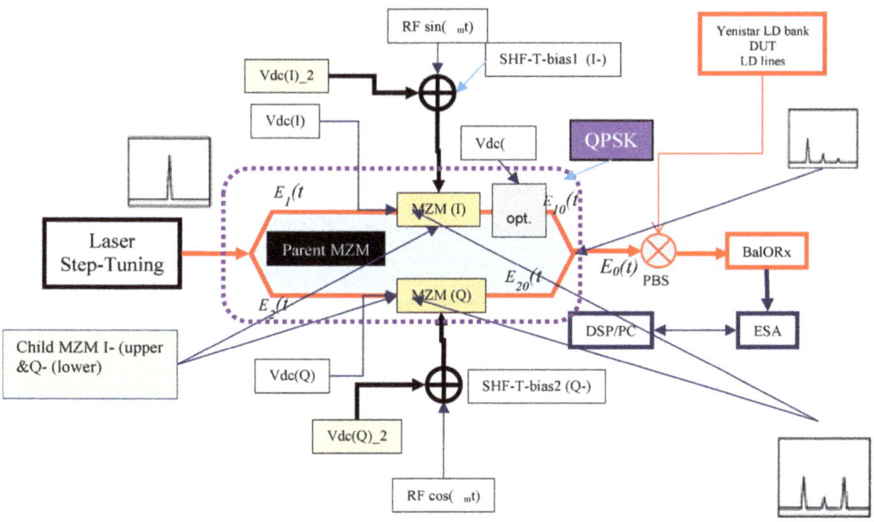

Figure 6.9. Circuit connection of an IQ modulator (EO-Space type) for CS-SSB modulation.

Therefore, a shift left to the lower sideband (LSB) or upper sideband (USB) can be generated depending on whether the phase shifter is added in the right arm or the left arm, respectively.

Analytically [1], excitation of an RF voltage is required at a frequency and amplitude level such that the modulation depth is sufficient for an alternative π phase shift of the MZIM biased at the maximum (max) or minimum (min) point such that single sideband suppression can be generated. Figure 6.10 depicts the schematics of the generator of such a CS-SSB using either a DD-MZM or a QPSK with two MZMs and a phase shifter Φ. Optionally, it is possible when using a polarization rotator QPSK, such as a Fujitsu FTM7980, a dual polarized QPSK, to generate and switch the optical carriers in the two arms into or out of the carrier suppression. These CS-SSB modulators can be used in optical network units (ONUs). For the DD-MZM shown in figure 6.10(a), the upper sideband of the optical field in the upper arm can be written as

$$E_{\text{DD-MZM}_{\text{USB}}}(t) = \text{Re}\left\{\frac{1}{\sqrt{2}}E_0 e^{j(\omega_0 t+\Delta\varphi(t))}\right\} = \frac{1}{\sqrt{2}}E_0\cos(\omega_0 t + \Delta\phi(t)) \tag{6.14}$$

$$E_{\text{DD-MZM}_{\text{USB}}}(t) = \frac{1}{\sqrt{2}}E_0\{(\cos(\omega_0 t)\cos\Delta\phi(t) - \sin(\omega_0 t)\sin\Delta\phi(t))\}, \tag{6.15}$$

where the subscript USB stands for the upper sideband, E_0 and ω_0 are the amplitude and angular frequency of the input optical carrier, respectively. $\Delta\phi(t) = m\cos(\omega_{RF}t)$, with m and ω_{RF} the modulation depth and the RF angular frequency, respectively.

Then, by replacing $\Delta\phi(t)$ by $\cos(\omega_{RF}t)$ and the solution of equation (6.15) in terms of the Bessel functions of the zeroth-, first-, second-, and third-orders, $J_1, \ldots, J_3$, with higher order ($\geqq$ fourth-order) Bessel terms negligible:

The lightwaves output electric fields of the upper arm are

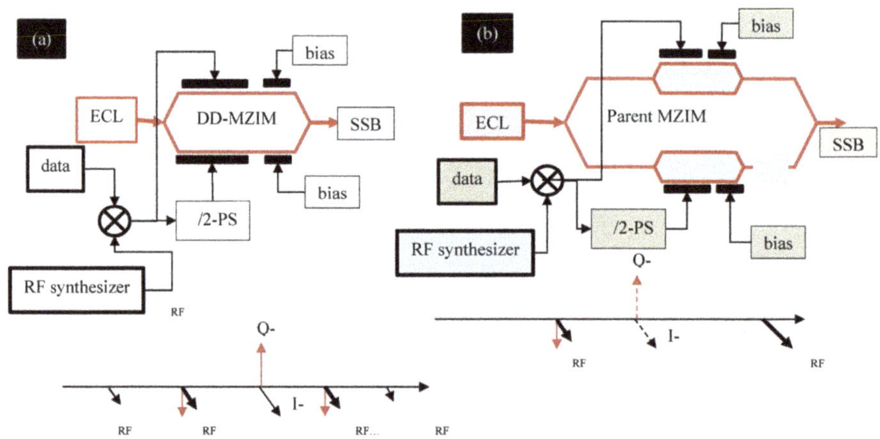

Figure 6.10. Generation of a CS-SSB by setting the biasing and PS conditions using (a) DD-MZM and (b) QPSK or DP-IQ modulators. I = in-phase; Q = quadrature; PS = phase shifter (delay line); SSB = single sideband; DD = dual-drive; MZIM = Mach–Zehnder interferometric modulator.

$$E_{\text{DD-MZM}_{\text{USB}}}(t) = \frac{1}{\sqrt{2}} E_0\{J_0(\text{m})$$
$$\begin{aligned} &\cos(\omega_0 t) - J_2(\text{m})[\cos(\omega_0 t + 2\varpi_{\text{RF}} t)] \\ &+ [\cos(\omega_0 t - 2\varpi_{\text{RF}} t)] \\ &+ J_1(m)[\sin(\omega_0 t - \varpi_{\text{RF}} t) - \sin(\omega_0 t - \varpi_{\text{RF}} t)] \\ &- J_1(m)[\sin(\omega_0 t + 3\varpi_{\text{RF}} t) + \sin(\omega_0 t - 3\varpi_{\text{RF}} t)]\}. \end{aligned} \tag{6.16}$$

For the lower arm of a DD-MZM, the phase shift is the opposite (negative) with respect to that in the upper arm, because of the electro-optic effect via interaction of the optical field and the RF electric field whose direction is in the opposite direction to that of the upper arm, with a phase shift of 90° radians applied to the phase shifter. Thus $\Delta\phi(t)$ can be expressed as

$$3\cos(\omega_{\text{RF}} t + \pi/2) = -m\sin(\omega_{\text{RF}} t). \tag{6.17}$$

Similar to the modulation of the upper arm, the lightwave electric field at the output of the lower phase modulated arm of the DD-MZM can be expressed as

$$E_{\text{DD-MZM}_{\text{LSB}}}(t) = \frac{1}{\sqrt{2}} E_0\{J_0(m)\cos(\omega_0 t)$$
$$\begin{aligned} &- J_2(m)[\cos(\omega_0 t) + 2\varpi_{\text{RF}} t + \cos(\omega_0 t - 2\varpi_{\text{RF}} t)] \\ &+ J_1(m)[\cos(\omega_0 t - \varpi_{\text{RF}} t) - \cos(\omega_0 t - \varpi_{\text{RF}} t)] \\ &- J_1(m)[\cos(\omega_0 t + 3\varpi_{\text{RF}} t) + \cos(\omega_0 t - 3\varpi_{\text{RF}} t)]\}. \end{aligned} \tag{6.18}$$

The phase of the lower arm should be shifted by another 90° because of the $V_\pi/2$ dc bias. Hence we can see the positive sign $+\ \omega_{\text{RF}}$.

Some components of the lightwave field can be canceled at the output port of the DD-MZM, as shown in (i) of figure 6.10(a). Then, we choose a modulation depth (m) to suppress the central electrical component, i.e. the zeroth-order of the Bessel function $J_0(m) = 0$. Hence the CS-SSB signal can be generated with a modulation depth of $m \approx 2.4$, thence at this level of driving voltage, the MZM is overdriven or higher than $2V_\pi$. This puts difficulties on the microwave driving signals if the operating region is in the mmW spectral window. The modulation index m is defined as

$$\Delta\phi(t) = m\cos(\varpi_{\text{RF}} t), \tag{6.19}$$

where m and ω_{RF} denote the modulation depth and the angular frequency of the RF driving signal, respectively.

6.3.1.2 Generating a CS-SSB using a DP-QPSK

A PM-QPSK (figure 6.10(b)) consists of three single-arm MZMs, designated as MZM1 (the upper arm child MZM), MZM2 (the lower arm child MZM), and MZM3 (the parent MZM). An RF driving voltage signal is applied to the two child

MZMs (MZM1 and MZM2). Similarly to the analysis for a DD-MZM, the optical field at the MZM1 output is given by

$$E_{\mathrm{MZM_1}}(t) = \frac{1}{\sqrt{2}}E_0\{2J_1(m)\{\sin(\omega ot + \varpi_{\mathrm{RF}}t) + \sin(\omega ot - \varpi_{\mathrm{RF}}t)\} \\ - J_3(m)[\sin(\omega ot + 23t)] + [\sin(\omega ot - 3\varpi_{\mathrm{RF}}t)]\}. \tag{6.20}$$

For MZM2, the lower child MZM with a $\pi/2$ phase shift, the optical electric field at the MZM2 output is given by

$$E_{\mathrm{MZM_2}}(t) = \frac{1}{\sqrt{2}}E_0\{-2J_1(m)\{\cos(\omega ot + \varpi_{\mathrm{RF}}t) - \cos(\omega ot - \varpi_{\mathrm{RF}}t)\} \\ - 2J_3(m)[\cos(\omega ot + 3\varpi_{\mathrm{RF}}t) - +\cos(\omega ot - 3\varpi_{\mathrm{RF}}t)]\}. \tag{6.21}$$

Similarly, the phase at the MZM2 output can be shifted by another $\pi/2$ due to the $V_\pi/2$ dc bias. The optical electric field components at the output port of the DP-QPSK are shown in panel (ii) of figure 6.10. We can observe that the term of the positive sign $+\omega_{\mathrm{RF}}$ generated by the DP-QPSK is dominant and independent of the driving modulation index.

This implies that we do not need to over-drive the modulator in order to achieve carrier wavelength suppression as in the case of the DD-MZM.

6.3.1.3 Implementation

Figure 6.11 shows the structure and hardware implementation of a CS-SSB modulator. The obtained spectra are given in figure 6.12 and indicate the suppression of the carrier and one sideband. The upper sideband or lower sideband can be selectively suppressed. Both sidebands can be scanned across the signal spectrum so that the data channels of the sub-RF carrier multiplexed channels can be detected simultaneously. A typical assignment of the selected sideband is given in figure 6.13. Under coherent reception a CS-SSB local oscillator (LO) can be generated using such a technique as shown in figure 6.13. The implementation of the coherent reception using a hybrid coupler is shown in figure 6.14 in which component details are given.

6.3.2 Linearization by suppression techniques with a paralleling MZM and biases

6.3.2.1 Remarks

Linearization of the optical device of an RPoF system aims to obtain an RPoF whose spurious free dynamic range (SFDR) and spur-free third to fifth harmonic frequencies are as far as possible from the fundamental so as not to interfere with the RF carrier or multi-carriers in a sub-carrier multiplexing system and the channels (data) in 6G bands (>60 GHz). The linearization of RPoF can be implemented using two major techniques, as follows.

The first method is linearization by extending the transfer transmittance linear region of an MZIM. Commonly, in a bare MZIM with or without chirping, dual-drive, or QPSK, the SFDR and third harmonic spur (IMD3) can only be within

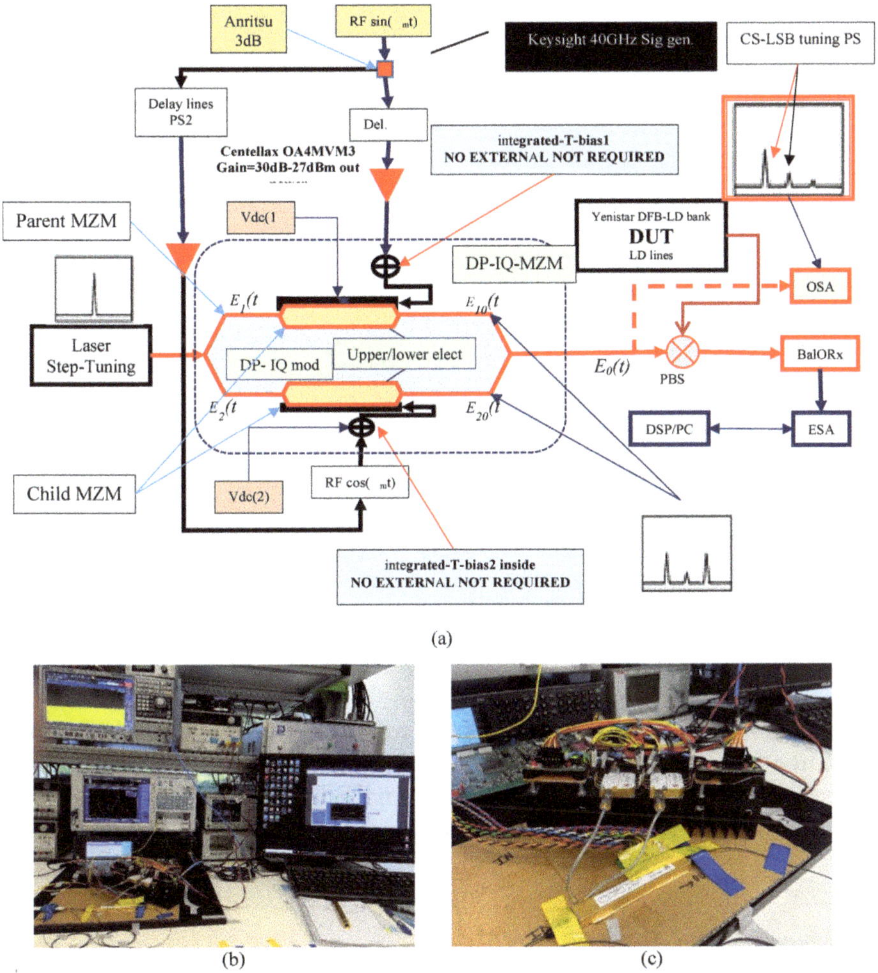

Figure 6.11. DP-IQ modulator, ABC platform (20200807), and RF amplifiers Centellax OM4MVM3 with 30 dB gain (a three stage FET amplifier) and output RF amplifier at max 27 dBm. (a) CS-SSB set-up platform for CS-SSB using an DP-IQ modulator and 30 dB gain RF amplifier. (b) A photograph of (a). (c) Centellax Amplifiers type OA3MVM3—gain = 30 dB; output power max = 23 dBm; BW = 50 GHz.

one decade of frequency. A micro-ring resonator (MRR) and phase modulator can be added to one branch of the MZIM and another phase modulator section on the other branch, as shown in figure 6.15. The phase control and coupling switch can offer some 20 dB improvement in the SFDR [2–5]. This linearization technique suffers from narrow passband characteristics to lower than those of a passband of the RF modulated optical carrier, even though the SFDR can be improved.

Second, linearization by suppression of the second and third harmonics can be implemented by considering the modulation format DP-QPSK by modulation

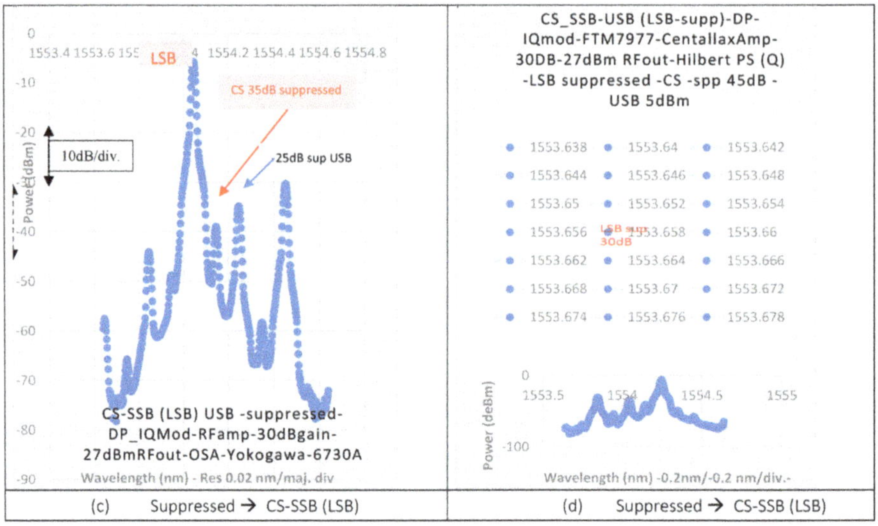

Figure 6.12. Spectra of (a) a CS-SSB with suppression of the upper sideband (USB). (b) A CS-SSB with suppression of the lower sideband (LSB), OSA set at a span of 1554.8–1553.4 or 1.4 nm full scale, 20 pm resolution.

over two parallel MZIMs (parent and children) that are manipulable to cancel the third harmonics and also allow suppression of NLFD harmonic free distortion and make it possible to increase the RF signals, i.e. 6G carrier-channels can go much farther, thus lowering the OPEX and making it more attractive for network operators.

6.3.2.2 Linearization by suppression

In this section we present the VPI simulation model of RPoF using a modulation technique. Two parallel MZIMs are used and biased in a minute differential mode by adding or subtracting the term 'dBias' to the biased levels of the two modulators or branches of an MZIM, controlled by sweeping this differential bias. It is noted here that an MZIM with dual-drive can be considered as two parallel phase modulators and interference. dBias is the differential bias of a very small amount added or subtracted from the dc bias in the differential and common modes to the biases of the two MZMs. Further dBias is also used to recover the clock signal.

We now can come to the conditions for depletion of the second harmonic distortion (SHD), third harmonic distortion (THD), IMD2, IMD3, and SFDR:

- The output optical fields of the DP-QPSK. *Notes*: two orthogonal polarized lights, a field of $\pi/4$, the intensity is by sqrt(2), power of 2 → equivalent feedforward linearization of the compensation of nonlinearity in the Volterra transfer function (see figure 6.16) of the RPoF operating under nonlinear regions, here as EM field analyses.

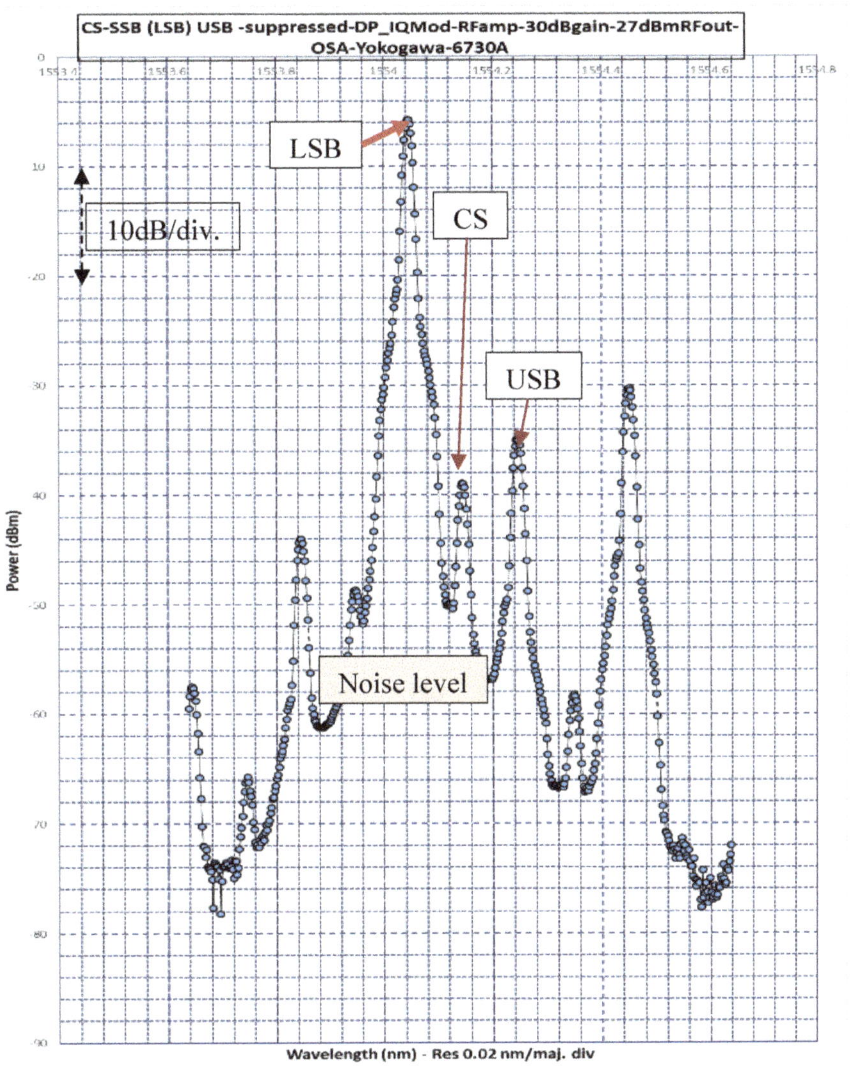

Figure 6.13. Typical assignment of modulated sidebands and carrier suppression in a CS-SSB modulation scheme.

- The summation of two optical field beats in the PD (single or hybrid—single preferred if IQ modulation is not required).
- The MZIMs of the QPSK modulator need to be biased relative to each other, to have $\varphi_{\text{im}}(t) = \pi V_{\text{im}}(t)/V_{I-\pi} = \beta_I V_m(t)$—modulation index, $\beta = \pi/V_\pi$.
- Detected current (6.20) and (6.21) under the square of the optical field sum → product terms → Bessel functions [sin$x$(sin$y$)].

Now if we set $\varphi_{i3} = k\pi + \pi/2$ (the phase of the IQ MZMs) and $\varphi_{i2} = (2n+1)\pi - \varphi_{I1}$ (the phases at the quadrature point of the IQ- and the PS-QPSK); $k = 0, 1, 2, 3, \ldots, N$.

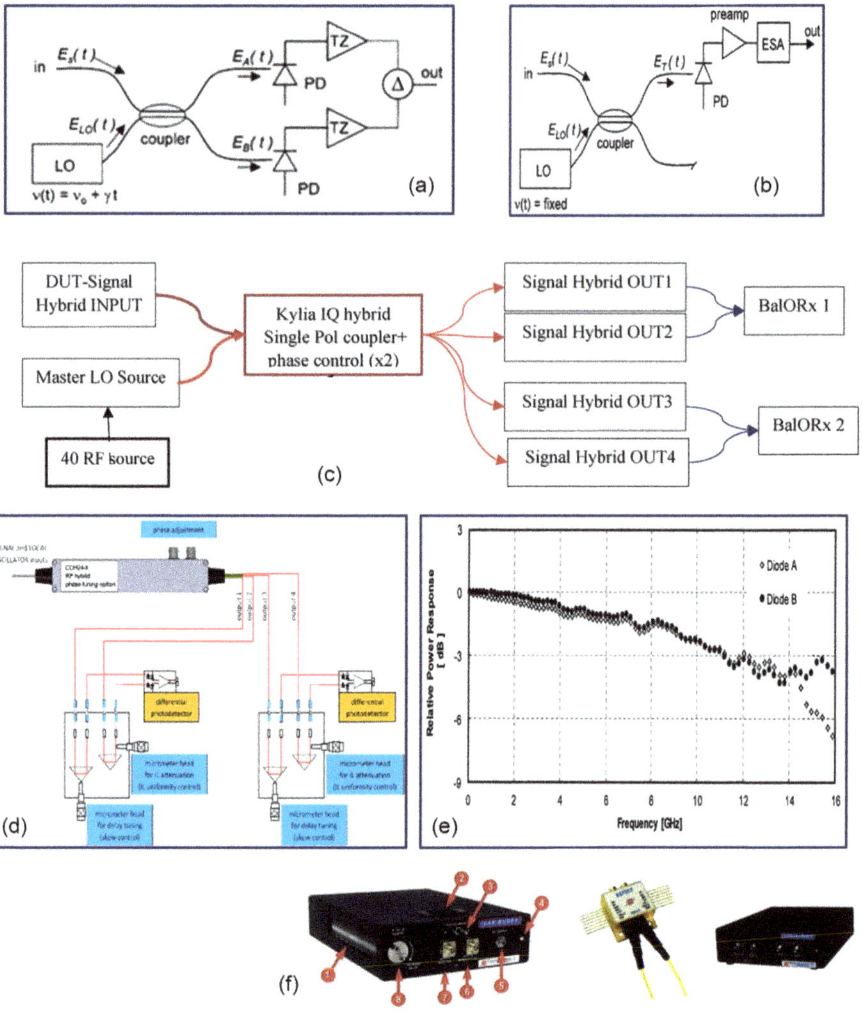

Figure 6.14. Schematic details of an MLO CS-SSB: (a) balanced receiver in COSA with a scanning LO; (b) un-balanced heterodyne receiver with fixed LO (non-scanning)—requires a broadband TIA and PD. (c) COSA reception structure for polarization multiplexed signals using a hybrid coupler and balanced receiver structure with a CS-SSB LO laser. (d) Details of the hybrid coupler with phase tuning (see also figure 6.40), for in-phase and quadrature optical paths. (e) Trans-impedance responses $S_{21}(\omega)$ of the optical receiver. (f) Images of the receivers by a Discovery semiconductor (DSC) and balanced photodetectors (BPDs).

Then the PD current is proportional to the responsivity R, $J(.)$, the Bessel functions, and β, the phase factor proportional to the dc bias level).

The conditions for suppression of all distortion coefficients for the depletion of IMD2, IMD3, and SHD are as follows:

- Carefully adjusting the biases of the PM-DP-MZM to make $\varphi_{i3} = k\pi + \pi/2$ and $\varphi_{i2} = (2n + 1)\pi - \varphi_{i1}$, where k and n are integers.

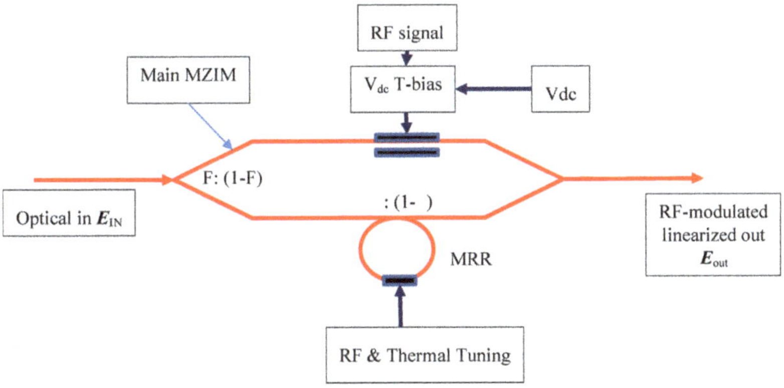

Figure 6.15. Schematic of of high-linear optical modulator, MRR-MZM in an integrated optic structure. MRR = micro-ring resonator with a cross coupling coefficient κ.

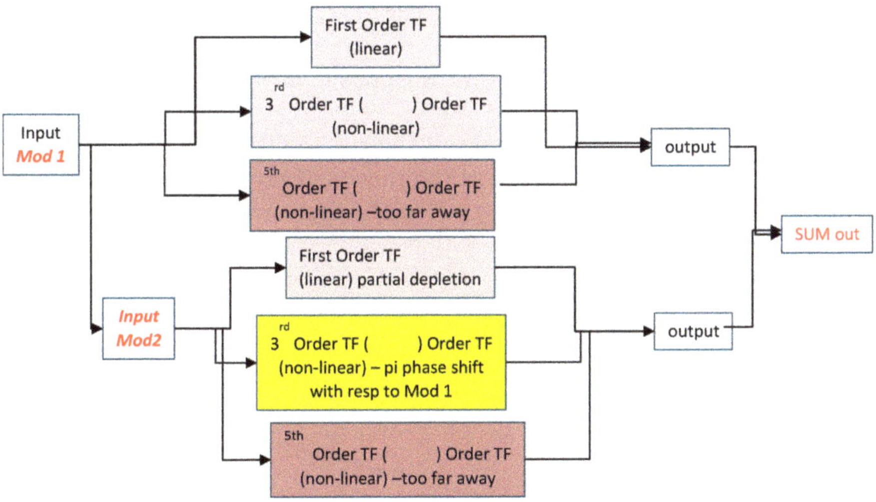

Figure 6.16. Volterra transfer functions of a nonlinear RPoF system operating in the nonlinear region.

- Adjusting the input RF amplitude-power and the microwave attenuator to tune β_1 and β_2 to satisfy (6.20) and (6.21) so that the inter-modulation due to the beating of these fields can be cancelled to each other.

Figure 6.17 shows the structure of a dual polarization QPSK and the arrangement of biases for minimization of the nonlinear IMD. A polarization emulator must be used for testing the effects of the polarization random rotation of the SSMF and thus its influence of the linearization after transmission through a length L of SSMF. Figure 6.18 illustrates a phasor diagram on which the static dc phases and the swinging of the RF modulated dynamic phase can rotate.

6.3.3 Simulation of nonlinear dispersion (NLD) suppression in RPoF

6.3.3.1 Simulation platform for RPoF with polarization rotation

The VPI TransmissionMaker platform is used to simulate the RPoF and demonstrate the suppression of nonlinear distortion and intermodulation effects. The simulation circuit is given in figure 6.19. A laser diode (an ECL with a narrow linewidth of 10 KHz) model is used to feed optical signals into two child MZMs which are driven by two RF signal branches in which a phase shifter is incorporated. Polarization rotators are added along the two arms of the child MZMs and then combined to a polarization beam combiner and detected by a single PD. In a later model we will use coherent reception. On the far right of the figure a two-tone analyzer is used to analyze the IMD effects and correspondingly the excitation of the RF two-tone oscillators, as shown in the far left. The RPoF parameters are obtained, such as the SFDR, and the signal polarization direction can be rotated by sweeping the rotation angle. The power of the first-order, second-order, and third-order spur and SFDR are obtained, as shown in figure 6.20. At the polarization angle of 13.58° the powers are peaked, indicating the best performance of the RPoF using a pulse position modulation (PPM)-QPSK modulator. The polarizer acts as a component resolver to split the power of one polarized channel to the opposite of that of the other channel in the opposite direction, hence mitigating the IMD effects.

At this optimum polarization angle of the rotator the power of the excitation RF tones is swept and the variation of the power of the SFDR, third-order spur, second-order spur, and IIP3 with respect to the first-order RF tone is depicted in figure 6.21.

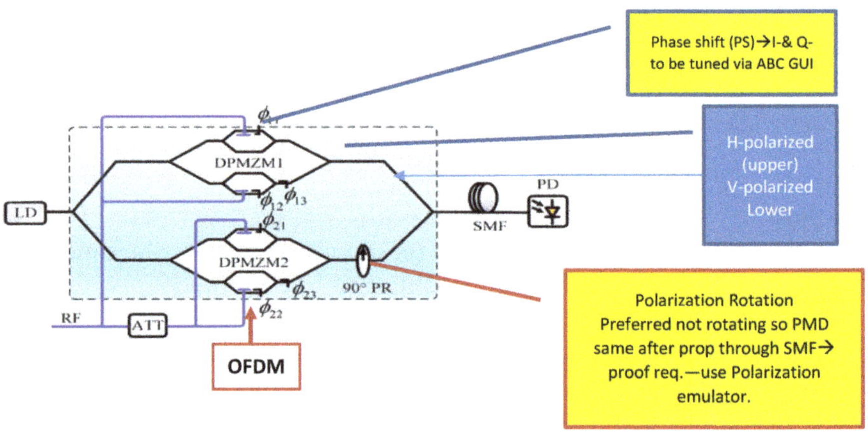

Figure 6.17. Dual polarization a QPSK modulator and the arrangement of the biases and polarization rotation. LD = laser diode; PD = photodetector; SMF = single mode fiber; PR = polarization rotator; DP-MZM1, 2: MZM1 and MZM2 of the DP-MZM.

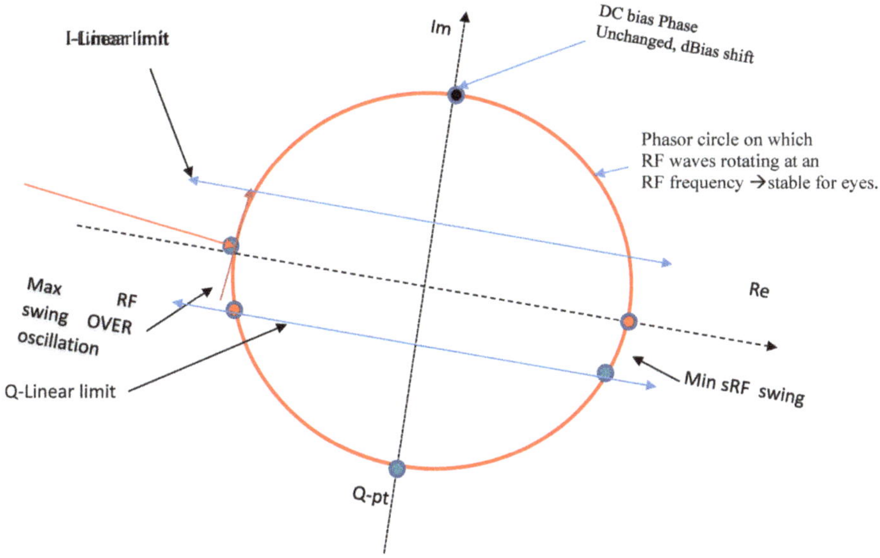

Figure 6.18. Biasing scheme for both MZIMs as represented on a phasor diagram, specifically with differential dBias.

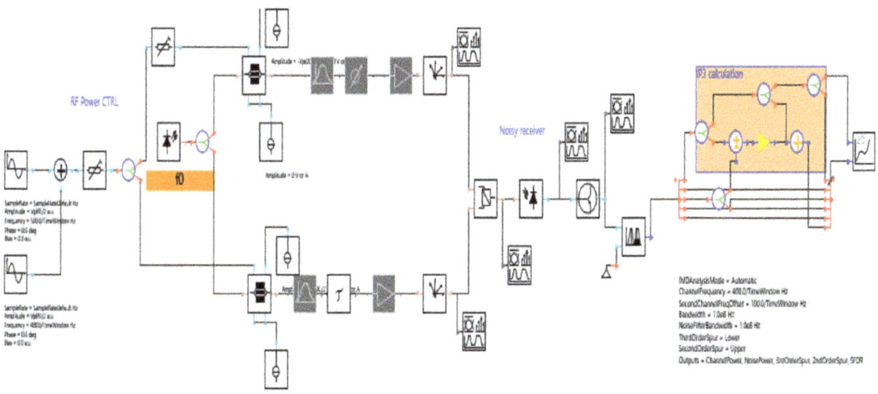

Figure 6.19. The VPI TransmissionMaker© model for a PM-QPSK RPoF transmission system.

6.3.3.2 Simulation platform for RPoF with biasing Q+ and Q−

6.3.3.2.1 Structure and bias

The model structure is shown in figure 6.22(b) and its corresponding model set up on VPI PHotonics GmbH platform given in figure 6.23. The biasing is set such that the biasing in the positive and negative slope of the linear region of the MZMs is as follows:

- The positive increase and negative increase can counter each other's non-linear distortion effects.
- Polarization rotation to adjust the coupling efficiency or optical amplitude ratio cancellation of the NLD with the RF attenuation to see the amplitude ratio of the RF waves.

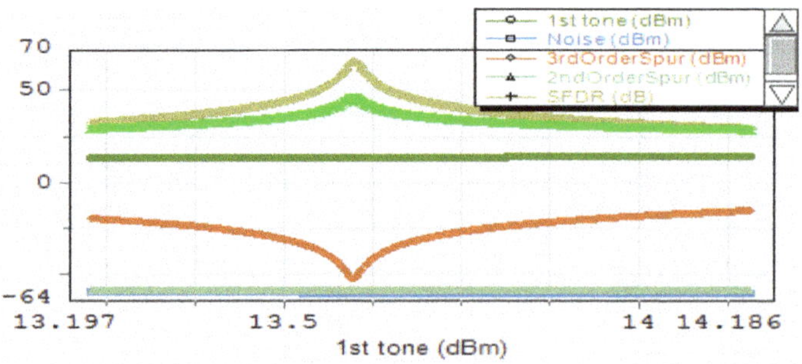

Figure 6.20. Variation of the RF power of the RF signals, first-order tone, noise floor level, third-order spur, second-order component, and SFDR as a function of the polarized angle.

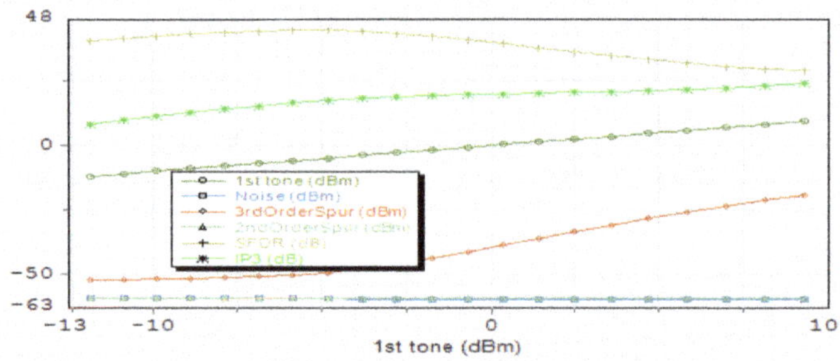

Figure 6.21. Variation of the power of the SFDR, third-order spur, second-order spur, and IIP3 with respect to the first-order RF tone.

- Now the attenuator is set as 3 dB with a sweeping polarization rotation angle, in practice applying voltage to the polarization rotator.
- If no 7980 PM-MZM by Fujitsu is available then use two MZMs and a PC as shown to control the polarization and a hybrid coupler
- This model shows that RPoF systems can be affected by intermodulation distortions (IMDs) resulting from the nonlinear nature of the optical link.
- In this example, a laser is split into two paths and MZMs modulated by five orthogonal frequency division multiplexing quadrature amplitude modulation (OFDM-QAM) radio channels (2 GHz carrier frequency, 100 MHz channel spacing).
- When an MZM is biased at Q+ or Q− then the magnitude of output optical intensity follows a negative or positive sloping of the transfer transmittance curve, respectively.
- The second MZM output goes into the polarization beam combiner (PBC) so as to mitigate the NLD effect.
- Second- and third-order IMD are generated by increasing either the modulation index of the laser or the fiber length.
- The error vector magnitude (EVM) of the central radio channel is displayed against the ratio third-order spur/carrier.

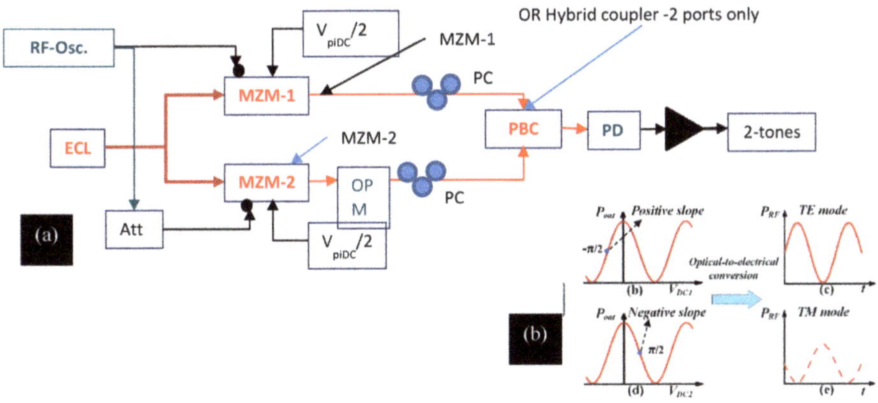

Figure 6.22. Schematic of the RPoF (a) using a QPSK with (b) biasing in the positive and negative slopes of the transfer transmittance of the child MZMs.

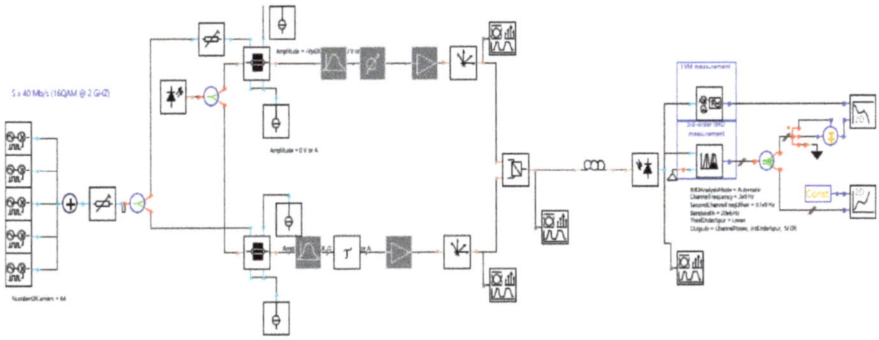

Figure 6.23. The model of the QPSK modulator at opposite biasing and two analyzers with five sub-carriers OFDM data channels and an RF carrier at the transmitter.

- The carrier, third-order spur (IMD3), and SFDR are displayed against the modulation index (respectively fiber length) as well.
- Relevant models: C:\VPI_Models_All\VPI-Models\ProjMdls\2021Proj_Prt1_Mdls-working\ROF System\NLD-mQAM-OFDM.
- And located at C:\VPI_Models_All\VPI-Models\ProjMdls\2021Proj_Prt1_Mdls-working\ROF System\NLD-mQAM-OFDM (figure 6.23).
- Single carrier OFDM: carrier frequency 2.5 GHz, RF at 20 GHz, IF at 15 GHz → processing at 5 GHz + carrier frequency.
- The adjacent channel power ratio (ACPR) is also measured = Pad (dBm) – Pref (dBm) over the bandwidth of the channel (the same bandwidth).

The RF carrier is assigned 20 GHZ and the OFDM channel a 1.0 GHz bandwidth. Five RF sub-carriers are injected into the optical MZMs which are biased in the positive and negative directions of the transmittance transfer curve. The RF-OFDM spectrum of one channel is shown in figures 6.24(a) and (b). The channel is coherently down-converted to the baseband, as shown in figure 6.25.

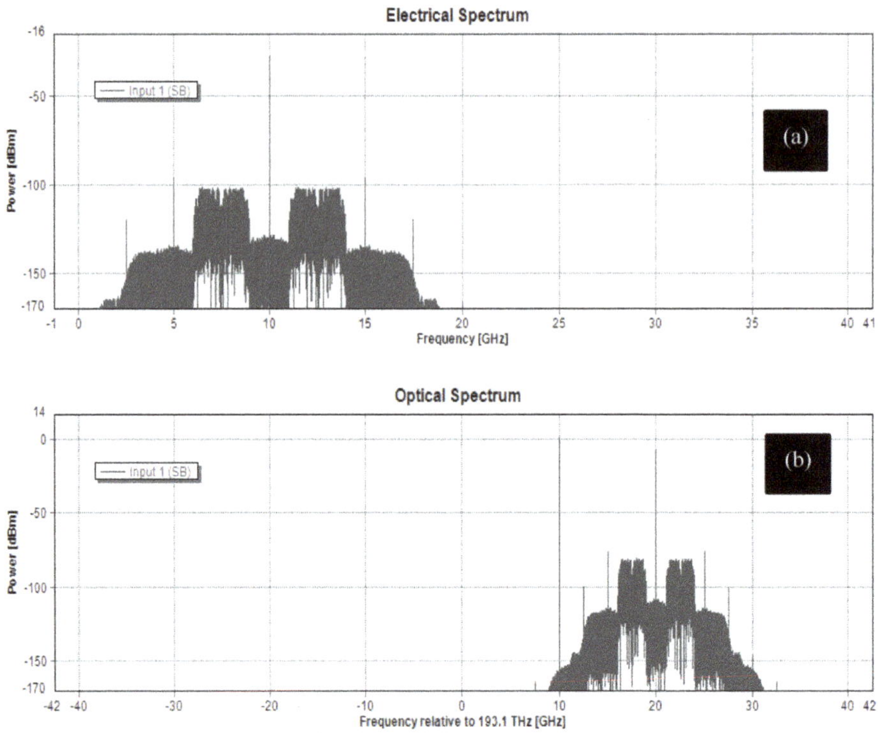

Figure 6.24. The spectra of one channel reception of (a) an electrical RF channel after down conversion and (b) from an RF 20 GHz modulated channel.

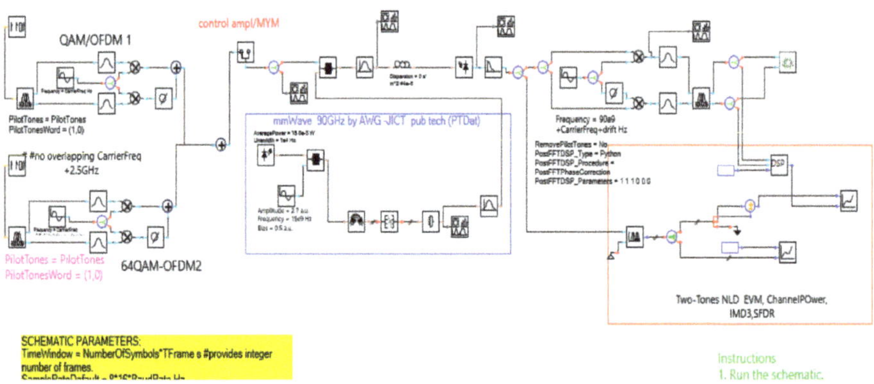

Figure 6.25. RPoF with QPSK and a coherent reception model. A local oscillator laser is inserted in the lower middle section. A two-tone analyzer in at the far right and five times sub-carrier multiplexing of the RF-OFDM on the far left.

This can be done in coherent reception or envelop detection. The IMD parameters are also measured, as well as the adjacent channel power ratio. The OFDM channel guard band is 100 MHz.

The modulation is 258QAM for RF carriers and the OFDM data channels remain the same.

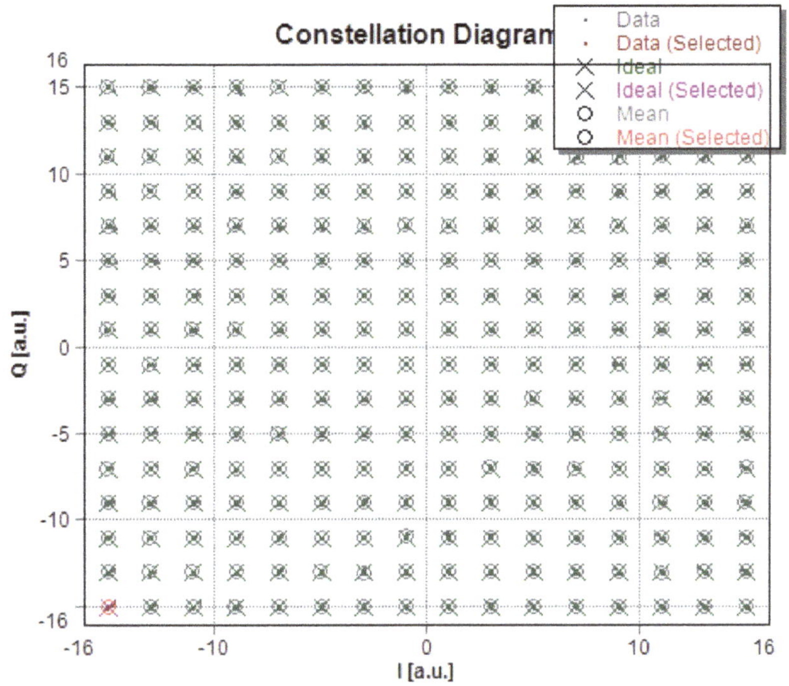

Figure 6.26. Constellation of a sub-carrier of the OFDM with 16 sub-carriers. Note: For decoding OFDM sequence: DMT (discrete multi-tone) model is used instead of OFDM modified model.

The system parameters are as follows:

- Data bit rate: 2.5 Gbps.
- RF_c = 20 GHz.
- IF = 15 GHz–<5 GHz processing region.
- 256QAM-OFDM; sub-carrier number = 16; OFDM coding and decoding.
- Fiber distance = 10 km; EVM < 5%; SFDR ~ 25–30 dB → 80–90 dB $Hz^{2/3}$; soft error rate (SER) < 1e-5; at this error rate it is fine for forward error coding (FEC) in practice 1e-3 OK so 1e2 for margin.

Figure 6.26 depicts the constellation of the detected RF signals, which is modulated by 2576QAM. If direct detection is used then the algorithm to recover the OFDM is set for discrete multi-tone (DMT) (figures 6.27 and 6.28).

6.4 Distribution of multi-radio-photonic carriers in passive optical networking

As described in previous sections, the CS-SSB using DD-MZM or DP-IQ modulators can be implemented with all the modulators which are commercially available. The modulation bandwidths of MOD1 and MOD2 are 9 GHz and 12 GHz, respectively, and the insertion losses are 5 dB and 3.5 dB, respectively. The V_π of MODa and MODb are 2.6 V at complementary driving and 5 V, respectively.

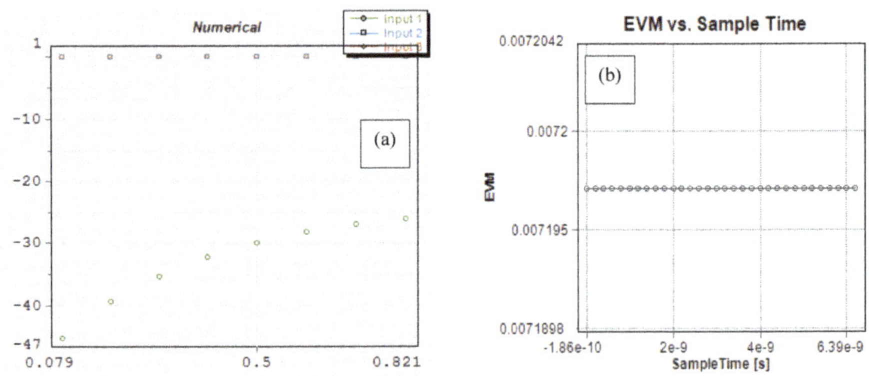

Figure 6.27. (a) EVM and SFDR with input 1 and 2 power. (b) An EVM of < 0.007 or 0.7%.

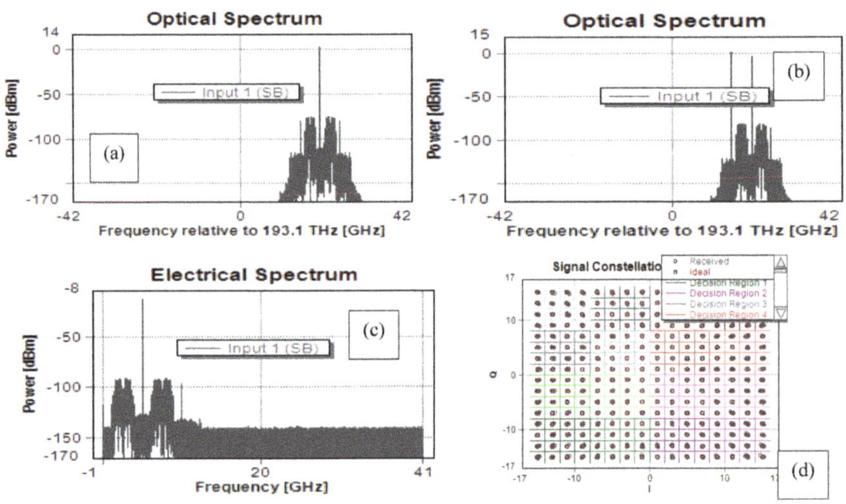

Figure 6.28. (a) Spectrum of one OFDM channel modulating an RF carrier. (b) At the output of the optical transmitter. (c) The optical spectrum in front of the optical receiver. (d) The down-converted spectrum, the IF, of 10 GHz with the constellation of the received channel.

The DP-MZM, in this configuration, can also be termed as a differential quadrature phase shift keying (DQPSK) modulator.

The RF bandwidths are 12 GHz and 7 dB, respectively, with a half-wave voltage V_π of 5.5 V. For the case of CS-SSB MOD1, a shared DD-MZM (MODa) could be located at the remote node [2] to simplify the optical network unit (ONU). However, active components, such as high speed electronic circuits, are required at the remote node, as shown in figure 6.29. A bank of optical lasers is used to launch into a multi-band optical-wireless distribution system. A particular wavelength laser is coupled and multiplexed via an array waveguide grating (AWG) whose diffraction grid spacing meets the grid spacing of 100 or 50 GHz in the C- and L-bands.

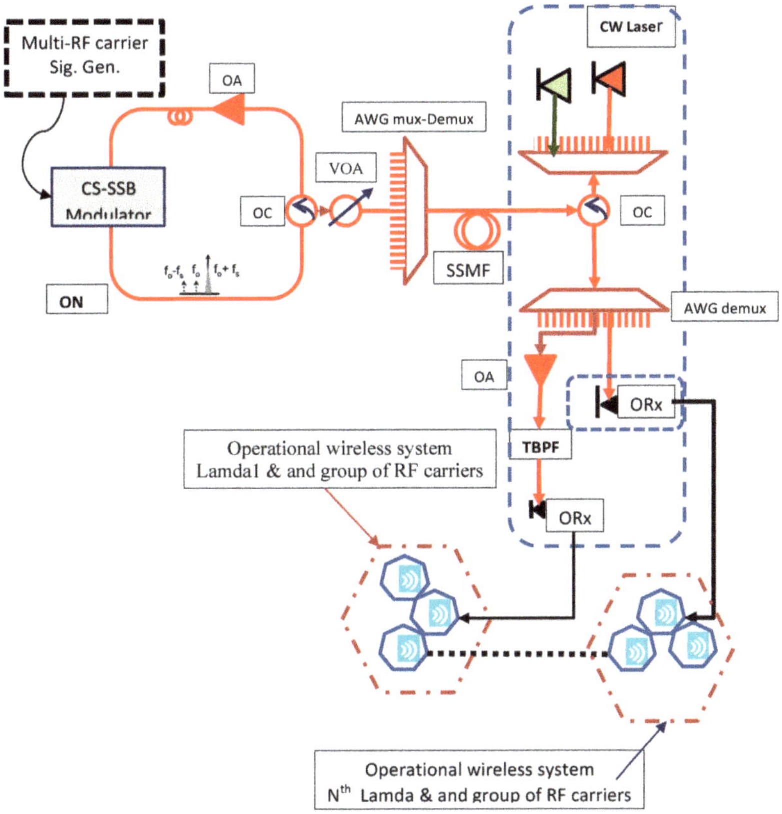

Figure 6.29. Carrier distributed PON using CS-SSB modulation for multi-band antenna testing. The antennae are operating in different spectral windows of 5G, 6G, and 7G networks. TOBPF = tunable optical bandpass filter; OC = optical circulator; PC = phase shifter; OA = optical amplifier. Insets: (i) CS-SSB MOD1; MODa = dual-drive MZM; MODb = single-arm MZM; PS = phase shifter; (ii) CS-SSB MOD2; DP-MZM = dual-paralleled MZM. This is applicable to RPoF.

The multiplexed channels are circulating and entering into a CS-SSB incorporated in an optical loop so that the RF carriers match to a particular RF band of the 5G or 6G radiation, for example, the mmW or μW regions. These RF carrier optical waves can then be circulated via an optical circulator (OC) and re-enter the optical reception multi-systems and be filtered by an optical bandpass filter whose central passband wavelength matches that of the RF wireless-optical carrier. This optical wave is then detected and converted to the electrical signal domain and is then amplified and fed into a wireless antenna system for transmitting to the 5G or 6G micro-cell. Therefore, several wireless systems can share an optical band and/or multi-band optical regions in massive optical-wireless networking.

The operation of a CS-SSB can be described again briefly here. Applying V_π to both MZM1 and MZM2 (the child MZMs) results in suppression of carriers in both arms. Then $V_{\pi/2}$ is applied to the phase shifter so that the Q-point is obtained. The PS of $\pi/2$ is applied as $v_{\mathrm{RF}}(t)$ to both MZMs (1, 2) or MZIM(I–) and MZM(Q–)

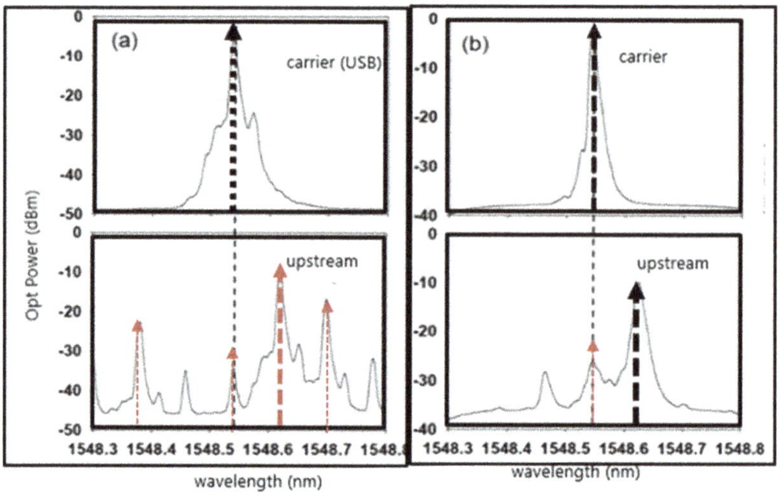

Figure 6.30. Optical spectra of the CW carrier (carrier RB) and upstream signal and signal-RB of (a) CS-SSB (i.e. MZM (I−)) and (b) CS-SSB MOD2 (i.e. MZM (Q−)). *Note*: biasing at the quadrature point in both child MZMs.

(PS shifted) with a modulation index of 2.4 to generate CS-SSB carriers. The multi-band optical channels can be circulating through the CS-SSB modulator. Thus, modulated RF bands can be injected into the optical carriers using systems of multi-band RF generators or transmitter sub-systems. Optical amplifiers (OA) are required to compensate for optical loss of the circulating loop.

Figure 6.30 shows the optical spectra of the CW carrier and up-link data. The spectral overlap between the CW carrier and the up-link signal is obviously reduced due to the CS-SSB modulation.

Interferometric noise is significantly improved because of reduced spectrum overlap. Due to the different modulation mechanisms of the two CS-SSB MODs, the frequency components of the generated CS-SSBs are different.

Referring to figure 6.30(a) and (b), it is observed that the suppression of the carriers is very much similar or CS-SSB signals can be generated by the two modulators. For the CS-SSB MOD1, are observed. One can over-drive the DD-MZM to suppress the center wavelength with a sufficiently high peak-to-peak amplitude of the RF signals, that is by setting the zeroth-order of the Bessel function $J_0(m) = 0$, the power levels of the other frequency components will increase.

6.5 Multiplexed RPoF up-link

An RPoF system employing a number of RF-shifted optically multiplexed carriers can offer a significant increase in information capacity of the transmission and distribution of wireless-optical networks. Figure 6.31 shows such a configuration for multiplexed RPoF systems and the distribution of the up-link to wireless antennae sites.

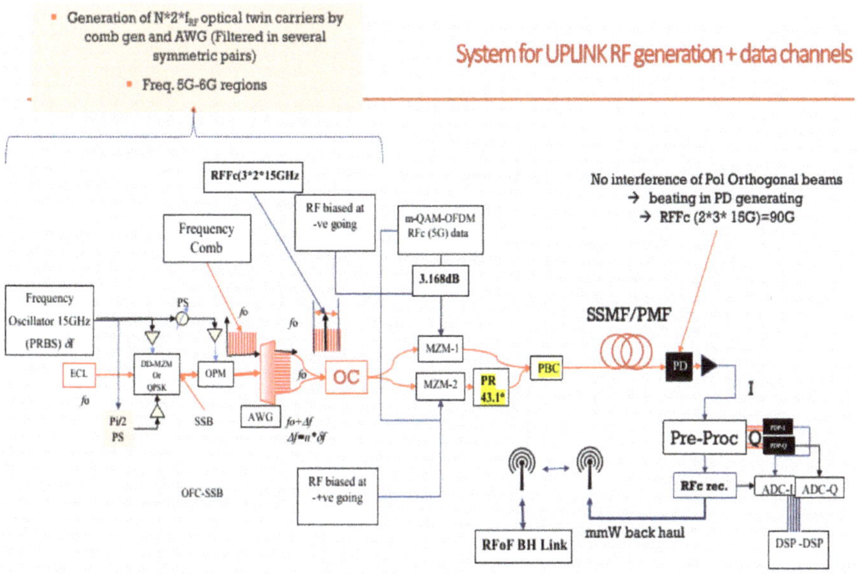

Figure 6.31. RPoF in up-link and down-link configurations with a dual parallel MZIM and polarization rotation with several optical carrier multiplexing.

6.6 Radio-photonic networks and optical to the antenna

Wireless signaling to users and the backbone of the network is critical. This section gives a generic overview and describes the technologies for such networking. Due to the interaction and interfacing between the radio frequency carriers over which data channels are embedded and transported over the photonic networks as the backbone, the term radio-photonic networking (RPN) is used.

6.6.1 Visions of 5G–7G and RPOF networking

The first products for 5G mobile phones were installed in 2019, in which the frequency region 1 (FR1) of 2–6 GHz was extensively deployed, but with 6G and 7G allocated in the FR3 of the μW and mmW spectral windows. Full roll-out of the 5G, 6G, and 7G technology will be developed progressively over the years to come. The current strategy includes the deployment of connections for other application technologies still in their early stages, for example the IoET, autonomous automotive crowds, etc. 5G–7G networking is a continuation in progress rather than a fully defined target standard. It requires a targeted delay time of millisecond latency and 99.9999% reliability with a high fault tolerance for network security in association with cloud computing. Currently it is more aspirational than attainable. Nonetheless, 5G, 6G, and 7G are shaping the future of human communications.

5G–7G are dominated by the words wireless and mobile, and then photonics. Photonic technologies become more and more dominant when RF carriers move higher and higher into the mmW and even THz regions. However, photonics will remain the data-backbone of the transport and distribution nodes of the global

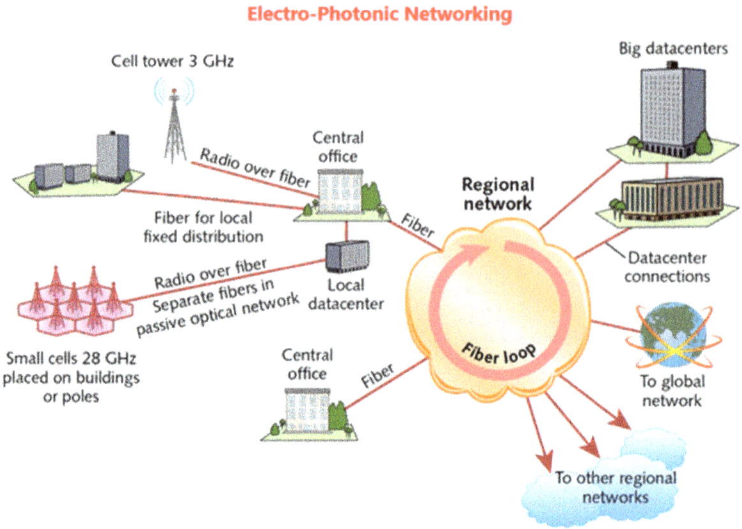

Figure 6.32. Vision of the progress of 5G radio-photonic networking (RPN). Copyright (2014) IEEE. Reprinted with permission from [6].

network, embedded in submarine, terrestrial, metro, and access areas, as shown in figure 6.32, including long-haul and short-haul cables, inter- and intra-connections of data centers (dc), and transmission lines running to wireless transmitters and to buildings and everything else for 5G, 6G, and 7G networks. 5G networking pushes fiber optics and photonics technologies all the way to the home, to the buildings, etc, i.e. fiber to the X (FTTx) in Europe, Asia, and North America. Currently, at the beginning of the third decade of the twenty-first century, some 40% of homes are linked by fiber in the USA, China, Europe, and Northeast Asia. It is expected that fiber line subscription rates will increase to 80% and 90% in Southeast Asia and China. The shift of fiber information connected societies is tending towards Asia where technological dominance is expected towards end of the twenty-first century.

6.6.1.1 Major targets of 5G, 6G, and 7G

The top four major targets for 5G, 6G, and 7G are as described in the following sections.

6.6.1.1.1 Enhanced mobile broadband (e-MBB)

Data rates of 10 Gbps or more are technologically possible over distances of more than 100–500 m using multiple input multiple output (MIMO) beam steering and antenna gain to compensate for channel propagation loss at high frequencies (μW and mmW). This is most applicable in advanced technologies such as augmented or virtual reality or ultra-high density (UHD) video. The UHD bit rate services via radio/wireless are for delivering all the bandwidth required by information hungry clients/customers.

6.6.1.1.2 Low delay latency and high reliability intercommunications

Ultra-high reliable service with 99.9999% fault tolerance and a low delay time in the millisecond range can be satisfied. Both are needed to track and control autonomous vehicles as well as everything else in their vicinity, including other vehicles, bicycles, pedestrians, construction sites, and sports stadia. Thus massive sensing networks and inter-communication networks are essential. Reliability and agility are vital for other network devices, such as home medical equipment and systems.

6.6.1.1.3 Massive IoET

Transporting data channels among the IoET, machine-to-machine, a large pool of device types is expected to be used with ultra-high densities of groups of hundreds of thousands per km^2 in dense population regions. Several home and building devices are already connected. Many more gadgets and devices are envisaged to function in an automatic mode (automata) with minimum human intervention, if at all. All the above targets are to be achieved in the near future for connected smart societies via RPoF in the μW and mmW spectral regions.

6.6.1.2 Radio-wireless broadband spectrum

5G will expand its bandwidth with new technology and by expanding the part of the radio spectrum allocated to radio-wireless mobile communications. Different parts of the spectrum serve different needs. Frequencies below 1 GHz offer wide-area coverage but limited bandwidth for IoET low data rates. A spectral range of 1 to 6 GHz, used for satellite links, mobile phones, and WiFi, offers more bandwidth, but the transmission distance decreases with increasing frequency.

The biggest source of >6 GHz for European frequency regions (FRs) is spreading into the μW and then mmW regions 27.5–28.35 GHz, 37–40 GHz, and 56–71 GHz. More data channels or wider bandwidths per channel can be allocated, but noting that the transmission range will be more limited. Agile interactive gaming and UHD video networking are expected to be major uses.

Over the first three decades of the twenty-first century, Asian and European cellular network areas have been 2 to 50 km in diameter, with the largest in rural areas. Microcells of 0.4 to 4 km are in metro areas for high density populations. Vacuum transmission losses are much higher in the 28 GHz band, so 5G and 6G use microcells and picocells tens of meters across, and femtocells which are about the same size but at lower power. Thus it requires that the receiving and transmitting antennae must be very close to the of homes.

All these cells require back-haul connections to the network, generally through a central office that connects signals from many cell towers. As of 2019, about 90% of cellular back-haul was through fiber cable and 6% via fixed wireless microwave. That mix is evolving, as 5G's need for more bandwidth is making copper inadequate and fiber transmission a must.

Figure 6.32 shows a simplified version of the emerging RPN network. Small cells in the 28 GHz band will distribute broadband signals short distances, with large cells operating over larger areas at frequencies below 6 GHz. Wireless signals reaching antennas or base stations will be transmitted through fiber or other links to central offices, either in radio format or digitally. Equipment at the central office will decode radio signals and process others. Those requiring low latency will be routed to a local data center; others will be routed to other destinations at remote data centers or network cores. Fiber will transport much of the data, but it will compete with fixed wireless microwave and free-space laser links. These systems are now in development, as described below, with the choice depending on cost/performance benefits.

6.6.1.3 'Pluggable' modules for upgrading

While 5G, 6G, and 7G hardware does not reach directly into the network core, they bring significant data traffic that requires major expansion of transmission capacity into multi-Tbps. Growth in the fiber transmission system can be contributed between data centers (interconnection between data centers (DCI)) and the data center networks' (DCNs) external fiber core networks. Fibers are abundant inside DCs and are connectable to outside gateways and the transmission distances are short, no longer than 2.0 km, and often shorter ~1200 m. Every router or switch port has its own dedicated fiber, and if a new switch or router is added additional fiber can be added to terminate the new ports. This will require increasing port densities and large-scale deployment of new high speed optics.

Advances in SiP integrated silicon photonic circuits have steadily decreased the cost per bit for electronics, but the cost per bit of photonics has not decreased as rapidly. Thus, as the bit rate increases, the cost of the photonics becomes a major portion of the hardware cost. Photonics account for 10% of the hardware cost in a 10 Gbps DCN, but at current rates the photonics exceed half the hardware cost when data rates reach 400 Gbps.

International telecoms network (ITN) operators' approach to controlling the capital expenses (CAPEX), mainly minimizing the photonics costs, is to focus on the 'pluggable' photonic modules, such as transmitter optical sub-assembly (TOSA) and receiver optical sub-assembly (ROSA), and even bidirectional OSA (BOSA). These modules are standardized packaging photonics and electronics for OE and EO conversion (see figure 6.32). One end of the module is an electronic module that plugs into the electrical interface on the router, while the other is an photonic/optic adapter connecting to the output fiber.

The modules generate optical signals in formats that optimize performance over specific distances, such as 500 m or 2 km. Data center transmission (DCT) is limited to about 10 km, so the pluggables use direct detection (DD) of '1s' and '0s' rather than coherent transmission to limit costs. However, the coherent solutions would offer much higher sensitivity and thus longer distances in the case that distributed networking to a wider area is essential, such as in DCs distributed to mobile edges. Coherent photonic solutions are much more flexible for networking, in particular and most likely in the approach to minimize the latency and ultra-wideband (UWB) data channels in which the RF is increasing towards the higher spectral RF region, the higher end of the mmW range.

Common pluggable module C form-factor pluggables (CFPs) convert signals (TOSA and COSA) from electronic form into optical form and vice versa, with one end accepting electrical input and delivering optical output.

In the long-haul transmission network outside DCs, fiber is scarce, data rates are very high to up to 200 GBd, and repeater spacings can be more than 80 km with optical amplifiers and dispersion compensating fiber (DCF) incorporated at the span ends so as to transmit over several thousand of kilometers underseas and/or across continentals. In those systems, the preferred signal advanced modulation formats combine with dense wavelength-division multiplexing (DWDM) and coherent transmission to increase the capacity and distance of installed cables. Pluggable coherent optics do not bulk up the connections at routers or switches, and allow carriers to automate network operation.

6.6.1.4 Radio-photonics over fiber (RPoF) transport and access lines

At some point on the way from the network core to the antenna, digital data must be converted into radio signals for wireless transmission. This can be implementable by installing the radio-generation equipment at the base station, where it can convert output signals from data into radio waves and the analog signals received from the antenna from radio waves into data. However, that requires installing the equipment at radio transmitters, which can be expensive and can add to maintenance costs.

An alternative transport medium, the RPoF, converts the data into radio frequency signals at a central office or other regional facility and hence transforming the radio signals into a form that can be transmitted optically through a fiber to the base station. A centralized software-defined network (SDN) can create a cloud-radio access network to transport the signals to and from central offices and transmission stations. In practice, the biggest advantage of radio over fiber comes when antennas are receiving multiple radio signals that require sensitive equipment to process. The radio signals can be transmitted at the received frequencies, or converted to a lower intermediate frequency that is easier to transmit and can be processed further at a central office.

Integration of mmW electronics and photonic transmission lines for adverse operating environments is essential. The antennae must be of MIMO type so that there is beam steering and high gain to compensate for free air propagation loss.

6.6.1.5 How and how far could RPoF penetrate into user premises?

One major question that remains unanswered is the best way to deliver RF signals to the closest wireless broadband transmitters. To deliver signals reliably in the low mmW 12–28 GHz band, antennae probably need to be within 100 m. Concrete or brick walls can block such high frequencies or line-of-sight is essential. This distance is even shorter when the RF carrier frequency goes to the higher end mmW region, in particular in the free-license band centered at the 58.6 and 7 GHz channel band.

Would it be better to bring the signals inside the house, in the form of fiber to the home (or premises for apartment dwellers)? The best approach would be to combine building out 5G, 6G, and 7G fiber back-haul with installed fiber to the home (FTTH or FTTx). 5G, 6G, and 7G technologies and FTTx are synergistic because of the density of high-frequency antennae that is needed. ‘Between 65% and 96% of fiber costs for 5G, 6G, and 7G back-haul networks may not be needed with the rolling out of optimized and future-proof converged fiber network (CFN)’.

We can base this argument on models of urban, suburban, and rural areas that include various density levels for wireless cells. It can be shown that the total costs could be slashed by building a single network to deliver both FTTH and 5G 28 GHz broadband wireless services. 'The extra investment needed to immediately make an FTTH network ready for 5G is only 1% to 7%'.

If only about 15% of households subscribe to fiber broadband, fibers pass by about 40% of homes. If a carrier could add fiber drops to 5G antennas to an existing FTTH network, it could save a lot in installation costs. Conversely, if the carrier had to install fibers close to homes anyway to install 5G antennas, it could recover some of the cost by adding FTTH connections.

However, fixed wireless links and free-space optics offer alternatives that could slash installation costs, at least in some areas. Fixed wireless has been used for previous generations of wireless base stations, and free-space optics had found some applications. However, both require line-of-sight operation and can be attenuated by weather.

Fixed wireless transmission in the radio and optical bands has been used for a long time, and offers a way to avoid high construction costs for transmission, but has suffered serious limitations from poor performance, licensing requirements, and high costs. 5G has opened up new possibilities in the radioband, including MIMO transmission that gives more concentrated beams and high-frequency transmission, although the latter is limited by distance. Free-space optics transmission was also vulnerable to group-velocity dispersion and chirp at long distances. Thus, combining radio and optical free-space links could improve reliability, but would add to the costs. Although wireless links could cut costs, they also face serious performance challenges.

A new alternative technology that uses free-space laser communication is putting constellations of small satellites into low earth orbit. In this scheme, ground stations will transmit signals up to the satellites, which will relay them around the world using intersatellite laser links to a point where they can be transmitted down to their destination. The goal is to bring broadband Internet to regions poorly served by terrestrial and submarine fiber cables.

6.6.1.6 Low latency and other challenges

The quest for millisecond (ms) latency poses several major challenges because it is a serious goal for applications from autonomous cars or pools of cars to home and medical monitoring systems. The speed of light in silica fiber glass is about 2e5 km s^{-1} at the 1550 nm region, placing a hard minimum on the latency of cloud computing centers (CCCs). Just the round trip to a data center a thousand kilometers away takes 5 ms (fiber speed = 5 ns km^{-1}), and that does not include the processing time at either end. That time period is too long when steering a car at freeway speeds or in reacting to a possible medical event.

Reducing that latency will require building a mesh of data centers across the country over global continents. They would not have to be big, but they would have to be fast. Several research groups in both the academic and industrial environments are looking at ways to ramp up the speed, but there are advantages and disadvantages to such technological breakthroughs. In one experiment, it was shown that it should be possible using a pair of WDM channels at different spectral windows distributed over the C- or L-band to transmit 25 Gbps or higher-speed

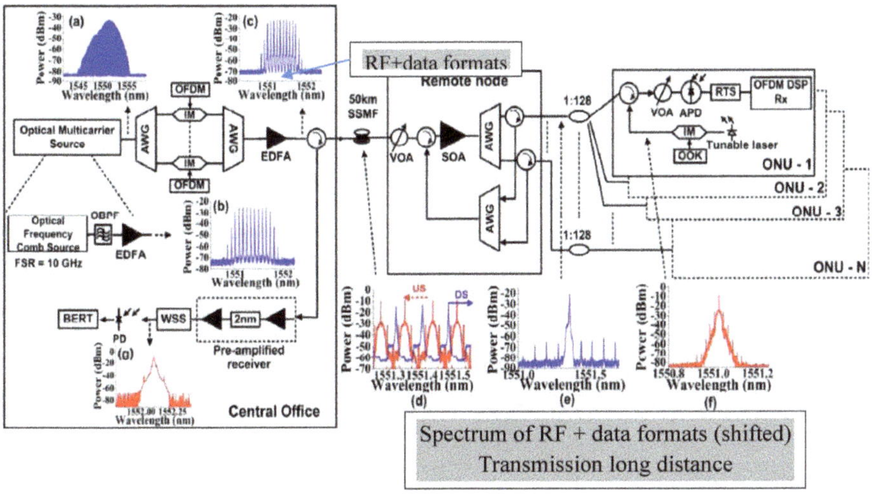

Figure 6.33. Multi-carrier modulation for RPoF transmission and networking for massive frequency antenna delivery networking. OBPF = optical bandpass filter; optical AWGs for multi-carrier modulated demultiplexing. ONU = optical network unit located at an antenna site. Antennae delivery networking. (Modified from [6] for RPoF.)

packets in parallel with latency below 400 μs over 20 km with a data rate of 50 Gbps. More challenges can be expected to emerge as 5G, 6G, and 7G technologies mature.

6.6.1.7 Remarks

In chapters 2–5 we have described the photonics transmission technologies for long-haul of several thousand kilometers to a few hundred meters, for inter-communication from global submarine, to terrestrial, to fiber to the home, and DCI and DCN. This section and the following section outline the main networking for the overall RPN scenario.

6.6.2 Multi-carrier RPoF networking

As described in chapter 5, comb lasers can be considered as a multi-carrier optical source. These sources are highly compact and can be employed as an integrated laser of multiple carriers in the RPoF transmission of a distributed network, such as in wireless access networking and testing environments for multiple antennae production environments. The comb source is demultiplexed into several optical lines or groups of optical lines and is further modulated by an optical modulator over which the RF carrier and data channels are modulated and fed to another optical modulator on which its data channels are converted into optical domain, as shown in figures 6.33 and 6.34.

6.7 Appendix

6.7.1 FTM-7961 diagram

See figure 6.35.

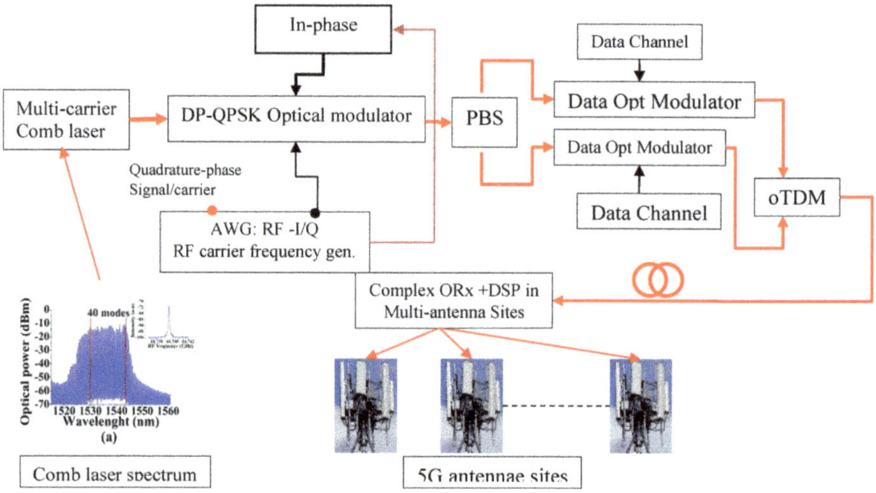

Figure 6.34. Multi-frequency RF antenna networking using an model locked laser and optical IQ modulator with or without polarization multiplexing for doubling capacity. AWG = arbitrary waveguide grating for multi-band optical filtering.

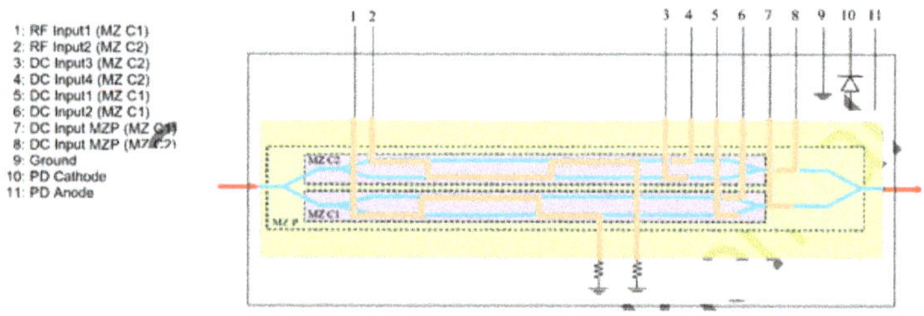

Figure 6.35. FTM-7961 structure and pin connections.

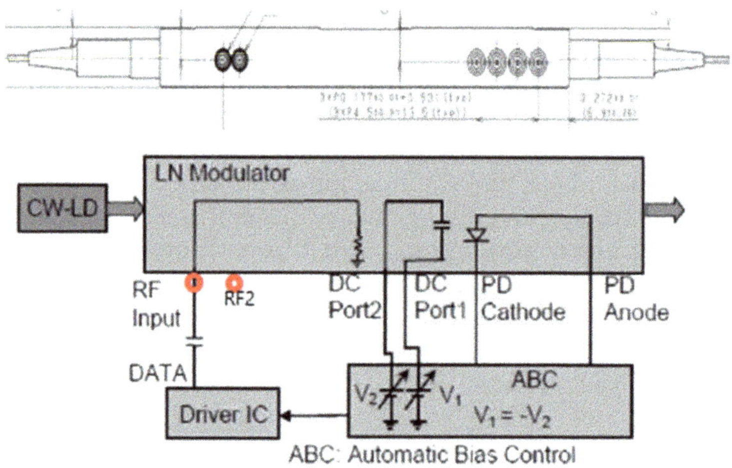

Figure 6.36. FTM-7937EZ structure and pin connection for a dual-drive MZM. The condition for CS-SSB swinging voltage $V_{p\text{-}p} = 2V_\pi$ or in dual-drive $V_\pi = 1.8$ V → $V_{p\text{-}p}$ → 24 dBm (3 dB down from 27 dBm for $V_{p\text{-}p} = 7.0$ V).

Specifications – SHF S807 B

Parameter	Unit	Symbol	Min	Typ	Max	Conditions
Absolute Maximum Ratings						
Maximum RF Input Power in Operation	dBm V	$P_{in\ max}$			4 1	 peak to peak voltage
Maximum RF Input Power without Power Supply	dBm V	$P_{in\ max}$			10 2	 peak to peak voltage
DC Voltage at RF Input	V				±9	
DC Voltage at RF Output	V				±7	
Supply Voltage	V		8		12	0.4 A, reverse voltage protected
Case Temperature[2]	T_{case}	°C	10	40	50	
Electrical Characteristics (At 40°C case temperature, unless otherwise specified)						
High Frequency 3 dB Point	GHz	f_{HIGH}	55			
Low Frequency 3 dB Point	kHz	f_{LOW}			60	
Gain	dB	S_{21}	22	23		non-inverting measured at P_{in}=-27 dBm
Max. Gain Reduction	dB		-2.5	-3	-4	Control via software interface
Output Power at 1 dB Compression	dBm V	P_{01dB}	15 3.5	16 4		10 MHz...20 GHz peak to peak voltage
Output Power at 2 dB Compression	dBm V	P_{02dB}	17 4.5	18 5		10 MHz...20GHz peak to peak voltage
Output Power at 3 dB Compression	dBm V	P_{03dB}	19 5.6	19.5 6		10 MHz...20 GHz peak to peak voltage
3rd Order Intercept Point	dBm	IP_3	28			
Max. RF Input for Linear Operation	dBm V	$P_{in\ lin}$			-8 0.25	I.e. Pout ≤ P01dB peak to peak voltage
Max. Output Power Reduction	dB		1			P_{in} ≥ - 2 dBm Crossing might need to be readjusted by using the crossing control feature. Control via software interface
Crossing Control Range	%		-4		4	Control via software interface
Input Return Loss	dB	S_{11}		-12 -7	-10 -5	< 30 GHz < 50 GHz

Specifications

Parameter	Unit	Symbol	Min	Typ	Max	Conditions
Absolute Maximum Ratings						
Maximum RF Input Power in Operation	dBm V	$P_{in\ max}$			10 2	 peak to peak voltage
Maximum RF Input Power without Power Supply	dBm V	$P_{in\ max}$			10 2	 peak to peak voltage
DC Voltage at RF Input	V				±9	AC coupled input
DC Voltage at RF Output	V				±7	AC coupled output
Positive Supply Voltage	V	V_{DD}	3	5	6	reverse voltage protected
Positive Supply Current	A	I_{DD}		0.14	0.2	
Bias Control Voltage	V	V_{G1}	-5	+1	+5	will not exceed 0.02 A
Case Temperature	T_{case}	°C	10	30	50	

Parameter	Unit	Symbol	Min	Typ	Max	Conditions
Electrical Characteristics (At 30°C case temperature, unless otherwise specified)						
High Frequency 3 dB Point	GHz	f_{HIGH}	65			
Low Frequency 3 dB Point	kHz	f_{LOW}			70	
Gain	dB	S_{21}	10	11		inverting measured at P_{in}=-20 dBm @ 40 MHz
Gain Ripple	dB	ΔS_{21}		±0,5	±1	40 MHz...40 GHz
Output Power at 1 dB Compression	dBm V	P_{01dB}	11 2.2	12 2.5		10 MHz...35 GHz peak to peak voltage
Output Power at 2 dB Compression	dBm V	P_{02dB}	14 3.2	15 3.6		10 MHz...35 GHz peak to peak voltage
Output Power at 3 dB Compression	dBm V	P_{03dB}	15.5 3.8	16 4		10 MHz...35 GHz peak to peak voltage
Input Return Loss	dB	S_{11}			-10 -5	< 12 GHz < 40 GHz
Output Return Loss	dB	S_{22}			-10 -5	< 30 GHz < 45 GHz
Rise Time/Fall Time	ps	t_r/t_f			 6 10	20%...80%, 2 V ≤ Vout ≤ 3 V Deconvoluted [2,3] Full Setup [3]
Jitter	fs	J_{RMS}		 350 450	 500 600	2 V ≤ Vout ≤ 3 V Deconvoluted [2,3] Full Setup [3]
Group Delay Ripple	ps				±50	40 MHz...40 GHz, 100 MHz aperture
Power Consumption	W			0.7		V_{DD} = 5 V / I_{DD} = 0.14A
Mechanical Characteristics						
Input Connector						1.85mm (V) female[4]
Output Connector						1.85mm (V) male[3]

Figure 6.37. SHF-727 specifications.

Key Characteristics @ 25°C

$Vdd_{1,2}$= 7.0 V; $Vg_{1,2}$= -0.1.0 V; $Vb_{1,2}$= N/C, Zo= 50 Ω

Parameter	Description	Time Domain Min	Typ	Max
Amplitude (V)	Eye Amplitude	4.1	4.5	-
Jitter (ps)	Added RMS Jitter	-	0.5	0.6
Tr/Tf (ps)	Rise/Fall Time	-	6	9

		Frequency Domain 0.01-26 GHz			26-40 GHz			40-45 GHz		
Parameter	Description	Min	Typ	Max	Min	Typ	Max	Min	Typ	Max
S21 (dB)	Small Signal Gain	18	21	-	15	19	-	15	17	-
|S21_A-S21_B| (dBm)	Small Signal Gain Difference	-	0.5	-	-	0.75	-	-	1	-
S11 (dB)	Input Match	-	-12	-10	-	-11	-8	-	-10	-8
S22 (dB)	Output Match	-	-12	-10	-	-12	-8	-	-10	-6
P_{sat} (dBm)	Saturated Output Power	-	20	-	-	20	-	-	17	-

Figure 6.38. Centallax amplifier, OA4SMM4 small signal amplifier.

Key Characteristics @ 25°C

Vd_1=Vd_2=Vd_3= 7 V; Vg_1=Vg_2=Vg_3= -0.1 V; Zo= 50 Ω

V_{in}= 400 mV

Parameter	Description	Time Domain Min	Typ	Max
V_{out} (V)	Output Eye Amplitude[2,3]	7	7.5	-
Jitter RMS (ps)	Added Jitter (RMS method)[1,2]	-	<0.5	1
Tr/Tf (ps)	Rise/Fall Time	-	<8	10

		Frequency Domain 0.01-26 GHz			26-45 GHz		
Parameter	Description	Min	Typ	Max	Min	Typ	Max
S21 (dB)	Small Signal Gain	27	30	35	24	27	35
S11 (dB)	Input Match	-	-12	-10	-	-10	-8
S22 (dB)	Output Match	-	-12	-10	-	-10	-4

Product Highlights

- 7.5 Vp-p (23 dBm saturated output power)
- <0.5 ps added RMS jitter
- <8 ps rise / fall time
- 27 dB gain
- 3.4 W power dissipation
- Useful gain to 65 GHz
- Small size package

Figure 6.39. Centellax OM4MOM4, large signal power amplifier (27 dB gain) max power output = 23 dBm.

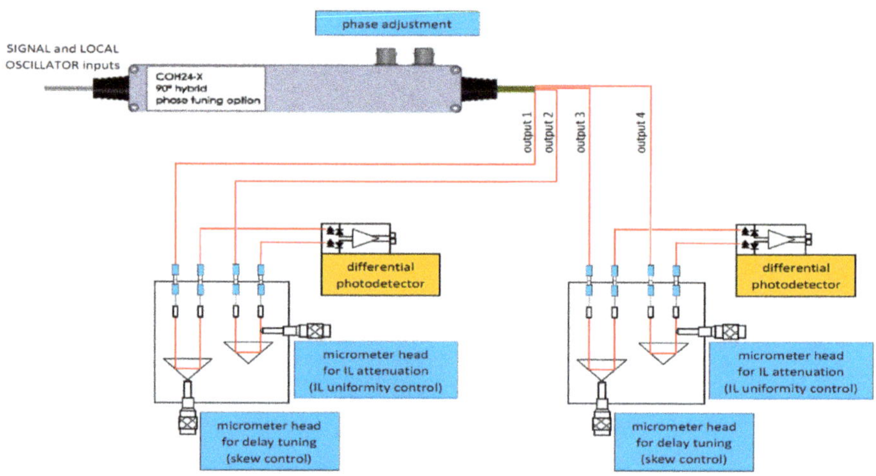

Figure 6.40. Kylia hybrid coupler COH24X.

Data Sheet
February 2016
20 GHz Linear AGC Balanced Optical Receivers with Optional Remote Command and Control Softwar

Electrical / Optical Specifications: (Conditions unless otherwise noted: $T_{AMBIENT}$ = 25°C, $V_{bd\,(A\,\&\,B)}$ = +5V, V_{cc} = +3.3V, 1550nm, 50 ohm loads, optical input power mismatch <2dB (1))

Parameter		Min	Typical	Max	Units
Responsivity	@ 1550 nm	0.50	0.60		A/W
	@ 1310 nm	0.55	0.65		
Wavelength Response Range		1064		1650	nm
Maximum Adjustable Bandwidth (-3dB) (2)			20		GHz
Minimum Adjustable Bandwidth (-3dB) (2)			7		GHz
Low Frequency Cut-off (-3dB)			50	100	KHz
Channel Match (DC Responsivity & Frequency Response)			1		dB
Maximum Adjustable Differential Conversion Gain (manual gain control mode, 10 Gb/s, -15dBm optical input power) (3)			2800		V/W
Minimum Adjustable Differential Conversion Gain (manual gain control mode, 10 Gb/s, 0dBm optical input power) (3)			30		V/W
Maximum Differential Output Voltage Swing (manual gain control mode, +2dBm optical input power)			1300		mVp-p
Differential Adjustable Output Voltage Swing in AGC Mode (4)			450	750	mVp-p
Optical NEP (Noise Equivalent Power)			40		pW/√Hz
Optical Sensitivity (23 Gb/s, NRZ-ASK, BER=1E-12, 2^{31}-1 PRBS)			-15		dBm Average
Optical Overload (23 Gb/s, NRZ-ASK, BER=1E-12, 2^{31}-1 PRBS)			+3		dBm Average
Total Harmonic Distortion (@ 1 GHz)			5		%
AGC Response Time (5)			350		ns
Peak Detector DC Output Voltage (with no optical input)			0.1		V
Peak Detector Output Response Slope (6)			3.75		V / Vp-p
Peak Detector Maximum DC Output Voltage			3		V
Peak Detector Output Resistance			5000		Ohms
Output RF Return Loss			10		dB
Optical Return Loss		27	35		dB
Optical PDL			0.05		dB
Vbd Bias (Photodiodes, A & B)		4.5	5	5.5	V
PD Dark Current			10	100	nA
Vcc Bias (Amplifier)		3.14	3.3	3.46	V
Icc Current (Amplifier)			80		mA
Power Consumption			270		mW

Figure 6.41. Operational parameters of the DSC Discovery Semiconductor optical receiver model DSC-412.

6.7.2 FTM-7937EZ/7938

See figure 6.36.

6.7.3 Electronic amplifiers

6.7.3.1 SHF-827 small signal broadband amplifier

See figure 6.37.

6.7.3.2 Centallax power amplifier and small signal amplifier

See figures 6.38 and 6.39.

Test Data Report

Customer:	Evotronic GmbH	Date :	12/07/2016
Part Number:	DSC710-89-FC/APC-K-3	PO :	3721013401
Description:	Balanced Optical Detector	WO :	5683
Tested by:	Matthew D'Angiolillo	Serial Number:	710491607
		Witnessed by:	Dan Mohr

Parameter	Specification	Measured Value		Units
		Diode A	Diode B	
-3 dB Bandwidth @ -3 dBm input optical power, Diode A = +5 V, Diode B = -5 V	Min: 8	11.2	11.1	GHz
Channel Match (< +/- 1.0 dB)	Min: 8	14.5		GHz
DC Responsivity @ 1550 nm	Min: 0.70	0.72	0.76	A/W
Detector dark current @ + 5 V	Max: 100	8	25	nA
Optical Return Loss @ 1550 nm	Typ: 27	38	27	dB

Figure 6.42. DSC710 test report of Q-lab DSC710.

6.7.4 Saturated output hybrid coupler

See figure 6.40.

6.7.5 Balanced optical receiver

6.7.5.1 DSC-412 specifications

See figure 6.41.

6.7.5.2 DSC710

See figure 6.42.

References

[1] Chow C W, Wang C H, Yeh C H and Chi S 2011 Analysis of the carrier-suppressed single sideband modulators used to mitigate Rayleigh backscattering in carrier-distributed PON *Opt. Express* **19** 10973

[2] Prescod A *et al* 2009 Super-linear modulator with extended bandwidth capability for broadband access applications *Proc. SPIE* **7234** 72340E

[3] Zhao F, Lu Y, Zhang L, Cai W, Wu S and Zhang Y 2020 Linearized microwave photonic link based on dual-driven Mach–Zehnder modulator *Opt. Eng.* **59** 016117

[4] Streshinsky M, Ayazi A, Xuan Z, Eu-Jin Lim A, Lo G-Q, Baehr-Jones T and Hochberg M 2013 Highly linear silicon traveling wave Mach–Zehnder carrier depletion modulator based on differential drive *Opt. Express* **21** 3818–25

[5] Zhu R 2015 Broadband linearization technologies for broadband radio-over-fiber transmission systems *PhD Thesis* Concordia University, Montreal

[6] Vujicic V *et al* 2014 Optical multicarrier based IM/DD DWDM-SSB-OFDM access networks with SOAs for power budget extension *European Conf. Optical Communications (Cannes, France)* paper We.1.6.6

Chapter 7

Electro-photonic RF oscillator

7.1 Introductory remarks

The demands are tremendous for pure RF carrier generation sources in the millimeterwave (mmW) and microwave (μW) regions as well as in the lower μW bands, that is, the frequency can reach the lower GHz region and even higher for 5G, 6G, and higher generations of wireless and optical networking.

In general, there are three possible approaches to the generation of such ultra-high frequency sources. First, traditional digital and analog electronics can be designed using integrated microelectronics such as the phase locked loop (PLL) and voltage control oscillator (VCO) and associated microprocessors [1]. Similarly, the natural oscillation in the μW and mmW frequency regions of yttrium iron garnet (YIG) crystal can be employed in an active resonant circuit to generate μW signals [2], to form a YIG tuneable oscillator (YTO) which is a complete analog oscillator as shown in [3]. The second approach is to derive the oscillation frequency (OF) from the photonic domain down-conversion to the μW and mmW domain, in particular in the hundreds of gigahertz band [4].

The third approach is a hybrid mode of photonic and electronic high frequency, the electro-photonic oscillation (EPO) generator, which is a closed loop composed of both electronic and photonic components. The generated signals are tapped and fed back to modulate or manipulate the optical waves. This chapter focuses on this technique. However, we also describe very briefly a modern technique for generating μW or mmW via the excitation of the vibration of the crystal structure YIG via an electromagnetic (EM) force. Furthermore, a control feedback mechanism needs to be used in the stabilization of the oscillation, thus an optical phase lock loop (OPLL) would be required. This OPLL is presented briefly in section 7.9.3. Various structures of EPOs are described, most rely mainly on the modulation technique so that either single sideband (SSB) or double sideband (DSB) modulated lightwaves are employed in the electro-photonic loop (EPL).

doi:10.1088/978-0-7503-3511-9ch7

For high speed electro-photonic signal processing (EOSP), photonics technological approaches can be very attractive to offer ultra-high speed and low phase noise. High speed EOSP incorporates interactions of electronic and photonic domain components so as to take advantage of both technologies. It is noted that for electronics, the high speed becomes a major obstacle but not for photonics, which can offer wideband over several terahertz. However, photonic processing cannot resolve low frequency and narrow band signals. A combination of these two photonic and radio wave technologies can offer advantages over analog RF systems, such as wireless networks in 5G, 6G, and 7G networking, in radar processing such as the Square Kilometre Array radio antenna, and in the μW and mmW spectral regions. Photonic signal processing (PSP) [5] can offer large instantaneous bandwidth (>60 GHz and even to 120 GHz), low insertion loss (0.2 dB km^{-1} in SSMF in the 1550 nm region), and a compact form factor. Furthermore, for integrated photonics using Si-on-insulator (SOI), electronic circuits such as drivers and lasers can be integrated discretely via flip chip bonding and coupling.

Further, EPO incorporated in a harmonic mode locked laser is presented so that the generation of optical comb lines and RF lines is possible and demonstrates both RF line and optical comb line generation in the same electro/photonic system.

7.1.1 The sampling needs of ultra-high frequency oscillators

As mentioned in chapters 5 and 6, there is a need for RF low phase noise oscillators in the μW and mmW regions that can cover several application systems in 5G–7G wireless and radio-photonic networking, from the pico-cell to macro-cell scales to large cellular structures. Further, in computer interconnections (CIs) and data center intercommunications (DCIs) between clusters of servers, on-chip optical interconnects (OCONs) can have the potential to overcome clock skew (the delay nonuniformity of channels). Furthermore, in the integrated circuit processing speeds or in the DCI approach 10–100 GHz, the phase noise of the data clock becomes highly critical. Phase noise reduction is thus essential. An EPO can offer phase noise for at least 20 dB improvement over its electronic counterparts. This allows the synchronization of ultra-high speed data bus on the chip and all high speed servers and transmission lines.

At the other end, optical networks (ONs) are employed extensively to transport Internet protocols (IPs) of voice, ultra-high definition video, and Internet data of 100 Gbps over thousands of kilometers and even much higher to 1.6 Tbps over a few kilometers in data center access networks (DCANs). The employed wavelength of the optical carriers are the assigned wavelength grid specified by the International Telecoms Union (ITU) [6].

Furthermore, the synchronization of routing and processing is highly important to meet the ultra-high speed demands. Thus low phase noise (PN) clocks or RF carriers are becoming key components for 5G, 6G, and 7G networks.

Low phase noise clocks (LPNCs) are also important in the analog-to-digital converters (ADCs) of ultra-high speed digital signal processing (UHS-DSP) based optical reception systems whose baud rate can reach 212 GBd. The sampling and hold (S&H) circuit of the UHS-DSP relies on a very low jitter clock to maximize the

ADC effective number of bits (ENOBs) and hence its signal-to-noise ratio (SNR) [7]. An interdependence between the timing jitter and SNR or carrier-to-noise ratio (CNR; in dB) is expressed as $20\log_{10}(1/2\pi f_{RF}t_j)$ and SNR (dB) = 6.02 × ENOB + 1.78 [8], where f_{RF} is the RF input frequency, and t_j is the 'clock' timing jitter. The jitter time can be approximated by $t_j \approx (\text{ENOB}/f_{Sa}) \times 2^{N\text{bits}+2})$, where the sampling rate $f_{Sa} = 2 \times f_{RF}$ is the Nyquist sampling rate of the ADC.

Therefore, when the ENOB is increasing from 8 to 14 bits, the jitter can reduce from 1.0 ps to about 10 fs, or 100 times shorter. For most of the ADCs for DSP based optical communication systems the sampling rate has now reached 200 GSa s^{-1}, approaching the requirement to obtain a higher ENOBs with a much more precise clock rate ($\ll$ fs) and lower phase noise [9].

7.1.2 Oscillator noise

Oscillator noise can be contributed by intensity noise, phase noise, timing jitter, and system influenced noise. Relative intensity noise (RIN) is the amplitude/power variation with respect to spectral components due to fluctuations of the laser emission carrier or RF carrier output. Furthermore, the random phase of the spectral components creates jitter, and hence phase noise.

Phase noise is due to fluctuations of the optical phase of the oscillator output whose dynamics are due to the constant distance of the resonator of the laser or the electronic oscillator circuits. In an EPO the resonator is composed of the RF part and the photonic section. It is desired that such a resonator can support only the selected mode. It is noted that several resonant modes are possibly generated in an unfiltered EPO and a resonant mode is selected by using a sharp bandpass filter (BPF) so that all the energy can be concentrated into this mode and give a high power-out oscillation mode.

Oscillator PN can lead to timing jitter and thus the CNR is the common parameter used to characterize oscillators in either the optical or RF domain. For an EPO, the PN is denoted as $N_P(f)$ and is defined as the relative SSB power spectral density (PSD) with respect to the EPO generated RF signal, in units of dBc Hz^{-1}. That is, the power of the RF signals is measured with a spectral window of $\Delta f = x$ Hz. The power distribution curve can then be normalized over a spectral resolution bandwidth of 1.0 Hz. PN is normally measured at a frequency offset from the carrier, assuming the $N_P(f)$ power follows a Lorentzian shape with respect to frequency, as shown in figure 7.1. EPO noise is dominated by different noise sources at different frequency offsets: (i) environmental noise in the 1–10 Hz range, mainly from environmental conditions such as temperature variations and vibrations, and follows a slope of f^{-4}; (ii) flicker noise over 10 Hz–1.0 kHz is phase noise due to the RF amplification essential in most EPO systems and this follows a slope of f^{-3}; (iii) in the 1 kHz–30 kHz range white noise is dominant and follows a slope of f^{-2}; and finally (iv) over the range of 30 kHz and higher, relative intensity noise (RIN) and PN are dominant, which follow a 'flat' noise floor.

For most applications in 5G–7G networks, timing jitter determines the 'clock' quality. Referring to a measured PN spectrum, as shown in figure 7.1, the total noise

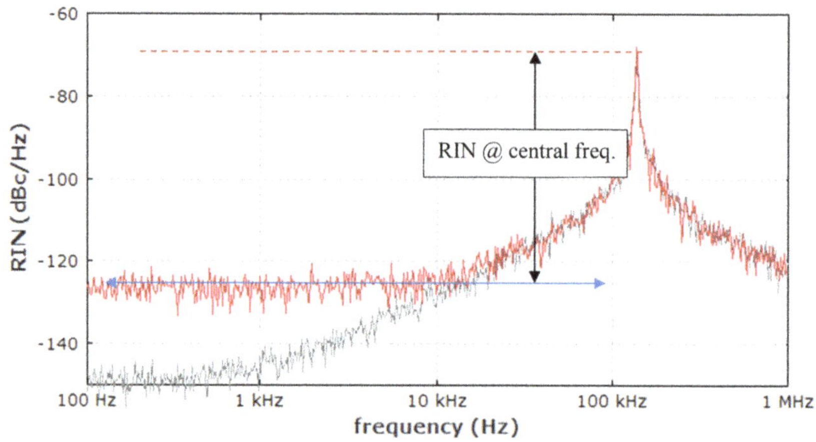

Figure 7.1. RIN of an EPO oscillation: RF detected power against frequency difference offset from the central RF oscillation frequency.

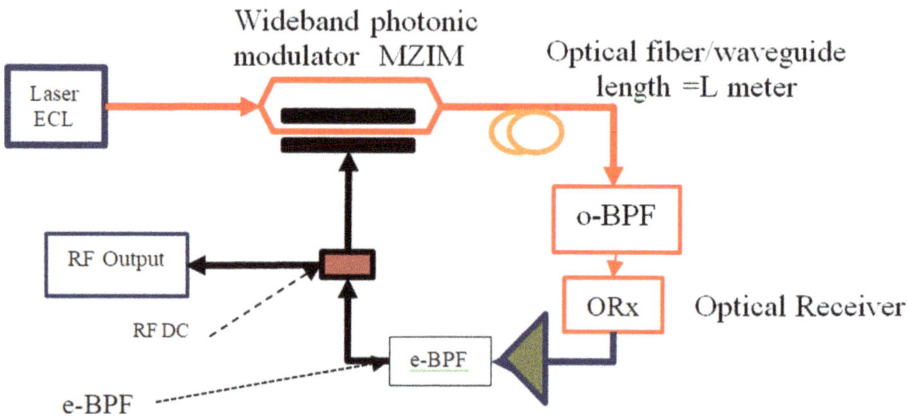

Figure 7.2. Basic EPO structure. o-BPF = optical bandpass filter, which can be inserted in the loop to minimize optical noise; similarly, an e-BPF = electrical BPF, can also be used to tune the RF frequency and RF noise. ECL = external cavity laser; MZIM = Mach–Zehnder interferometric modulator; dc = directional coupler; ORx = optical receiver.

can be derived by integrating the RIN(f) over the entire bandwidth of the data and carrier frequency data channel band.

We also give an account of an optical phase locked loop that can be placed in-line with the optical section of the EPO so as to obtain stabilized RF oscillation at the EPO output.

The EPO can use an external modulation laser (EML) or an external cavity laser (ECL) and an external modulation device, as shown in figure 7.2. A laser, typical an ECL, launches lightwaves into the photonic modulator. The RF modulation signal is connected to the modulator RF port to convert any RF noise or signals to an optical domain signal. The modulator output is then detected and the converted RF signal is amplified to a sufficient level to drive the modulator.

Initially, the MZIM is modulated with system noise which is derived from the laser and the recirculated lightwaves. Such modulated lightwaves are then filtered in the RF domain and/or in the optical domain, and it takes a very short time for the RF signal to reach the right level to modulate the optical modulator. The RF oscillation signal is derived by tapping via an RF directional coupler at the output of the RF amplifier. In the case that the optical power is not high enough, an optical amplifier can be used. However, as the output power of the DFB can now reach 15 dBm with a linewidth of less than 1.0 MHz being normal, this power level is reduced by the total insertion loss of the optical modulator of a few dB leading to the remaining power being injected into the optical receiver, a photodetector in cascade with a broadband transimpedance amplifier. This RF signal can then be further amplified by an RF power amplifier to about 23 dBm. This is sufficient for the RF power level to be tapped and fed into the optical modulator to generate the RF modulated lightwaves to recirculate in the electro-photonic loop.

In practice, in particular when these oscillators are deployed in wireless access networks, the stability of the RF frequency and the phase noise or fluctuations of the carriers in frequency is highly critical. Therefore, a feedback control would normally be conducted in this case to ensure that the RF oscillation is stabilized. Similar techniques were applied to the stabilization of lasers. This technique is described in a specific experimental platform in the appendix, section 7.9.3.

In general, an oscillator with a feedback loop can be represented by an open loop circuit with a gain of $G(\omega)$ and then a feedback loop with a gain of $H(\omega)$. Then the overall closed loop transfer function $\mathrm{TF}(\omega)$ [10] can be written as

$$\mathrm{TF}(\omega) = \frac{G(\omega)}{1 - G(\omega)H(\omega)}. \tag{7.1}$$

In the EPO the forward path is in the optical domain and the feedback is in the electrical domain (figure 7.3). The conversion from the electrical to optical domain occurs via the modulation of the photonic modulator, and the conversion from the photonic to electrical domain occurs via photodetection. The condition for the oscillation is that the denominator of equation (7.1) equals zero, that means that the denominator goes to zero and then the output reaches a very high value or is in resonance. As observed for a typical pole-zero pattern of the TF in the Laplace plane, the poles of of a lossy network can be in conjugate poles outside the vertical axis or the total closed loop gain $\mathrm{GH}(s)$ is less than unity. When it is lossless or $\mathrm{GH}(s)$ reaches the unity level, the conjugate pole pair is positioned on the imaginary axis and thus oscillation is obtained. The gain in an EPO is obtained mainly from the RF electronic amplifier and possibly an optical amplifier in the optical forward path. These gains supply sufficient energy for the oscillation modes. Figure 7.4 depicts a typical spectrum of the multi-mode RF output of an EPO.

7.2 EPO type 2: an integrated external modulation laser EPO

The external modulator EPO can be integrated to form an EML, as shown in figure 7.5(a) in which a DFB is integrated with an absorption modulator and isolated

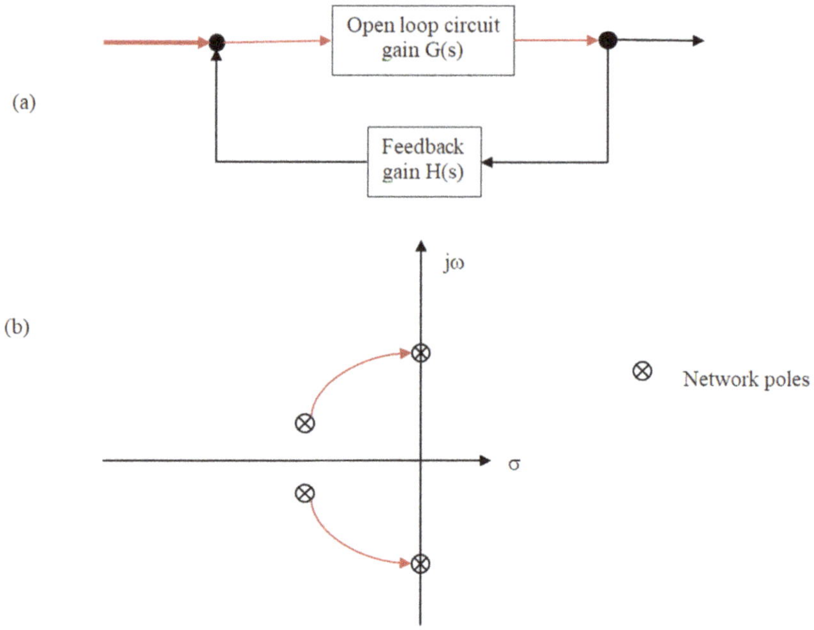

Figure 7.3. (a) Generic scheme of an oscillator under a closed feedback loop. *S* = Laplace variable. (b) Dynamic pole-zero patterns in the *S* Laplace plane for oscillation of the oscillation network.

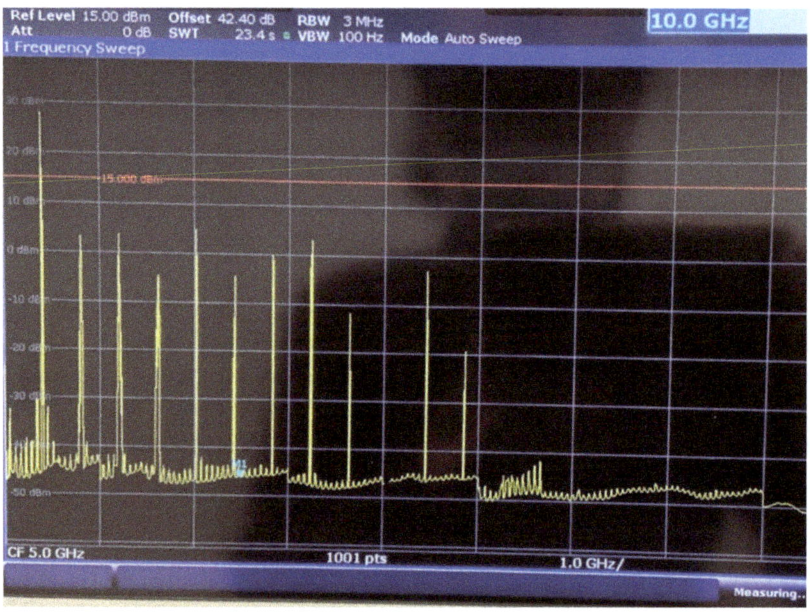

Figure 7.4. Typical spectrum of an EPO with significant phase noise suppression. Full scale span = 10 GHz or 1 GHz/division.

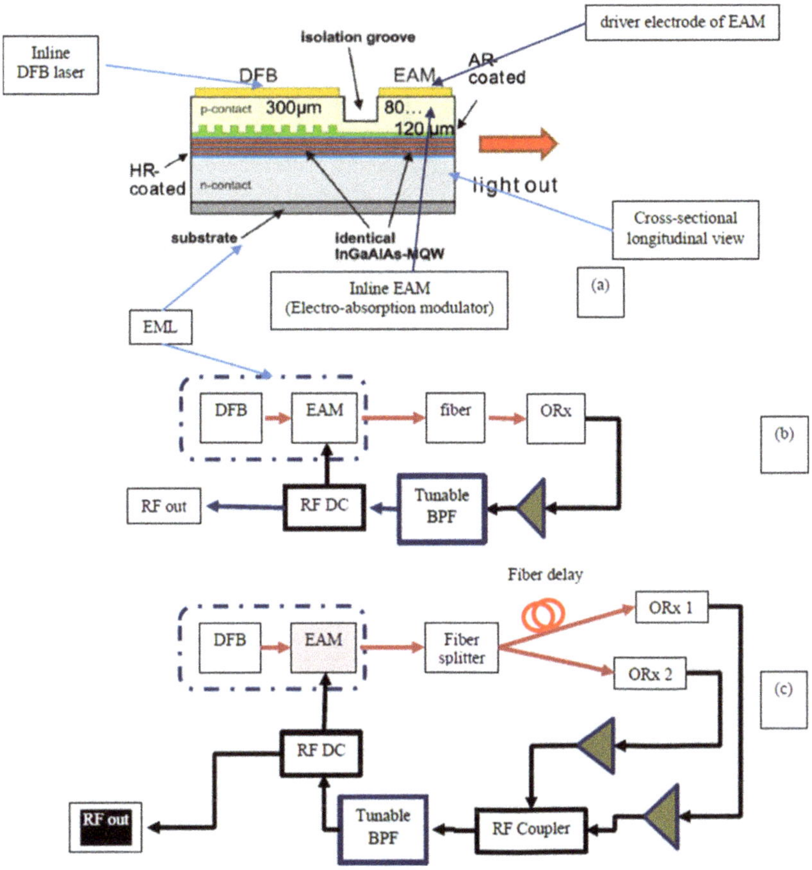

Figure 7.5. EPO using an integrated EML. (a) The integrated structure of an EML. (b) An EPO using an EML and RF feedback. (c) Double-loop EPO to suppress the phase noise. Black lines = electrical path. Red or light color = optical path.

in an etched slot. This integrated EML can replace the DFB laser and the Mach–Zehnder modulator (MZM) shown in figure 7.2. The RF generation via an opto-electronic receiver and feedback to drive the electrical amplifier (EA) section of the EML to generate the RF oscillation output signals via an RF directional coupler is shown in figure 7.5(b). A tuneable RF BPF can be inserted to tune the RF frequency of the oscillator. Similarly, the wavelength can be tuned to match the sensitivity of the receiver by inserting an optical BPF in the optical section of the loop. Figure 7.6(a) shows a typical spectrum of the RF modes of the electro-photonic cavity in the case that no RF BPF is inserted in the loop. Figures 7.6(b) and (c) show the filter RF oscillation mode generated in principle and in a practical experiment. A phase noise level of $-140\,\mathrm{dBc\,Hz^{-1}}$ can be observed with a carrier power of about -90 dBc. This level can be amplified up with some penalty with its noise figure. It is observed that the suppression of other cavity modes is not completely depleted. The phase noise can thus be improved by using two optical cavity paths with an optical splitter, as shown in figure 7.5(c).

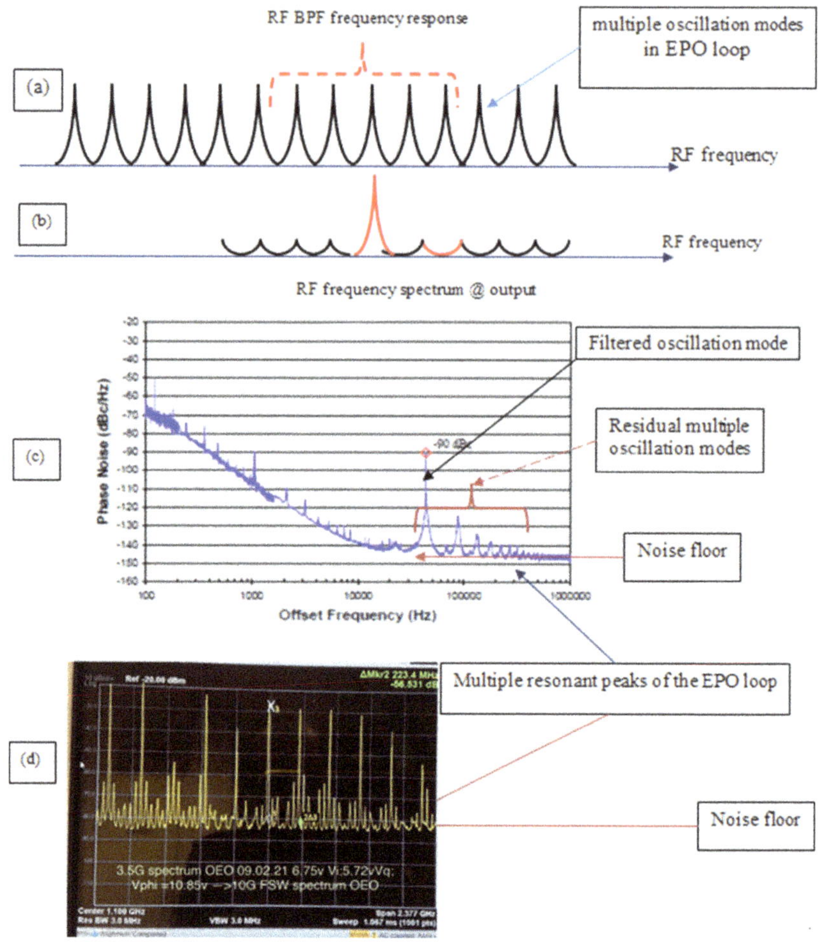

Figure 7.6. A typical spectrum of the generated oscillator of an EOP as detected at the optical spectral window of the laser and the optical BPF. (a) RF cavity mode of the EOP without an RF BPF and (b) with a BPF as extracted/coupled out. (c) Experimental single sideband noise and signal spectrum [11]. Note that the frequency is in the $\log_{10}$ scale. The vertical scale is a scaled unit (/Hz) from the bandwidth of an electrical spectrum analyzer. (d) Free running (without RF BPF) RF spectrum.

The operational principles of this type of EPO are as follows:

- First the noise and filtered noise in the RF domain at the output of the BPF are building up after the lightwaves which circulate a number of rounds, through the electro-photonic loop.
- The frequency mode spacing is determined by the loop length L whose delay time is given as $T = L/(c/n_{\text{eff}})$ with c the speed of light in a vacuum and n_{eff} the effective refractive index of the optical guided medium at the operating wavelength, which is about 1.448 at 1550 nm. The electronic delay of the electrical path is assumed to be much smaller than that of the optical path. Thus for a 0.5 m fiber pig-tail of the optical modulator the resonant frequency

of the zeroth order mode is about 200 MHz. Thereafter the higher order resonant mode is located at a multiple of this fundamental mode. If 10 GHz is to be generated then the length would be 200 m and thermal stability must be implemented for the fiber length.

- There are two loops which share a common lightwave source, which is an integrated EML, as shown in figures 7.5(a)–(c).
- An integrated EML (IEML) consists of a DFB laser and a external modulator which is normally of electro-absorption modulator (EAM) type. The laser and the modulator are separated by an etched slot, the distributed Bragg grating (DBR), which is formed by an etched periodic slot whose distance matches the optical wavelength of the desired optical frequency. The external EAM can thus modulate the lightwaves without affecting their linewidth.
- The two optical paths can be designed so that their different delay times can create a constructive electro-photonic wave. Thus, we can improve the RF CNR. The RF waves can be tapped via an RF directional coupler (RF-dc) as indicated in figure 7.5(c).
- The photoreceiver of figure 7.5 is of the direct detection type or the square-law is applied, that is, the RF current produced at the output of the photodetector is proportional to the output optical intensity of the IEML. Nonlinearity can be expected due to this nonlinear optical-to-electrical (OE) conversion process. This can be further improved by the low pass filter (LPF) self-coherent detection described in sections 7.4 and 7.5.
- Stabilization of the loop in the RF path can be implemented to ensure a high CNR and low phase noise.

7.3 Type 3: a dual parallel EPO with MZM complementary output ports

In general, a dual-loop EPO can be represented in figure 7.7 in which a photonic comb generator is coupled by an ECL in cascade with an optical booster amplifier so as to supply sufficient energy to all resonant modes of the EPO. The fiber and optical receiver are similar to all other cases of opto-electronic oscillators (OEOs).

This section describes the generation of RF oscillation waves by a dual-loop EPO which consists of a dual-output Mach–Zehnder interferometric modulator (DO-MZIM) [12] modulated by closed loop feedback RF signals via an optical receiver using a photodetector pair (PDP), as shown in figure 7.8. This model is analysed briefly [13] in this section and section 7.7.

The operation principles are based on the following:

(i) A dual loop formed by two complementary and orthogonal output ports of DO-MZIM sharing the same MZM modulator, that is, the MZIM has two coupled output ports which could be combined via an optical directional coupler or a polarization beam splitter/combiner.

(ii) A fiber and two input ports of PDP connected back-to-back to form a balanced photodetection (BPD) structure.

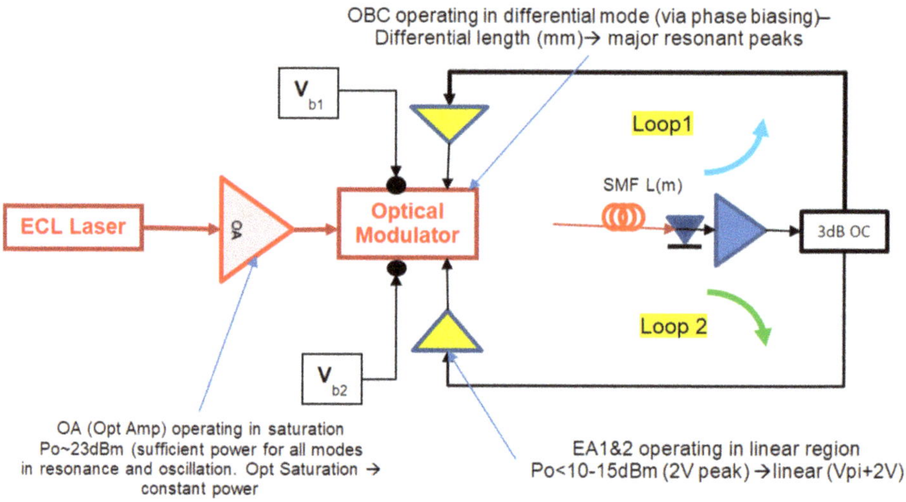

Figure 7.7. Generic structure of a dual-loop EPO. The optical path is indicated by colored lines and the electrical path by black lines.

The dual-loop (DL)-EPO structure operates in either the common or differential mode so that it can enhance its suppression capability of the side modes and achieve higher reduction of the generated RF phase noise. In contrast to the common single-loop EPO (SL-EPO) based on an optical loop consisting of a single-output modulator (SO-MZM), an optical coupler, and a PD with associated electronic amplifiers, the advantages of this DL-EPO are low noise and flexible selection of the RF output frequency. It is demonstrated that in the μW–mmW region RF oscillation can be achieved with a side-mode suppression ratio (SMSR) of higher than 70 dB, phase noise $\ll -110$ dBc Hz^{-1}, and a linewidth in the order of a few tens of hertz. The frequency region depends on the properties of the photonic and electronic components of the loops.

Microwave generators are used widely in electronic and optical communication systems. Their spectral purity and phase noise are critical in the performance of such electronic/photonic systems [14]. Opto-electronic oscillators (OEOs) are reported to generate high frequency microwave signals with high spectral purity and low phase noise [15]. With the spectral purity enhancement and the phase noise reduction of EPO, the performance of electronic systems can be improved [16]. Such a high spectral purity and low phase noise EPO can find wide applications in RPoF, ultra-high speed digital signal processing (UHS-DSP), etc. In order to generate high resolution microwave signals, a long optical fiber OEO can be used as the energy-storage component [17]. However, the long loop EPO leads to narrow mode spacing. Thus the selection of the side-mode for generation of the desired microwave signals is difficult by filtering in the electronic feedback loop.

The EPO is fundamentally a longitudinal multi-mode (LMM) system with moderate lower side-mode suppression properties. Each longitudinal mode falling

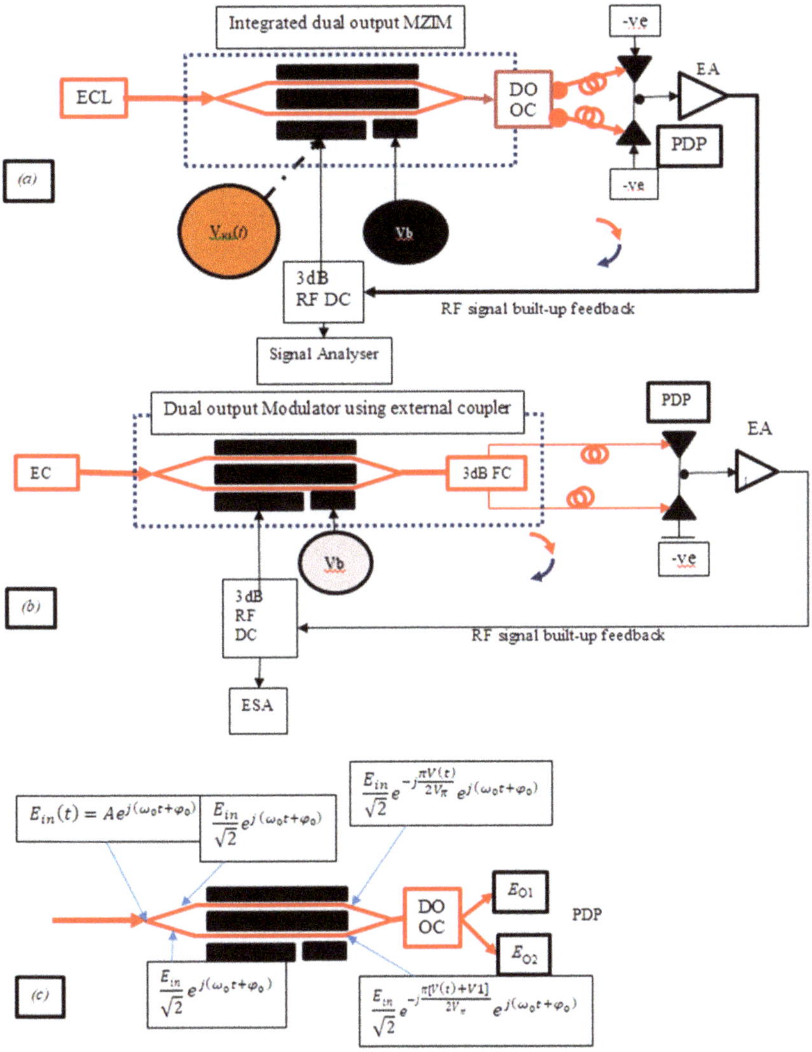

Figure 7.8. (a) Schematic of a DO-EPO. (b) Structure of a single-out (SO)-EPO (c) Field distribution of a DO-MZIM. ECL = external cavity laser; DO-MZIM = dual-output MZIM; BPD = balanced detector pair; EA = RF electrical amplifier; ES = RF electrical power splitter; ESA = electrical spectrum analyser; OC = optical coupler; $V_{1,\mathrm{dc}}$ = bias voltage in path 1. DO-OC = dual output optical coupler.

within the RF filter bandwidth could possibly be excited into the resonant mode. To optimize such suppression of the non-resonant mode, a number of structures are proposed using a multi-loop EPO to achieve the narrow-spacing property for the EPO [18–20]. It can be demonstrated that the EPO longitudinal modes could be suppressed efficiently by the interference of the fields of the two optical paths due to the two OE hybrid feedback loops with different lengths, hence differential phases. The use of dual loops in EPO can obtain high suppression.

The low extinction ratios (ERs) of the PBS and PBC offer a low level of either constructive or destructive interference, thus the beating noise between the two optical paths leads to high suppression of phase noise. A tuneable frequency quadrupling dual-loop EPO based on an optical polarization modulator (OPolM) could be used in addition to the six proposed structures given in sections 7.2–7.6.

To eliminate the interference effects due to different polarization states in each polarization loop, controlling the polarization states of all optical paths is required. This increases the complexity of the EPO. A dual-loop EPO using two distinct wavelength carriers can be used to reduce the beating noise between the two optical paths. However, this is more complex and comes at a higher cost.

Balanced detection is used in optical systems so as to suppress the common mode noise [10, 21] due to quantum shot noise generated in the PD [22]. Balanced detection is also used in EPOs to effectively decrease the phase noise in the EPO [23, 24]. The EPO phase noise remains high at about −120 dBc Hz^{-1} at a 10 kHz offset.

In this section a DL-EPO scheme is described and examined from the point of view of differential resonance/interference that offers a higher SMSR and further suppression of RF phase noise.

In contrast to a common single-output EPO (SO-EPO) based on a single-output MZIM, plus an optical coupler, and BPD, the dual-loop EPO (DL-EPO) using DO-MZIM and BPD can be formed by two 180° output ports of DL-MZIM and two input ports of the BPD, without any additional optical coupler or PBS devices. The DL-EPO scheme could be used to suppress both the side modes and RF phase noise. We note here that the two output ports of a DL-MZIM are complementary to $\pi/2$ via the $\pi/2$ phase factor when the lightwaves are cross-coupled to the output port, but not the through port.

A DL-EPO structure is depicted in figure 7.8(a) in which a continuous wave (CW) lightwave source of angular frequency ω_0 is coupled to an integrated photonic DO-MZIM. The modulator is modulated by the feedback electrical RF signal. The MZIM is biased at the Q+ or Q− point or at quadrature in either the positive or negative slope of the transmittance transfer curve of the MZIM. The DO-MZIM thus operates at the half-value of the V_π, the half-wave voltage, such that the oscillation is operating in the linear region. A voltage control-gain (VCG) amplifier is used to ensure that the oscillation amplitude is limited to the linear range of the MZIM.

The two output ports of the DO-MZIM are connected to a 1 km long SSMF (path 1) and 6 km long SSMF (path 2). Then the output lightwaves transmitted through path 1 (P1) and path 2 (P2) are received by a pair of identical photodetectors (PDPs) connected back-to-back (B2B) in a balanced mode, thus balanced photo-detection (BPD). The multi-longitudinal-mode (MLM) electrical microwave signals are generated via the BPD. The desired-frequency RF signal is selected by filtering out the RF component by a sharp roll-off electrical bandpass filter (EBPF). The filtered and detected signal is amplified by a high gain RF electrical amplifier (RF-EA), which is then split by a 3 dB electrical power splitter (3 dB-ES). One part is further amplified by a variable gain amplifier (VGA) with a limited amplitude below the nonlinear level of the transmittance transfer curve, the power versus applied voltage characteristics, which is then applied to modulate the DO-MZIM to form the electronic-photonic feedback loop. The other part is delivered to an electrical spectrum analyser (ESA) or a signal analyser to characterize the electrical signal analyser, e.g. a Keysight X-series analyser [25] for measurement of the spectrum and phase noise.

The integrated DO-MZIM can be implemented by an SO-MZIM and a 2 × 2 fiber coupler (FC). The optical fields at different points of the DO-MZIM are indicated in figure 7.8(c).

As shown in figure 7.8(b), the single-output (SO, Y-coupled) MZIM and the dual-output (DO, X-coupled) DO-MZIMs form two different kinds of dual-loop EPOs with BPD. For the traditional SO-EPO scheme, as shown in figure 7.2 and figure 7.8(b), the output optical signals of the SO-MZIM from the two arms are first combined by the Y-coupler and then split into two parts by an additional 50:50 optical coupler. For the case of the DO-EPO shown in figures 7.8(a) and (b), the Y-coupler of the single-output MZIM can be replaced by an integrated 3 dB directional coupler (dc). Thus there are two different $\pi/2$ phase outputs.

The optical fields distributed in the DO-MZIM are shown in figure 7.8(c). The CW of an ECL DFB-LD is given as

$$E_{\text{in}}(t) = Ae^{j(\omega_0 t+\varphi_0)}, \tag{7.2}$$

where A, ω_0, and φ_0 are the optical E-field amplitude, optical angular frequency, and optical initial arbitrary phase.

The corresponding modulated RF signal is given as

$$V_{\text{RF}}(t) = V_0 \sin(\omega_{\text{RF}} t), \tag{7.3}$$

where V_0 and ω_{RF} are the RF signal amplitude and angular frequency.

The RF driving voltage can be applied complementarily to the two arms of the DO-MZIM leading to the two arms in opposite phase shift. The output Y-coupler of a traditional non-dc coupled modulator can be replaced by a 3 dB fiber directional coupler.

Assuming that the input Y-coupler and optical dc output are ideal with no loss, the output optical fields at the two optical ports of the DO-MZIM, E_{O1} and E_{O2}, can be expressed as

$$E_{O1} = \frac{E_{\text{in}}}{\sqrt{2}} e^{-j\frac{\pi[V(t)+V1]}{2V_\pi}} + \frac{E_{\text{in}}}{\sqrt{2}} e^{-j\frac{\pi V(t)}{2V_\pi}+\frac{\pi}{2}} \tag{7.4}$$

$$E_{O1} = \frac{\sqrt{2}}{2} A e^{j\left(\omega_0 t+\varphi_0+\frac{\pi}{4}\right)} \left\{ J_0(\beta) + \sum_{n=1}^{\infty} J_{2n}(\beta) - 2\sum_{n=1}^{\infty} J_{2n-1}(\beta)\cos([2n-1]\omega) \right\} \tag{7.5}$$

$$E_{O2} = \frac{E_{\text{in}}}{\sqrt{2}} e^{-j\frac{\pi[V(t)+V1]}{2V_\pi}+\pi/2} + \frac{E_{\text{in}}}{\sqrt{2}} e^{-j\frac{\pi V(t)}{2V_\pi}}$$

$$E_{O2} = \frac{\sqrt{2}}{2} A e^{j\left(\omega_0 t + \varphi_0 + \frac{\pi}{4}\right)} \left\{ J_0(\beta) + 2\sum_{n=1}^{\infty} J_{2n}(\beta)\cos(2n\omega) \right.$$
$$\left. + 2\sum_{n=1}^{\infty} J_{2n-1}(\beta)\cos([2n-1]\omega) \right\},$$

where

- $\beta = \frac{\pi V_0}{2V_\pi}$, with V_π is phase factor due to the applied V_o with respect to the RF half-wave voltage of the DO-MZM;
- $J_n(\beta)$ is the nth-order Bessel function of the first kind; and
- $V(t) = V_{dc} + V_{RF}(t)$ is the sum of the dc and RF voltages applied to the two arms of the DO-MZIM.

It is noted that the phase shift of $\pi/2$ in equation (7.4) is contributed by the phase shift of the cross-coupling in the optical directional coupler.

Neglecting the high-order sidebands, the spectrum of the output voltage of the BPD can be assumed to follow a Lorentzian shape spectrum. In the spectral domain the RF generated oscillation signal can be written as

$$\begin{aligned} V_{BPD}(t) &= \frac{1}{2}A^2 \mid E_{out1}(t)E^*_{out1} \mid . \mathcal{R}. \mid Z \mid - \frac{1}{2}A^2 \mid E_{out2}(t)E_{out2} \mid . \mathcal{R}. \mid Z \mid \\ &= 4\rho\mathcal{R}A^2 J_0(\beta)J_1(\beta)\cos(\omega t), \end{aligned} \tag{7.6}$$

where

- R is the PD responsivity in units of A W^{-1} or the number of electrons which can be generated by the PD absorption of one photon energy; and
- Z is the load impedance of the BPD and $|Z|$ is the resistance of the impedance at the mid-band.

Assuming a linear operating region, the loop electrical voltage can be expressed as

$$\widetilde{V}_{out}(\omega) = \sum_{i=0}^{\infty}\left[g_1 e^{j\omega\tau_1} + g_2 e^{j\omega\tau_2}\right]\widetilde{V}_i(\omega) = \frac{\widetilde{V}_o}{1 - \left[g_1 e^{j\omega\tau_1} + g_2 e^{j\omega\tau_2}\right]}, \tag{7.7}$$

in which

- g_1 is the path gain of the optical-electronic (OE) path 1;
- g_2 is the path gain of OE path 2;
- ω is the angular frequency of the generated RF oscillation signal; and
- τ_1 and τ_2 are the loop delay times of loops 1 and 2.

Thus the corresponding RF power can be described as

$$P_{out}(\omega) = \sum_{i=0}^{\infty}[g_1 e^{j\omega\tau_1} + g_2 e^{j\omega\tau_2}]\widetilde{V}_i(\omega); \tag{7.8}$$

and hence the generated RF oscillation spectrum can be obtained:

$$P_{osc}(\omega) = \frac{G_A^2 \mid V_{BPD} \mid^2 / 2R}{1 + \mid g_1 \mid^2 + \mid g_2 \mid^2 + 2 \mid g_1 \mid\mid g_2 \mid \cos(\Phi_1 - \Phi_2) - 2 \mid g_1 \mid \cos(\Phi_1) + \mid g_2 \mid \cos(\Phi_2)}, \quad (7.9)$$

where $\Phi_i(\omega) = \omega\tau_i + \phi_i$, with $i = 1, 2$ (i.e. path 1 and path 2 of the MZM). In equation 7.8 Φ_i is the phase factor of the complex gain g_i and G_A is the amplifier voltage gain. For the EPO structure of figure 7.8(a), the oscillation frequency must satisfy the resonance conditions

$$\Phi_i(\omega) = 2k\pi$$

$$\Phi_i(\omega) = 2m\pi, \quad (7.10)$$

in which k and m are the integers, $k = 0, 1, 2, \ldots, m = 0, 1, 2, \ldots$.

Using equations (7.10) and (7.8), the RF oscillation power can be written as

$$P(\omega) = \frac{G_A^2 |V_{BPD}|^2 / 2R}{1 + |g_1|^2 + |g_2|^2 + 2\,|g_1|\,|g_1| - 2\,|g_1| - 2\,|g_2|} \quad (7.11)$$

where $\mid V_{BPD} \mid$ is the voltage dropped across the PD output resistance R. Equation (7.11) indicates the factors that influence the RF oscillating mode power. However, the SMSR and phase noise of the EPO link cannot be derived from (7.11). The power spectral density of the EPO can be given, to follow the Lorentzian shape, as [3]

$$S_{RF}(\Delta f') = \frac{\text{NSR}}{\left(\frac{\text{NSR}}{2\tau}\right)^2 + (2\pi\tau\Delta f)^2} = \frac{1}{\text{SNR}\left[\left(\frac{1}{2\tau.\text{SNR}}\right)^2 + (2\pi\tau\Delta f)^2\right]}, \quad (7.12)$$

where Δf is the frequency offset from the oscillation frequency f_{RF}, τ is the loop total delay time, and NSR = 1/SNR is the inverse of the SNR of the EPO given as

$$\delta = \text{NSR} \equiv S_N \frac{G_A^2}{P_{osc}} \text{ or SNR} = \frac{P_{RF,\,osc}}{G_A^2 \rho_N} S_N(df), \quad (7.13)$$

where $S_N(df)$ is the total noise-spectral density referred to the input to the oscillator (i.e. the input port of the amplifier or at the output of the PD), i.e. the summation of the thermal noise, the quantum shot noise (due to average optical power), shot noise due to the biasing current, and the laser's relative intensity noise (RIN).

G_A is the gain of the electronic RF amplifier in linear scale, given as

$$S_N(df) = \frac{4k_B T}{R} + 2qI_{ph}\mathcal{R} + N_{RIN} I_{ph}^2 \mathcal{R}, \quad (7.14)$$

where k_B is the Boltzmann constant, T is the operating temperature in kelvins, NF is the RF amplifier noise figure, q is the electronic charge, I_{ph} is the photocurrent flowing across the PD load resistor, N_{RIN} is the laser RIN, R is the PD series resistance, $\mathfrak{R}$ is the PD responsivity, I_{ph}^2 is the photocurrent produced by the average PD fiber pig-tail optical power, i.e. in front of the PD, and $P_{\mathrm{RF,\ osc}}$ is the RF wave power at a steady state level.

From equation (7.12) the spectral density of the RF oscillating mode follows a Lorentzian shape with a full width at half-maximum (FWHM) Δf_{FMHM} given by

$$\Delta f_{\mathrm{FWHM}} = \frac{\delta}{2\pi\tau^2} = \frac{G_A^2 \rho_N}{2\pi\tau^2 P_{\mathrm{osc}}}. \tag{7.15}$$

Thus the power spectral density of the oscillating mode of equation (7.12) can be rewritten as

$$S_{\mathrm{RF}}(f') = \begin{cases} \dfrac{4\tau^2}{\delta} & |f'| \ll \dfrac{\Delta f_{\mathrm{FWHM}}}{2}; \text{ in-band power} \\ \dfrac{\delta}{(2\pi f')^2} & |f'| \gg \dfrac{\Delta f_{\mathrm{FWHM}}}{2}; \text{ out-of-band power.} \end{cases} \tag{7.16}$$

Equation (7.16) indicates that (i) the oscillation component power spectral density is proportional to the SNR, (ii) the side-mode power spectral density is inversely proportional to the SNR of the EPO link, and (iii) the EPO phase noise can be identified as the side-mode power spectral density.

Hence, the SMSR is inversely proportional to the SNR and the phase noise is proportional to the NSR. The noise performance of the X-coupled and Y-coupled link is given in [26]. It indicates that the DO-EPO link has a lower NSR than the SO-EPO link. Thus the side-mode suppression ratio can be given as

$$\text{SMSR (DO-EPO phase noise)} > \text{SMSR (SO-EPO)}. \tag{7.17}$$

A platform of DO-EPO is shown in figure 7.8(a). A distributed feedback laser diode (DFB-LD; Pure Photonics PP-550L, with a linewidth of 10 kHz) is used as the ECL. The lightwave from the ECL with a wavelength of 1550.53 nm and a power of 16.0 dBm is modulated by the feedback electrical signal via the DO-MZIM of 3 dB bandwidth of 22 GHz. The two optical signals of the DO-MZIM pass through a 1 km fiber (path 1, standard SMF-G652) and a 6 km fiber (path 2, standard SMF-G652), respectively. It is noted here that there is a $\pi/4$ phase shift between the optical fields of the two ports when the optical waves are cross-coupled from one port to the other cross port of an integrated or fiber optical directional coupler (ODC). Then, the two path optical signals are fed in to a pair of PDs (a Finisar-XPDV3120, with a 3 dB bandwidth of 65 GHz and a responsivity of 0.8 A W^{-1} in the 1550 nm spectral region). The desired RF signal is filtered out by an EBPF. The RF waves are obtained via a 3 dB RF splitter. The RF signal is thus driving the DO-MZIM to form a close EPO loop after a further amplification by an electronic amplifier (SHF-807, with a gain of 23 dB and a bandwidth of 65 GHz). The RF splitter is placed between the RF amplifier and the DO-MZIM to obtain the RF oscillation

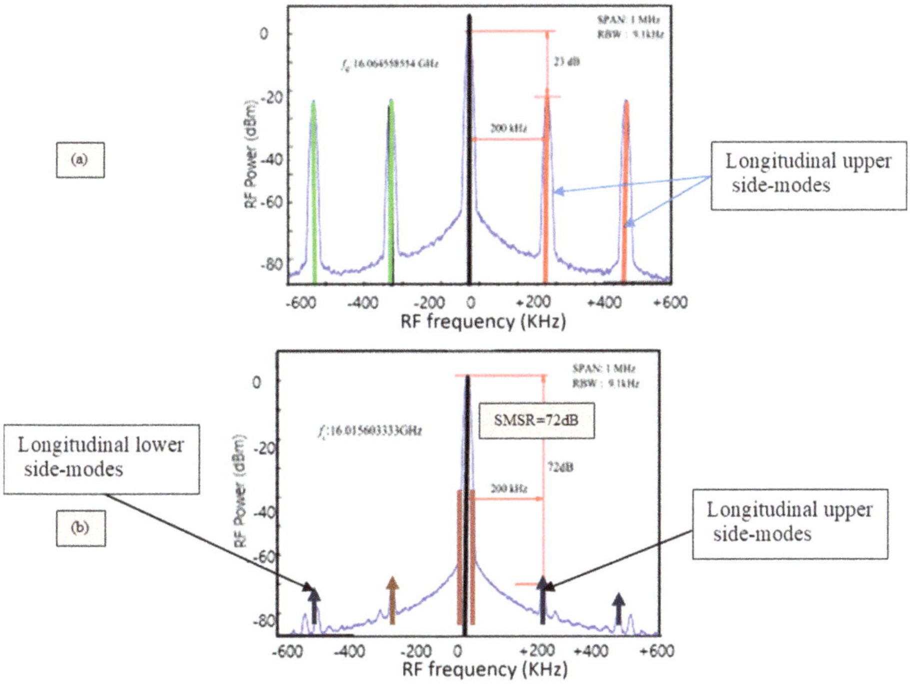

Figure 7.9. Oscillation from a DO-MZIM EPO system. (a) The RF power spectrum of the single-loop EPO with 1 km fiber length. (b) The RF power spectrum of the DO-EPO. Resolution bandwidth (RBW) = 9 Hz; full span = 1.0 MHz. The centered frequency is at 16 GHz, 10 GHz, or 20 GHz as the resonant mode.

component output and is measured by an Rhode and Schwartz FSW spectrum analyser (ESA) or a Keysight X-series signal analyser of 43.6 GHz full span.

First, to verify the DO-EPO scheme a contrast is made between the SL-EPO and the DO-EPO in the spectrum and phase noise. As shown in figure 7.8(a), disconnecting either path, there is an integral single-loop EPO. First, when only path 1 is connected to the single-loop EPO with 1 km loop length, it has a 23 dB SMSR and 200 kHz mode spacing, shown in figure 7.9(a). This can be implemented by open the loop path 2 or effectively disconnect the coupling of the output 2 of the DO-MZIM and replace with an index matching unit to avoid any reflection back into the loop, hence the random oscillation or instability.

Second, when only path 2 is connected, the single-loop EPO with a 6 km loop length has an 8 dB SMSR and 33.3 kHz mode spacing, shown in figure 7.9(b). Finally, when both paths are connected, there is a DO-EPO with balanced detection, which could be naturally combined without any additional optical coupler or PBS devices. The RF power spectrum of the DO-EPO with balanced detection is shown in figure 7.2(c), the SMSR is 72 dB and the mode spacing is 200 kHz. Compared with the 1 km and 6 km loop length single-loop EPOs, the SMSR of the DO-EPO with balanced detection is improved by 49 and 64 dB. Meanwhile, we could also note that the use of the double loops of the DO-EPO can improve the ability of the free spectral range (FSR) to avoid hopping (or multiple frequency oscillation) [27]. An experiment of the SO-EPO is also performed based on the configuration shown in figures 7.9(b) and 7.10(b). The spectrum of the SO-EPO formed by a single-output MZM, a 50:50 optical coupler, and a BPD is demonstrated in figure 7.9(a) indicating that the SMSR of the SO-EPO is 27 dB, much lower than that obtained for the DO-EPO. This can be explained by the fact that the Y-coupled balanced detection link in the SO-EPO may increase the noise induced by temperature fluctuation, vibration, and other factors, whereas the X-coupled balanced detection link in the DO-EPO may reduce the noise induced by those. As a comparison, the phase noise of the SL-EPO with different loop lengths, SO-EPO, and the DO-EPO with balanced detection is measured.

Figure 7.10(b) shows the resonant peaks and phase noise of the single-loop EPO with different loop lengths and the DO-EPO with balanced reception. The phase noise of the single-loop EPO with 1 km and 6 km loop lengths is −94.3 and −106.6 dBc Hz^{-1} at an offset frequency of 10 kHz when the oscillating frequency is 16 GHz. The measured phase noise of the 16 GHz RF oscillation of the DO-EPO is −133.2 dBc Hz^{-1} at an offset frequency of 10 kHz. As can be seen from figure 7.3(a), it is obvious that the phase noise of the DO-EPO with balanced detection is much lower than that of the single-loop. Compared with the 1 km and 6 km loop lengths of the single-loop EPO, the phase noise of the dual-loop EPO with balanced detection has been improved by 38.9 and 26.6 dBc Hz^{-1} at an offset frequency of 10 kHz. Figure 7.10(b) also shows the phase noise comparison between the single EPO and dual EPO. More details of a single resonant mode and side modes under the lowest resonant condition are depicted in figure 7.11. Figure 7.12 shows the case of setting or tuning of the modulation conditions such that all major resonant modes are excited. The excitation is very close to full suppression of phase noise. The phase noises are all suppressed to about −140 dBm Hz^{-1} when the full resonance is reached.

The phase noise of the SO-EPO is −110.4 dBc Hz^{-1} at an offset frequency of 10 kHz and is 21.8 dBc Hz^{-1} worse than the result in the DO-EPO [23]. This indicates that the X-coupler of the DO-MZM is a principal factor for the reduction

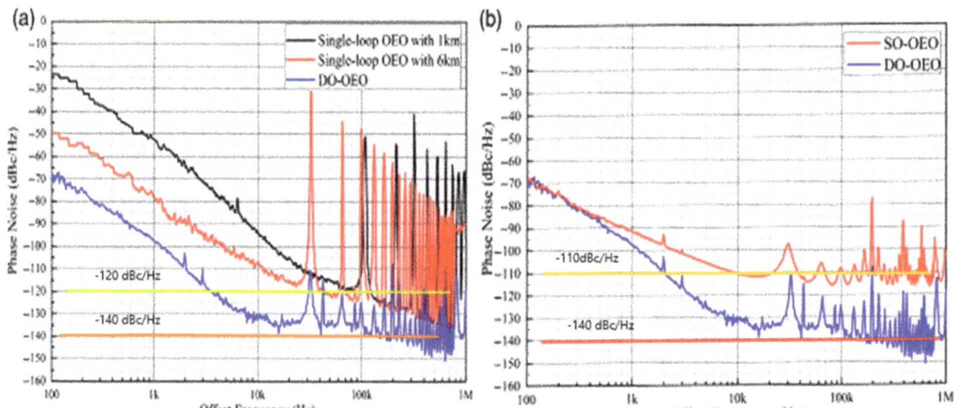

Figure 7.10. Experimental result. (a) Phase noise curve comparison of the single-loop EPO with 1 km fiber length, the single-loop EPO with 6 km fiber length, and the DO-EPO. (b) Phase noise of SO-EPO and the DO-EPO. PN = phase noise.

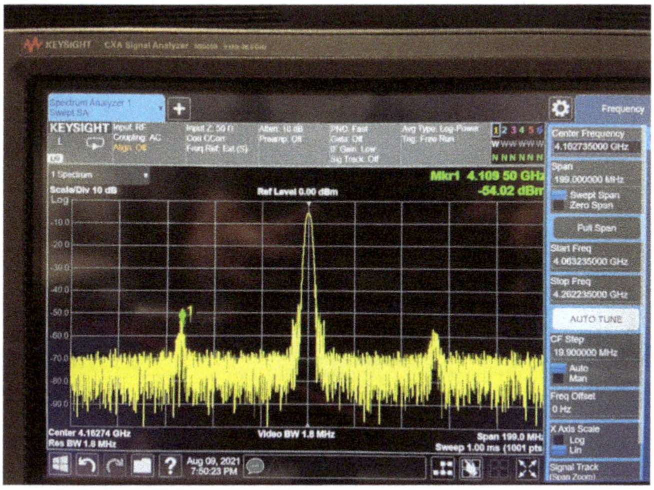

Figure 7.11. Detailed oscillation mode and other side modes at the lowest resonance and thus more noisy.

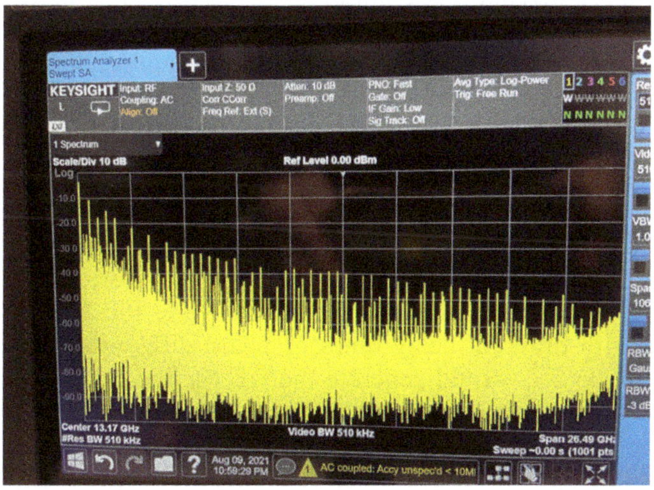

Figure 7.12. Oscillation mode of a DO-MZIM under excitation of all major peak longitudinal modes. Excitation at near resonance, that is phase noise is not fully suppressed.

of the EPO phase noise. As a consequence, we could note that the use of the DO-EPO with balanced detection can maintain the lower phase noise performance than the use of the single-loop EPO and the SO-EPO. The frequency drift of the DO-EPO with balanced detection is less than 17 kHz in laboratory environment conditions within 60 min. The frequency drift performance can be further improved if a phase locked loop module can be used in the dual-loop EPO.

The dual-loop EPO is analysed and demonstrated, including the use of a balanced opto-electronic receiver. The system employs only a DO-MZIM and a BPD. The phase noise can be highly suppressed effectively by more than 20 dB in the DO-EPO

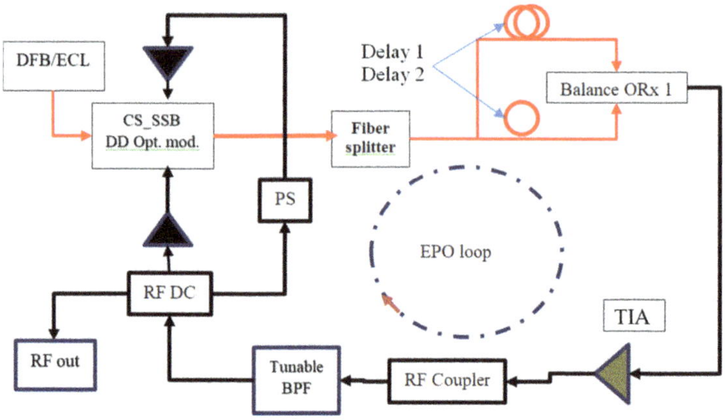

Figure 7.13. Self-homodyne reception electro-photonics oscillator.

with balanced detection compared to a single-loop EPO. A microwave RF signal with an SMSR of 72 dB and phase noise of -133.2 dBc Hz^{-1} at 10 kHz frequency offset can be achieved. Furthermore, the DO-EPO scheme can be characterized by a simple structure, low insertion loss, and polarization independence.

7.4 EPO type 4: a self-homodyne reception dual-loop EPO

The SSB can be used to generate lower phase noise, as shown in figure 7.13, using the MZDI at the output of the SSB modulator. For the EOE the SSB with a low level optical carrier must be used so that the beating of the SSB component and the original carrier in the PD can generate the RF carrier whose frequency equals the difference of these two optical carriers. Alternatively, the MZDI generates two optical carriers with different delay times or phase differences and the beating of these two optical carriers would generate the RF oscillation. For high frequencies the delay time difference must be small enough, for example for 10 GHz the delay difference time is 100 ps. For this difference, this delay must be implemented in an integrated photonic optical waveguide such as silicon nitrate (Si_3N_4) or in SOI.

7.5 EPO type 5: a single sideband carrier suppressed (CS-SSB) external modulation laser (EML)

Figure 7.14 shows the schematic of an EPO using a dual-drive optical MZIM and carrier suppression to generate an SSB with a double optical path cavity loop. The optical spectrum of a CS-SSB laser can be depicted in figure 7.15, which are obtained by modulating a DD-MZM with two orthogonal RF signals as show in figure 7.14 and a dual polarization DP-QPSK modulator as depicted in figure 7.16. Thence the optical and RF paths of the loops are the same as given in figure 7.5. We observe that the phase noise of the EPO using the CS-SSB can be reduced by another 30 dB. Thus, the phase noise can reach the minimum level of around -150 dBc Hz^{-1} and the CNR can be improved to >80 dBc (figures 7.16 and 7.17).

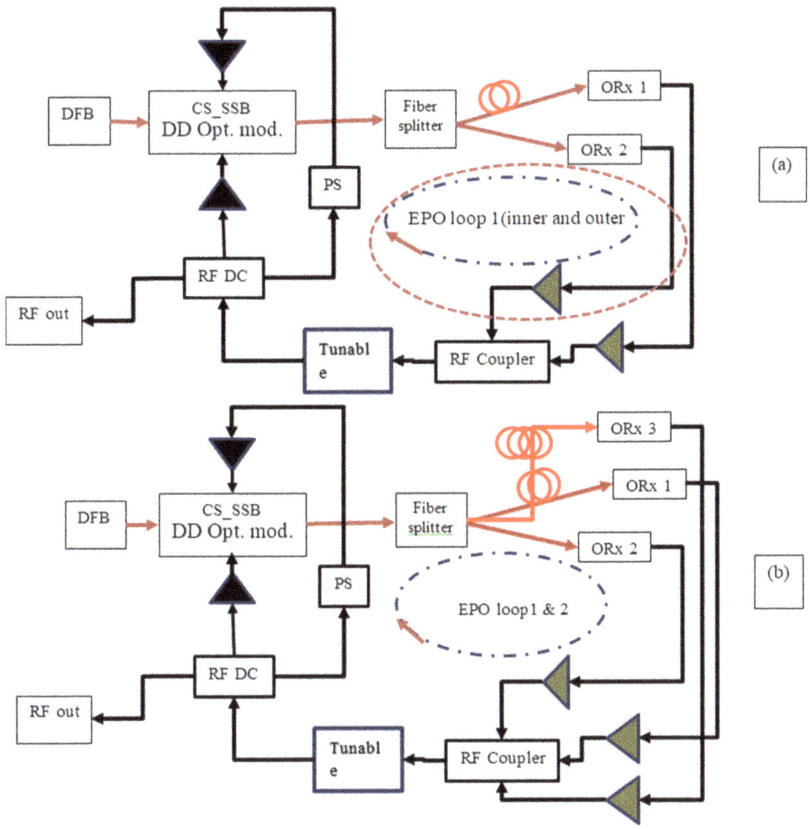

Figure 7.14. Schematic of an EPO using a dual drive optical MZIM and carrier suppression to generate an SSB with a double optical path cavity loop. (a) A dual-loop with a delay difference in the optical lines. (b) Triple loop with delay lines. DFB = distributed feedback laser; CS-SSB = carrier suppressed; DD = dual drive.

7.6 EPO type 6: an optical tunable EPO (OTEPO)—a master–slave heterodyne-coherent receiving tunable EPO

We can appreciate the importance of the tunability of the EPO that can produce the RF oscillation output due to the beating of the two lasers in the PD. The RF output is proportional to the product of the amplitudes of the two lasers and the oscillation at the frequency difference of the two lasers. This was done by tuning the RF BPF, as shown in figure 7.13. However, the electrical tuning of the BPF can be difficult. An alternative technique is to apply coherent heterodyne detection by injecting a second laser into the loop, as shown in figure 7.18. This second laser is assigned as the master laser whose frequency can be tuned so that the frequency difference of the two lasers is equal to that of the desired RF frequency. This is now possible due to the advanced properties of modern single frequency lasers. It is noted that the bandwidth of the PD and the transimpedance amplifier (TIA) must be wide enough to cover the RF frequency band as desired. The local oscillator or the slave laser, is shown in figure 7.18. This LO frequency is mixing with the input lightwaves whose frequency is shifted by an amount with respect to that of the master laser by a frequency amount equal to that of the RF oscillation which we desire to generate for the EPO. This thus can be better than that of the e-LPF-heterodyne EPO described in section 7.4, as the tuneable laser can now be tuned to less than 1.0 kHz.

7.7 Type 7: an optical polarization modulator-based EPO (OPolM-EPO)

A microwave down-converter can offer the possibility of sampling in the lower frequency region of RF signals to the baseband, and an intermediate frequency (IF) spectral window is desired in many microwave-radio systems so as to minimize the complexity of reception and processing in RPoF systems, in particular for 6G and 7G mmW and μW wireless communication systems [28]. Conventional μW systems use electronic mixers to implement frequency down-conversion, but the electronic mixers exhibit a narrow bandwidth and strong unwanted IMD. RF generated with down-conversion to an IF spectral region can be conducted by transferring the RF-shifted optical carrier to an appropriate intermediate optical position, thence an OE conversion produces the appropriate RF signals in the electrical domain, hence requiring only a simple low-speed photodetector (PD).

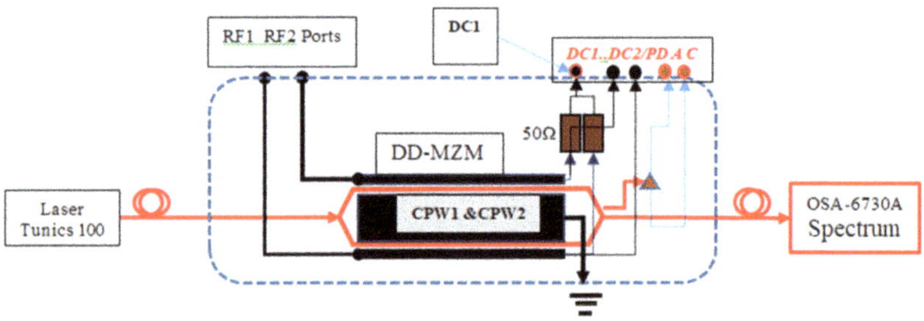

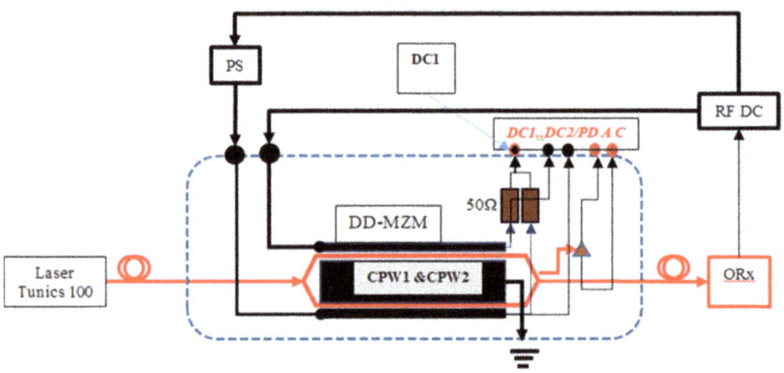

Figure 7.15. A schematic arrangement of a CS-SSB laser using a dual polarized QPSK optical modulator. (a) DD-MZM. (b) Core loop DD-MZM for the EPO structure. (c) DP-QPSK modulator. (d) Closed loop to generate the EPO. ORx = optical receiver; PS = phase shifter; dc = directional coupler/splitter.

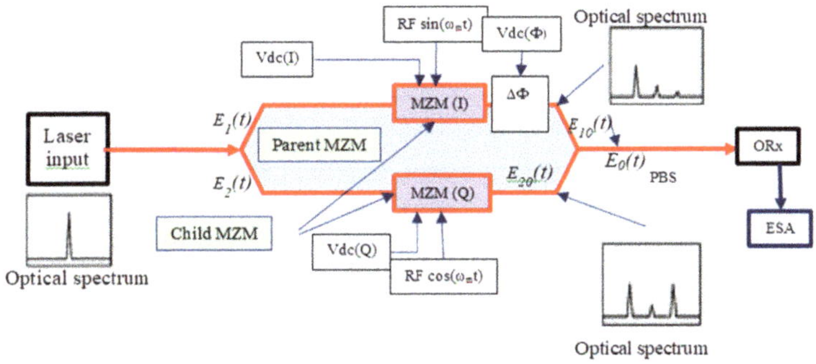

(c)

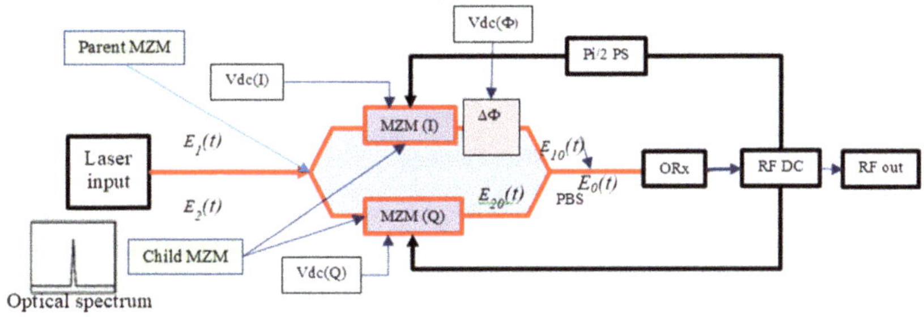

(d)

Figure 7.15. (Continued.)

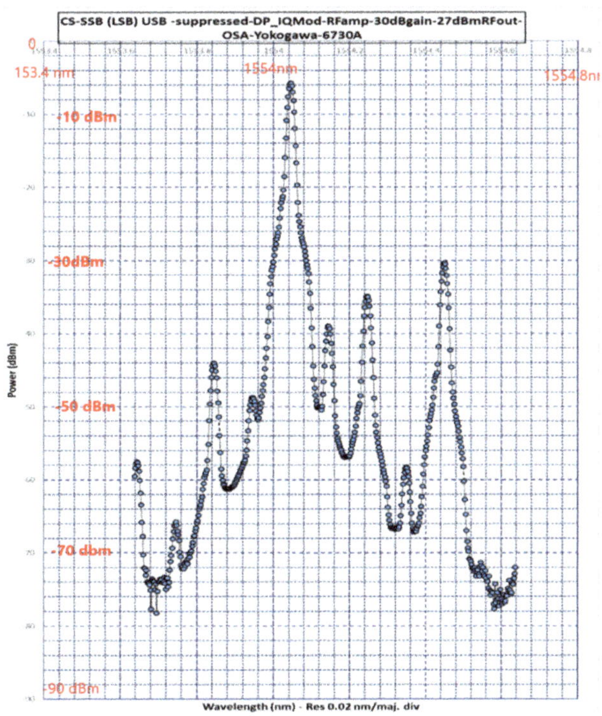

Figure 7.16. Optical spectrum of a CS-SSB laser as measured using an optical spectrum analyzer Yokogawa 6730.

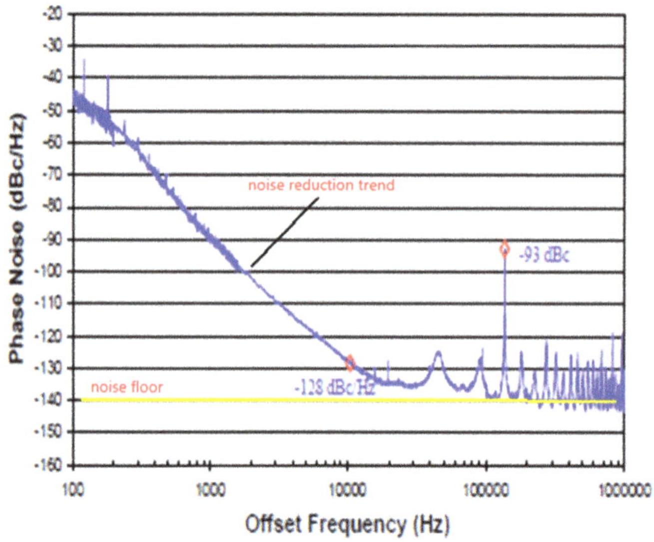

Figure 7.17. Phase noise in dBc Hz^{-1} as a function of frequency of a triple loop EPO. Plots are normalized from plots obtained by an ESA. The main peak of the RF oscillation frequency at 12 GHz and −93 dBc at 10 Hz bandwidth sweeping.

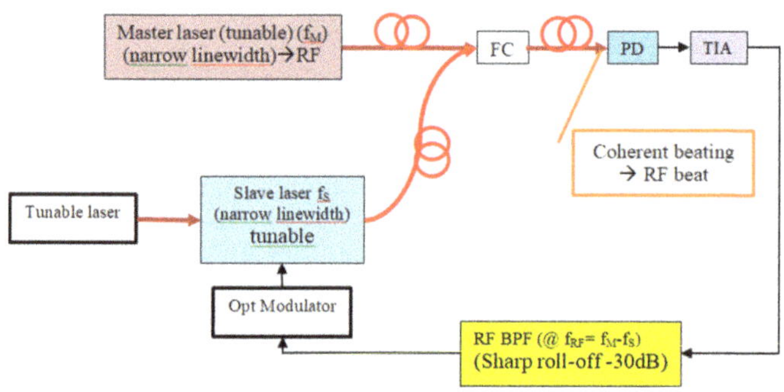

Figure 7.18. The master–slave configuration in an electro-photonic coherent receiving loop for RF oscillation generation. PD = photodetector; TIA = transimpedance amplifier.

To generate such a high-quality local oscillator (LO), an EPO is a candidate with great potential. The EPO is an opto-electronic feedback loop in which a photonic-modulation device is integrated to convert the RF to the photonic domain, as described in sections 7.2–7.6. A high-Q factor of the EPO cavity can produce a high-spectral-purity oscillation signal with a phase noise as low as −140 dBc Hz^{-1} at 10 kHz offset from its central resonant frequency.

7.7.1 Principles of operation

The structure of the photonic microwave down-converter based on a frequency doubling-EPO (FD-EPO) is shown in figure 7.19. It is then adapted for a novel EPO structure, depicted in figure 7.20.

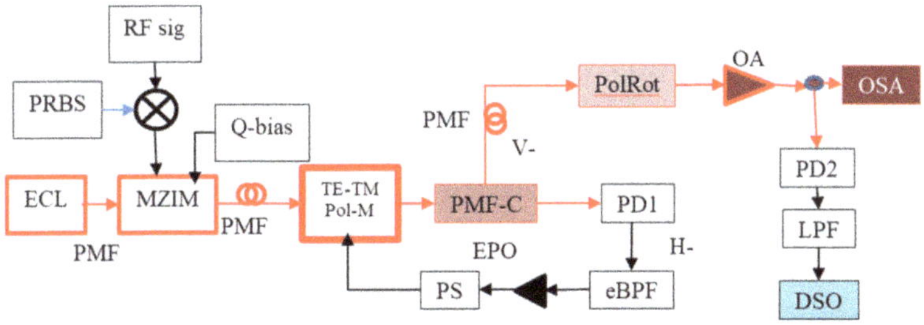

Figure 7.19. Schematic of a photonic radio microwave IF down-converter using a PolM based forward-decision EPO. ECL= external cavity laser; PC = polarization controller; PMF= polarization maintaining fiber; MZIM = Mach–Zehnder intensity modulator; OPolM = optical polarization modulator; PBS = polarization beam splitter; RF sig = RF signal generator; OA = optical amplifier; PD = photodetector; EA = electrical amplifier; eBPF = electrical bandpass filter; PRBS = pseudo-random bit sequence; PS = phase shifter; LPF = low pass filter; OSA = optical spectrum analyser; DSA = digital signal analyser. Adapted from [29].

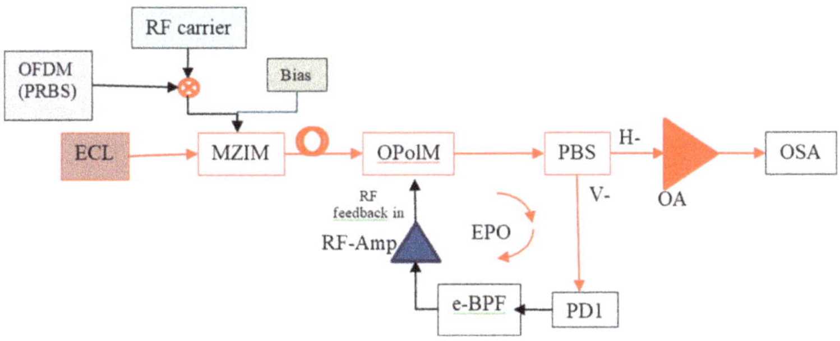

Figure 7.20. Schematic of an EPO integrated in an RoF to suppress nonlinear harmonics and noise. PBS = polarization beam splitter; e-BPF = electric bandpass filter; PolM = polarization modulator; MZIM = Mach–Zehnder interferometric modulator; OA = optical amplifier (see also figure 7.21).

The RF carrier signal is coupled with a pseudo-random bit sequence (PRBS) with an OFDM data channel. They are then used to modulate the MZIM, hence RF modulated lightwaves are generated and transmitted into the opto-electronic (OE) loop. An external cavity laser (ECL) is used whose output field is modulated via the MZIM. The polarization states of the lightwaves coupled to the MZIM must be manipulated or switched, using a polarization controller (PC), for example, to maximize the lightwave energy propagating in the optical section of the OE loop. Alternatively, a polarization mode converter operating at high speed can be used to switch or change the polarization states of the optical guided fields.

The polarization modulator (PolM) is coupled to a polarization beam splitter (PBS). This PolM in association with the PCs, can be considered equivalent to a two-output-port intensity modulator (IM). An EPO using the radio-photonic converter DO-MZIM is described in section 7.3. One output of the IM is detected by a PD, then is amplified and fed back to the RF port of the PolM, hence forming a closed loop EPO.

An EPO system requires a 'positive' feedback loop whose closed loop gain should be close to unity, as given in equation (7.2), so that RF oscillation waves can be obtained. The polarization converter can be adjusted so that the PBS output is equivalent to a linear intensity modulation device. Photodetector PD_1 is used to convert optical signals to the RF electrical domain. The oscillation output frequency can be selected by an EBPF whose passband can be narrow with a sharp roll-off.

An EA is also incorporated to ensure a loop gain higher than unity. It is preferred that an automatic gain-controlled (AGC) amplifier is used so that the output RF signal level can fall into the linear range of the MZIM so that no nonlinear harmonics are generated. This EPO structure can be considered as a feed-forward radio-photonic processed oscillator which can be termed as an FF-EPO. A PC can be used here to adjust to the output field of the PBS. Thus, this action is equivalent to intensity modulation biased at the quadrature transmission point (QTP).

Alternatively, one can structure an equivalent forward-decision electro-photonic oscillator (FD-EPO) by adjusting the PC to allow output 1 of PBS_2 to behave equivalently to an intensity modulation device such as an MZIM which is biased at the minimum transmission point ($Q_{\min}$), that is a π radians phase shift at the initial transmission point. At this operating bias point, the lightwaves at the output have a spectrum whose frequency shifts by an RF frequency doubling of the RF carrier. In addition, suppression of the central optical carrier is obtained.

The intermediate frequency (IF) down-conversion process can be as follows. (i) When an RF signal at a frequency f_m is injected into the FD-EPO, an oscillation of $f_m/2$ is produced at the initial stage at the output of the EPO. (ii) Due to the broadband properties of the noise of the EPO structure, an $f_m/2$ component exists. This component is then be amplified by the electronic amplifier. (iii) The amplified and resonant $f_m/2$ component is then fed back into the PolM based MZIM, thus forming the closed opto-electronic loop. This then modulates the input laser field. (iv) Under this modulation, the electrical modulating signal and the injected RF signal are mixed to produce further new RF spectral components. Thus the positive feedback forms an RF-photonic resonance loop that generates oscillation waves at the frequency of $f_m/2$. A high-quality RF carrier with low phase noise at $f_m/2$ can thus be obtained, this is the IF.

Once the EPO output is oscillating at $f_m/2$, the RF modulated optical signal is multiplied by the optical intensity transfer function of the PolM based IM, given as

$$T(\omega) = \alpha\left(1 + \cos\left[\beta \sin\frac{\omega_m t}{2} + \varphi_b\right]\right), \tag{7.18}$$

in which ω_m is the RF angular frequency, β is the phase modulation index ($[V_{\text{RF}}(t)]\pi/V_\pi$), α is the transmittance factor, φ_b is the bias phase factor ($V_b.\pi/V_\pi$), and V_{RF} is the RMS magnitude of the radio microwave sinusoidal carrier. The function sin[cos(x)] is equivalent to Bessel functions of the first order.

7.7.2 EO polarization mode converter

The 3 dB bandwidth of a 20–25 GHz EOPM [30, 31] operates at a high switching speed of the state of polarization (SOP) of the lightwaves. It can be constructed by electro-optic [32], acousto-optic [33], or magneto-optic [34] effects. Functioning

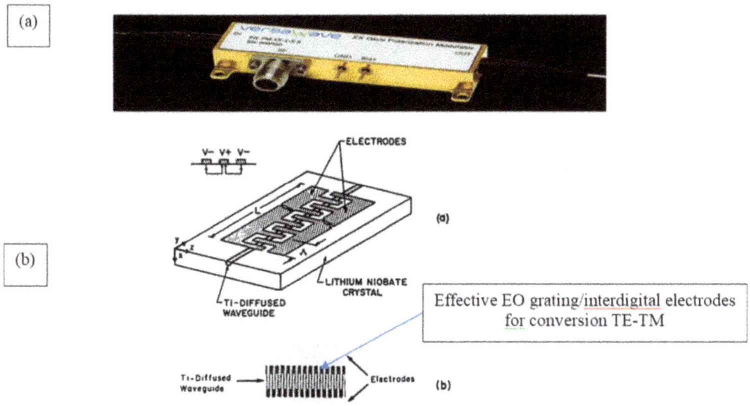

Figure 7.21. Polarization modulator: (a) packaged device and (b) PolM schematic structure using diffused waveguides and interdigital electrodes in $LiNbO_3$.

equivalently to a high speed electrically variable wave plate, the modulators can switch the SOP of linearly polarized lightwaves to an orthogonal linear polarization state which is similar to passing through a circular polarizer. The range and degree of the change in the SOP can be varied by either dc bias or RF drive voltages.

For an EO polarization converter or modulator, an interdigital electrode (IDT) is deposited on the surface across the optical waveguide, as shown in figure 7.21(b). When the RF signal is applied to the electrodes it generates an effective EO grating via the change of the refractive index of the guided medium. If the spacing distance producing the EO frequency matches the difference of the effective refractive index of the two polarized modes, the transverse electric (TE) and magnetic modes (TM) of the optical waveguide, then a polarization conversion between these TE and TM modes occurs. This polarization conversion/modulation efficiency depends on the applied RF signal magnitude and the splitting ratio from TE to TM, hence the polarization modulation efficiency and speed. No biasing is required.

Unlike the TE–TM mode converter based on lithium niobate, the OPolMs exhibit low birefringence and thus a low differential group delay of polarized lightwaves. This leads to more flexibility in EPO system design. Furthermore, for such an EPO using an OPolM for RF waves employed in an RPoF system, suppression of the harmonics and nonlinear distortion or IMD can be obtained, as shown in figure 7.21. The OPolM based EPO can have properties of low drive voltage, ultra-wide bandwidth, and compact size in an integrated optic structure.

7.8 Type 8: EPO generation integrated in-line of an HMLL

A harmonic mode locked laser (HMLL) [35, 36] can be formed by the resonance of an optical circulating loop in which an optical modulator is integrated. The modulator is modulated by an RF signal, as shown in figure 7.22. An EPO structure is integrated within the HMLL, as indicated in figure 7.22. The optical comb laser lines are derived from the HMLL, detected by a PD, then amplified and split 50:50, driving two ports of two parallel MZIMs. The amplifiers for driving the MZIMs are

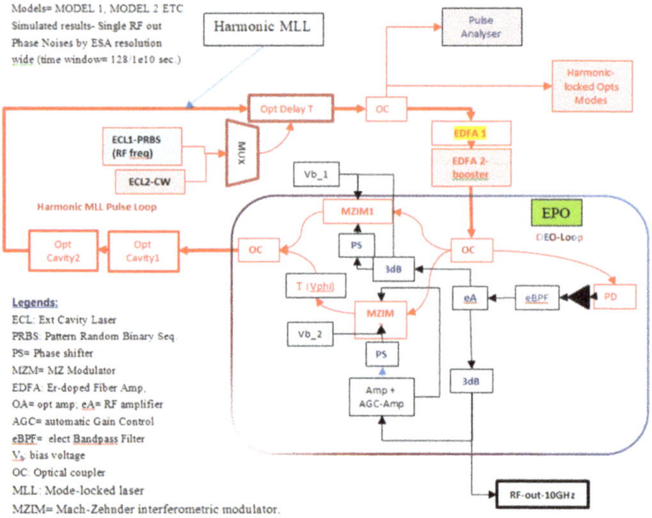

Figure 7.22. Schematic of a recirculating comb laser.

cascaded with an AGC stage whose output level is limited to the maximum linear range of the MZIM transmittance transfer curve so that no nonlinear or IMD products can be generated.

This section provides the following:

- The structure of a basic mode locked fiber laser (MLL) which is composed of an optical loop incorporating an opto-electronic oscillator, that is the optical signals circulate in the optical closed loop and are then detected by an optical receiver. The produced RF spectra are then filtered by an RF bandpass filter, and then drive an optical modulator embedded in the mode locked optical lasing.
- An optical time delay is inserted to ensure that the input lightwaves can be isolated and no interference of the RF modulating MZM can occur.
- The optical loop forms an MLL with an optical fiber amplifier (Er-doped fiber amplifier) and resonant loop. This laser and the EDFA lock the harmonics of the optical pulses which are directly related to the RF harmonic shifting frequency. Thus the RF frequency components of the OEO can also be in harmonics. The noise of the EDFA is sufficiently wide so that locking of RF components can occur.

Integrated within the MLL are a number of EPO types:

- OEO type 1: (a) A single MZM with biasing in the linear range and both RF input ports are driven by the same RF signal. (b) A single MZM but biased in inverted voltage and RF signal driven in complementary format so that a single sideband can be produced.
- OEO type 2: A dual MZM in the form of a CS-SSB and no phase shift so both SSBs can be formed and a phase can be added to form the real and complexed parts of the SSB signals to travel as optical waves in the loop. A phase of $\pi/4$ is applied to the phase shifter in the Q-branch of the QPSK modulator.

The EPO performance includes: (i) RF and optical spectra which indicate the output RF waves as selected or tuned by the EPO and (ii) phases noise output which agrees with experimental results.

The best performance for a DD-MZM is with an AGC RF to limit the driving amplitude to the DD-MZM—noise at −150.5 dBm Hz^{-1}.

The best performance for a QPSK is with SSB and CS biasing so that phase noises can be reduced to the level of −152 dBm Hz^{-1}.

A VPI simulation model of this harmonic MLL-EPO is given in figure 7.23. This MLL-EPO can produce both the comb laser lines and the RF oscillation signal. The RF harmonics is determined by the electrical bandpass filter in the EPO loop. This is also the frequency spacing of the short pulses of the MLL output. Two Fabry–Perot optical filters (OFs) are used to determine the FSR of the HMLL.

The VPI models include an MLL loop and integrated OEO to modulate its MZIM ports and coupling out the RF oscillation. The MLL loop acts as the cavity of harmonics of the RF frequency. There is an OA EDFA and MZM loop locking to the harmonics of the RF.

There is a tapped optical signal to close the EPO feedback loop for MZIM modulation, whose modulation frequency is 10 GHz of the RF output signal as set by the EBPF.

The EPO reaches a steady state when the HMLL pulse sequences attain their stability. Initially random noise is built up and then such noise is fed through an EBPF which filters off all other bands apart from the RF oscillation component. Then there is building up of the RF oscillation and the harmonics of the MLL comb lines or equivalently a periodic sequence of short pulses at the output of the MLL.

The simulation models are assigned with the following parameters:

- Model name: OEO-MLL_SSB_CS_Q+Q-CW_Noises-built-up_Juni03. Folder location: Lenovo C:\VPI_Models_All\VPI-Models\ProjMdls\2021Proj_Prt1_

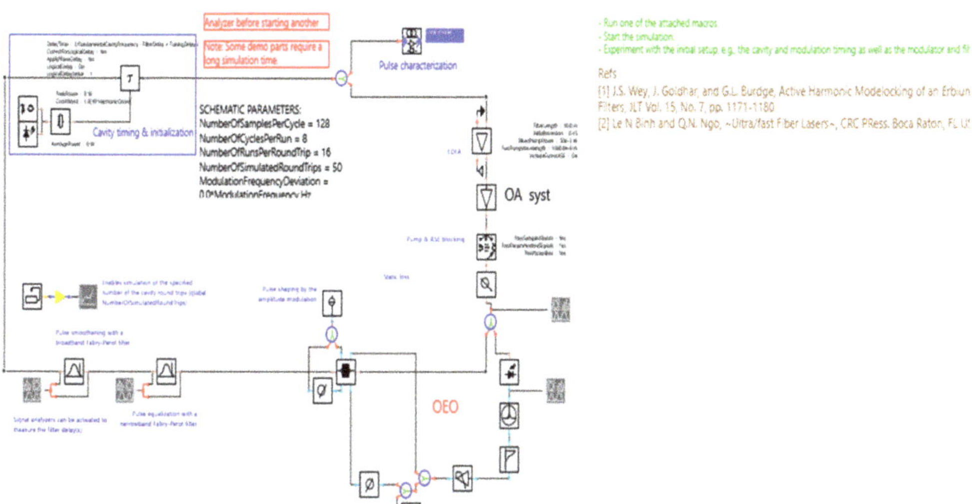

Figure 7.23. VPI TransmissionMaker© model of the harmonic MLL-EPO structure of figure 7.22.

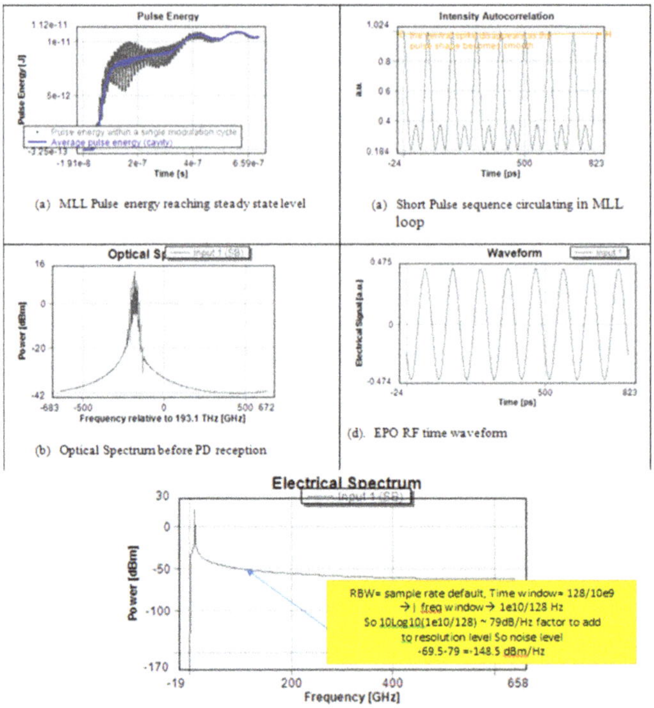

(a) MLL Pulse energy reaching steady state level

(a) Short Pulse sequence circulating in MLL loop

(b) Optical Spectrum before PD reception

(d). EPO RF time waveform

(e.) spectrum and phase noise level of -148 dBm/Hz

Figure 7.24. Simulated outputs of an HMLL incorporating an EPO: (a) harmonic pulse sequence in dynamic reaching the steady state level; (b) sequence of mode locked pulses; (c) optical spectrum of optical signals before detection by a PD and then amplifier and driving MZMs of the EPO; and (d) and (e) time waveform of RF oscillation and electrical spectrum with a low phase noise level.

Mdls-working\OEO. The EPO uses a dual parallel MZIM structure with differential bias levels of the two MZIMs so that the oscillation can be obtained, similar to the case of a clock recovery mechanism.

- MZM: $V_{\pi dc} = 5$ V; $V_{\pi RF} = 5$ V; biases Q+ = 2.5 and Q− = −2.5 V (carrier suppressed); eAmp; VGC set at $V_{\pi RF}/2$ or −12 dBW; EA gain = 20 dB; VGC gain = 20 dB; VGC level = 30 dB; EDFA1 gain = 20 dB; EDFA2 saturable gain = 20 dB and 25 dB; $V_{\pi} \sim \pi/2$ shift phase.

The simulated pulse sequences and the RF oscillation outputs are given in figures 7.24(a)–(e). This shows the effectiveness of the generation of both the MLL comb lines via the short pulse sequence in the optical domain and the RF oscillation signals, and very low phase noise. A number of EPO structures integrated in harmonic lasers are listed in table 7.1 with their performance in RF phase noise. The RF carrier is filtered and generated at 10 GHz, plotted in figure 7.25. Single drive MZIM and dual drive MZIM structures under different biases and driving conditions are given. The EPO employs dual parallel optical phase modulators and achieves −211 dBm Hz^{-1}, the best performance of all structures.

Table 7.1. EPOs with different modulators and biasing features.

OEO type	Phase noises (dBm Hz^{-1})	Biasing V (MZM-I)	Biasing V (upper MZM)	Biasing V (MZM-Q)	Biasing V (lower MZM)	Biasing V_{ϕ} (Q–) (°)	dBias	AGC (dBW EA) $V_{\pi RF}$ = 5 V (−12.5 dBW = 2.5 V peak)	LD LD +PRBS	RF sig EBPF (10 GHz) −90° PS between RF1–RF2
Single MZM										
MLL-OEO1, single MZM	−146		Q+		Q–	No	0	−12.5		DSB
MLL-OEO1, single MZM, no PRBS, CW laser	−138		Q+		Q–	No	0	−12.5		DSB
MLL-OEO1, single MZM, no PRBS, CW laser, AGC limit	−143		Q+		Q–	No	0	−10	No	PRBS
SSB										
Dual MZM-DSB										
QPSK, EA-AGC FB DSB	−152	Q+/Q+		Q+/Q+		90	0	−10	No	PRBS
QPSK, EA-AGC FB DSB	−152	Q+/Q+		Q+/Q+		0	0	−10	No	PRBS
QPSK, EA-AGC FB	−157	Q+/Q+		Q+/Q+		0	0	−12.5	CW LD only	DSB
QPSK, EA-AGC FB	−148	Q+/Q+		Q+/Q+		45	0	−10	CW LD only	DSB
QPSK, EA-AGC FB	0	Q+/Q–		Q+/Q–		90	0	−10	CW LD only	SSB
QPSK, EA-AGC FB, 20 GHz	−138	Q+/Q–		Q+/Q–		90	0	−10	CW LD only	SSB
QPSK, EA-AGC FB, 10 GHz	−148	Q+/Q–		Q+/Q–		180	0	−10	CW LD only	SSB

(*Continued*)

Table 7.1. (*Continued*)

OEO type	Phase noises (dBm Hz^{-1})	Biasing V (MZM-I)	Biasing V (upper MZM)	Biasing V (MZM-Q)	Biasing V (lower MZM)	Biasing V_ϕ (Q–) (°)	dBias	AGC (dBW EA) $V_{\pi RF}$ = 5 V (−12.5 dBW = 2.5 V peak)	LD LD +PRBS	RF sig EBPF (10 GHz) −90° PS between RF1−RF2
Dual MZM, SSB, dBias										
QPSK, EA-AGC FB, ±dBias	−160.2	Q+/Q− ±dBias		Q+/Q− ±dBias		90	±dBias	−10	CW	SSB
QPSK, EA-AGC 10 dB, ±dBias	−191	Q+/Q− ±dBias		Q+/Q− ±dBias		90	±dBias	−10	CW	SSB
QPSK, EA-AGC 10 dB, ±dBias	−180.9	Q+/Q− ±dBias		Q+/Q− ±dBias		90	±dBias	−10	CW	CS-DSB
QPSK, EA-AGC 20 dB, ±dBias	−171.95	Q+/Q− ±dBias		Q+/Q− ±dBias		90	±dBias	−10	CW	CS-DSB
Single DD-MZM										
DD-MZM, dBias, RF-DSB	−150.5		Q+/Q−		Q+/Q−	N/A	0	−10	CW	CS-DSB
DD-MZM, dBias, RF-DSB	−159.5		Q+/Q−		Q+/Q−	N/A	±dBias	−10	CW	In-phase DSB
DD-MZM, dBias, RF-DSB In-phase DSB	−121		Q+/Q−		Q+/Q−	N/A	NO	−10	CW +	PRBS MLL
DD-MZM, dBias, RF-DSB In-phase DSB	−137		Q+/Q−		Q+/Q−	N/A	NO	−10	CW +	PRBS MLL
Single DD-MZM + BalORx In-phase DSB	−163		Q+/Q−		Q+/Q−	N/A	NO	−10	CW	+PRBS MLL
Dual DD-MZM + BalORx	−187	Q+/Q− ±dBias		Q+/Q− ±dBias		90	±dBias	−10	CW	SSB

Dual DD-MZM + coherent BalORx	−201	Q+/Q− ±dBias	Q+/Q− ±dBias	90	±dBias −10	CW	SSB
DD-OPM (No.211)	−211	OPM1 Q +- dBias	OPM2 Q- +dBias	90	±dBias −10	CW	SSB

Notes	DD-MZM	Dual drive MZIM	dBias	Differential bias
	Q+, Q−	Quadrature points with positive/ negative slope	SSB, DSB CS-DSB	Single/dual sideband Carrier suppressed DSB
	V_ϕ	Voltage applied on the Q-branch of the QPSK modulator	CW	Continuous wave
	MLL	Mode locked laser	PRBS	Pseudo-random binary sequence
	LD	Laser diode	OPM	Optical phase modulator
	AGC	Automatic gain control		

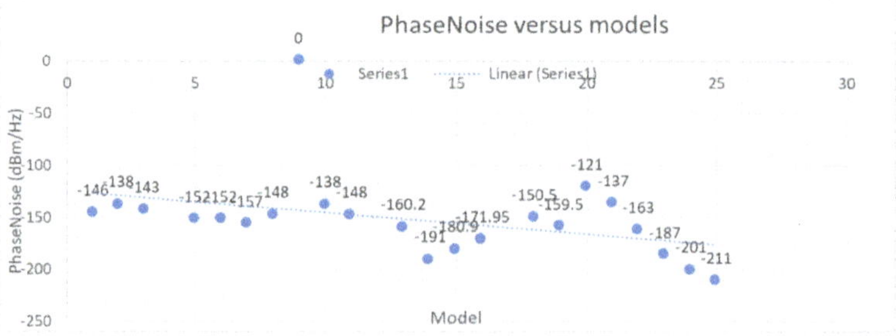

Figure 7.25. EPO phase noise versus different modulation structure. The numbering indicates the phase noise level of the EPO. See table 7.1 for the EPO types and phase noise performances.

7.9 Optical phase lock loop for an EPO

All the above EPO structures require optical phase locking so that the generation of the RF oscillation can be stabilized. This section gives an overview of the optical phase lock loop (OPLL) for EPO structures. Coherent receivers offer far more efficient detection and sensitivity of optical signals under various phase, amplitude, or frequency modulation schemes and techniques compared to simple direct detection. The critical component of a practical coherent receiver is the phase lock loop (PLL) in the optical domain, the OPLL. The PLL has found widespread applications in the electrical domain. However, the OPLL has yet to reach maturity due to its difficult set-up and problems with stability. This section thus aims to improve the performance of the OPLL with the development of a novel field programmable gate array (FPGA) based low pass filter (LPF), a digital phase locking state loop.

The OPLL in coherent reception can be highly applicable in EPO and RPoF systems and offer the following advantages [37]: (i) high sensitivity due to mixing a high power LO and the signal or effective amplification, thus a possible 10–20 dB can be achieved over IM/DD reception; (ii) frequency selectivity by either homodyne or heterodyne mixing; (iii) tuneable optical receivers, in particular with DSP association and related algorithms; and (iv) the capability for many modulation formats (MFs) such as frequency shift keying (FSK), phase shift keying (PSK), minimum shift keying (MSK), etc.

Currently, in association with the digital processing technique [38], the incorporation of OPLL would offer ultra-high stability for the EPO and in turn for the RPoF. The OPLL is considered as an efficient method for decoding various MFs, such as multi-level PSK (M-ary PSK), etc, of the RF carriers or sub-carrier multiplexing RPoF systems, and remote locking of the LO laser to a reference carrier for maximizing the receiver sensitivity.

In EPO the OPLL can be inserted via the optical coupling of the output port of the optical modulator (RF-shifted optical carrier) and then with phase locking, as shown in figure 7.26. A blocking state diagram of the OPLL sub-system is shown in figure 7.27.

Note that the LPF in figure 7.27 uses a first-order analog active filter for the PLL and forms a second-order closed loop transfer function. Alternatively, in a DSP

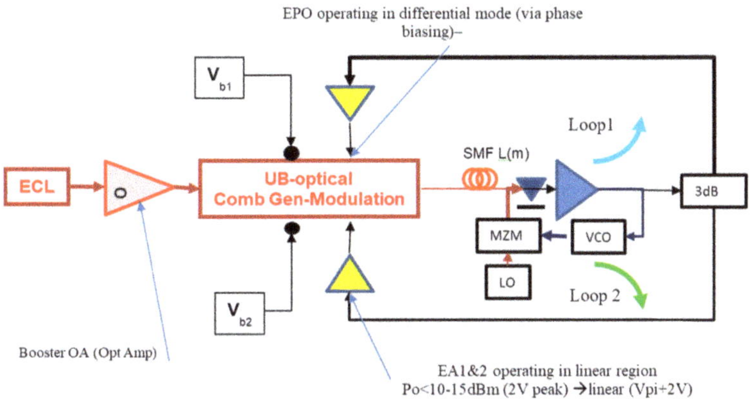

Figure 7.26. Generic structure of a dual-loop EPO incorporating coherent reception for an OPLL (VCO + MZM + LO). The optical path is indicated by the colored lines and the electrical path by the black lines. UB = ultra-wideband; LO = local oscillator; VCO = voltage control oscillator; MZM = Mach–Zehnder modulator; SMF = single mode fiber; ECL = external cavity laser; OA = optical amplifier.

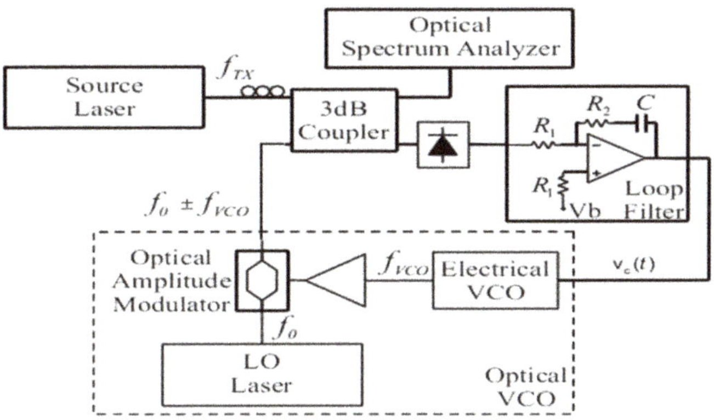

Figure 7.27. A sub-carrier OPLL (SC-OPLL), an analog feedback control technique.

based optical receiver such filtering can be implemented in the soft domain. This type of PLL is unconditionally stable to any step change in phase and reaches a steady state for any step change in frequency. However, with this simple filter design there exists a trade-off in the case where the bandwidth of the filter is to be as narrow as possible and have sharp roll-off to reduce the filtering noise contribution to the EPO RF oscillation at the output, and simultaneously be wide enough for the locking state on the initial error signal and 'pull in' to the locking state on the received signal [39].

By digitizing the LPF and combining a control system to bring the error signal within the pull-in region of the suitably narrow digital LPF the trade-off situation disappears. A further benefit of a digital implementation is the flexible and simple implementation of a monitoring system. A particularly efficient example of implementing a digital LPF onboard an FPGA was demonstrated by Linn [40].

The natural frequency of the PLL is generally relatively low, so the PLL itself is equivalent to an LPF.

This section thus focuses on the digital LPF which acts to condition the signal as obtained at the PD or PDP such that the LO can be adjusted to a matching or locking state in the frequency and phase of the incoming RF-shifted optical signal at the output of the MZIM or optical comb generator, shown in figure 7.26. Experimentally, an analog LPF of a simple first-order active filter loop, constructed using a single operational amplifier, is implemented for comparison with the performance of the FPGA-digital based phase locking. However, the LPF can be seen to be the 'active' or controlling component of the loop and thus this project aims to 'smarten' the LPF, thereby solving some of the observed problems of a traditional OPLL.

The proposed LPF can be implemented onboard an FPGA consisting of a first-order active digital filter which also performs other functions, such as conditioning the input signal, locking state monitoring, and sweeping for the incoming signal over the full range of the voltage controlled oscillator (VCO). The locking range is much greater than the bandwidth of the internal filter, thereby simplifying the structure.

7.9.1 Optical coherent detection in an RPOF with an OPLL

The principle behind coherent detection is remarkably simple, as described in chapter 4. A narrow linewidth laser is used as a local oscillator and the frequency and phase are tuned to equal those of the received signal. Any phase information is then available by simply applying a low pass filter.

Obviously, the complexity in the EPO coherent reception system is to tune or lock the local oscillator to the incoming laser and RF-shifted signal, and moreover maintain that locking state. To solve this problem the PLL has been developed, as shown in figure 7.28.

7.9.1.1 General PLL theory

The PLL shown in figure 7.29 consists of three main components forming an opto-electronic feedback loop, phase detector (PhD), and loop filter (LPF) as a function of the Laplace variable and the VCO.

Beginning with the PhD, two input mixing signals are compared and a signal is returned which is proportional to the phase error. This signal is then conditioned by an LPF before being fed into the VCO. The frequency (and thus phase) output of the VCO is proportional to the input voltage level. The input voltage applied to the VCO, in turn, reduces the phase error from the phase detector, thus stability is reached when the phase errors are gradually reduced until they reach the targeted RF oscillation.

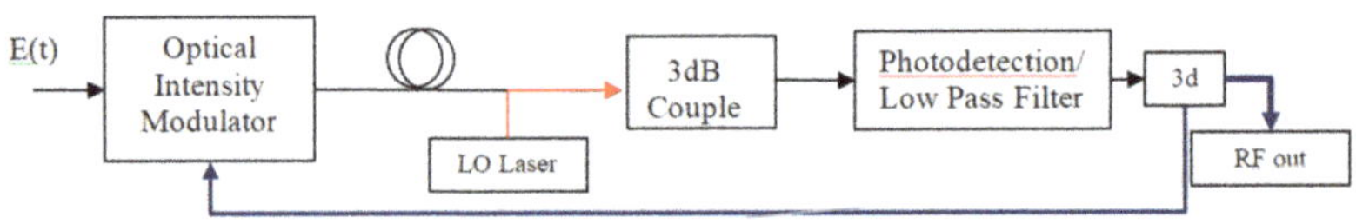

Figure 7.28. Typical arrangement of an EPO with coherent reception in the feedback path.

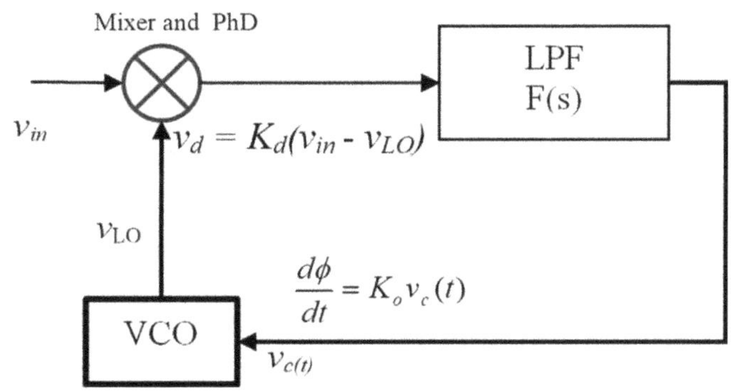

Figure 7.29. Generic PLL locking state diagram based on the analog technique. s is the Laplace variable and given as $s = j\omega$.

7.9.1.1.1 Phase detector (PhD)

The PhD is generally a signal multiplier or coupler, however, other types exist, such as the sequential phase detector used in some digital PLLs. In this chapter we are concerned with a multiplier type PD.

If we consider two signals $s_{\text{in}}(t)$ and $s_{\text{LO}}(t)$, assumed to be 90° apart, and multiply them together the output of the multiplier will be

$$s_3(t) = K_m A_{\text{in}} A_{\text{LO}} \sin[\omega_{\text{in}} t + \phi_{\text{in}}(t)]\cos[\omega_{\text{LO}} t + \phi_{\text{LO}}], \tag{7.19}$$

where K_m is a gain factor associated with the multiplier, which is typically less than 1 for a passive coupler. Also, by applying a common trigonometric, this output signal can be written as

$$\begin{aligned} s_3(t) = \frac{K_m A_{\text{in}} A_{\text{LO}}}{2} \Big[& \sin(\omega_{\text{in}} t + \phi_{\text{in}}(t) + \omega_{\text{LO}} t + \phi_{\text{LO}}(t)) \\ & + \sin(\omega_{\text{in}} t + \phi_{\text{in}}(t) - \omega_{\text{LO}} t - \phi_{\text{in}}(t)) \Big] \end{aligned} \tag{7.20}$$

$$\begin{aligned} s_3(t) = \frac{K_m A_{\text{in}} A_{\text{LO}}}{2} \Big[& \sin\left((\omega_{\text{in}} + \omega_{\text{LO}})t + (\phi_{\text{in}} + \phi_{\text{LO}}(t))\right) \\ & + \sin\left((\omega_{\text{in}} - \omega_{\text{LO}})t + (\phi_{\text{in}}(t) - \phi_{\text{LO}}(t))\right) \Big]. \end{aligned} \tag{7.21}$$

If ω_{LO} is set as equal or asymptotically close to ω_{in}, the above equation can be reduced to

$$s_3(t) = \frac{K_m A_{\text{in}} A_{\text{LO}}}{2} \Big[\sin\left(2\omega_{\text{in}} t + (\phi_{\text{in}}(t) + \phi_{\text{LO}}(t))\right) + \sin\left(\phi_{\text{in}}(t) - \phi_{\text{LO}}(t)\right) \Big]. \tag{7.22}$$

Note that the second sinusoidal term is only a function of the phase error and the first term contains a frequency of $2\omega_{\text{in}}$ that can be filtered off

$$s_3(t) = \frac{K_m A_{\text{in}} A_{\text{LO}}}{2} \Big[\sin\left(\phi_{\text{in1}}(t) - \phi_{\text{LO}}(t)\right) \Big]. \tag{7.23}$$

This is the phase error signal from the PhD and is received at the input of the LPF. When the phase error is zero, $\phi_1(t) - \phi_2$ is also zero. However, due to the system being configured as above, we recall that $\phi_1(t)$ is the phase of a sinusoidal function and ϕ_2 of a co-sinusoidal function, thus the actual phase difference between s_1 and s_2 is $\frac{\pi}{2}$. This can be chosen so as to match the behavior of the loop locking state in quadrature with the received signal.

7.9.1.1.2 Loop filter

In its most basic form the LPF is passive, effectively a low pass filter, but commonly an active filter is used. This type of filter not only incorporates low pass filtering but also features an integrator. This LFP is indeed an integrator and outputs a voltage signal level with an idealized magnitude response.

The transfer function of this filter for large open loop gain can be written as

$$F(s) = -\frac{s\tau_2 + 1}{s\tau_1}, \tag{7.24}$$

where s is the variable of the Laplace transform.

The magnitude response of this filter ($s = j\omega$) can be intuitively determined be considering that if ω is small (low frequencies) the magnitude grows without bound according to $\frac{1}{s\tau_1}$. If ω is large (high frequencies) the magnitude response is constant according to $\frac{s\tau 2}{s\tau 1} = \frac{\tau 2}{\tau 1}$. The negative sign in front of the filter introduces a phase reversal of $\frac{\pi}{2}$.

7.9.1.1.3 Voltage controlled oscillator

The VCO is determined by the applied output signal level of the LPF. The frequency generated at the output of the VCO is described by

$$\omega_{\mathrm{LO}} = \omega_c + K_o v_c(t), \tag{7.25}$$

where ω_c is the VCO central frequency, K_o is the VCO sensitivity factor in units of Hz V^{-1} and v_c is the controlling voltage output level of the LPF. Also note that the phase can be expressed as the integral of frequency over time, thus the output of VCO in terms of phase is

$$\varphi_{\mathrm{LO}}(t) = \int_{-\infty}^{\infty} \omega_c + K_o v_c(t)dt = \omega_c \int_{-\infty}^{\infty} dt + \int_{-\infty}^{\infty} K_o v_c(t)dt = \int_{-\infty}^{\infty} K_o v_c(t)dt. \tag{7.26}$$

Note we can discard the constant central frequency component ω_c due to the convention of phase wrapping around every 2π radians (or every complete cycle), thus it does not contribute to the output phase.

7.9.1.1.4 Second-order PLL

By rewriting the key loop equations, equation (7.21) gives the output of the phase detector which can be simplified by assuming the output is linear. This is true for small $\Delta\varphi$, that is, when the loop is closely locked:

$$s_3(t) = K_d[\phi_{\text{in}}(t) - \phi_{\text{LO}}(t)]; \quad \text{with} \quad K_d = \frac{K_m A_{\text{in}} A_{\text{LO}}}{2}, \tag{7.27}$$

and taking the Laplace transform of equation (7.27) leads to

$$S_3(s) = K_d[\phi_{\text{in}}(s) - \phi_{\text{LO}}(s)] \tag{7.28}$$

Furthermore, the output of the LPF is given by

$$V_c(s) = F(s)S_3(s), \tag{7.29}$$

where $F(s)$ is the LPF transfer function. We can also take the Laplace transform of equation (7.25) to obtain

$$\phi_{\text{LO}}(s) = \frac{K_o V_c(s)}{s}. \tag{7.30}$$

By substituting equation (7.21) into equation (7.22) and then substituting the result into equation (7.26) we can write the closed loop transfer function of a PLL as

$$\frac{\phi_{\text{LO}}(s)}{\phi_{\text{in}}(s)} = H(s) = \frac{K_o K_d F(s)}{s + K_o K_d F(s)}. \tag{7.31}$$

If we substitute the first-order LPF developed previously into equation (7.29) we can write

$$H(s) = \frac{\left(\dfrac{K_o K_d (s\tau_2 + 1)}{\tau_1}\right)}{s^2 + s\left(\dfrac{K_o K_d \tau_2}{\tau_1}\right) + \dfrac{K_o K_d}{\tau_1}} \tag{7.32}$$

$$\text{or } H(s) = \frac{2\xi\omega_n s + \omega_n^2}{s^2 + 2\xi\omega_n s + \omega_n^2}, \text{ where } \omega_n = \sqrt{\frac{K_o K_d}{\tau_1}} \text{ and } \xi = \frac{\tau_2 \omega_n}{2}. \tag{7.33}$$

Furthermore, we can write the phase error response

$$\frac{\phi_e(s)}{\phi_{\text{in}}} = \frac{\phi_{\text{LO}} - \phi_{\text{in}}}{\phi_{\text{in}}} = 1 - H(s) = \frac{s^2}{s^2 + 2\xi\omega_n s + \omega_n^2}. \tag{7.34}$$

Equation (7.31) is in second-order canonical form and the parameters ω_n and ξ are known as the natural frequency and damping factor and are important parameters in characterizing the loop. Using the final value theorem to determine the steady state phase error after a step change in the input $\Delta\varphi_{\text{in}}$ is

$$\lim_{t\to\infty} \phi_e(t) = \lim_{s\to 0} \frac{s\Delta\phi_{\text{in}}}{s + K_o K_d F(s)} = 0. \tag{7.35}$$

That is, the tracking of the second-order loop can be out any step'in-phase'state applied to the input, this is an important result and demonstrates why the second-order loop is so important.

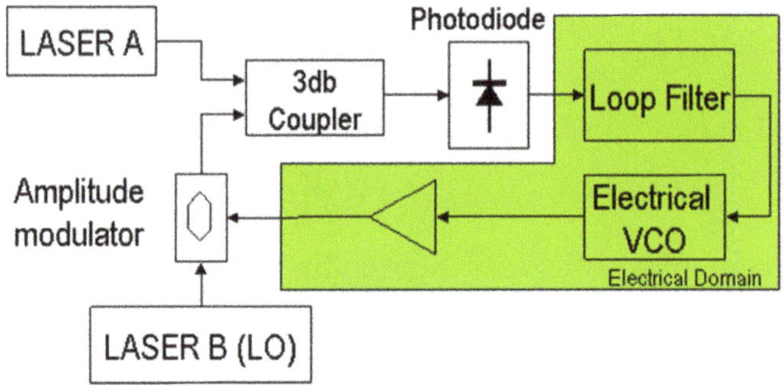

Figure 7.30. An OPLL in an RPoF block diagram with an RF modulated carrier and an RF tunable (modulating MZM) local oscillator and integrated feedback control loop.

For a ramp in the phase (a step change in input frequency $\Delta\omega_{\text{in}}$) the final value theorem results in a steady state error of

$$\lim_{t\to\infty}\phi_e(t) = \lim_{s\to 0}\frac{\Delta\omega_{\text{in}}}{K_o K_d F(s)} = \frac{\Delta\omega_{\text{in}}}{K_o K_d F(0)}. \tag{7.36}$$

This is also an important result as it provides the constant dc voltage need to provide an offset from the VCO's central frequency. $F(0)$ is the LPF's dc gain.

Plotting the magnitude responses given by equation (7.33) and equation (7.34) over a normalized frequency range $\omega/\omega_{\text{in}}$ furthers our understanding of the second-order loop. Figure 7.30 shows an OPLL for an RPoF block diagram with the RF modulated carrier (laser beam A), an RF tunable (modulating MZM) LO, and an integrated feedback control loop (IFBCL). The PLL shows a low pass filtering effect on phase inputs in which a loop high pass filter provides the phase error. In other words, the loop can track low frequency errors but high frequency errors pass through unaffected. This principle is used to allow data to be encoded in the incoming signal. The PLL will in effect 'pass through' this high frequency data while tracking any low frequency drift in the local oscillator.

7.9.1.2 OPLL

The OPLL can also be designed based on the described dynamic principles. Stability limitations in today's tuneable lasers are no longer a hurdle, for example the Pure Photonics PPLC-550 can be tuned via a software interface without much difficulty [41] with preservation of a narrow linewidth. Alternatively, an external modulator can be used, as shown in figure 7.26. This type of loop is known as sub-carrier modulated locking, where the loop actually locks onto a modulation sideband rather than the LO emitted frequency, as shown in figure 7.30.

The 3 dB optical coupler acts as the multiplier PD. The optical intensity modulator (MZIM) can be biased at the minimum transmission point (min-point) of the MZIM transfer characteristics so as to suppress the carrier and generate two sidebands whose frequencies can be related to the output of the VCO. With a stable VCO, the tuning properties and thus the sidebands can follow these tracking

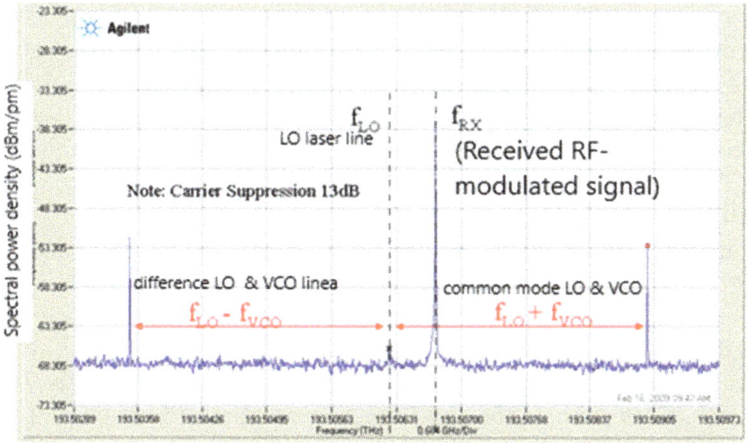

Figure 7.31. Input to the photodiode as seen by a high resolution spectrometer.

states and provide a stable RF oscillation frequency that, in turn, is used to lock onto the received signal. Figure 7.31 shows the spectral components of the VCO under a state of locking.

Thus, the frequency locking condition is

$$f_{\mathrm{LO}} \pm f_{\mathrm{VCO}} = f_{\mathrm{RX}}. \tag{7.37}$$

An important issue unique to the OPLL is the LO linewidth. According to Kazovsky [42], a typical second-order PLL requires a factor of at least 0.8 pW K^{-1} Hz^{-1}, the LO power per laser linewidth. In addition, the loop bandwidth should be a few magnitude orders higher than the linewidth of the received signal. Other effects such as the laser RIN can influence the OPLL but are not considered here.

7.9.2 Digital loop bandpass filter (DLBPF)

In this and subsequent sections the term DLBPF is used to refer to the entire 'smart' filter system (SFS), while the term 'digital filter' refers to the digital filter which is incorporated in the LPF. The digital filter can be very stable and is integrable in a common micro-controller. The functional locking states of the SFS can be implemented in an FPGA system which can consist of a sweeping unit, a proportional-integral-derivative (PID) controller and a digital filter. In a blocking state diagram three distinct data pathways can be formulated, that is the sweep unit, the PID controller, and the digital filter. Only one path is chosen to go to the output via the 3:1 multiplexer (MUX). A description of each pathway is described by referencing the start-up procedure. Initially the loop is not locked and the filter has no indication of where the received may be, hence the sweep unit is activated and ramps up and down until a beat signal is detected. Upon the frequency counter recognizing a beat signal the sweep unit will be disengaged and the PID controller will coarsely tune the signal using the frequency counter output as an error signal. This brings the signal into the locking state in the range of the digital filter, and once this occurs the output is switched to the output of the digital filter. The frequency counter is continually monitoring the locking state and signal in case action needs to be taken to correct the locking state. The control unit controls the MUX and provides diagnostic information to the outside world, and its inputs are the frequency counter and external switches which can be used for manual control.

7.9.2.1 OPLL hardware details

The hardware set-up used to create the OPLL consists of the components shown in table 7.2.

7.9.2.2 Optical power budget

The optical power at the output of the coupler is approximately −5 dBm, or effectively is 3 dB less than the input power, as shown in table 7.3. Ideally the electrical power available out of the OE converter should be larger; this could be achieved by using a higher LO power or an OE converter with a higher gain.

7.9.2.3 Simulation

A MATLAB Simulink model was constructed of the 'smart' filter and was tested in a regular PLL without simulating the optical system. The PID pathway was not

Table 7.2. Experimental OPLL hardware details.

Component	Brand/model	Remarks
Input laser	Radians Innova Intun 1500 or Pure Photonics PPCL 550	Tunable ECL
Local oscillator laser	Photonetics Tunics 1550 or Photonic PPCL 550	Tunable ECL
High resolution spectrometer (HRS)	Agilent 83 453 HRS	Optical HRS
Optical/electric (OE) converter	Tektronix SA-64	Photodiode
OE pre-amplifier	Mini-Circuits ZFL-500LN+	0.1 to 500 MHz, min. gain 24 dB
OE amplifier	Mini-Circuits ZX60–14012L+	300 kHz to 14 GHz (wideband), min. gain 9 dB
Low pass filter	Mini-Circuits PLP-5+	5 MHz anti-alias filter
FPGA	Altera DE1 board	Can use a smaller dedicated chip
Voltage controlled oscillator (VCO)	Mini-Circuits ZX95–3080-S + wider bandwidth SHF-807 or 824 (65 GHz)	Range: 2920 to 3080 MHz (~200 MHz), tuning sensitivity average 90 MHz V^{-1}, dc-30 GHz
VCO amplifier	Mini-Circuits ZVA-183+	700 MHz to 18 GHz (wideband), min. gain 24 dB
Amplitude modulator	Pirelli PIRMZM 153X or Fujitsu QPSK FTM-7980	Bias at 3.1 V

Table 7.3. Optical power budget.

Component	Power	Port
Input laser	−1 dBm	At the output
Local oscillator laser	5 dBm	At the output
Amplitude isolator–polarization controller	−6.2 dBm	At the output
Coupler	−4.9 dBm	At the output

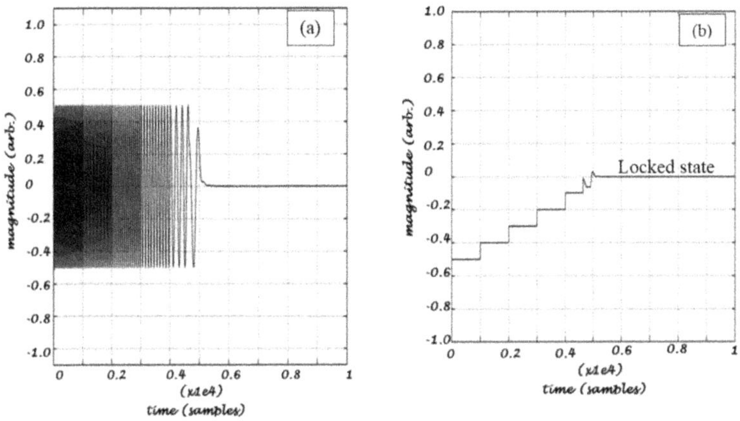

Figure 7.32. (a) LPF input (10 μs). (b) VCO drive voltage or LPF output (10 μs). The horizontal axis is the number of samples and is equivalent in time. Copyright (2014), reprinted from [43] by permission of Taylor & Francis Ltd.

included in the final model. The model considered quantization error and the limited bandwidth available to the interface board. However, several aspects such as fixed-point arithmetic overflow and electrical noise sources are not considered in this work.

The simulation was performed with an initial frequency offset of 50 MHz and analog LPF parameters set as $\tau_1 = 10\ \mu$s and $\tau_2 = 50\ \mu$s which describe the following digital filter (a sampling rate of 10 MHz was used):

$$H(z) = 5.5\frac{1-0.8182z^{-1}}{1 - z^{-1}}. \tag{7.38}$$

In the first LPF of the plot the smart filter is sweeping the local oscillator searching for the received signal. Incrementally it brings the beating 'error' signal closer, which results in the input frequency being reduced, as can be seen in figure 7.32. During this time the frequency counter is constantly monitoring the input frequency. When this is below a set threshold (10 MHz in this case) the control unit enables the digital filter and halts the sweeping unit. The action of the digital filter can be seen in the right-hand LPF of the plots. Figure 7.32(a) shows that the filter continues to reduce the frequency of the input signal until it reaches a null level, and at this point the loop can be said to be in a locked state.

Figure 7.32(b) shows the voltage applied to the VCO. Initially it consists of regular steps generated by the sweep unit, however, once the LPF is enabled the signal quickly settles to the required voltage of 0.5 V that produces $\Delta f = 50$ MHz for a VCO sensitivity of 100 MHz V^{-1}. Figure 7.33(a) shows the input from the ADC and shows just how little information is needed to complete the locking state in the almost ideal conditions offered by the simulation environment. Figure 7.33(b) shows the output of the digital filter, particularly the integral action settling at a steady state offset voltage.

7.9.2.4 OPLL platform

The experimental set-up can be seen in figure 7.34(a). The input signal is mixed with an externally modulated local oscillator laser via a fiber coupler with a polarization controller. The beating signals in the photodetection diode are then amplified to the right level, digitally converted, then low pass filtered in a FPGA. The output level is then fed to a VCO whose output then acts as a sinusoidal signal to be modulated, thus shifting the optical carrier of the local oscillator towards the frequency of the

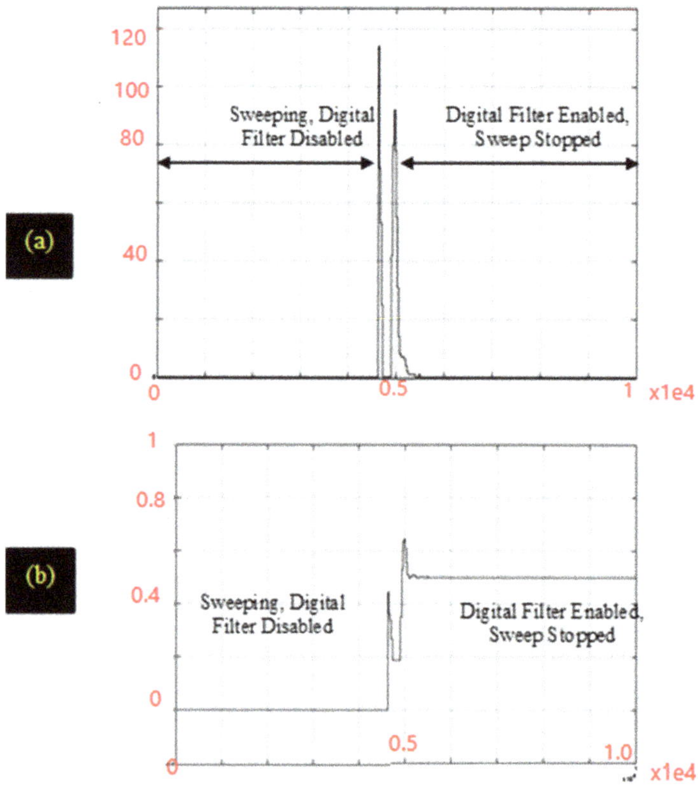

Figure 7.33. (a) Output from the ADC (10 μs full scale). (b) Output from the digital filter (10 μs full scale). Copyright (2014), reprinted from [43] by permission of Taylor & Francis Ltd.

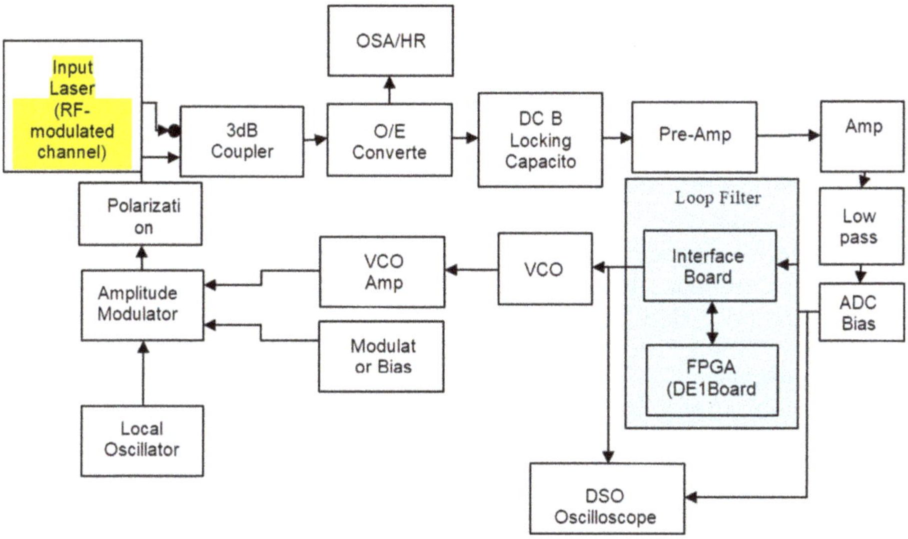

Figure 7.34. OPLL platform schematics. OSA = optical spectrum analyzer; HRS = high resolution spectrum analyzer; ADC = analog-to-digital converter.

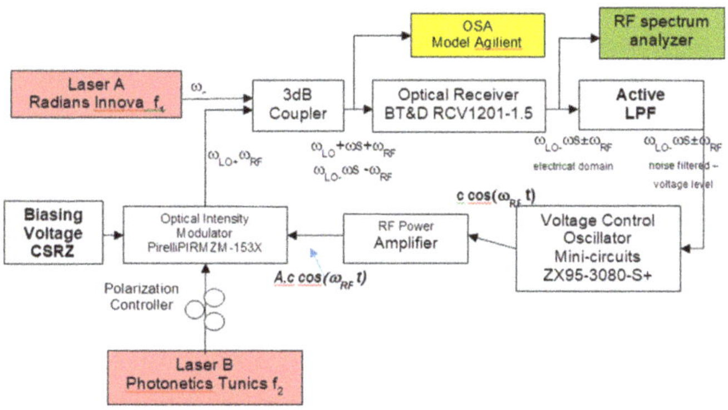

Figure 7.35. OPLL detailed platform with lasers and photonic components.

optical carrier of the optical signals. The FPGA allows the tuning and change of the strategy for tuning and locking the LO to the carrier. Figure 7.34(b) shows the prototype set-up in the laboratory of the OPLL. Figure 7.35 shows a schematic of the set-up, the evolution of the frequency of the LO, and a reference laser for testing the functionalities of the OPLL before feeding the real channel. The coherent detection and frequency shifting with the digital feedback loop allows the locking and then digital control to generate the locked LO, which would be fed to the coherent receiver $\pi/2$ hybrid coupler or to a comb generator whose outputs would be mixed with the super-channel and then de-multiplexed to be detected coherently by a parallel balanced receiver and DSP interconnecting network for the optical receiving structure. This topic is described briefly in the next section.

7.9.2.4.1 *Noise sources*

Various noise sources can affect the performance of the PLL and the primary sources are observed where electrical noise arises from the amplifiers after the OE converter. Furthermore, noise was observed to be superimposed on the interface board from the FPGA when programmed and switching at 50 MHz. This could be solved by ensuring that the interface board is electrically isolated from the FPGA using optical isolators.

Noise power is a major difficulty for any type of system, in particular in a system incorporating a feedback loop as the loop gain may be continually amplifying the noise that would overflow the filter responses.

7.9.2.4.2 *Limitations*

The greatest limitation observed with locking state quality maintained over time is the range of the VCO. The VCO employed was designed for an electrical PLL and had an approximate deviation from the central frequency of about 200 MHz. It was noticed that over time, in particularly when the temperature changes rapidly (e.g. dusk) the lasers become separated by a greater frequency deviation than the VCO range and thus the locking state is irrecoverably lost.

It was also observed that occasional high frequency jumps occurred which the filter was not able to counter due to its limited bandwidth. This is consistent with the operation of a PLL as discussed in section 7.2.2, that is high frequency errors are passed through, however, since no data was being transmitted these deviations are undesirable. A solution could be to upgrade the ADC sampling rate to allow a larger filter bandwidth, however, this increases the complexity of the interface board.

Finally, greater optical power would allow more aggressive filtering and require less amplification, hence the OPLL noise reduction. This could be achieved with an LO power of about 16 dBm.

7.9.2.4.3 *Simulation: analog feedback control loop*

Two lasers are mixed via the optical fiber coupler (OC) and then detected by an optical receiver with a sensitivity of 0.6 A W^{-1}. The receiver output is then filtered by an active LPF of 500 kHz corner frequency. A 20 dB gain dc amplifier is used to boost the signal level. The filtered electrical signal then drives the VCO. The signal output of the VCO of 3 dBm is further amplified by an RF power amplifier to drive the MZIM.

The feedback loop of the optical PLL is formed so as to produce any different quantity in frequency and phase between laser A and laser B (see figure 7.36) which can be stably tuned to the loop and corrected by the VCO. The components of the OPLL platform are listed in table 7.4. The component parameters such as output power, sensitivity, and gain/loss are also given in table 7.4.

The LO is modulated via an MZIM [44]. The modulated sideband is produced at the VCO output [45]. To match the frequency and phase errors between the LO and the received optical carrier signal, both the optical signals and the LO laser are mixed in an optical coupler. The coupler output total fields are then received in the optical receiver to produce a mixing product electrical signal [46]. This product output is then filtered by an LPF, then fed into the VCO.

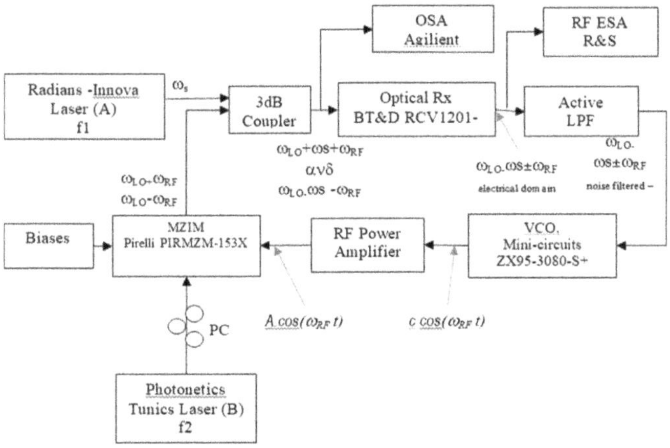

Figure 7.36. Schematic of the OPLL experimental test-bed. PC = polarization controller; VCO = voltage control oscillator; LPF = low pass filter; MZIM = Mach–Zehnder interferometric modulator.

Table 7.4. Technical parameters of the OPLL components.

Devices	Power output (dBm)	Gain/loss	Sensitivity	Insertion loss
Laser A lightwave carrier	−1.5		Narrow linewidth (<100 kHz)	
Laser B local oscillator	16 dBm		Linewidth < 1 kHz	
Optical 3 dB coupler (50:50)		$0.35\cos(\omega_{LO} - \omega_s - \omega_{RF})t$		3 dB
Optical modulator				6 dB
Polarization controller				0 dB
Electrical power amplifier SHJF-807	16 dBm	21 dB		
Optical receiver			0.67 A W^{-1}—responsivity	
Active LPF			3 dB corner frequency at 500 kHz, gain 20 dB	
Voltage control oscillator			3 dBm at 0–5 V 100 MHz V^{-1}	

The OPLL locking procedure can be as follows. (i) Two laser sources A and B are selected and arranged for mixing in a 3 dB fiber coupler. The beating signal produced in the ORx is analysed in an RF spectrum analyser. (iii) A RF function generator is manually controlled to near locking states of the LO frequency and that of the RF-received signal. (iv) Closing the OPLL loop with VCO and LPF circuits.

7.9.2.4.4 Close-loop OPLL

A high resolution OSA is connected to one port of the 3 dB optical coupler, as shown in figure 7.36. The carrier signal is tuned such that the frequency difference between the carrier signal and one sideband of the LO is less than 400 MHz. Figure 7.37 shows the spectra of the LO and carrier signal. The LO carrier is suppressed about 13 dB compared to its sideband. Figure 7.38 shows the spectra of the carrier signal and the LO under high resolution operation. Their frequency difference is lower than 120 MHz. Then, the OPLL is closed and the LO sideband is locked to the carrier signal (figure 7.37). Under lab conditions, the OPLL is able to maintain the lock in less than 30 min.

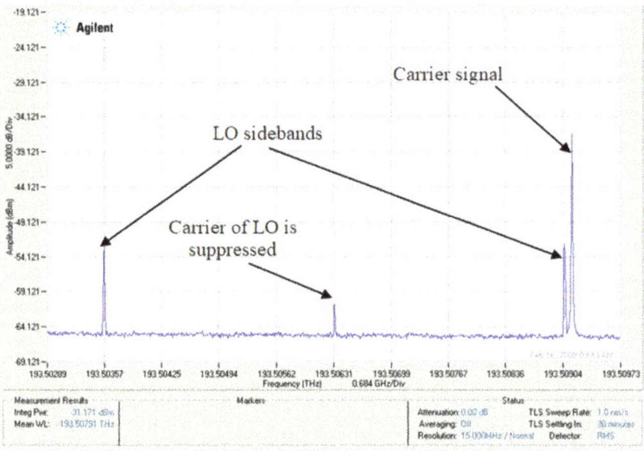

Figure 7.37. Spectrum components of the LO and the carrier signal. Observed on a high resolution optical spectrum analyzer (Agilent). Wide mode vertical axis 5 dB div^{-1}. Horizontal axis 684 MHz div^{-1} [43].

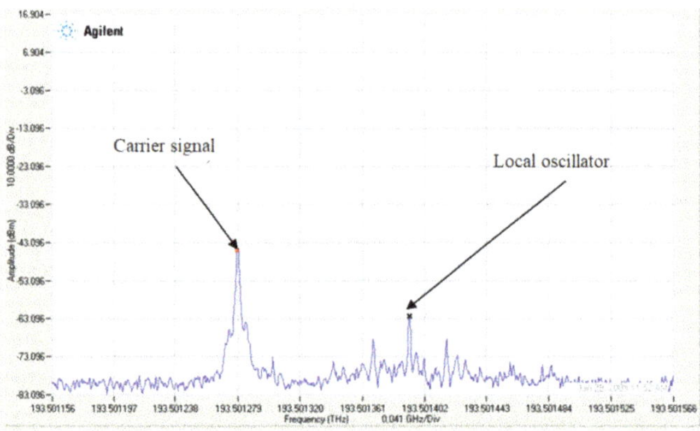

Figure 7.38. The frequency of the carrier signal and LO before being locked. The frequency error is 113.2 MHz. Observed on a high resolution OSA (Agilent type). Vertical axis 10 dB div^{-1}. Horizontal axis 41 MHz div^{-1} [42].

7.9.2.4.5 Phase error and LPF time constant

The time constants of the LPF can be denoted as R_1C and R_2C which can be termed as τ_1 and τ_2, respectively. The ratio of τ_1 and τ_2 changes the cut-off frequency of the active low pass filter. Therefore, R_1 and R_2 can be varied by either manually or digitally controlled tuning. The phase error can be estimated as

$$\sigma_{\mathrm{PE}} = \frac{V_{\mathrm{RMS}}}{K_d}, \tag{7.39}$$

where V_{RMS} is the RMS voltage measured at the electrical output port of the optical receiver, and σ_{PE} is the standard deviation of the phase error.

Locking of the LO and the signal carrier can be tracked and converged by the LPF-VCO and MZIM optical modulator loop. The time constant of the LPF is critical and further work is required to extend the locking range of the difference between the two laser sources.

The demonstrated locking loop for the OPLL is influenced and applicable for locking and stabilizing the EPO.

Some remaining issues of OPLL to be resolved are: (i) the rise time and fall time as well as the roll-off of the loop filter and effects on the phase locking; (ii) the phase noise of the local oscillator and its impact on the phase locking; and (iii) the quantum shot noise of the receiver due to the high level power of the local oscillator—a noise analysis of the receiver should be performed.

7.9.3 OPLL in an EPO receiver

EPO RF carriers are composed of a number of sub-carriers. Naturally, the number of sub-channels must be greater than 1 as at this rate the electronic and digital circuits would not be operating due to limited electronic speed (figures 7.39 and 7.40).

Figure 7.41 shows the parallel processing of the detected sub-channels in the analog domain which are then sampled by the bank of ADCs to obtain a digitalized sampled sequence of the sub-channels. Interconnecting DSP processors then process the digital sampled sequences of the sub-channels.

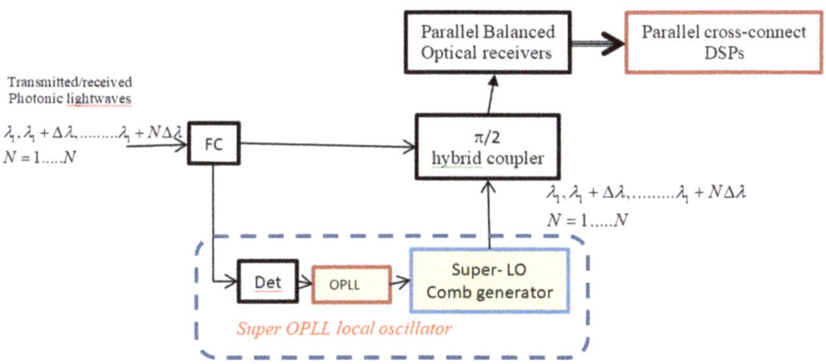

Figure 7.39. Schematic of the generation of a comb source to act as the LO of the super-channel.

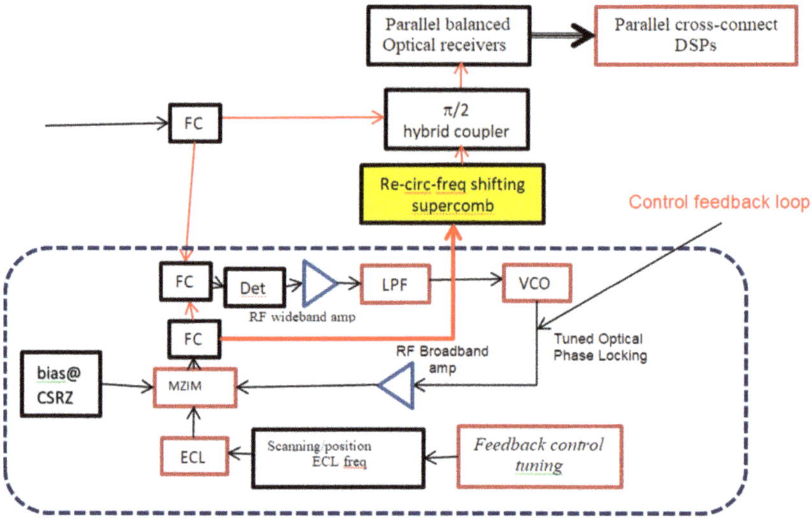

Figure 7.40. Super LO using an OPLL and comb generation. FC = fiber coupler; LPF = low pass filter; PD = photodetector; VCO = voltage controlled oscillator; MZIM = Mach–Zehnder interferometric modulator; ECL = external cavity laser.

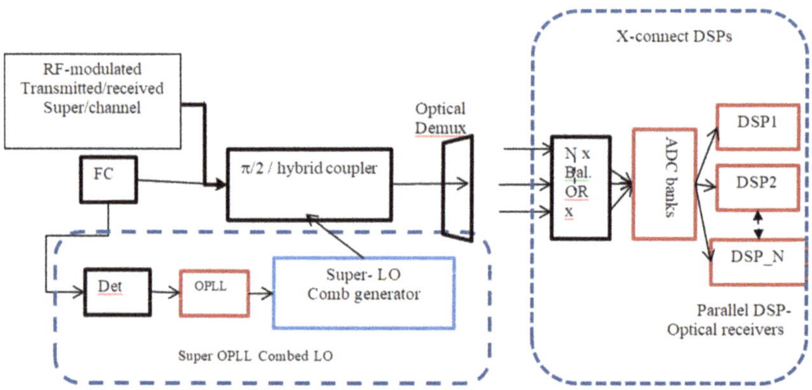

Figure 7.41. A super LO and DSP based super-channel coherent receiver system.

If optical phase locking techniques for EPO generation of the oscillation RF signals allow the signals to be generated and matched to the local oscillator of the receiver, then this would reduce the complexity of the receiving system significantly for any desired RF oscillator for RPoF systems. Frequency shifting can be a recirculating optical loop in which the lightwave is frequency shifted once every time they pass through the frequency shifter, hence multiple comb lines, or a nonlinear-conversion optical device though which several harmonics of the fundamental RF component can be generated. The frequency spacing in the comb lines can thus be adjusted to match the spacing of the sub-channels of the super-channel. The main issue is the matching of a reference frequency between the received channels and the

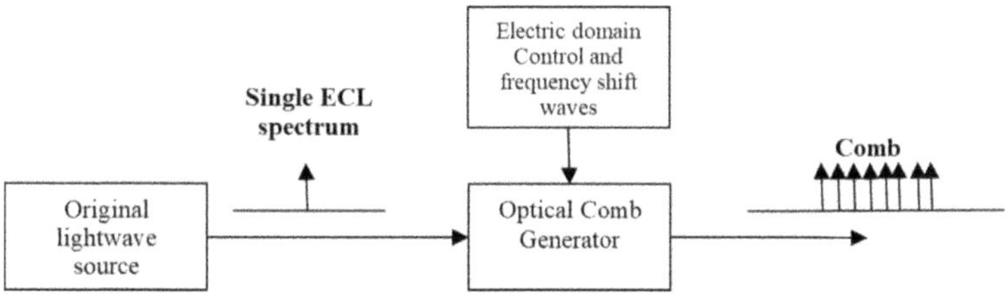

Figure 7.42. Schematic block diagram of a generic comb generator.

first referenced comb line. These can be used in multi-carrier RPoF systems, as described in chapter 4 and chapter 5 (figure 7.42).

7.9.4 Remarks

Simulated results are presented supporting the operation of the 'smart' digital LPF leading to the physical implementation of such a filter. An experimental optical phase lock loop is demonstrated for further exploration of the LPF and to confirm its 'real-life' viability. Various design considerations pertaining to the hardware implementation of the LPF have been discussed, leading to the development of an interface board between the 'smart' filter implemented onboard an FPGA and the OPLL. This section demonstrates a case study for the generation of local oscillators for intradyne reception systems, in particular the super-channel coherent receiver.

Experimental results have been presented displaying the correct operation of the LPF and ultimately it can be concluded that a successful proof of concept is demonstrated with the modified OPLL maintained in a locking state for at least one hour with limited spurious beats in the locking state. Further work remains to incorporate the OPLL into a coherent receiver and perform data transmission experiments. Ultimately an OPLL is useful in the control of the stability of the RF output oscillation of an EPO.

References

[1] Analog devices Inc.

[2] Stein W, Huber F, Bildik S, Aigle M and Vossiek M 2017 An improved ultra-low-phase noise tunable YIG oscillator operating in the 6–12 GHz range *47th European Microwave Conf. (EuMC)* pp 767–70

[3] VHF Communications, A simple approach to YIG oscillators web page, http://www.vhfcomm.co.uk/pdf/A%20Simple%20Approach%20tyo%20YIG%20Oscil.pdf

[4] Lucas E, Brochard P, Bouchand R, Schilt S, Südmeyer T and Kippenberg T J 2020 Ultralow-noise photonic microwave synthesis using a soliton microcomb-based transfer oscillator *Nat. Commun.* **11** 374

[5] Binh L N 2018 *Photonic Signal Processing, Second Edition: Techniques and Applications Optical Science and Engineering* vol 1 2nd edn (Boca Raton, FL: CRC Press)

[6] Binh L N 2017 *Advanced Digital Optical Communications Technology and Engineering* (Boca Raton, FL: CRC Press, Taylor and Francis)

[7] Redmayne D, Trelewicz E and Smith A 2006 Understanding the effect of clock jitter on high speed ADCs *Design Note* 1013 Linear Technology, Milpitas, CA

[8] Kester W 2004 *Analog–Digital Conversion* (Norwood, MA: Analog Devices)

[9] Binh L N 2009 *Digital Optical Communications* (Boca Raton, FL: CRC Press, Taylor and Francis)

[10] Desoer C A and Vidyasagar M 1975 *Feedback Systems: Input-Output Properties* (New York: Academic)

[11] Eliyahu D and Maleki L 2003 Low phase noise and spurious level in multi-loop opto-electronic oscillators *Proc. 2003 IEEE Int. Frequency Control Symp. and PDA Exhibition, Jointly with the 17th European Frequency and Time Forum (Tampa, FL)*

[12] Leonberger F J 1982 Guided-wave electrooptic modulators *Fiber-Optic Rotation Sensors and Related Technologies* ed S Ezekiel and H J Arditty (Springer Series in Optical Sciences, vol 32) (Berlin: Springer)

[13] Fan F, Hu J, Zhu W, Gu Y, Han X and Zhao M 2017 Dual-loop optoelectronic oscillator based on a compact balanced detection scheme *Opt. Eng.* **56** 026107

[14] Yao X S and Maleki L 1994 High frequency optical subcarrier generator *Electron. Lett.* **30** 1525–6

[15] Zou X and Liu X 2016 Opto-electronic oscillators (EPOs) to sensing, measurement, and detection *IEEE J. Quantum Electron.* **52** 1–16

[16] Yao X S and Maleki L 1996 Optoelectronic microwave oscillator *J. Opt. Soc. Am.* B **13** 1725–35

[17] Yao X S and Maleki L 2000 Multiloop optoelectronic oscillator *IEEE J. Quantum Electron.* **36** 79–84

[18] Yang J *et al* 2007 An optical domain combined dual-loop optoelectronic oscillator *IEEE Photonics Technol. Lett.* **19** 807–9

[19] Zhu D and Pan S 2012 Tunable frequency-quadrupling dual-loop optoelectronic oscillator *IEEE Photonics Technol. Lett.* **24** 194–6

[20] Jia S *et al* 2015 A novel optoelectronic oscillator based on wavelength multiplexing *IEEE Photonics Technol. Lett.* **27** 213–16

[21] Devgan P S *et al* 2012 Cancellation of photodiode-induced second harmonic distortion using single side band modulation from a dual parallel Mach–Zehnder *Opt. Express* **20** 27163–73

[22] Hu J *et al* 2016 Microwave photonic link with improved phase noise using a balanced detection scheme *Opt. Commun.* **370** 1–5

[23] Zang J *et al* 2013 Dual-loop optoelectronic oscillator for generation of stable and ultralow timing-jitter electrical and optical clock *Proc. Opto Electronics and Communications Conf. and Photonics in Switching* pp 1–2

[24] Jiang C *et al* 2015 Improvement of phase noise in a widely tunable optoelectronic oscillator based on balanced detection *Proc. Optoelectronics and Communications Conf.* pp 1–3

[25] Keysight Technologies X-Series Signal Analyzers

[26] Nichols L T *et al* 1997 Optimizing the ultrawide-band photonic link *IEEE Trans. Microwave Theory Tech.* **45** 1384–9

[27] Kim J-Y *et al* 2012 Dual-loop dual-modulation optoelectronic oscillators with highly suppressed spurious tones *IEEE Photonics Technol. Lett.* **24** 706–8

[28] Brès C-S, Zlatanovic S, Wiberg A O J and Radic S 2011 Reconfigurable parametric channelized receiver for instantaneous spectral analysis *Opt. Express* **19** 3531–41

[29] Zhu D and Pan S 2012 High-performance photonic microwave downconverter based on a frequency-doubling optoelectronic oscillator *IEEE J. Lightw. Technol.* **30** 3036–42

[30] Versawave polarization modulator type

[31] Binh L N 2009 Acousto-optic tunable wavelength and selective packet filter by simultaneously launching of acousto-optic waves of different frequency for TE–TM mode conversion in diffused optical and acoustic waveguides *Opt. Eng.* **48** 054603
[32] Heismann F and Aferness R C 1988 Wavelength-tunable electrooptic polarization conversion in birefringent waveguides *IEEE J. Quantum Electron.* **24** 83–93
[33] Binh L N, Livingstone J and Steven D 1980 A wide-band acousto-optic TE–TM mode converter using a doubly confined structure *IEEE J. Quantum Electron.* **16** 964–71
[34] Mu Q *et al* 2019 Tunable magneto-optical polarization device for terahertz waves based on InSb and its plasmonic structure *Photon. Res.* **7** 325–31
[35] Lam H Q, Shum P, Lam H Q, Shum P, Binh L and Gong Y 2006 Polarization-dependent locking in SOA harmonic mode-locked fiber laser *IEEE Photonics Technol. Lett.* **18** 2404–6
[36] Binh L N and Ngo N Q 2011 *Ultra-fast Fiber Lasers: Principles and Applications with MATLAB® Models* 1st edn (Boca Raton, FL: CRC Press)
[37] Kazovsky L 1986 Balanced phase-locking loops for optical homodyne receivers: performance analysis, design considerations, and laser linewidth requirements IEEE J. Lightw. Technol. **4** 182–95
[38] Binh L N 2017 *Digital Processing: Optical Transmission and Coherent Receiving Techniques* (Boca Raton, FL: CRC Press)
[39] Gardner F M 1979 *Phase Locking State Techniques* 2nd edn (New York: Wiley-Interscience)
[40] Linn Y 2007 Efficient loop filter design in FPGAs for phase locking state loops in high-datarate wireless receivers—theory and case study Wireless Telecommunications Symp. pp 1–8
[41] Tunable Laser Products, Pure-Photonics Inc. https://pure-photonics.com/
[42] Kazovsky L 1985 Decision-driven phase-locked loop for optical homodyne receivers: performance analysis and laser linewidth requirements *IEEE J. Lightw. Technol.* **3** 1238–47
[43] Binh L N 2014 *Digital Signal Processing: Optical Transmission and Coherent Reception Techniques* (Boca Raton, FL: CRC Press, Taylor and Francis)
[44] Pirelli Inc. PIRMZM-153X http://pirellilabs.com/web/default.page
[45] Mini-Circuits Inc. ZX95–3080-S+
[46] Optical receiver Active BT&D model no. RCV1201-1.5, BT&D Technology Inc. UK

Chapter 8

SMF and RF transfer function

This chapter, an appendix on optical fibers, gives an account of the analytical expressions of the fiber RF transmittance transfer function under both linear dispersive and nonlinear effect single mode fibers (SMFs) for broadband operation. It is applicable in the scenarios of several multiplexed radio-photonics on fiber (RFoF) systems and optical wavelength carriers. Near future 5G–7G networks will witness the extensive deployment of RPoF to dense pico-cell structures so as to distribute and monitor millimeterwave (mmW) channels for transport over SMF transmission lines.

Under the linear operation of photonic or electronic devices, the operating signal level must fall into the region in which the variations of other related parameters follow a linear relationship, the small signal conditions.

Under nonlinear (NL) operating conditions, both chromatic dispersion (CD) and self-phase modulation (SPM) effects due to nonlinear phase variations can be considered valid for optical double-sideband (DSB) modulation, optical single sideband (SSB) modulation, and the chirp phenomena of optical lightwaves from the transmitters. At the reception end, coherent reception gives a linear product between the signals and that of a local oscillator, while direct detection follows a square-law process and hence nonlinearity. The measurement of the transfer function of the fiber can be expressed in terms of the function of the detected amplitude over the amplitude of the excitation waves into the fiber as a function of the RF carrier transported through the fiber length.

8.1 Introductory remarks

Planar optical waveguides and circular dielectric fibers are composed of a guiding region, a slab or circular core, embedded between a substrate and a superstrate or cladding in a circular casing, having identical or different refractive indices. Lightwaves are guided by the confinement of the fields of the transverse electric (TE) or transverse magnetic (TM) components. Unlike the planar waveguiding structure, in the circulator fiber these two modes are randomly aligned unless it is a polarization maintaining type.

doi:10.1088/978-0-7503-3511-9ch8

Lightwaves with oscillation solutions and penetration into the fiber cladding region are evanescent waves. The number of oscillating solutions which satisfy the boundary conditions is the number of modes which can be guided. The guiding of lightwaves in an optical fiber is similar to that of the planar waveguide except that the lightwaves are guided through a circular core embedded in a cladding region. Fibers are commonly fabricated and manufactured using molecular organic chemical vapor deposition (MOCVD) techniques or a flow of chemical deposition on the wall of a pure silica tube that is then collapsed, and then drawn into fiber lines under a high-power microwave heater.

This chapter treats optical fibers as a perfect circular optical waveguide that can support a single mode with two polarized modes or a few modes under different polarizations. Although it is termed as a single mode fiber, it supports two polarized modes which can rotate randomly, thus there is polarization mode dispersion since there is random birefringence of the fiber core under different stresses along the long length of the fiber. This dispersion or the polarized mode delay difference between the polarized modes is considered critical for high speed or dense multiplexing systems.

The step-index (SI) and graded index multimode (GIMM) optical fibers find very limited applications in transmission systems and networks for long-haul applications. These types of fibers were developed extensively in the early 1970s. These fibers offered only a limited length bandwidth product of about 900 MHz km^{-1} due to the interferences of different modes which are estimated at around a few thousand for an SI with a 0.1 index difference and a profile core of 50 μm. Thus, only 1.0 km of this fiber supports only up to 900 MHz signal bandwidth, hence it is impossible for such a fiber to support ultra-high-speed channels.

Single mode optical fibers can be designed and realized with a very small difference in the refractive indices between the core and cladding regions (about 0.03%) and a core diameter of about 8 μm. Thus, such guiding phenomena are called 'weakly' guiding, indeed there is only 70% of mode energy concentrated in its core. The common single mode optical fiber (SMF) is the standard SMF or denoted as SSMF. This development was intensively debated and agreed by the optical fiber communications technology community during the late 1970s. The SMF leads to the standard SMF (SSMF), or SMF-28 coded by Corning Inc. Since the successful fabrication of such SMFs, there has been extensive development of coherent reception optical systems and long-haul intensity modulation and direct detection (IM/DD) transmission systems in the 1980s. However, the span length between transport repeaters reaches only 40 km, limited by the fiber loss of 0.35 dB K^{-1} m^{-1} in the 1550 nm spectral window of silica fiber. This loss was then compensated by the invention of optical amplifiers in the late 1980s.

The discovery of optical amplification in rare-earth elements doped in SMFs has transformed the design and deployment of optical fiber communications systems and networks in the last decade and the coming decades of the twenty-first century. The optical loss of the fiber and the optical components, both in fibers and integrated optical devices, in optical transport networks can be compensated for by using these in-line fiber optical amplifiers (OAs).

Therefore, the pulse broadening of optical signals during transmission and distribution in the networks becomes much more important for the system, in particular, the expected distribution of ultra-high RF carriers to the dense micro- and pico-cells of the 6G and 7G wireless signals from the fiber back-haul and front-haul to antenna sites.

Recent advances in ultra-high sampling rate analog-to-digital converters (ADCs) and digital-to-analog converters (DAC), to 200 GSa s^{-1}, and DSP processors allow the incorporation of ADC-DSP or DAC-DSP into coherent reception systems using only optically amplified SSMF spans and no dispersion compensating fiber (DCF), hence only one OA is required per span. This minimizes the accumulated amplification stimulated emission (ASE) noise at the receiver front-end. Thus under DSP based reception coherent systems with polarization division multiplexing (PDM) the aggregate bit rate (BR) per channel can reach 100 Gbits/s (Gbps) or 400 Gbps employing the QPSK or 16QAM modulation formats, and multi-sub-carrier multiplexing RPoF.

Furthermore, due to recent demonstrations of the use of DSP based coherent reception of modulated lightwaves and multiple-input multiple-output (MIMO), plus the training sequence, processing techniques have enhanced the sensitivity of optical receivers significantly. The transmission distance and the capacity of optical communication systems can thus be increased significantly [1]. MIMO techniques offer some possibilities for the use of different guided modes through a single fiber, for example few-mode fibers which can support more than one mode (up to 16) but not too many as in the case of multimode types. Each mode would carry a number of multiplexed channels. They are spatially demultiplexed in the optical domain at the fiber output and then detected and converted back to the electrical domain. These demultiplexed signals are then digitally processed using MIMO techniques. Thus the conditions under which a circular optical waveguide can operate as a few-mode fiber (FMF) are also mentioned in this chapter.

We focus the understanding of lightwave propagation in optical fibers with a focus on the practical aspects for the structure of minimum dispersion fibers or for a specified dispersion factor under the assumption that the optical field distribution follows a Gaussian distribution profile. Knowing the field distribution, one would be able to obtain the propagation constant of the single guided mode and hence the spot size (or energy concentration) of this mode within the core of the fiber.

The basic concept of optical dispersion related to the chromatic dispersion in SMFs can be addressed by using the definition of group velocity and group delay.

In current 5G Internet networking, the frequency regions 2 (FR2) and 3 (FR3) are employing the RF carriers in the microwave (μW) and mmW so that wideband composite data channels can be exploited. Thus in order to carry these RF carrier modulated data, the transport medium, the optical fiber SSMF, must be used so as to minimize the harmonic and inter-channel distortion and attenuation of these RF modulated optical channels.

The technical parameters of two typical optical fibers (Corning types) G652, commonly known as standard single mode optical fiber, and G655, known as large effective area fiber, can be found by referring to Corning [2].

In addition to understanding the physical phenomena of guided optical waves and the interference of the modulated wave spectral components, the fiber transmittance transfer function in the linear domain and its distortion due to nonlinear effects are important for RF transport systems.

8.2 Optical fiber: general properties

8.2.1 Geometry and index profile

An circular fiber consists of two concentric dielectric layers. The inner layer, or core, has a refractive index of $n(r)$ and radius a. This core region is normally fabricated by chemical deposition of Ge-doped silica on the wall of a pure silica tube, which is then collapsed. The outer layer of pure silica, or cladding, has an index of n_2 with $n_2 < n(r)$. Depending on the concentration of the impurity GeO_2 in the silica material the distribution of the core index can have a step size, changing gradually, or a graded index distribution.

An SMF core diameter of about 4–9 μm and 125 μm for cladding are typical for silica-based SMFs. A schematic diagram of the structure of a uniform circular SMF and the multimode and refractive index profiles are shown in figures 8.1(a)–(d). The fiber faces are etched so that the exposure of the cross-sections of the fibers can be identified clearly. The silica fibers are etched in a hydro-peroxide solution so that the core region doped with impurities will be etched faster than that of pure silica, thus the exposure of the core region as observed. The refractive index profile can be stepped or graded.

The refractive index $n(r)$ can be expressed by

$$n^2(r) = n_2^2 + \mathrm{NA}^2 s\left(\frac{r}{a}\right), \tag{8.1}$$

where NA is the fiber core numerical aperture, $s(\frac{r}{a})$ is the fiber core profile distribution for any profile shape ($s = 1$ at maximum) with a scaling parameter with respect to the fiber core radius a.

For a step-index profile

$$s\left(\frac{r}{a}\right) = \begin{cases} 1 & r \leqslant a \quad \text{hence} \\ 0 & r > a \, \text{ref\_index} \end{cases} n^2(r) = \begin{cases} n_1^2 & r \leqslant a \\ n_2^2 & r > a \end{cases}. \tag{8.2}$$

For a graded index profile, a power-law-index and the Gaussian profile can be assumed.

For the power-law profile, $s(r/a)$ can be expressed as

$$s\left(\frac{r}{a}\right) = \left\{1 - \left(\frac{r}{a}\right)^{\alpha}\right\} \text{for } r \leqslant a \text{ and } = 0 \text{ for } r < a, \tag{8.3}$$

with α as the power exponent. Thus the index profile distribution $n(r)$ can be expressed in the usual way, by substituting $\mathrm{NA}^2 = n_1^2 - n_2^2$:

$$n^2(r) = \begin{cases} n_1^2\left[1 - 2\Delta\left(\frac{r}{a}\right)^{\alpha}\right] & \text{for } r \leqslant a \\ n_2^2 & \text{for} \quad r > a \end{cases}, \tag{8.4}$$

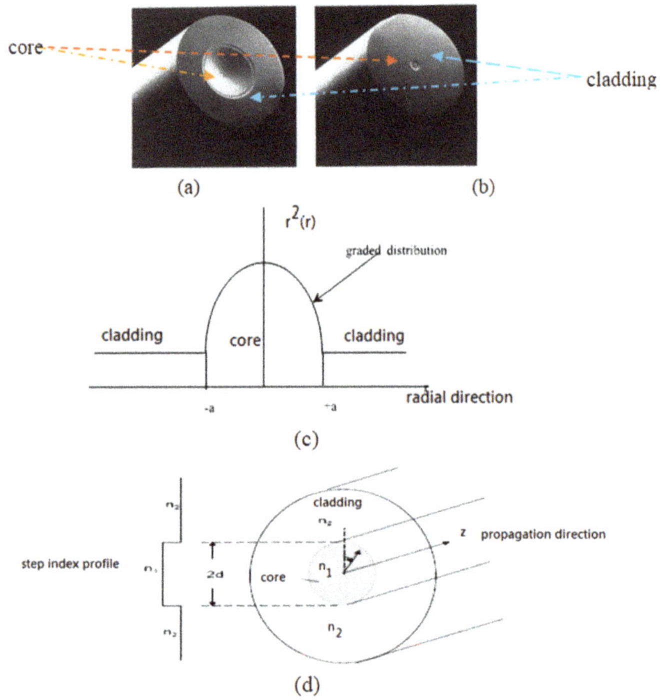

Figure 8.1. Photographs of a core-etched circular fiber: (a) multimode and (b) single mode. (c) Refractive index profile of a graded fiber core. (d) Fiber of step-index profile core. Copyright (2016), reprinted from [3] by permission of Taylor & Francis Ltd.

with $\Delta = \frac{\mathrm{NA}^2}{n_1^2}$ as the relative refractive difference with a small difference between that of the cladding and the core regions. The profile shape can have three special distributions: (i) $\alpha = 1$, the profile function $s(r/a)$ is linear or follows a triangular distribution; (ii) $\alpha = 2$, a quadratic or parabolic profile; and (iii) $\alpha = \infty$, a step type profile.

For the Gaussian type we have

$$s\left(\frac{r}{a}\right) = e^{-\left(\frac{r}{a}\right)^2}. \tag{8.5}$$

8.2.2 Weakly guiding mode

The electric and magnetic fields $E(r, \phi, z)$ and $H(r, \phi, z)$ of the circular SMF can be expressed in a well-known radial coordinate system as

$$\left[\frac{\delta^2}{\delta r^2} + \frac{1}{r}\frac{\delta}{\delta r} + k^2 n_j^2\right]\varphi(r) = \beta^2\varphi(r), \tag{8.6}$$

where $n_j = n_1$, n_2 and $\varphi(r)$ is the TE spatial field distribution

$$E_x = \psi(r)e^{-i\beta z} \text{ and } H_y = \left(\frac{\varepsilon}{\mu}\right)^{1/2} E_x = \frac{n_2}{Z_0} E_x, \tag{8.7}$$

where E_y, E_z, H_x, and H_z are negligible, $\varepsilon = \varepsilon_0 n_2^2$, and $Z_0 = (\varepsilon_0 \mu_0)^{1/2}$ is the vacuum impedance.

A Gaussian shape of the guided mode can be assumed based on experimentally measured distributions. Hence the solution of the wave equation (8.6) can be obtained in a straightforward manner. Then the eigen-propagation constant of the guided wave can be evaluated and optimized to achieve the best fiber structure.

The refractive index difference between the core and cladding of the SMF is found for the best loss and dispersion properties and is minute, only about 3%. Under this scenario the guide mode field penetrates into the cladding region with 30% power or only 70% of mode power contained within the core region. Under this circumstance the guiding of the mode is termed as a weakly guiding phenomenon.

8.2.2.1 Step-index SMF

The field spatial function $\varphi(r)$ would have the form of the Bessel functions given by equation (8.6) as

$$\varphi(r) = \begin{cases} A\dfrac{J_0(ur/a)}{J_0(u)} & 0 < r < a \text{ for the core} \\ A\dfrac{K_0(vr/a)}{K_0(v)} & r > a \text{ for the cladding} \end{cases}, \tag{8.8}$$

where J_0andK_0 are the Bessel functions of the first kind and modified second kind, respectively, and u and v are the two effective propagations along the core axis and the perpendicular direction to this axis of propagation towards the cladding region, given as

$$\frac{u^2}{a^2} = k^2 n_1^2 - \beta^2 \tag{8.9a}$$

$$\frac{v^2}{a^2} = -k^2 n_2^2 + \beta^2. \tag{8.9b}$$

Using the Maxwell's equation relation, we can find that E_z can take two possible orthogonal solutions as

$$E_z = -\frac{A}{kan_2}\begin{pmatrix}\sin\phi \\ \cos\phi\end{pmatrix}\begin{cases} \dfrac{uJ_1\left(u\dfrac{r}{a}\right)}{J_0(u)} \text{ for } 0 \leqslant r < a \\ \dfrac{vK_1\left(\dfrac{vr}{a}\right)}{K_0(v)} \text{ for } r > a \end{cases}. \tag{8.10}$$

The terms u and v must simultaneously satisfy two equations:

$$u^2 + v^2 = V^2 = ka(n_1^2 - n_2^2)^{1/2} = kan_2(2\Delta)^{1/2} \tag{8.11}$$

$$u\frac{J_1(u)}{J_0(u)} = v\frac{K_1(v)}{K_0(v)}. \tag{8.12}$$

This equation is obtained under constraints of boundary continuity conditions at the core–cladding interface $r = a$ (E_z is tangential and continuous at the interface). Equation (8.12) is commonly known as the wave eigenvalue equation whose solution leads to the propagation constant or effective index of the guided mode.

8.2.2.2 SMF and FMF conditions

Conditions for a single mode (SMF) and few modes (FMF) in a fiber can be obtained. In practice $\Delta \ll 1$, this longitudinal component is negligible compared to the transverse component, thus it is a transversely polarized mode. The fundamental mode is dominant and is designated as the LP_{01} mode (LP = linearly polarized) that follows a Gaussian distribution.

Furthermore, when the number of guided modes is higher than two polarized modes, they do form a set of modes over which information channels can be simultaneously carried and spatially demultiplexed at the receiving end so as to increase the transmission capacity, as illustrated in figure 8.2 [4, 5]. Such few-mode fibers are employed in the most modern optical transmission systems, and a schematic is shown in figure 8.2. A mode multiplexer acts as a mode spatial mixer and likewise the demultiplexer splits the modes LP_{11} and LP_{01} into individual modes and then converts them to the LP_{01} mode field and then they are injected into the SMF to feed into the coherent receiver. Obviously, there must be mode spatial demultiplexing and then modulation and multiplexing back into the transmission fiber for transmission. Similar structures would be available at the receiver to separate and detect the channels. Note the two possible polarizations of the mode LP_{11}. Note that there are four polarized modes of the LP_{11} mode, only two polarized modes are shown in this diagram. The delay due to the propagation velocity, from the difference in the propagation constant, can be compensated easily in the DSP processing algorithm, similar to that due to the polarization mode dispersion (PMD). The main problem to resolve in this spatial mode multiplexing optical transmission system is the optical amplification for all modes so that long-haul transmission can be achieved, that is, the amplification in the multimode fiber structure.

The number of guided modes is determined by the number of intersecting points of the circle of radius V and the curves representing the eigenvalue solutions. Thus for a single mode fiber the V parameter must be less than 2.405 and for a few-mode fiber this value is higher, for example if $V = 2.8$ we have three intersecting points between the circle of radius V and three curves, and then the number of modes would be LP_{01} and LP_{11} and their corresponding alternative polarized modes are as shown in figures 8.2(b) and (c). For the single mode there are two polarized modes whose polarizations can be vertical or horizontal. Thus a single mode fiber is not a mono-mode but supports two polarized modes. The main issues also concern the optical amplification gain for the transmission of modulated signals in such few-mode fibers. These remain the principal obstacles.

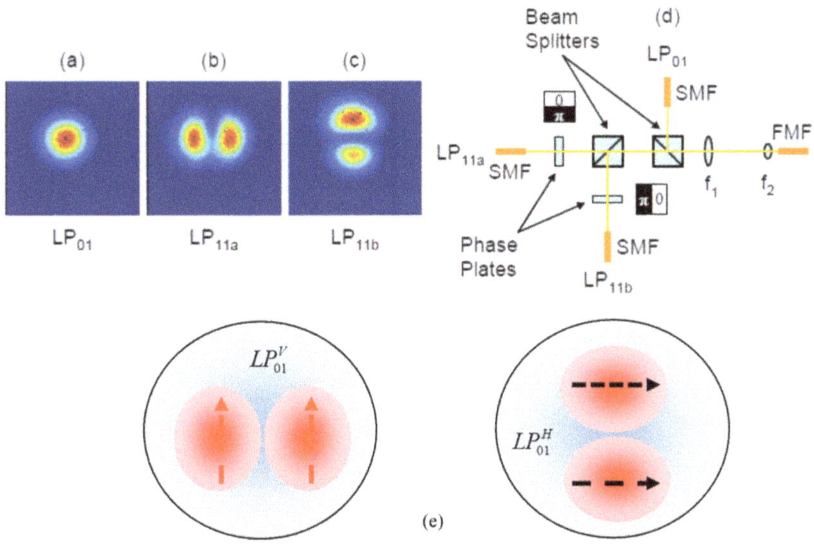

Figure 8.2. (a), (b), and (c) Intensity profiles distributed in the core of the first few order modes of a few-mode optical fiber employed for a 5 × 65 Gbps optical transmission system (reproduced with permission from [6]) and optical system arrangement for spatially demultiplexed and multiplexed modal channels. (d) schematic of an optical system to launch the input lightwave into a few-mode-fiber to excite individual mode. (e) Horizontal (H) and vertical (V) polarized modes $LP_{01}^{V,\,H}$. The polarization directions are indicated by arrows. Copyright (2016), reprinted from [3] by permission of Taylor & Francis Ltd.

Figures 8.3(a) and (b) illustrate the modes of the FMF. The rays of the three modes can be axially straight or skewed and twisted around the principal axis of the fiber. Thus there are different propagation times between these modes, hence dispersion of the modes. However, this dispersion property can be employed to compensate for chromatic dispersion effects in an SMF by cascading FMFs and SMFs [7]. Figure 8.3(c) shows a spectrum of the graphical solution for FMF modes. In figure 8.3(d) the regions of single operation and then the higher order and second-order mode regions as determined by the value of the V parameter are indicated.

Channel transmission through an FMF with spatial division multiplexing (SDM) and demultiplexing (SDD) optical devices can be implemented using a spatial demultiplexer and multiplexer using integrated optical devices in which appropriate polarization and waveguides support a few modes matching the fiber modes and their polarizations (see figure 8.2). For example, an SSMF is single-moded at a wavelength 1550 nm, so if a lightwave's source has wavelength 850 nm, then there would be more than one mode that can be guided in such fiber. Guided modes of up to LP41 can be guided and thus 16 partial modes can be guided to transport information. The bit rate can reach 100 Gbps per spatial channel resulting in 1.6 Tbps for the transmission over the FMF. However, the transmission would be less than 10 km. Another advantage of this SDM transmission is that at the receiver multiple DSP systems can be combined and processed under MIMO to optimize and reduce the bit error rate (BER) significantly (figure 8.4).

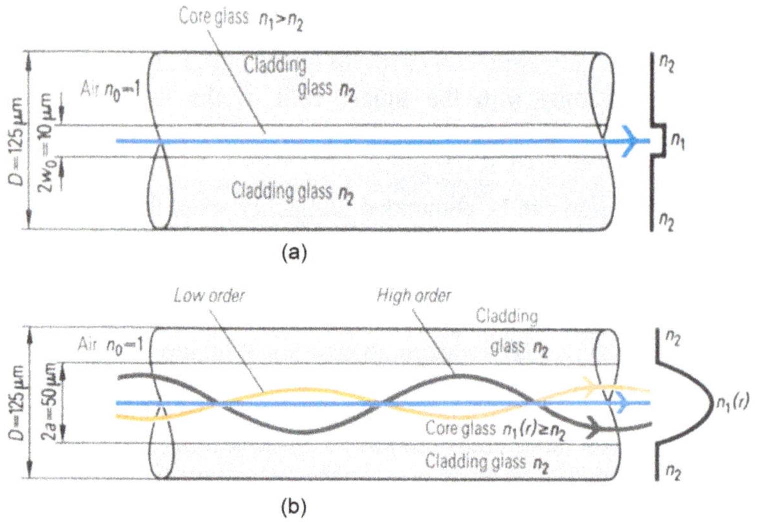

Figure 8.3. (a) Guided modes as seen by 'a ray' in the transverse plane of a circular optical fiber, the 'ray' model of a lightwave propagating in a single mode fiber. (b) The ray model of the propagation of different modes guided in a few/multimode graded index fiber. (c) Graphical illustration of the solutions for eigenvalues (propagation constant—wave number of optical fibers). (d) *b–V* characteristics of graded fibers. Copyright (2016), reprinted from [3] by permission of Taylor & Francis Ltd.

8.2.2.3 Weakly guiding and Gaussian approximation

We note again that E and H are approximate solutions of the scalar wave equation and the main properties of the fundamental mode of weakly guiding fibers that can be observed as follows.

The propagation constant β (in the z-direction) of the fundamental mode must lie between the core and cladding wave numbers. This means the effective refractive index of the guided mode lies within the range of the cladding and core refractive indices. Accordingly, the fundamental mode must be nearly a transverse electromagnetic wave as described by equation (8.7):

$$\frac{2\pi n_2}{\lambda} < \beta < \frac{2\pi n_1}{\lambda}. \tag{8.13}$$

The spatial dependence $\psi(r)$ is a solution of the scalar wave equation (8.6).

The main objectives are to find a good approximation for the field $\psi(r)$ and the propagation constant β. This can be found though the eigenvalue equation and Bessel's solutions, as shown in a previous section. It is desirable, if we can approximate the field to a good Gaussian shape, to obtain simple expressions to have a clearer understanding of light transmission in an SMF without going through graphical or numerical methods. Furthermore, the experimental measurements and numerical solutions for step and power-law profiles show that $\psi(r)$ is approximately Gaussian in appearance. We thus approximate that the field of the fundamental mode follows a Gaussian profile, thus its output also follows this shape, as

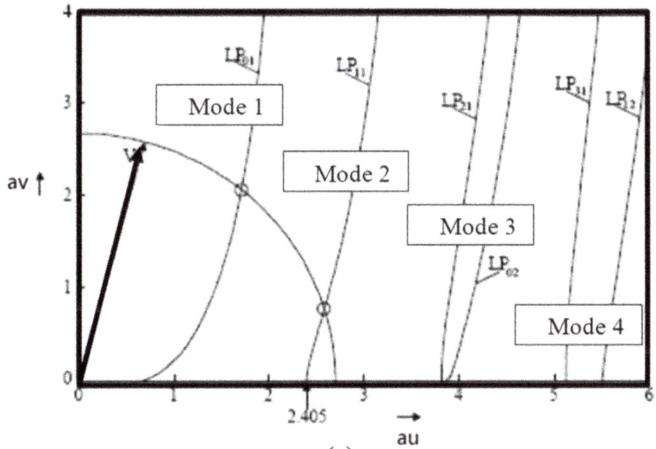

(c)

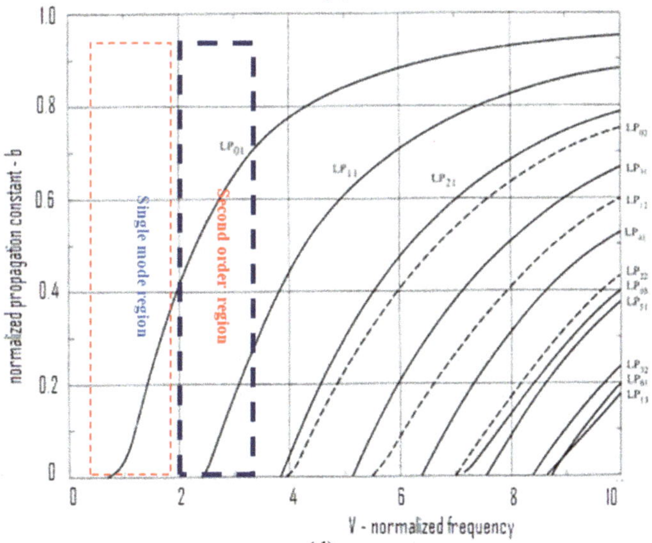

(d)

Figure 8.3. (Continued.)

$$\phi(r) \cong Ae^{-\frac{1}{2}(\frac{r}{r_0})^2}, \tag{8.14}$$

where r_0 is defined as the spot size, i.e. the size at which the intensity equals e^{-1} of the maximum. Thus if we have the wave equation (8.6) and solve for the propagation wave vector, we have

$$\beta^2 = \frac{\int_0^\infty \left[-\left(\frac{\delta\varphi}{\delta r}\right)^2 + k^2 n^2(r)\varphi^2\right] r\delta r}{\int_0^\infty r\varphi^2 \delta r}. \tag{8.15}$$

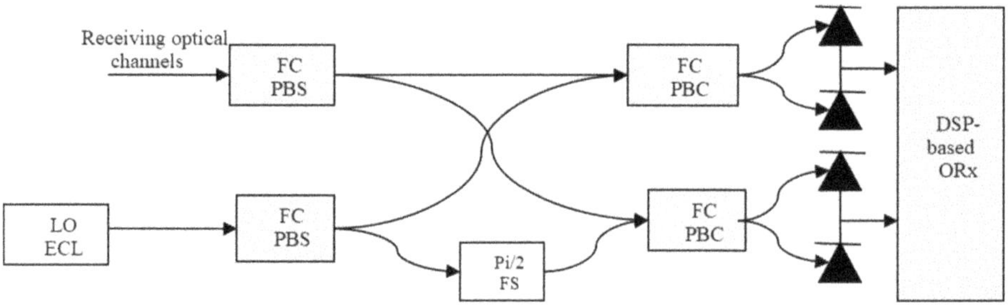

Figure 8.4. $\pi/2$ hybrid coupler for polarization demultiplexing and mixing with a local oscillator in a coherent receiver of a modern DSP based optical receiver for the detection of phase modulated schemes. FS = phase shifter (optical); PBC = polarization beam combiner; FC = fiber coupler; LO = local oscillator laser; ECL = external cavity laser; DSP = digital signal processor.

The spot size is then followed by substituting the $\psi(r)$ (Gaussian) in equation (8.14) into equation (8.15) then differentiating and setting $\frac{\delta^2\beta}{\delta r}$ at r_0 to zero; the propagation constant β of the fundamental mode gives the largest value of r_0. Therefore, knowing r_0 and β the fields E_x and H_y (equation (8.7)) are fully specified. For the step-index profile, substituting equation (8.15) and $\psi(r)$ into equation (8.14) leads to β in terms of r_0 by

$$V = \text{NA}.ka = \text{NA}\frac{2\pi}{\lambda}a. \tag{8.16}$$

The spot size is thus given by setting

$$\frac{\delta^2\beta}{\delta r_0} = 0. \tag{8.17}$$

Thus r_0 is given by

$$r_0^2 = \frac{a^2}{\ln V^2}. \tag{8.18}$$

Combining equations (8.1) and (8.8) we have

$$(a\beta)^2 = (akn_1)^2 - \ln V^2 - 1. \tag{8.19}$$

This expression is physically meaningful only when $V > 1$, that is only when r_0 is positive.

8.2.2.4 Cut-off properties

Similarly to the case of planar dielectric waveguides, from figure 8.3 we observe that when we have $V < 2.405$, only the fundamental LP_{01} exists. It is noted that for single mode operation the V parameter must be less than or equal to 2.405. However in practice $V < 3$ may be acceptable for single mode operation. Indeed, the value 2.405 is the first zero of the Bessel function $J_0(u)$. In practice one cannot really distinguish between the V value between 2.3 and 3.0. Experimental observation also shows that the optical fiber can still support only one mode. Thus designers do usually take the value of V as 3.0 or less to design a single mode optical fiber.

The V parameter is inversely proportional with respect to the optical wavelength which is directly related to the operating frequency. Thus if an optical fiber is launched with lightwaves whose optical wavelength is smaller than the operating wavelength at which the optical fiber is single mode, then the optical fiber is supporting more than one mode. The optical fiber is said to be operating in a few regions and then multimode regions when the total number of modes reaches a few hundred.

Thus one can define the cut-off wavelength for optical fibers as follows: the wavelength (λ_c) above which only the fundamental mode is guided in the fiber is called the cut-off wavelength, λ_c. This cut-off wavelength can be found by using the V parameter as $V_c = V|_{\text{cut-off}} = 2.405$, thus

$$\lambda_c = \frac{2\pi a \text{NA}}{V_c}. \tag{8.20}$$

In practice the fibers tend to be effectively single mode for larger values of V, say $V < 3$ for the step profile, because the higher order modes suffer radiation losses due to fiber imperfections. Thus if $V = 3$, from equation (8.11) we have $a < (3\lambda/2)$. In the case that $\lambda = 1\ \mu$m thence the NA must be very small ($\ll 1$) for radius a to have some reasonable dimension. Usually Δ is about 1% or less for SSMFs employed in long-haul optical transmission systems so as to minimize the loss factor and the dispersion.

8.3 Dispersion

Chromatic dispersion (CD) and nonlinear phase dispersion (NPD) affects the multi-quadrature amplitude modulation polarization division multiplexing (M-QAM PDM) of the RF carrier RPoF transmission over SMF and are critical for RF modulated optical carriers. So the modulation and generation of only one sideband is preferred for long distance RPoF to avoid the interference and destruction due to the two sidebands.

In the compensation of CD, we understand that a training sequence can be employed, e.g. the CAZAC training sequence [8], to mitigate such effects. However, if the transmission distance is long the complexity of the training sequence may be high, so if one can compensate parts of the linear CD by pre-emphasis at the transmitter it would ease some difficulties for the DSP at the receiver.

Similarly, nonlinear effects can be implemented at the receiver using back-propagating techniques or the inverse kernels of the fiber nonlinear Volterra transfer function. However, the nonlinear transfer function of the M-QAM can be found by examining the in-phase and quadrature phase components which are oscillating between 0 and π, and $\pi/2$ and $3\pi/2$. This is superimposed on the linear CD transfer function.

This section briefly summarizes the techniques for the determination of the fiber transfer function under CD and nonlinear distortion effects, then we suggest techniques of pre-emphasis to be applied to the DAC so as to equalize the CD and nonlinear impairments as much as possible so as to ease the complexity of the DSP at the coherent receiver.

8.3.1 Theoretical analysis

The evolution along the propagation path z of the 'small signal' intensity modulation (IM), or complex power $p(\omega, z)$ and phase rotation (PM) $\phi(\omega, z)$, during the propagation of the guided mode through the single mode optical fiber (SMF), taking into account both the chromatic dispersion and the nonlinearity (SPM) effects, is governed by the following set of differential equations [9–12]:

$$\frac{\delta \bar{p}(\omega, z)}{\delta z} = \beta_2 \omega^2 \bar{P}_0 \bar{\phi}(\omega, z); \bar{A}(\omega, z) = \sqrt{\bar{p}(\omega, z)} \tag{8.21}$$

$$\frac{\delta \bar{\phi}(\omega, z)}{\delta z} = -\left[\frac{\beta_2 \omega^2}{4\bar{P}_0} + \gamma e^{-\alpha z}\right] \bar{p}(\omega, z), \tag{8.22}$$

where $A(\omega, z)$ and $\phi(\omega, z)$ are defined as the normalized complex amplitude and phase, respectively, of the optical field in the Fourier domain, ω is the radial frequency of the RF or broadband signal, z is the distance along the propagation axis of the fiber, α is the attenuation coefficient in linear scale of the SMF, and β_2 is the first-order dispersion coefficient, i.e. the group delay factor as a function of the optical wavelength given by

$$\beta_2 = -\frac{\lambda^2 D(\lambda)}{2\pi c}, \tag{8.23}$$

where c is the velocity of light in a vacuum, $D(\lambda)$ is the dispersion factor of the fiber, typically taking values of 17 ps nm^{-1} km^{-1} for a silica SMF at the operating wavelength $\lambda = 1550$ nm.

γ is the nonlinear SPM coefficient defined by

$$\gamma = \frac{2\pi n_2}{\lambda A_{\mathrm{eff}}}; \text{ with } A_{\mathrm{eff}} = \pi r_0^2, \tag{8.24}$$

where n_2 is the nonlinear coefficient of the fiber, typically $n_2 = 3.2 \times 10^{-23}$ m_2 W^{-1} for SMF-28, and $A_{\mathrm{eff}} = \pi r_0^2$ is the effective area of the SMF, that is the area of the Gaussian mode spot of the guided mode in an SMF under the weakly guiding condition [13].

These equations are derived from the observer positioned on the moving frame of the phase velocity of the waves which is normally expressed by the nonlinear Schrödinger equation (NLSE):

$$\frac{\partial A(t, z)}{\partial z} = -\left[\alpha A(t, z) + \frac{j}{2}\beta_2 \frac{\partial A^2(t, z)}{\partial t^2}\right] + j\gamma \mid A(t, z)|^2 A(t, z). \tag{8.25}$$

In the coherent optical receiver and digital signal processing at the transmitter and after the detection at the receiver, under quadrature amplitude modulation (QAM), both the in-phase and quadrature phase components are recovered in the amplitude, which is proportional to the power of the optical signals arriving at the front of the optical waves of the optical hybrid coupler and the balanced photodetector pair. A schematic of the transmission is shown in figure 8.5 in which the transmitter can generate an optical sequence or near single frequency sinusoidal waves at frequencies reaching 30 GHz using a Fujitsu DAC sampling rate of 65 GSa s^{-1}. The optical modulator is a typical Fujitsu IQ modulator modulated by electrical signals output from the DAC and phase shifted in the RF domain by an electrical phase shifter. The RF phase can be set such that when they are $\pi/2$ shifted with respect to each other the suppression of one of the single sidebands can be achieved at the output spectrum. The main carrier can be suppressed by biasing the child Mach–Zehnder intensity modulators (MZIMs) at the minimum transmission point. Figure 8.6 shows the constellation of 16QAM under dispersion effects and the rotation of the 16QAM is effectively three distinct circles generated by the phase angle variation with respect to the transmission length. The three amplitudes are effectively the three circles of the constellation of the square 16QAM. In RPoF systems an M-ary QAM is used to modulate the RF modulated optical carrier. The data channels are superimposed on such optical carriers. Thus the constellation indicates the degree of distortion of the RF carrier under optical modulation. Figure 8.7 illustrates the experimental constellation obtained in a coherent digital receiver at the output of the sampling real-time oscilloscope. Once the carrier phase and the dispersion are compensated by the constant amplitude algorithm the phase diagram is reversed to its original state as at the transmitter plus noise contributed by the optical amplification and the shot noise and equivalent noise of the electronic amplifiers of the receiver.

The propagation equation in the spectral domain can be written using equation (8.25) as

$$\frac{\delta^2\bar{p}(\omega, z)}{\delta z^2} = -\left[\frac{\beta_2^2\omega^4}{4} - \beta_2\omega^2 + \gamma\bar{P}_0 e^{-\alpha z}\right]\bar{p}(\omega, z). \tag{8.26}$$

Subject to the initial conditions of

$$\bar{p}(\omega, 0) = \bar{p}_{\text{in}}(\omega) \text{ and } \frac{\delta\bar{p}(\omega, 0)}{\delta z} = \beta_2\omega^2\bar{P}_0 e^{-\alpha z}\phi(\omega, 0) = \beta_2\omega^2\bar{P}_0 e^{-\alpha z}\phi_{\text{in}}(\omega), \tag{8.27}$$

where the subscript ‘in’ indicates the input location, which is the starting of the propagation of the modulated optical waves through the SMF.

Now, by changing some variables with the setting of

$$x = 2\sqrt{B}e^{-\alpha z/2};\ B = -\frac{\beta_2\omega^2\gamma P_0}{\alpha^2}, \tag{8.28}$$

it can be rewritten as

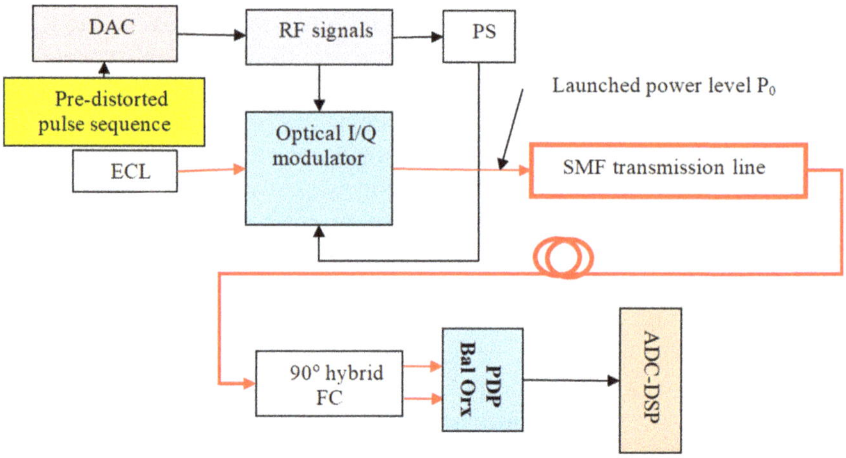

Figure 8.5. Digital-based optical transmitter and coherent reception with real-time sampling and digital signal processing. DAC = digital-to-analog converter; ADC = analog-to-digital converter; DSP = digital signal processing; PDP = photodetector pair; FC = fiber coupler; I/Q = in-phase/quadrature phase.

$$\left[x\frac{\partial^2}{\partial x^2} + x\frac{\partial}{\partial x} - (x^2 - \upsilon^2)\right]\bar{p}(\omega, z) = 0. \tag{8.29}$$

The solution of this equation is a combination of purely imaginary Bessel's functions L and K, and subject to the initial conditions of equation (8.27). Thus the evolution of the complex amplitude of the modulated optical waves along the propagation path is given by [14, 15]

$$\bar{p}(\omega, z) = \frac{2\sin h(\pi\upsilon)}{\pi}\left\{\begin{aligned}&\sqrt{B}\bar{p}_{\text{in}}(\omega)\left[\frac{\partial L_{i\upsilon}}{\partial x}\left(2\sqrt{B}\right)K_{i\upsilon}(x) - \frac{\partial K_{i\upsilon}}{\partial x}\left(2\sqrt{B}\right)I_{i\upsilon}(x)\right]\\&+\frac{\alpha B}{\gamma}\phi_{\text{in}}(\omega)\left[K_{i\upsilon}\left(2\sqrt{B}\right)L_{i\upsilon}(x) - L_{i\upsilon}\left(2\sqrt{B}\right)K_{i\upsilon}(x)\right]\end{aligned}\right\} \tag{8.30}$$

where $\upsilon = -\frac{\beta_2\omega^2}{\alpha}$. The first term on the RHS of equation (8.30) is the magnitude part and the second is the phase part, that is the in-phase and quadrature phase components of the QAM signal. Thus the in-phase and quadrature parts of the complex magnitude can be expressed as

$$\bar{p}_I(\omega, z) = \frac{2\sinh(\pi\upsilon)}{\pi}\left\{\sqrt{B}p_{\text{in}}(\omega)\left[\frac{\partial L_{i\upsilon}}{\partial x}\left(2\sqrt{B}\right)K_{i\upsilon}(x) - \frac{\partial K_{i\upsilon}}{\partial x}\left(2\sqrt{B}\right)I_{i\upsilon}(x)\right]\right\} \tag{8.31}$$

$$\bar{p}_Q(\omega, z) = \frac{\alpha B}{\gamma}\phi_{\text{in}}(\omega)\frac{2\sinh(\pi\upsilon)}{\pi}\left[K_{i\upsilon}\left(2\sqrt{B}\right)L_{i\upsilon}(x) - L_{i\upsilon}\left(2\sqrt{B}\right)K_{i\upsilon}(x)\right]. \tag{8.32}$$

In figures 8.8 and 8.9 the following parameters are used in the estimation of the frequency responses: propagation length $L = 100$ km of SSMF for the optical launched power levels, i.e. $z = 0$ at the fiber inputs of 0, +10, and +13 dBm (solid lines).

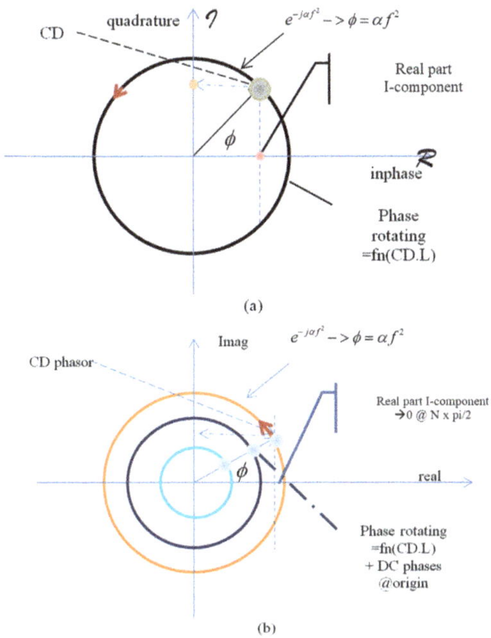

Figure 8.6. Phasor of the amplitude evolution along the propagation path, assuming the amplitude is not affected by attenuation: (a) the phasor and (b) the phase or the phase constellation of the M-QAM (M = 16 square QAM), at the three amplitude level.

The SSMF parameters are α = 0.2 dB km^{-1}, D = 17 ps/(nm.Km), $n_2 = 3.2 \times 10^{-23}$ m^2 W^{-1}, and effective area = 70 μm^2 at a wavelength of 1550 nm.

The behavior of the amplitude and phase of the complex power can be explained by referring to figure 8.8. The in-phase and quadrature phase components move along the horizontal and vertical axes within the ± 1 limits, meaning that as the phase rotating around the unit circles these components oscillating such that when the phase is $(2M + 1)$ ($M = 0, 1, 2, \ldots$) or an odd number of $\pi/2$ the in-phase component becomes nullified. Likewise the quadrature phase at $N\pi$ ($N = 0, 1, 2, \ldots$). In the case of a QAM scheme, e.g. 16QAM, there would be three amplitude levels of the phase constellation and these levels are rotating, as shown in figure 8.9; the initial phase is set by the initial position of the constellation point of a square 16QAM but the oscillation and nullified locations would be very similar to those of figure 8.8.

Under the linear operating regime, i.e. $\gamma = 0$, we can thus obtain the expressions for the complex power amplitude and phase, from

$$\bar{p}_I(\omega, z) = \cos \frac{\beta_2 \omega^2 z}{2}. \tag{8.33}$$

The values of the frequency at which the real part reaches a minimum level can be observed in figure 8.8. We can see that the frequency at which the in-phase and quadrature phase components vanish is

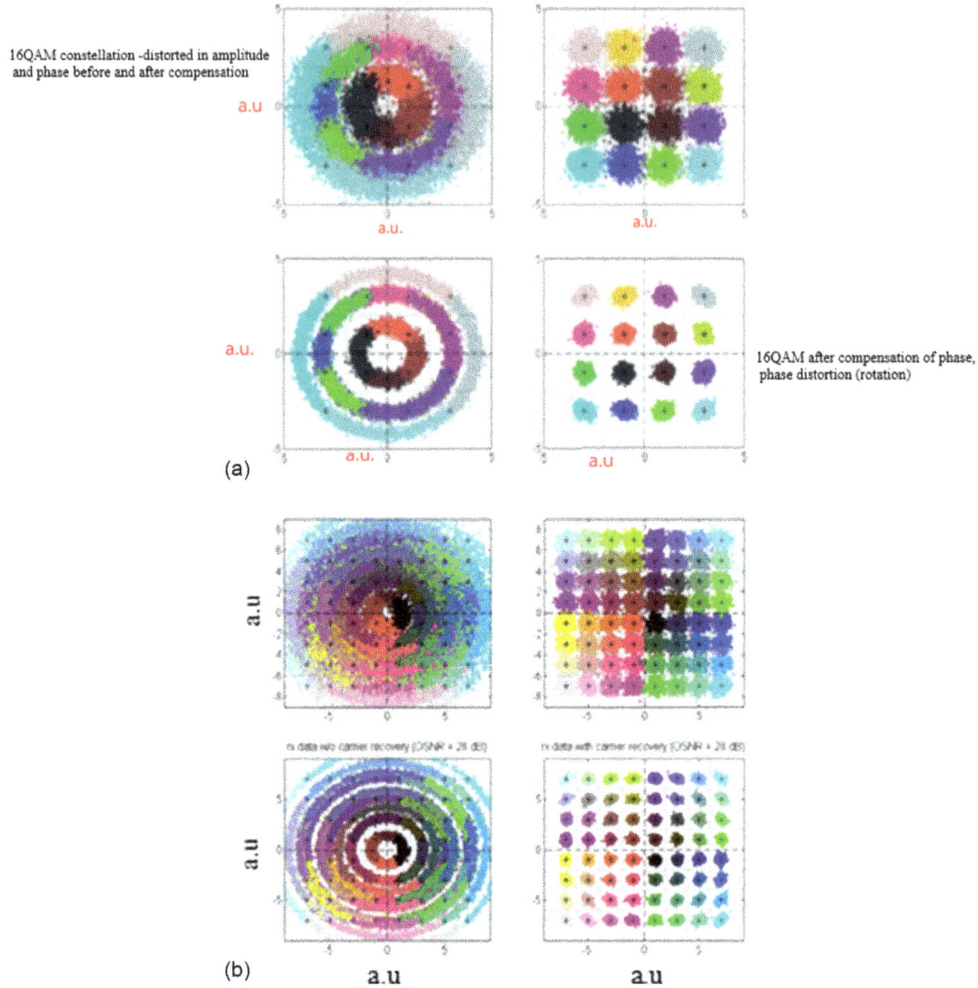

Figure 8.7. Constellation rotation of a 16QAM under dispersion and recovered after the RF carrier recovery.

$$f = f_0\sqrt{\frac{m}{2\pi LD(\lambda)c}}; m = 1, 3, 5. \ldots \tag{8.34}$$

8.3.2 SMF transfer function

8.3.2.1 Linear TF

The transfer function (TF) of the SMF in the frequency domain can be obtained from the fiber impulse responses which can be obtained from solving the NLSE with the delay term ignored [16]. The fiber TF is simply a quadratic phase function as a result of the removal of the nonlinear term of the NLSE and using a Taylor series approximation around the operating frequency (the central wavelength). The frequency and impulse responses of an SMF. Equation (8.35) gives the time domain impulse response $h(t)$ and its frequency transfer function $H(\omega)$ as

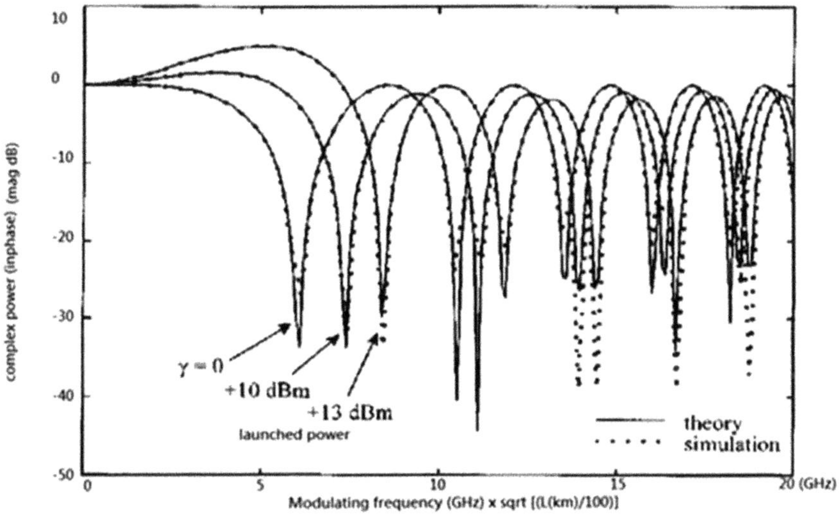

Figure 8.8. Variation of the magnitude of the real part of the complex amplitude component with frequency, the intensity frequency response of an SMF-28 of $L = 100$ km (at $z = 100$ km) under coherent detection with a normalized amplitude of the intensity of the optical waves under linear ($\gamma = 0$) and nonlinear operating conditions. (Reproduced with permission from [5].) The frequency scale can be multiplied by a factor of $1/\sqrt{(L}$ (km)/100) to values for different transmission lengths in kilometers. For example for $L = 1000$ km then the scaling factor is $1/\sqrt{(10)}$.

$$h(t) = \sqrt{\frac{1}{j4\pi\beta_2}} \exp\left(\frac{jt^2}{4\beta_2}\right) \quad \leftrightarrow \quad H(\omega) = \exp(-j\beta_2\omega^2), \tag{8.35}$$

where β_2 is the group velocity dispersion (GVD) parameter, i.e. the second-order differentiation of the propagation constant. $f(t)$ is the input rectangular pulse sequence and β_2 is proportional to the length of the fiber.

The output $g(t)$ is the dispersed waveform. The fiber transmittance transfer function in equation (8.35) is an exact analogy of diffraction in optical systems (see item 1, in table 2–1 of [17]). Thus the quadratic phase function also describes the diffraction mechanism in one-dimensional optical systems, where the distance x is analogous to time t. Thus the step response $s(t)$ and $H(\omega)$ can be expressed in terms of Fresnel cosine and sine integrals as

$$s(t) = \int_0^t \sqrt{\frac{1}{j4\pi\beta_2}} \exp\left(\frac{jt^2}{4\beta_2}\right) dt = \sqrt{\frac{1}{j4\pi\beta_2}} \left[C\left(\sqrt{1/4\beta_2}\, t\right) + jS\left(\sqrt{1/4\beta_2}\, t\right) \right], \tag{8.36}$$

with $C(t) = \int_0^t \cos\left(\frac{\pi}{2}\tau^2\right) d\tau$ and

$$S(t) = \int_0^t \sin\left(\frac{\pi}{2}\tau^2\right) d\tau, \tag{8.37}$$

where $C(t)$ and $S(t)$ are the Fresnel cosine and sine integrals.

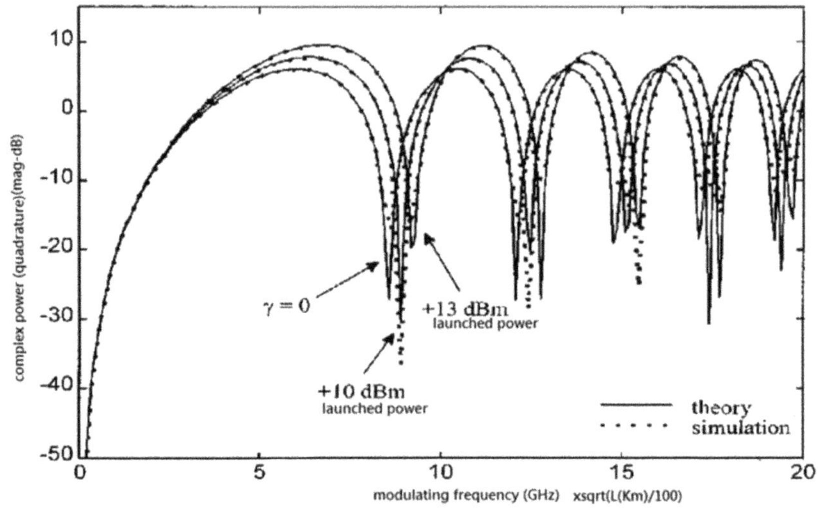

Figure 8.9. Variation of the magnitude of the quadrature part of the complex amplitude component (the in-phase component) with frequency, the QAM in-phase frequency response of the QAM signal SMF of L = 100 km (at z = 100 km) under coherent detection with a normalized amplitude of the intensity of the optical waves under linear ($\gamma = 0$) and nonlinear operating conditions. (Copyright (1991) IEEE. Reproduced with permission from [4].)

Figures 8.10(a)–(c) depict, respectively, the square pulse, the impulse response, and step response of an SMF. The symmetric over-damping of the response due to a rectangular pulse agrees with that of the step response. The impulse response also shows the corresponding nulls or chirping effects on the impulse and step responses. The response chirp occurs mainly near the edge of the pulses when it is in the near field region, about a few kilometers for SSMF. In this near field distance, the accumulation of nonlinear effects is still very weak and thus this linear chirp effect dominates the behavior of the single mode fiber. The nonlinear Volterra transfer function (VTF) presented in the next section would thus have a minimum influence. This point is important for understanding the behavior of lightwaves circulating in short length fiber devices in which both the linear and nonlinear effects are to be balanced, such as active mode locked soliton and multi-bound soliton lasers [18, 19]. In the far field the output of the fiber is Gaussian-like for the square pulse launched at the input. In this region the nonlinear effects would dominate over the linear dispersion effect as they have been accumulated over a long distance.

A linear time variant system, such as the single mode fiber, would have a transfer function of

$$H(f) = |\,H(f)|\; e^{-j\alpha(f)}, \tag{8.38}$$

where $\alpha = \pi^2\beta_2 L = \frac{-\pi D L\lambda^2}{2c}$ is proportional to the length L and the dispersion factor D (λ) (ps nm^{-1} km^{-1}). The phase of the frequency transfer response is quadratic in terms of the frequency, thus the group delay is a linear function of frequency, as shown in figure 8.11. The frequency response in the amplitude term is infinite and is a constant while the phase response is a quadratic function with respect to the frequency of the

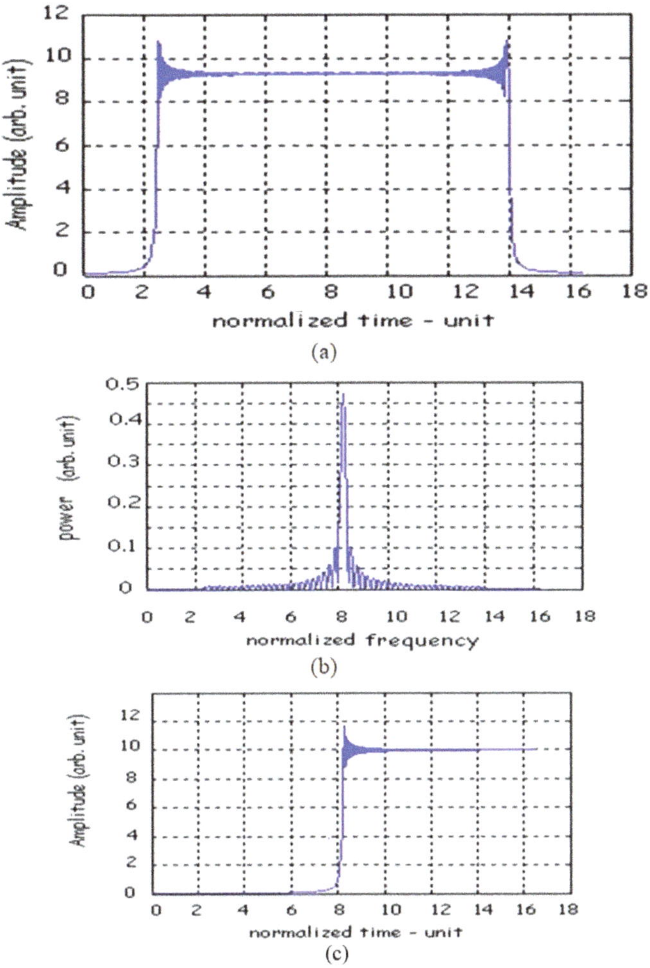

Figure 8.10. SMF temporal responses due to (a) rectangular pulse excitation through an SMF, (b) the impulse response, and (c) the step response. Note that the horizontal scale is in normalized units of time. Copyright (2016), reprinted from [3] by permission of Taylor & Francis Ltd.

baseband signals. The carrier is chirped accordingly, as observed in figures 8.12 and 8.13. The chirping effect is very significant near the edge of the rectangular pulse and almost nil at the center of the pulse, in the near field region of less than 1.0 km of standard single mode fiber. In the far field region the pulse becomes Gaussian-like. Thus the response of the fiber in the linear region for a Gaussian pulse input to the fiber. The output pulse is also Gaussian by taking the Fourier transform of the input pulse and multiplying by the fiber transfer function. Then an inverse Fourier would indicate that the output pulse shape follows a Gaussian profile.

This leads to a rule of thumb for consideration of the scaling of the bit rate and transmission distance as ‘Given that a modulated lightwave of a bit rate B can be transmitted over a maximum distance L of single mode optical fiber with a BER of error-free level, then if the bit rate is halved then the transmission distance can be

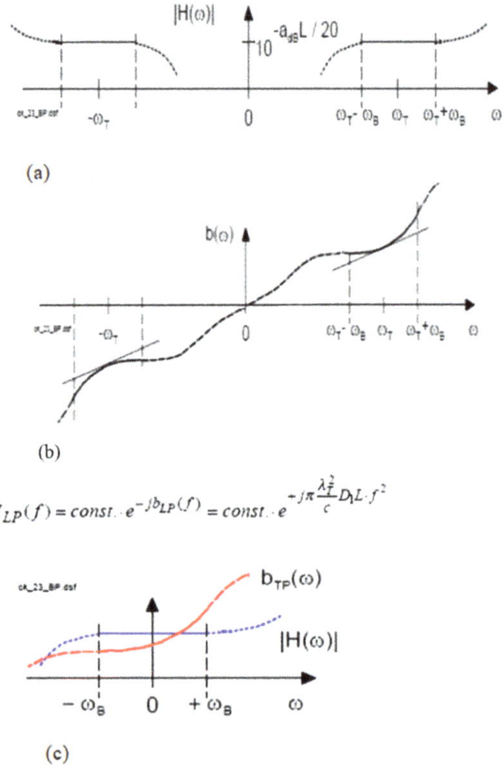

Figure 8.11. Frequency response of an SMF: (a) magnitude; (b) phase variation in the bandpass regime; and (c) baseband equivalence. Copyright (2016), reprinted from [3] by permission of Taylor & Francis Ltd.

increased by four times'. For example, for 10 Gbps amplitude shift keying modulation format signals can be transmitted over 80 km of standard single mode optical fiber then at 40 Gbps only 5 km can be transmitted for a bit error rate (BER) of 10^{-9}. Figure 8.11 shows a typical frequency response in the magnitude phase and the low PS property. Ideally one could see that the amplitude response of the fiber is constant throughout all frequencies if no attenuation or constant attenuation throughout all frequency ranges. Only the phase of the lightwaves is altered, that is, the chirping of the carrier. In the optical receiver, a DSP could be used to determine the exact dispersion and hence the rotation of the constellation by an angle such that it can be recovered to its original position. The chirp of the carrier can be seen in figure 8.12 in which the chirp is much less at the center but heavily chirped near the edge of the pulse. The step responses shown in figures 8.12(b) and indicate the damping oscillation of the step pulse which is due to the chirp of the carrier as we also observe from the calculated impulse and step responses given in figure 8.10. Figure 8.13 shows the Gaussian-like impulse response when the transmission distance is large. This is the typical pulse shape in long-haul non-dispersion compensating transmission. These dispersive pulse sequences are then coherently detected and processed by the digital signal processors. The number of samples must

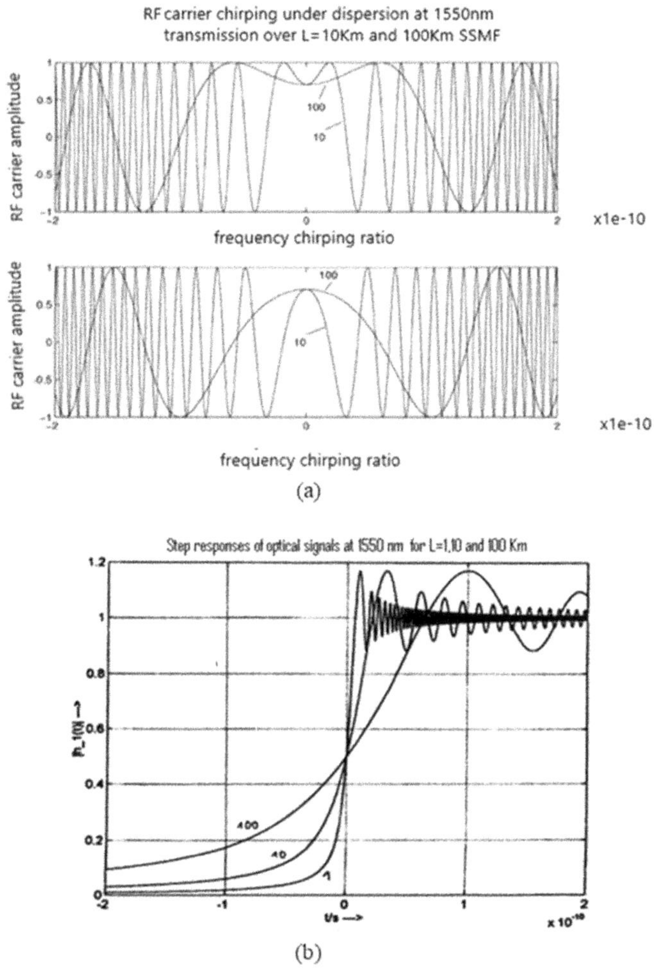

Figure 8.12. (a) Carrier chirping effects and (b) step response of a single mode optical fiber of $L = 1$, 10, and 100 km. *Note*: over-damping. Copyright (2016), reprinted from [3] by permission of Taylor & Francis Ltd.

be long enough to cover the dispersive sequence and the number of taps of the finite impulse response (FIR) filter must be high enough to ensure that the whole dispersive pulse is covered and falls within the filter length. Thus the longer the transmission length the higher the number of taps of the FIR. This is the first step in the DSP based optical receiver, by using the constant modulus amplitude (CMA) algorithm to compensate for the dispersion effects before compensating and recovering the phase constellation of the modulated and transmitted channels. It is noted that the chirp of the pulse is much higher if we measure the pulse response in the near field region than that if obtained in the far field of the output lightwaves, as observed in the step responses given in figures 8.12(b) and 8.13. Thus if a Gaussian pulse is launched and propagates in the single mode fiber we would obtain a Gaussian pulse shape at the output (see figure 8.13) provided that the fiber length is sufficiently long, typically more than 20 km which is commonly met in real transmission systems.

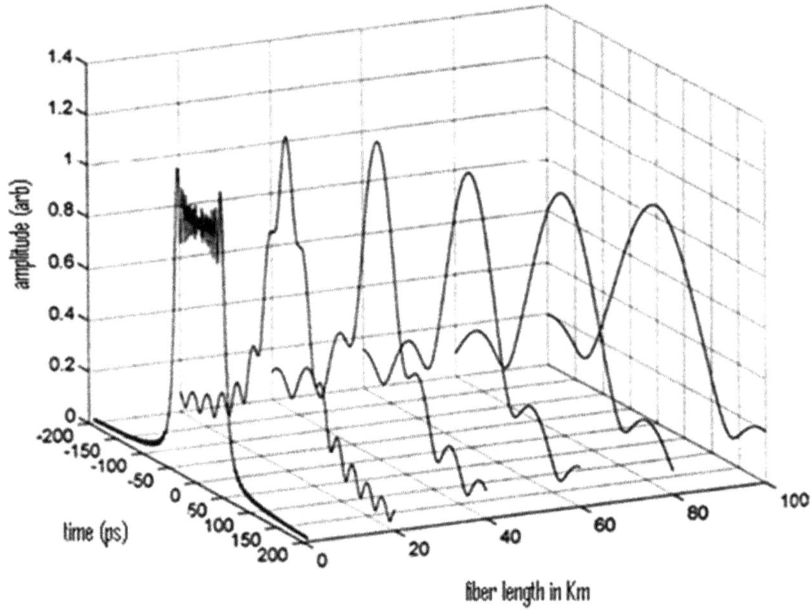

Figure 8.13. Pulse response from the near field (~< 2 km) to the far field (>80 km). Copyright (2016), reprinted from [3] by permission of Taylor & Francis Ltd.

8.3.2.2 Remarks

The analytical expression for the in-phase and quadrature phase components can be obtained and thus the frequency responses of these components can be identified. Experimental proof can be conducted by launching into the optical IQ modulator using a DAC with a broadband signal, the impulse in the time domain, and obtaining the pulse sequence at the output of the fiber under different average optical power at the input, then taking the fast Fourier transform (FFT) of the sampled waveform would give us the frequency response. The detection of the time domain waveform by a coherent receiving DSP system will allow us the determination of the response and hence the design of pre-emphasis for equalization of CD and nonlinear impairments of the fiber transmission system, in particular under M-QAM with $M > 4$.

Thus in order to conduct pre-distortion of the transmitted pulse at the transmitter we have to satisfy the condition for both the in-phase and quadrature phase components of the M-QAM modulated channels by

$$H_{I\_\text{SMF-CD}}(j\omega).\ H_{C\text{Deq}}(j\omega) = 1 \rightarrow H_{C\text{Deq}}(j\omega) = \text{equalizer TF}, \tag{8.39}$$

which can be rewritten as

$$H_{C\text{Deq}}(j\omega) = H^{-1}_{I\_\text{SMF-CD}(j\omega)}. \tag{8.40}$$

This can be achieved by taking the inverse of equations (8.31) and (8.32), then performing an inverse FFT to obtain the time domain discrete sequence to be fetched into the DAC to pre-distort the pulse sequence.

Thus it is important to measure the nonlinear transfer function using the DAC and sinusoidal excitation signal. Once we obtain the frequency response we can use the zero forcing technique to obtain the equalized filter characteristics and then conduct an inverse FFT to obtain the time domain response for driving the DAC. This should be a principal task in DAC based 16QAM PDM Nyquist super channel transmission systems.

8.3.3 Fiber TF under intensity modulation

As observed from figure 8.6 the carrier magnitude and phase as represented by a phasor do not change and rotate around a constant radius circle, so is it correct that its magnitude is unchanged? Not quite, as the diagram illustrates, from a single frequency component of the lightwave the components would travel at different speeds due to the frequency under the modulation, in fact, in a modulated lightwave spectrum. Each spectral component travels at a different velocity due to the frequency dependence of the propagation constant $\beta(\omega)$.

Thus one can measure the fiber transmittance TF using amplitude modulation (AM), as shown in figure 8.14. The MZIM is modulated by the RF waves obtained from the output of an RF synthesizer. Sweeping of the RF frequency can be scanned across the frequency range of interest.

The output waves at the SMF end are then detected by a PD and monitored in an RF spectrum analyzer. The RF amplitude can be plotted as function of the scanned frequency. Alternatively, instead of using the sinusoidal waves from the synthesizer a step function can be obtained from the signal analyzer, in which spectral and time domain reflectometry (TDR) measurement can be obtained. The $S_{21}(f)$ parameter can be interpreted to give the fiber TF(f).

Analytically, for the power received at the output of the fiber, the SMF can be obtained as follows.

The optical field at the MZIM output can be written as

$$E_i(t)\alpha \cos\left[b\frac{\pi}{2} + m\cos(\omega_m t + \varphi_0)\right]\cos(\omega_c t), \tag{8.41}$$

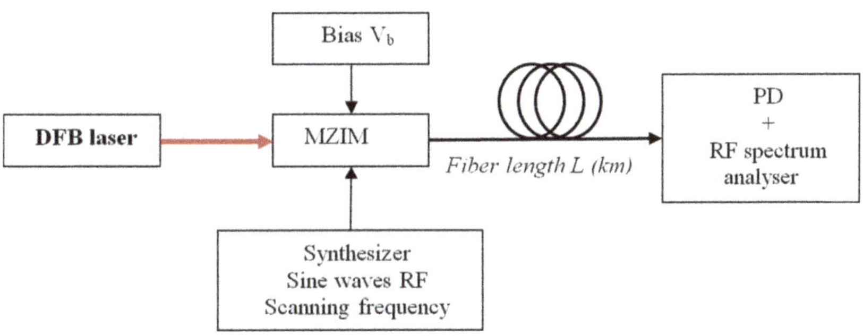

Figure 8.14. Measurement set-up for the fiber transfer function. DFB = distributed feedback; PD = photodetector; RF = radio frequency.

where b is the phase ratio between the bias voltage dc (V_{dc}) and V_π of the MZIM, V_π is the half-wave voltage at which the total phase change of π is imposed on the optical fields passing through the electrodes, m is the modulation index defined as the ratio between the level amplitude of the RF wave over V_π, and ω_m and ω_0 are, respectively, the angular frequencies of the modulating RF wave and optical waves.

The lightwave field amplitude at the fiber output can be given as [20]

$$E_0(t) \, \alpha \, J_0(m)\cos(\omega_m t + \varphi_0) - J_1(m)(\cos(\{\omega_0 - \omega_m\}t - \varphi_1) + \cos(\{\omega_0 + \omega_m\}t + \varphi_2)) + \ldots, \tag{8.42}$$

where φ_0, φ_1, and φ_2 are the phase delays of the spectral components due to CD from the wavelength dependence of the LP mode propagation constant $\beta(\omega)$ which can be approximated by a Taylor power series of the central frequency of the baseband signal. The phase variation of the lightwaves is thus a function of the traveled distance z and the dispersion factor.

At the PD (p–i–n type; p = p-doped, i = intrinsic, n = n-doped) the amplitude of the converted current is proportional to the power of the RF modulated dispersive optical waves given as

$$i_m(t) \, \alpha P_0(t) = \cos^2\left[\frac{\varphi_1 + \varphi_2}{2}\right] = \cos\left[\pi LcD\left(\frac{f_m}{f_0}\right)^2\right], \tag{8.43}$$

where D is the dispersion factor of the fiber in units of ps $(\mathrm{nm.km})^{-1}$, c is the velocity of light in a vacuum, and L is the total length of the fiber. Thus from this expression we can see that the detected current vanishes at the following discrete frequencies:

$$f = f_m = f_c\sqrt{\frac{N}{2LDc}}; N = 1, 3, 5. \ldots \tag{8.44}$$

The measured frequency response is shown in figure 8.15(a) and the analytical one in (b). We can see that the SSMF over a length L acts like a low pass filter associated with the first null point of the transfer function.

The transfer function of a direct detection and intensity modulation plotted against different transmission lengths of 100 and 200 km is depicted in figures 8.16(a) and (b), respectively. It is shown that the first null point occurs at about 8 GHz for 100 km while it narrows down to about 4 GHz when L is extended to 200 km. Thus the passband of the transfer functions narrowed significantly as a function of the transmission length. The dips in the fiber frequency transfer function would create depletion of the optical carrier, hence that of the RF over fiber signals using DSB modulation, or the limited length of transmission of RFoF.

The null points can be observed at 8.7, 14.8, and 19.1 GHz for a 50.4 km length of SSMF. The variation of the null order as a function of the fiber length is shown in figure 8.17. Thus with 50 km length all null frequencies occur below 20 GHz for the SSMF. The first-order $N = 1$ null point of the transfer function under direct detection is the most important for the pre-emphasis technique to shift the passband to a higher frequency (figure 8.18).

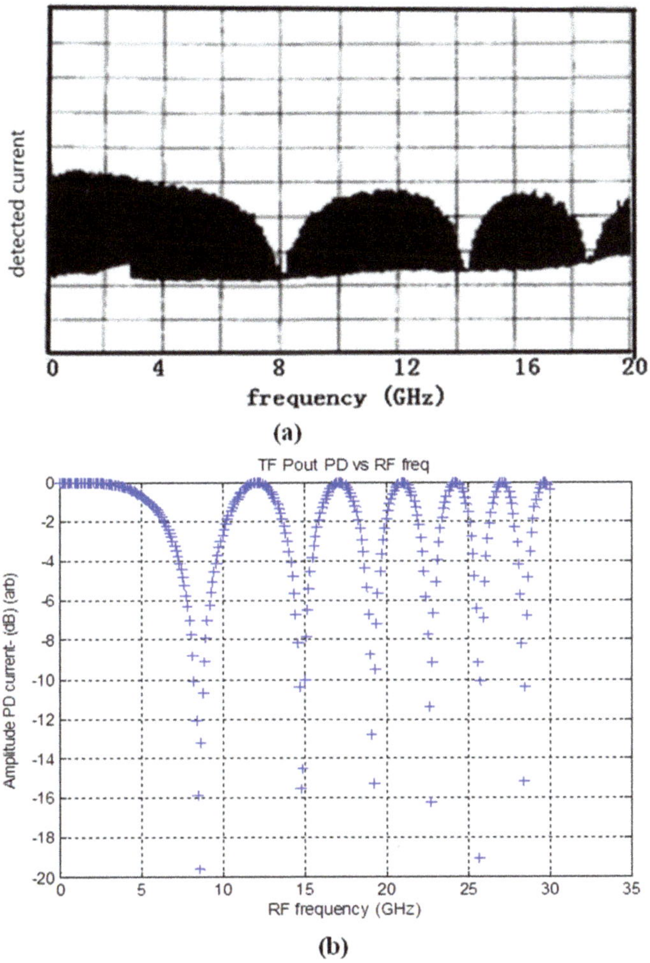

Figure 8.15. Variation of the amplitude of the detected current with respect to the frequency of the launched RF waves modulating the MZIM. (a) The experimental results for $L = 50$ km with a detected photodetector (PD) current as a function of a pure sinusoidal RF frequency modulating the optical waves. (b) Analytical results.

8.3.4 Fiber nonlinear transfer function (NLTF)

8.3.4.1 Nonlinear effects

Figure 8.19 illustrates the family of the linear and nonlinear effects on the SMF.

The elastic nonlinear effects are due to the change of the refractive index of the core region by the intensity of the lightwaves, hence the corresponding change of their phases and interference effects, or to the different propagation times of the components of the modulated spectra. The fiber refractive index (RI) is dependent on both the operating wavelengths and lightwave intensity. This intensity-dependent phenomenon is known as the Kerr effect and is the cause of fiber nonlinear effects.

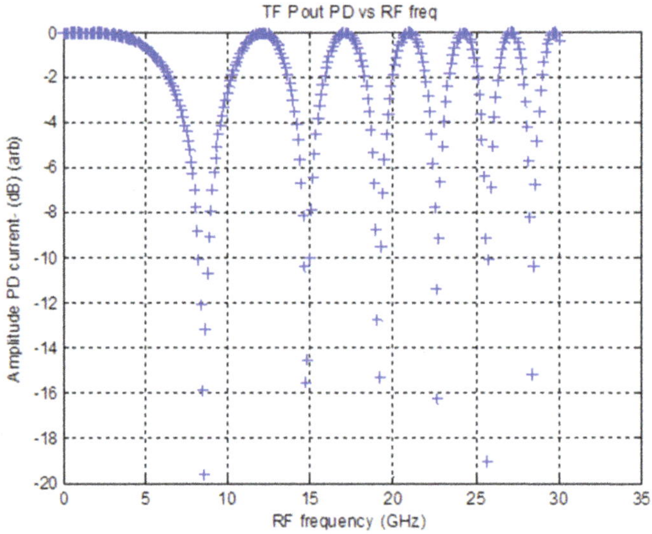

(a)

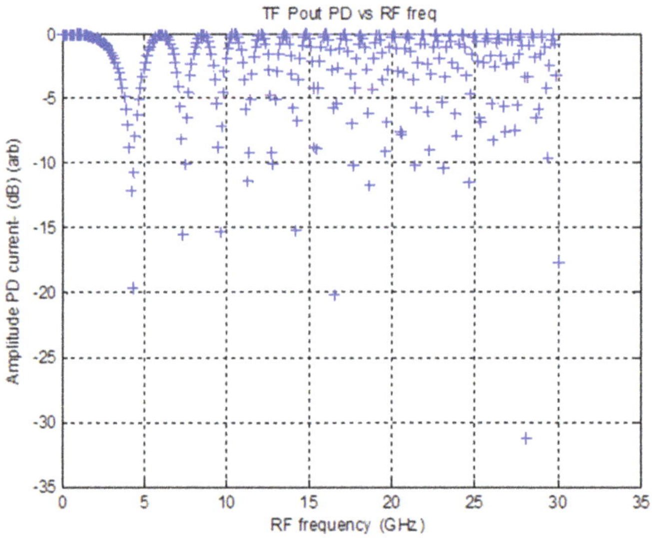

(b)

Figure 8.16. The transfer function of the SSMF under intensity modulation and direct detection for (a) L = 100 km, (b) 200 km SSMF with D = 17e-6 s m^{-2}. Note that the null N = 1 moves from the RF frequency 8 GHz to about 4 GHz when L changes from 100 to 200 km.

On the other hand, the non-elastic nonlinear effects are due to the electronic vibration of the atoms of the impurities doped in the fiber core. The electronic vibrations then transfer the energy of the lightwaves to other components whose frequency would be some distance away, for example 100 nm away or a few THz by Raman scattering or only 100 MHz amplitude modulation by Brillouin scattering.

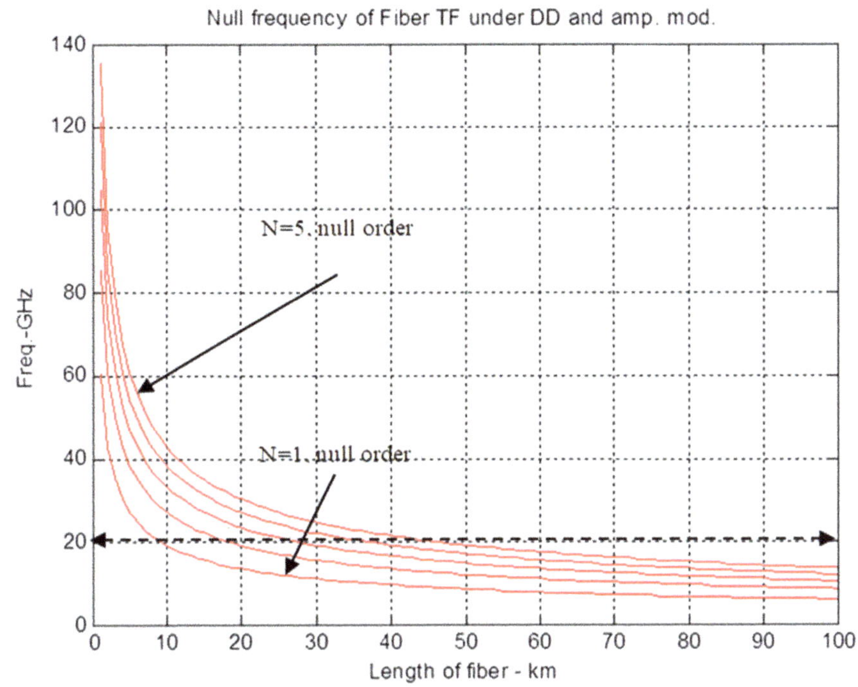

Figure 8.17. Variation of the null frequency as a function of fiber length in unity steps [3].

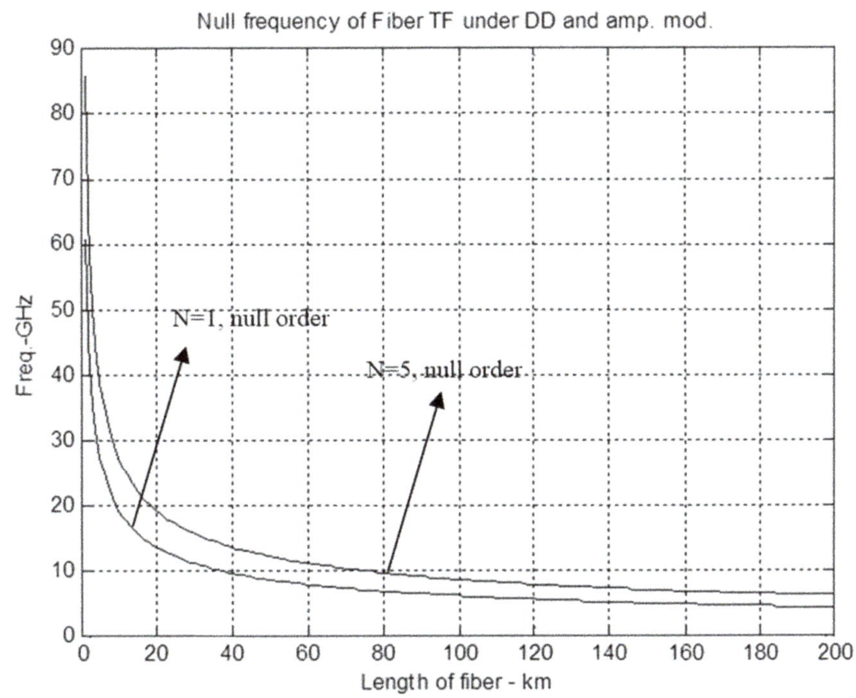

Figure 8.18. The null frequency variation as a function of fiber length, for $N = 1$ and $N = 2$. Copyright (2016), reprinted from [3] by permission of Taylor & Francis Ltd.

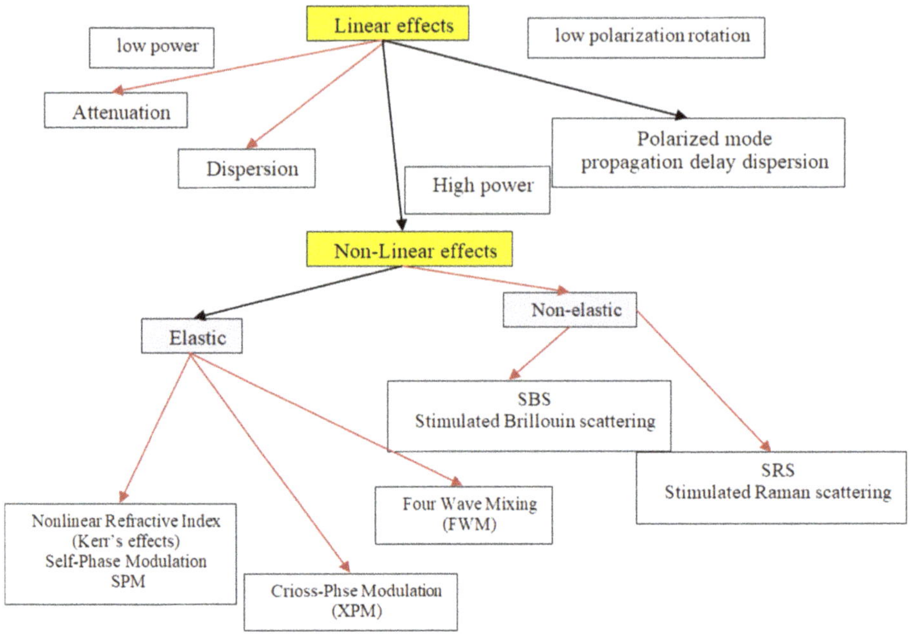

Figure 8.19. Linear and nonlinear fiber effects due to signal power propagation in single mode optical fibers.

8.3.4.1.1 SPM and XPM effects

The power dependence of the RI can be expressed as

$$n' = n + \bar{n}_2(P/A_{\mathrm{eff}}), \tag{8.45}$$

where P is the optical average power of the guided mode, $\bar{n}_2$ is the fiber nonlinear coefficient, and A_{eff} is the effective area of the fiber or the $1/e$ area of the guided mode field distributed across the fiber cross section.

Fiber nonlinear effects include intra-channel SPM, inter-channel cross phase modulation (XPM), four-wave mixing (FWM), stimulated Raman scattering (SRS), and stimulated Brillouin scattering (SBS). SRS and SBS are degrading factors and become noticeably large with very high optical power.

However, in multi-carrier multiplexed systems, SRS generates FWM products, that is, a mixing of three equally spaced optical carriers creates the fourth components which can be located in the other signal bands or effectively three-wave intermodulation (IMD) products. This can be critical in sub-carrier multiplexing (SCM) in radio-photonics over fiber (RPoF) in wavelength multiplexed (SCM-WDM) transmission, and cross-talk can be generated in many RF subcarriers in several wavelength bands. Thus the SRS mixing of three optical waves generates the fourth IMD waves whose frequency satisfies the momentum phase matching conditions, or is located at equal frequency spacing, thus creating interference with the other waves located at that of the generated waves of equally spaced RF frequency grids.

The SBS generates low frequency components in the order of a few hundreds of MHz and thus jittering in the narrow data bands in the RPoF, and degrades the system performance severely with the generation of ghost waves.

The SRS and SBS effects can be minimized by assigning the optical carriers in the spectral regions of SMFs whose local dispersion factors are high so that the momentum phase matching conditions are not satisfied. Thus we should not use the 1300 ± 10 nm of silica fiber spectral windows at which the dispersion factor is very low or even zero.

In terms of XPM, or similar to the intermodulation (IMD) effects in audio and radio systems or amplifiers, the nonlinearity generates the modulation products of two adjacent waves to the third waves from such two-wave mixing. XPM effects are considered to be negligible in a dense wavelength division multiplexing (DWDM) system under the following scenarios: (i) a highly locally dispersive system and (ii) large channel spacing. However, XPM should be serious in RPoF deploying a non-zero dispersion-shifted fiber (NZ-DSF), that is the zero crossing of the dispersion factor is shifted from 1300 nm to another different wavelength window where local dispersion values are small.

SPM effects are normally the dominant nonlinear effect for systems employing transmission fibers with high local dispersion, for example the SSMF and DCF. The effect of SPM is normally coupled with the nonlinear phase shift ϕ_{NL} defined as

$$\begin{aligned} \phi_{\mathrm{NL}} &= \int_0^L \gamma P(z)dz = \gamma L_{\mathrm{eff}} P \\ \gamma &= \omega_c \bar{n}_2 / (A_{\mathrm{eff}} c) \\ L_{\mathrm{eff}} &= (1 - e^{-\alpha L})/\alpha, \end{aligned} \tag{8.46}$$

where ω_c is the lightwave frequency, L_{eff} is the effective fiber length, and α is the SMF attenuation factor ~ 0.17 to 0.2 dB km^{-1} in the 1550 nm spectral window.

The temporal variation of the nonlinear phase ϕ_{NL} comes from the generation of new spectral components far apart from the lightwave carrier ω_c, thus spectrum broadening, $\delta\omega$, which can be derived from the time dependent nonlinear phase shift as

$$\delta\omega = -\frac{\partial \phi_{\mathrm{NL}}}{\partial T} = -\gamma \frac{\partial P}{\partial T} L_{\mathrm{eff}}. \tag{8.47}$$

Equation (8.46) indicates that $\delta\omega$ is proportional to the temporal derivative of the average signal power. Furthermore, due to the new spectral components generated at the rising and falling edges of optical pulses the chirping is substantially larger.

The NLSE propagation equation can be represented as

$$\begin{aligned} &\frac{\partial A(z,t)}{\partial z} + \frac{\alpha}{2} A(z,t) + \beta_1 \frac{\partial A(z,t)}{\partial t} + \frac{j}{2}\beta_2 \frac{\partial^2 A(z,t)}{\partial t^2} \\ &- \frac{1}{6}\beta_3 \frac{\partial^3 A(z,t)}{\partial t^3} \\ &= -j\gamma\, |A(z,t)|^2 A(z,t) - \frac{1}{\omega_0}\frac{\delta}{\delta t}(|A|^2 A) - T_R A \frac{\delta(|A|^2)}{\delta t} \end{aligned} \tag{8.48}$$

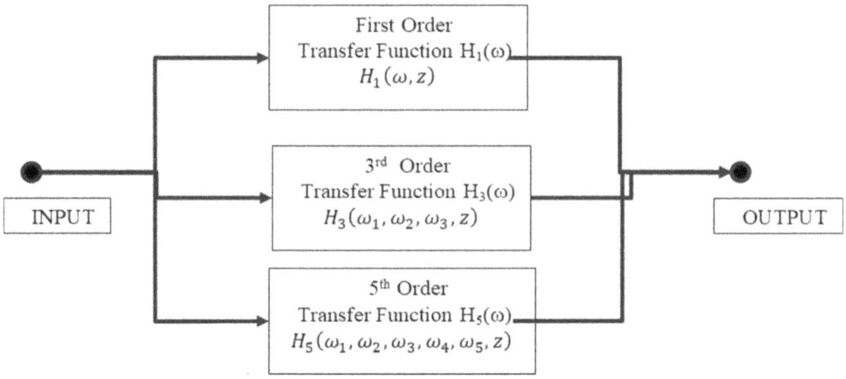

Figure 8.20. Schematic representation of the fiber nonlinear transfer functions with kernels of the first-, third-, and fifth-order angular frequencies ω_1, ω_2, ω_3, ω_4, and ω_5.

in which we have ignored the pure delay factor involving β_1. The last term on the RHS represents the Raman scattering effects.

8.3.4.2 Nonlinear transfer functions

The weakness of most of the recursive methods in solving the NLSE is that they do not provide much useful information to help the characterization of nonlinear effects. The Volterra series model can provide a system's nonlinearities and its frequency transfer function can be represented by a number of parallel transfer functions of first-, third-, and fifth-order kernels, as shown in figure 8.20 [21]. These kernels involve the fundamental, second harmonics, third harmonics, and intermodulation terms which can be found in [22].

The second-order as well as third transfer function creates the IMD products between ω_1, ω_2, and ω_3. This is critical for SCM RPoF systems in which several multiplexed RF carriers exist. That is, the frequency spacing in the SCM RPoF can be only about 500–600 MHz and in an optical grid of 50 GHz.

The detailed expressions of these VTFs can be found in [15].

The VTF expressions allow us to evaluate each effect individually, in particular the nonlinear effects so that for the RPoF system we can manage their distortion products by a mitigation algorithm in a post-compensation algorithm DSP based optical receiver. These techniques are outlined in the next section.

8.3.5 Pre-emphasis/pre-compensation of the fiber TF for SCM RPoF systems

Here we can state that when the optical signals travel through a length of SSMF they would be filtered as the fiber behaves like a low pass filter associated with the first null point of the transfer function. Under the intensity modulation and direct modulation this low pass filter will limit the bit rate of the modulated signals. Thus, in order to reduce this effect, one can employ the pre-emphasis technique, that is creating a bandpass filter whose passband falls on the roll-off region of the fiber low passband, as illustrated in figures 8.21 and 8.22. In the case of an RPoF the RF sub-carrier multiplexed optical channels are important and the distortion due to

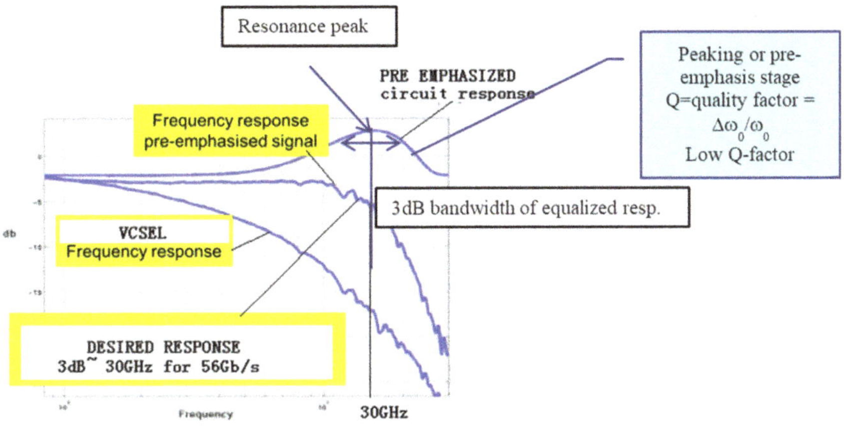

Figure 8.21. Pre-emphasis by equalization in the frequency domain responses.

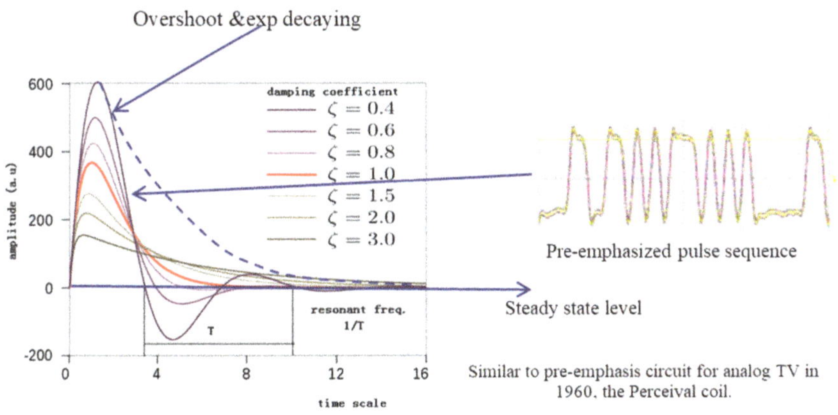

$$\varsigma = \frac{\alpha}{2\pi f_0}; \quad f_0 = \frac{1}{T} = resonance\_freq; \quad \alpha = damping\_attenuation$$

Figure 8.22. Principles of pre-emphasis in the time domain.

nonlinearity is critical, in particular the third- and fifth-order harmonics and intermodulation effects.

Figures 8.23 and 8.24 show a digital system that can create a pre-emphasized signal so that it acts as a pre-distortion effect opposite to the dispersion due to the linearity effects of the fiber or electronic amplifier, hence the RPoF system is distortionless after the transmission. Similarly, a simple analog pre-emphasis circuit can be used, as shown in figure 8.25.

8.4 Concluding remarks

This chapter describes briefly the physical propagation phenomena and dynamics of signals of both broadband and multi-carrier multiplexed channels of RPoF, through

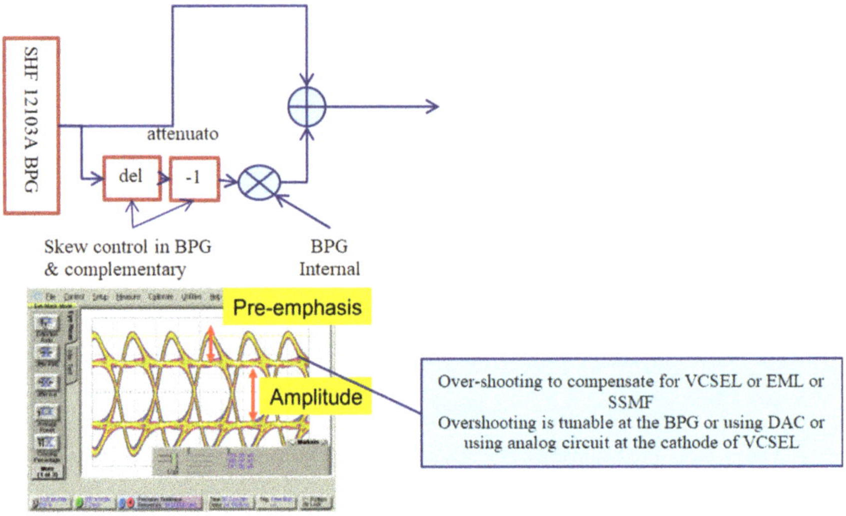

Figure 8.23. Generating pre-emphasized signals so as to use the fiber transfer function as a compensator. EML = electro-absorption modulated laser or external modulated laser; VCSEL = vertical surface emission laser.

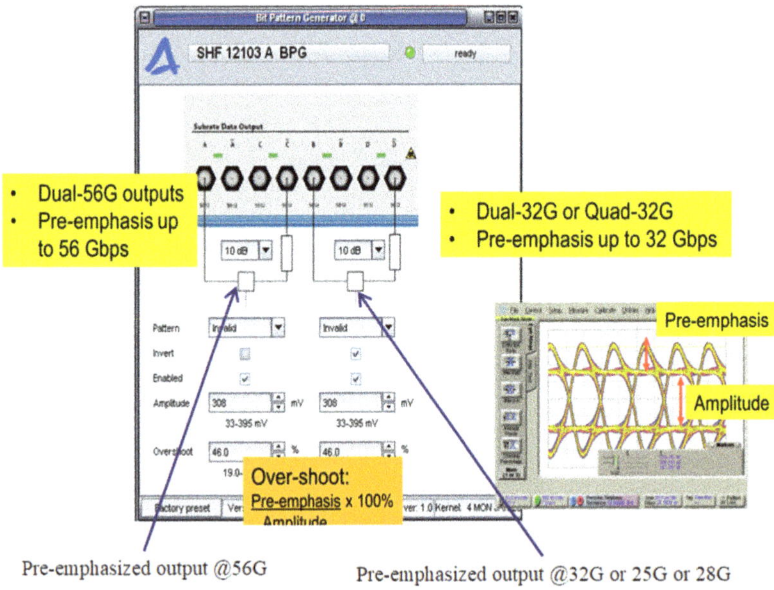

Figure 8.24. Pre-emphasis signals generated by the BPG SHF12103A model.

an SMF. The signal under linear and nonlinear distortion effects traveling through SMF transmission lines is given. The analytical and schematic representations of these effects are summarized together with some brief techniques for compensating such distortion effects on transmitted signals, in particular the RF modulated signals transported by optical waves.

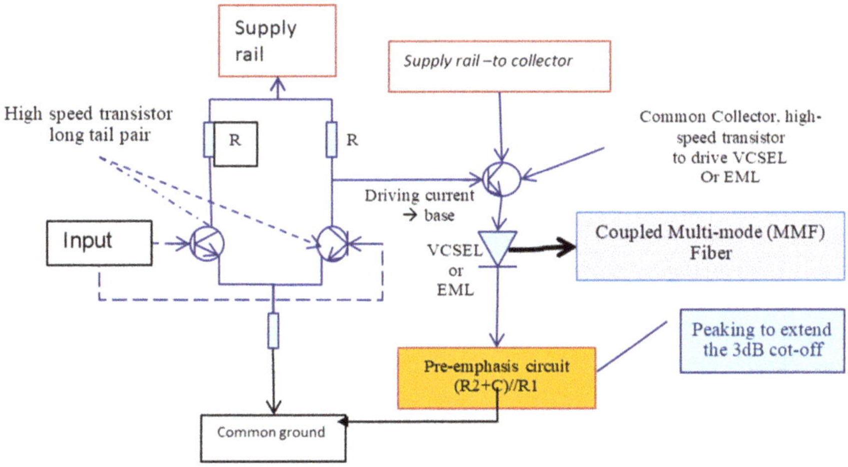

Figure 8.25. Pre-emphasis circuit in association with a source circuit. VCSEL = vertical cavity surface emission laser; EML = external modulation laser.

References

[1] Xia C, Bai N, Ozdur I, Zhou X and Li G 2011 Supermodes for optical transmission *Opt. Express* **19** 16653

[2] Corning Corning® SMF-28® Ultra Optical Fiber

[3] Binh L N 2016 *Advanced Digital Optical Communications* 2nd edn (Boca Raton, FL: CRC Press, Taylor and Francis)

[4] Safaai-Jazi A and McKeeman J C 1991 Synthesis of intensity patterns in few-mode optical fibers *IEEE J. Lightw. Technol.* **9** 1047

[5] Salsi M *et al* 2011 Transmission at 2 × 100 Gb/s, over two modes of 40 km-long prototype few-mode fiber, using LCOS based mode multiplexer and demultiplexer *Opticl Fiber Conf. and National Fiber Optic Engineers Conf. (Los Angeles, CA)*

[6] Randel S, Ryf R, Sierra A, Winzer P J, Gnauck A H, Bolle C A, Essiambre R-J, Peckham D W, McCurdy A and Lingle R Jr 2011 6 × 56-Gb/s mode-division multiplexed transmission over 33-km few-mode fiber enabled by 6 × 6 MIMO equalization *Opt. Express* **19** 16697

[7] Poole C D, Wiesenfeld J M, DiGiovanni D J and Vengsarkar A M 1994 Optical fiber-based dispersion compensation using higher order modes near cutoff *IEEE J. Lightw. Technol.* **12** 1746–58

[8] Pittala F, Zanaty A, Vaquero Caballero F J, Goeger G and Ye Y 2016 Training-aided 6 × 6 MIMO frequency-domain equalizer for optical SDM transmission systems *IEEE Photonics Conf. (Waikoloa, HI)*

[9] Cartaxo A V T, Wedding B and Idler W 1998 New measurement technique of nonlinearity coefficient of optical fibre using fibre transfer function *ProcECOC98 (Madrid, Spain)* pp 169–70

[10] Cartaxo A V T, Wedding B and Idler W 1998 Influence of fiber nonlinearity on the phase noise to intensity noise conversion in fiber transmission: theoretical and experimental analysis *IEEE J. Lightw. Technol.* **16** 1187

[11] Ramos F and Martí J 2000 Frequency transfer function of dispersive and nonlinear single-mode optical fibers in microwave optical systems *IEEE Photon. Tech. Lett.* **12** 549

[12] Agrawal G 1995 *Nonlinear Fiber Optics* 2nd edn (San Diego, CA: Academic)
[13] Binh L N 2012 *Guided Wave Photonics* (Boca Raton, FL: CRC Press, Taylor and Francis)
[14] Ramos F and Martí J 2000 Frequency transfer function of dispersive and nonlinear single-mode optical fibers in microwave optical systems *IEEE Photon. Tech. Lett.* **12** 549
[15] Dunster T M 1990 Bessel functions of purely imaginary order, with an application to second-order linear differential equations having a large parameter *SIAM J. Math. Anal.* **21** 995–1018
[16] Binh L N 2009 *Digital Optical Communications* (Boca Raton, FL: CRC Press)
[17] Proakis J G 2001 *Digital Communications* 4th edn (New York: McGraw-Hill) pp 185–213
[18] Binh L N 2009 *Digital Optical Communications* (Boca Raton, FL: CRC Press)
[19] Binh L N and Nguyen N 2009 Generation of high-order multi-bound-solitons and propagation in optical fibers *Opt. Commun.* **282** 2394–406
[20] Schmuck H 1995 Comparison of optical millimeter wave system concepts with regard to chromatic dispersion *Elect. Lett.* **31** 1848–9
[21] Binh L N 2013 *Advanced Optical Digital Communications* (Boca Raton, FL: CRC Press, Taylor and Francis)
[22] Ha T T 1981 *Solid State Microwave Amplifiers* (New York: Wiley) p 256

www.ingramcontent.com/pod-product-compliance
Ingram Content Group UK Ltd.
Pitfield, Milton Keynes, MK11 3LW, UK
UKHW051905100926
8556IPUK00005B/774

9 780750 335096